Lecture Notes in Networks and Systems   1331

Series Editor

Janusz Kacprzyk, *Systems Research Institute, Polish Academy of Sciences, Warsaw, Poland*

Advisory Editors

Fernando Gomide, *Department of Computer Engineering and Automation—DCA, School of Electrical and Computer Engineering—FEEC, University of Campinas—UNICAMP, São Paulo, Brazil*

Okyay Kaynak, *Department of Electrical and Electronic Engineering, Bogazici University, Istanbul, Türkiye*

Derong Liu, *Department of Electrical and Computer Engineering, University of Illinois at Chicago, Chicago, USA*

   Institute of Automation, Chinese Academy of Sciences, Beijing, China

Witold Pedrycz, *Department of Electrical and Computer Engineering, University of Alberta, Alberta, Canada*

   Systems Research Institute, Polish Academy of Sciences, Warsaw, Poland

Marios M. Polycarpou, *Department of Electrical and Computer Engineering, KIOS Research Center for Intelligent Systems and Networks, University of Cyprus, Nicosia, Cyprus*

Imre J. Rudas, *Óbuda University, Budapest, Hungary*

Jun Wang, *Department of Computer Science, City University of Hong Kong, Kowloon, Hong Kong*

The series "Lecture Notes in Networks and Systems" publishes the latest developments in Networks and Systems—quickly, informally and with high quality. Original research reported in proceedings and post-proceedings represents the core of LNNS.

Volumes published in LNNS embrace all aspects and subfields of, as well as new challenges in, Networks and Systems.

The series contains proceedings and edited volumes in systems and networks, spanning the areas of Cyber-Physical Systems, Autonomous Systems, Sensor Networks, Control Systems, Energy Systems, Automotive Systems, Biological Systems, Vehicular Networking and Connected Vehicles, Aerospace Systems, Automation, Manufacturing, Smart Grids, Nonlinear Systems, Power Systems, Robotics, Social Systems, Economic Systems and other. Of particular value to both the contributors and the readership are the short publication timeframe and the worldwide distribution and exposure which enable both a wide and rapid dissemination of research output.

The series covers the theory, applications, and perspectives on the state of the art and future developments relevant to systems and networks, decision making, control, complex processes and related areas, as embedded in the fields of interdisciplinary and applied sciences, engineering, computer science, physics, economics, social, and life sciences, as well as the paradigms and methodologies behind them.

Indexed by SCOPUS, EI Compendex, INSPEC, WTI Frankfurt eG, zbMATH, SCImago.

All books published in the series are submitted for consideration in Web of Science.

For proposals from Asia please contact Aninda Bose (aninda.bose@springer.com).

Esteban Mauricio Inga Ortega ·
Nuria García Herranz ·
Vladimir Espartaco Robles-Bykbaev ·
Eduardo Gallego Diaz
Editors

Systems, Smart Technologies, and Innovation for Society

Proceedings of CITIS 2024

 Springer

Editors
Esteban Mauricio Inga Ortega ⓘ
Universidad Politécnica Salesiana
Cuenca, Ecuador

Nuria García Herranz ⓘ
Universidad Politécnica de Madrid
Madrid, Spain

Vladimir Espartaco Robles-Bykbaev ⓘ
Universidad Politécnica Salesiana
Cuenca, Ecuador

Eduardo Gallego Diaz ⓘ
Universidad Politécnica de Madrid
Madrid, Spain

ISSN 2367-3370   ISSN 2367-3389 (electronic)
Lecture Notes in Networks and Systems
ISBN 978-3-031-87064-4   ISBN 978-3-031-87065-1 (eBook)
https://doi.org/10.1007/978-3-031-87065-1

This work was supported by Universidad Politécnica Salesiana del Ecuador.

This Springer imprint is published by the registered company Springer Nature Switzerland AG
The registered company address is: Gewerbestrasse 11, 6330 Cham, Switzerland

If disposing of this product, please recycle the paper.

Preface

This book is composed of the papers written in English and accepted for presentation and discussion at the X International Conference on Science, Technology, and Innovation for Society (CITIS 2024). This conference had the support of the Universidad Politécnica Salesiana (UPS). It will take place at Guayaquil, Ecuador, during July 18–19, 2024.

The International Conference on Science, Technology, and Innovation for Society (CITIS 2024), in its *tenth* edition, offers the national and international academic community a unified communication platform, aimed at covering the theoretical and practical problems of greater impact in the modern Society through Engineering. Main topics are related to the application of science, technological development and innovation to the industry, environment, information and telecommunications, and mobility.

The Program Committee of CITIS 2024 was composed of a multidisciplinary group of 89 experts and those who are intimately concerned with Information and Telecommunication (Infotelecom), Mobility, Sustainability and Environment, and Industry. They have had the responsibility for evaluating, in a "double-blind review" process, the papers received for each of the main themes proposed for the conference: A) Information and Telecommunication; B) Mobility; C) Sustainability and Environment; and D) Industry.

CITIS 2024 received 188 contributions from 5 countries around the world. The papers accepted for presentation and discussion at the conference will be published by Springer (this book) and will be submitted for indexing by Scopus and/or Google Scholar, among others.

Eduardo Gallego Diaz
Nuria García Herranz
Vladimir Espartaco Robles-Bykbaev
Esteban Mauricio Inga Ortega

Organization

Honorary Committee

P. Marcelo Farfán, sdb. Ph.D. — Chancellor of the Universidad Politécnica Salesiana, Ecuador

P. Juan Cárdenas Tapia, sdb. Ph.D. — Rector of the Universidad Politécnica Salesiana, Ecuador

Fernando Pesántez Avilés, Ph.D. — Vice rector general of the Universidad Politécnica Salesiana, Ecuador

Raúl Alvarez Guale, Ph.D. — Vice Rector of the Universidad Politécnica Salesiana's Guayaquil-Ecuador campus

Esteban Inga Ortega, Ph.D. — Vice Rector for Research of the Universidad Politécnica Salesiana, Ecuador

Angela Flores Ortiz, Ph.D. — Academic Vicerrector of the Universidad Politécnica Salesiana, Ecuador

Vladimir Robles Bykbaev, Ph.D. — Vice Rector of Graduate Studies of the Universidad Politécnica Salesiana, Ecuador

PC Chairs

José Manuel Aller Castro, Ph.D.
Fabricio Espinoza, Ph.D.
Juan Diego Valladolid, Ph.D.
Mónica Karel Huerta, Ph.D.
Juan Lata, Ph.D.
Lenin Cevallos, Ph.D.
René Vinicio Sánchez Loja, Ph.D.
Diego Francisco Carrión Galarza, Ph.D.

Advisor Committee

José Luis Verdegay — Universidad de Granada, Spain

Riccardo Cristoforo Barberi — Università della Calabria, Italia

Carlos Lozano Garzón — Director IEEE ComSoc, Latin American Region, Colombia

Roger Cltet Martínez — Universidad Internacional de Valencia, Spain

José Ignacio Huerta Cardoso — Tecnológico de Monterrey, Mexico

Manuel Dïaz-Madroñero	Universitat Politècnica de València, Spain
Hugo G. Espinoza	Griffith University, Australia
Vicente Parra Vega	CINVESTAV, Mexico
Francisco Jurado	Universidad de Jaén, Spain

International Scientific Committee

Humberto Michinel	University of Vigo, Spain
Carlos Lozano-Garzón	IEEE ComSoc, Latin American Region, Colombia
Vicente Parra-Vega	Center for Research and Advanced Studies, Mexico
Hugo G. Espinosa	Griffith University, Australia
Sandro Silveira	Instituto Federal de Educação, Ciência e Tecnologia do Ceará, Brazil
José Luis Verdegay	Universidad de Granada, España
Wilson Castro-Silupu	Universidad Nacional de Frontera, Peru
Pablo Díaz-Núñez	University of Manchester, UK
Luis Fernández-Ramírez	Universidad de Cádiz, Spain
Josefa Mula	Universidad Politécnica de Valencia, Spain
José A. Alvarado-Contreras	Universidad de Los Andes, Venezuela
Pedro Fernández de Córdoba	Universidad Politécnica de Valencia, Spain
Luis Fernando Mulcue-Nieto	Universidad Autónoma de Manizales, Colombia
Víctor Ancajima-Miñán	Universidad Nacional de Piura, Peru
Luís Guimarães	Universidade do Porto, Portugal
Nervo Xavier Verdezoto	Cardiff University, UK
Daniel Gamermann	Universidade Federal do Rio Grande do Sul, Brazil
Gilberto Reynoso-Meza	Pontificia Universidade Católica do Paraná, Brazil
José Ignacio Huerta-Cardoso	Tecnológico de Monterrey, Mexico
Higinio Sánchez	Universidad de Cádiz, Spain
Francisco Jurado	Universidad de Jaén, Spain
Salvatore Patera	Universidad de Salento, Italia
Ricardo Silva	Foundation for Living, Wellness and Health, E.U.A
Julian Triana-Dopico	Arizona State University, E.U.A
Miguel Angel Díaz-Rodríguez	Universidad de los Andes, Venezuela
Ignacio Castillo-Velázquez	Universidad Autónoma de Ciudad de Mexico
R. S. Ajin	Idukki District Disaster Management Authority, India
Joan Vázquez-Molina	Research Engineer, Tenneco, Belgium
Simón Jesús Fygueroa-Salgado	Universidad de Pamplona, Colombia

Reinaldo Ramirez C.	Universidad Politécnica Salesiana, Ecuador
Tzinnia Soto	Monterrey Institute of Technology and Higher Studies, Mexico
Christopher Reyes López	Universidad Politécnica Salesiana, Ecuador
Héctor Asa De León	Autonomous University of Zacatecas, Mexico
Luis Alberto Escalera Velasco	Technological Institute of Aguascalientes, Mexico
António Jiménez Carrascosa	Paul Scherrer Institute, Switzerland
Marcelo Estrella Guayasamin,	Universidad Politécnica Salesiana, Ecuador
Juan Carlos Lata García	Universidad Politécnica Salesiana, Ecuador
Karen A. Guzmán-García	ENUSA, Spain
Héctor René Vega-Carrillo	Universidad Autónoma de Zacatecas, Mexico
Guillermo Eduardo Campillo-Rivera	Universidad Autónoma de Zacatecas, Mexico
Joel Vázquez-Bañuelos	Universidad Autónoma de Zacatecas, Mexico
António Baltazar-Raigosa	Universidad Autónoma de Zacatecas, Mexico
Ángel García-Durán	Universidad Autónoma de Zacatecas, Mexico
Carina Oliva Torres-Cortés	Universidad Autónoma de Zacatecas, Mexico
Claudia Angélica Márquez-Mata	Universidad Autónoma de Zacatecas, Mexico
Roberto López Chila	Universidad Politécnica Salesiana, Ecuador
Álvaro López-Cazalilla	University of Helsinki, Finland
William Ipanaque Lalama	Universidad de Piura, Peru
Raúl Montenegro	Universidad de Córdoba, Argentina
Guillermo Gutierrez Montoya	Universidad Don Bosco, El Salvador
Humberto Javier Michinel Álvarez	Universidad de Vigo, Spain
Hugo G. Espinosa	Griffith University, Australia
Agustín Yagüe Panadero	Universidad Politécnica de Madrid, Spain
Jairo Castillo Calderón	Universidad Nacional de Loja, Ecuador

Contents

Information and Telecommunications (Infotelecom)

Mobility

Sustainability and Environment

Industry

About the Editors

Professor Eduardo Gallego Diaz is Full Professor of Nuclear Engineering at UPM in the Department of Energy Engineering of the Technical School of Industrial Engineers, where he has taught in the areas of Radiological Protection and Nuclear Safety since 1985. He is Industrial Engineer specializing in Energy Techniques (1982) and PhD in Industrial Engineering (1990) from the UPM. He has been Advisor of 10 PhD Theses and over a hundred Final Degree or Master Theses. He is Director of the University Master's Degree in Nuclear Science and Technology at the UPM since 2011.

Currently, his main lines of research are neutron detection, dosimetry and spectrometry—being responsible for the Neutron Measurement Laboratory of the UPM—and support for analysis and decision-making in nuclear emergencies, to evaluate protection measures for the population, having been Co-Founder and Member of the Management Board of the European Platform NERIS on Nuclear and Radiological Emergencies since its creation in 2012 until 2022.

He is Member of the Board of Directors of the Spanish Nuclear Society (2020–2024). He has been President of the Spanish Society for Radiological Protection (2013–2015) and previously Vice President (2010–2012) and Treasurer (2000–2004). At the International Radiation Protection Association (IRPA), he has been Member of the Executive Council (2008–2016) and Vice President (2016–2020). He belongs to the International Commission on Radiological Protection (ICRP) since 2013, as Member of Committee 4.

Nuria García Herranz is Associate Professor of Nuclear Engineering at UPM, Department of Energy Engineering, where she has taught in the areas of Nuclear Technology and Nuclear Fission Reactors since 2005. She is Energy Engineer (1995) and PhD in Nuclear Engineering (2000) from UPM. She has been Advisor of 6 PhD theses, four of them awarded or finalists with the European Nuclear Education Network PhD prize. She has graduated about 45 BSc and MSc students.

Her main research areas include Reactor Physics (neutronics and multi-physics for both current reactors and Generation-IV fast reactors), Inventory and Criticality Safety, together with Sensitivity and Uncertainty analysis due to Nuclear Data. Innovation in education in Reactor Physics is also one of her active areas of work.

Her scientific contributions have been published in 37 papers in indexed journals, 67 peer-reviewed international conference proceedings and around 75 presentations in workshops and national conferences. She has participated in 12 European projects (principal investigator in 3 of them) and in 17 national research projects.

She has internationally recognized expertise in Reactor Physics, being Spanish Delegate of the Working Party on Scientific Issues of Reactor Systems (WPRS) of the OECD/NEA Nuclear Science Committee since 2021, and Spanish Delegate of the Expert Group on Physics of Reactor Systems (EGPRS) since 2020. She is Member of the Editorial Board of Annals of Nuclear Energy since 2021.

Vladimir Espartaco Robles-Bykbaev (Senior Member, IEEE) received the degree in computer science from Universidad Politécnica Salesiana, Ecuador, in 2006, the M.S. degree in Artificial Intelligence, Pattern Recognition, and Digital Imaging from the Polytechnic University of Valencia, Spain, in 2008, and the Ph.D. degree in Information and Communication Technologies from the University of Vigo, Spain, in 2016. Since 2012, he has been Principal Professor with the Computer Science Department, UPS. Since 2016 he collaborates as Head of the smart technologies area at the UNESCO Chair on Support Technologies for Educational Inclusion at the UPS. His research interests include the application of artificial intelligence techniques for improving the educational inclusion of children, youth and older adults as well as the rescue, and preservation of cultural heritage of Andean people.

Esteban Mauricio Inga Ortega is currently Research Vice-Chancellor at Universidad Politécnica Salesiana. In 2001, he graduated as Electronic Engineer at the Salesian Polytechnic University—Cuenca Campus—and has more than 21 years of experience in university teaching. In 2008, he received his Master's in Education and Social Development from the Universidad Tecnológica Equinoccial-Quito/Ecuador. At the beginning of 2017, he received his Master's in Engineering from the Universidad Pontificia Bolivariana de Medellín—Colombia. In March 2018, he obtained his Ph.D. degree from the Universidad Pontificia Bolivariana de Medellín—Colombia with honorable mention: Magna Cum Laude awarded when the thesis exceeds the expectations of what was proposed in the project, by the unanimous concept of the jurors and scientific production is demonstrated. He has held positions as Head of the Curricular Area, Director of the Electrical Engineering career, Coordinator of Dual Technical and Technological Training, Coordinator of the Master's Degree in Electricity, mention in Electrical Power Systems. Currently, he is Director of the Master's Program in ICT for Education offered by the Salesian Polytechnic University, Research Coordinator of the Quito Campus, Coordinator of the Research Group on Smart Grids (GIREI) of the Salesian Polytechnic University at the Quito Campus, Mentor and Coordinator 2022–2025 of the "Red-IUS en redes eléctricas y ciudades inteligentes (RECI)," Associate Editor of the Ingenius Journal, Senior Member of IEEE (IEEE Communications Society-ComSoc and IEEE Education Society) and Accredited Researcher by the Senescyt. In terms of scientific publications, he has generated more than 71 articles indexed in ISI Web of Science/Scopus and more than 30 articles indexed by Scielo, latindex, and other indexing; he has coordinated and directed more than ten research projects, his number of citations in scholar google exceeds 1400 citations. He currently directs doctoral and Master's theses in Italy, Colombia, Chile, and Ecuador and has produced more than 20 theses within the Salesian Polytechnic University. Electrical Engineering and Biomedical Engineering professor, where he teaches Research Techniques, Numerical Methods Programming, and Wireless Networks. He is also Professor of the Doctorate Program in Computer Science and Master's programs in Educational Innovation, Special Education, Electricity, and Information and Communication Technologies for Education.

Information and Telecommunications (Infotelecom)

Towards a Learning Experience on the Pre-Assembled Mobile Robot Platform in the First Year of Engineering

Ruben Puma-Rodriguez[1], José Jaime-Carriel[1], Mariana Pintado-Cuji[1], and Joe Llerena-Izquierdo[1,2,3]

[1] Universidad Politécnica Salesiana, Guayaquil, Ecuador
{rpuma,jjaime,mpintado,jllerena}@ups.edu.ec, jllerenai@uoc.edu
[2] Universitat Oberta de Catalunya, Barcelona, Spain
[3] Grupo de Investigación en Enseñanza-Aprendizaje de las Ciencias para la Ingeniería GIEACI, Guayaquil, Ecuador
https://gieaci.blog.ups.edu.ec/

Abstract. The use of emerging technologies in the classroom is an open field that is being researched in order to demonstrate their implication in learning. The aim of this paper is to present an experience of a work placement in the subject of programming in the first year of engineering studies at a polytechnic university in the city of Guayaquil, Ecuador. In order to determine how the use of a pre-assembled mobile robot platform, irobot, allows the consolidation of programming concepts and algorithmic strategies. An empirical-analytical, quantitative, quasi-experimental research methodology is used. A population of 90 students from five courses took part. The survey technique is applied at the end of a work session with the irobot platform, to integrate an algorithmic logic experience in two phases. It is discussed that the training of the professors to generate activities with robotics elements that allow the integration of algorithmic tasks is still a challenge. It is concluded that an experience with robotics elements in a platform such as irobot, achieves a greater reach in high enrolment courses, in addition to the preparation of activities that involve tasks with algorithmic development allow a better understanding of programming concepts. Finally, 90.9% of the students agree that using this platform, irobot and similar, motivates students' performance in the study of programming.

Keywords: Educational robots · Programming concepts · Learning experience · Irobot platform

1 Introduction

The use of emerging technologies in the classroom is an open field that is being investigated for its implications for learning[11,14]. There is very little research on classroom experiences with the use of robots by professors [4,7]. The existing

© The Author(s) 2025
E. M. Inga Ortega et al. (Eds.): CITIS 2024, LNNS 1331, pp. 3–11, 2025.
https://doi.org/10.1007/978-3-031-87065-1_1

gap in education stems from the ability of professors to learn and teach the use of robotics elements to translate them into activities that generate an educational experience [2,12,17]. Learning experiences with the use of educational robots, in turn, require a high level of hardware and software infrastructure [3,11,20]. Especially in the first year of engineering, where students are in a transition between high school and college, the initial courses have the characteristic of being in high demand for enrollment [1,10]. To minimize this impact, the use of robotic platforms as well as programmatic tools can become an ideal alternative to this educational challenge [15,19].

The integration of different disciplines such as the learning of algorithmic concepts allows the promotion of new ways of teaching and learning with the use of educational robots integrated in virtual platforms [5,19].

The use of robotics elements such as irobot, allows students a significant experience in the development of logical procedures in the movement of a robot [8,16], represented as an electronic device that, in addition to the excitement of seeing movements ordered by instructions by the student, awakens the interest and motivation to learn an algorithmic logic to achieve the right movement [6].

The aim of this paper is to present a work experience of programming practices in the first year of engineering studies at a polytechnic university in the city of Guayaquil, Ecuador. In order to determine how the use of a pre-assembled mobile robot platform, irobot, allows the consolidation of programming concepts and algorithmic strategies. (see Fig. 1).

Fig. 1. Groups of students practising on the irobot platform to strengthen the learning of sequential algorithmic structures.

2   Methods and Materials

This work uses a quantitative, quasi-experimental, empirical-analytical research methodology. A population of 90 students of the first year of engineering studies of a polytechnic university in the city of Guayaquil, Ecuador, participated. Students who belong to five courses of practices of the subject of Programming in an extracurricular schedule participate in the study. The survey technique is applied at the end of a work session with the pre-assembled mobile robotic platform, irobot, to integrate an algorithmic logic experience[1]. The survey is composed of two demographic questions, one question about aspects of the platform, and ten questions structured using a five-level likert scale to determine satisfaction with the irobot platform. The results have a confidence level of 90% and a margin of error of 9%.

It is intended to establish an experience that integrates algorithmic logic in a mobile robotics platform to solve problems of movements of a robot. For this, a two-phase planning is organized (see Fig. 2).

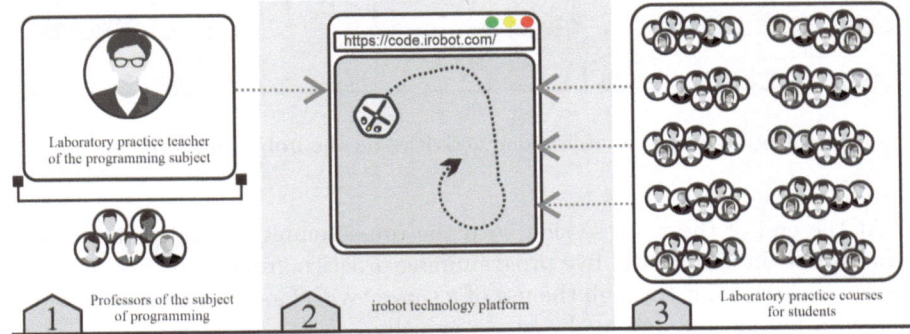

Fig. 2. Work process planning on the irobot platform.

The first phase is composed of three stages. In the first stage, an organized work planning is developed between the professors of the five courses of the Programming subject with the laboratory practice professor in charge. The role of the professor of practices is to accompany students in learning the subject of programming using good problem solving practices, use learning strategies for the study of programming and integrate the use of educational digital tools in practical activities to enhance learning. The second stage focuses on the recognition of the pre-assembled mobile robotic platform, irobot, and its available tools for the integration of programming concepts and introducing algorithmic strategies to solve robot motion problems. In the third stage, a work session is organized with the irobot platform for the development of a laboratory practice and the completion of the survey.

For the practical sessions of the Programming using irobot course, they are developed in two moments. In the first moment (A), students must solve a problem focused on the movement of the robot to perform a given path using a set of defined and ordered instructions. The student performs the observation of the applicability of his proposal and supervises the robot tracing. This work is done individually. In the second moment (B), the students must solve the problem of the supervised movement of the robot by means of work teams, with solution level, in 20 steps, then in 8 steps and finally in 6 steps. With this, it

---

[1] https://code.irobot.com.

is intended to establish the best algorithmic solution among the same students through collaborative work (see Fig. 3).

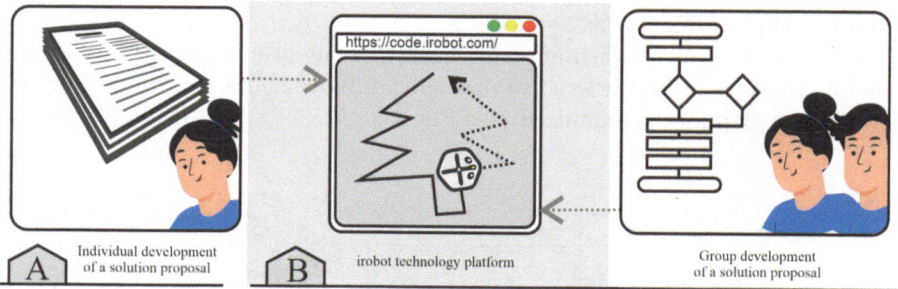

Fig. 3. Planning of learning activities on the irobot platform.

At the end of the work session with the pre-assembled mobile robotic platform, irobot, in each of the five programming practice groups, the evaluation of the day is carried out through the use of a survey with Google Forms technology. This data is collected and analyzed statistically.

3   Results

The results of the survey in the demographic questions show a participation of 23% of female students and 77% of male students. Although the representation of the female gender is lower, its representativeness is significant to promote proposals that motivate their inclusion in the field of engineering and avoid stereotypes [9,13,18]. It is also evident that 75% of students are between 18 and 20 years of age, while 25% are over 20 years of age.

In the question regarding the most relevant aspects of the irobot platform, a relevance scale from 1 to 5 was established to determine, being 1, Not relevant at all, 2, Not very relevant, 3, Somewhat relevant, 4, Relevant and 5 Highly relevant.

The first aspect (A1) to determine in the irobot platform is its ease of use. The second aspect (A2) is to know if the platform is useful for learning algorithmic logic. The third aspect (A3) is to know if the platform is intuitive. The fourth aspect (A4) is to know if the platform requires little logical knowledge from the participant. The fifth aspect (A5) is to recommend to the professors the use of this platform to learn algorithmic logic. The results are presented below see Table 1.

The results show that aspect A1 has a relevance of 89%. Aspect A2, reaches a relevance of 89%. Aspect A3, reaches a relevance of 75%. Aspect A4, reaches a relevance of 77%. Aspect A5, reaches a relevance of 88%. It is noteworthy to observe that the intuitive aspect of the platform, although it reaches a moderate

Table 1. Relevant aspects of the platform.

irobot platform		1	2	3	4	5
A1	Easy to use	0%	2%	9%	27%	62%
A2	Usefulness for learning algorithmic logic	0%	2%	9%	34%	55%
A3	Intuitive	0%	2%	23%	25%	50%
A4	Requires little logical knowledge	0%	7%	16%	34%	43%
A5	Recommend the use of the platform to the professors	2%	2%	8%	26%	62%

percentage, it is understood that it requires more cognitive load at the beginning for learning. Finally, it is noteworthy that the aspect to recommend this platform to the professors achieves 88%.

The results to the ten questions on student satisfaction with the use of the irobot platform for hands-on experience show, for question Q1, have I found that learning irobot requires a minimum amount of time? 45.5% agree completely, 38.6% agree, 13.6% neither agree nor disagree, and 2.3% disagree. In other words, 84.1% agree that learning the irobot platform requires a minimum of time.

For question Q2, have I noticed that learning irobot has allowed me to consolidate the algorithmic logic of movements?, 54.5% agree completely, 31.8% agree and 13.6% neither agree nor disagree. In other words, 86.3% agree that the irobot platform allows the student to consolidate the algorithmic logic of a problem.

For question Q3, does learning irobot make me excited to learn programming?, 50.0% strongly agree, 29.5% agree, and 20.5% neither agree nor disagree. In other words, 79.5% indicated that the platform makes students enthusiastic about learning programming.

For question Q4, has learning with irobot allowed me to understand how to integrate programming logic into electronic device simulators?, 45.5% strongly agree, 34.1% agree, and 20.5% neither agree nor disagree. In other words, 79.6% understand the integration of programming logic in electronic device simulators.

For question Q5, would you recommend other students to learn irobot to complement programming studies?, 56.8% completely agree, 25.0% agree and 18.2% neither agree nor disagree. That is, 81.8% would recommend learning the platform to complement programming studies.

For question Q6, do simulators such as irobot allow reinforcing knowledge in the field of programming logic?, 54.5% completely agree, 29.5% agree, and 15.9% neither agree nor disagree. In other words, 84.0% agree that robotic tools such as those found on the irobot platform strengthen knowledge in the field of programming logic.

For question Q7, should the use of simulators such as irobot be increased to expand programming knowledge?, 63.6% completely agree, 27.3% agree and 9.1% neither agree nor disagree. That is to say that 89.9% indicate that the use of this platform allows them to expand their knowledge in the field of programming.

For question Q8, should simulators such as irobot be used to foster engagement in studying algorithms?, 63.6% completely agree, 25.0% agree, and 11.4% neither agree nor disagree. That is, 88.6% of the students indicate that the use of this platform fosters the commitment to study algorithms (see Fig. 4).

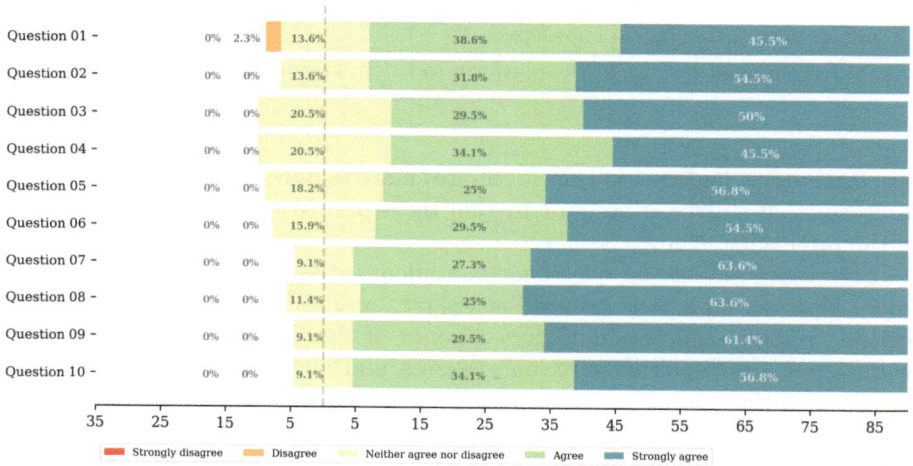

Fig. 4. Percentage of the results of the survey developed for students.

For question Q9, should simulators such as irobot be used to motivate the performance of programming students?, 61.4% completely agree, 29.5% agree and 9.1% neither agree nor disagree. That is to say that 90.9% agree that using this platform and similar motivates students' performance in the study of programming.

For question Q10, does knowing simulators such as irobot motivate me to improve academic performance?, 56.8% completely agree, 34.1% agree and 9.1% neither agree nor disagree. In other words, 90.9% of the students indicated that including this type of platform in the laboratory practices motivates them to improve their academic performance.

4   Discussion

This research paper discusses that the creation of educational activities with the use of environments that integrate robotics elements and that allow to generate meaningful learning is still being studied [2] and require greater input from the scientific community. Although the development of activities of a subject, such as Programming, elaborated collaboratively by a group of professors is a strength, the work led by a professor of practices, within a planning, where he integrates elements of robotics allows the monitoring of the study groups, to correct and improve didactic actions in each session [6,7]. With this, introducing a robotic platform such as irobot allows to enhance the concepts of a programming curriculum. Additionally, the preparation of professors for the use of robotics elements in collaborative environments applied in educational activities should be the attention of educational institutions as a challenge for the future of education [16].

5   Conclusions

This paper is an applied research work that develops a learning experience in an extracurricular laboratory session with the pre-assembled mobile robotic platform, irobot, which integrates algorithmic logic topics. A work strategy is established with five groups of students belonging to the first year of studies in a polytechnic university in the city of Guayaquil, Ecuador. It is evidenced that the accompaniment of a professor of laboratory practices, in work sessions that integrate digital tools such as robotic platforms create an ideal space to enhance elements of algorithmic reasoning. This work is of relevance because of its contribution to the existing literature.

It is concluded that an experience with robotics elements in a platform such as irobot, achieves a greater reach in high enrollment courses, also the preparation of activities that involve tasks with algorithmic development allow a better understanding of programming concepts. Finally, 79.5% of the students indicate that the platform excites students to learn programming by showing direct, animated and visual results and 90.9% agree that using this and similar platforms motivates students' performance in the study of programming.

Future work from this research focuses on the development of meaningful learning activities with the use of the irobot platform for a longitudinal study and its effectiveness on academic performance.

Acknowledgements. The authors would like to thank the authorities of the Universidad Politécnica Salesiana for their support for this work, which is linked to the research project "Design of a research training methodology for new university professors and administrative staff" (with the acronym FINVE+P) of the GIEACI group, with Resolution No. 032-002-2024-02-27.

References

1. Cedeño-Tello, A., Llerena-Izquierdo, J.: Homogeneity of engineering courses from an assignment management model in virtual learning environments. In: International Conference on Science, Technology and Innovation for Society, pp. 439–447. Springer (2023). https://doi.org/10.1007/978-3-031-24327-1_38
2. Chen, X., Cheng, G., Zou, D., Zhong, B., Xie, H.: Artificial intelligent robots for precision education. Educational Technology & Society **26**(1), 171–186 (2023). https://www.jstor.org/stable/48707975

3. Darmawansah, D., Hwang, G.J., Chen, M., Liang, J.C.: Trends and research foci of robotics-based STEM education: a systematic review from diverse angles based on the technology-based learning model. Int. J. STEM Educ. **10**(1), 12 (2023). https://doi.org/10.1186/s40594-023-00400-3

4. Fan, K.Y.D., Dimiduk, K.C.: Using the matlab-based irobot create simulator to engage introductory computer programming students in program development and observing computational errors. In: 2011 Frontiers in Education Conference (FIE), pp. S2G–1–S2G–6 (2011). https://doi.org/10.1109/FIE.2011.6143104

5. Greifenstein, L., Graßl, I., Heuer, U., Fraser, G.: "help me solve it" or "solve it for me": Effects of feedback on children building and programming robots. In: Proceedings of the 55th ACM Technical Symposium on Computer Science Education V. 1. pp. 401–407. SIGCSE 2024, Association for Computing Machinery, New York, NY, USA (2024). https://doi.org/10.1145/3626252.3630752

6. Greifenstein, L., Heuer, U., Fraser, G.: Hint cards for common ozobot robot issues: Supporting feedback for learning programming in elementary schools. In: Proceedings of the 55th ACM Technical Symposium on Computer Science Education V. 1, pp. 408–414. SIGCSE 2024, Association for Computing Machinery, New York, NY, USA (2024). https://doi.org/10.1145/3626252.3630868

7. Gökçe, H., Gökçe, Z., Bektas, O., Saylan Krmzgül, A.: Robotic coding perceptions of middle school students. J. Educ. Future **1**(25), 31–44 (2024). https://doi.org/10.30786/jef.1274671

8. Lee, H.J., Yi, H.: Development of an onboard robotic platform for embedded programming education. Sensors **21**(11) (2021). https://doi.org/10.3390/s21113916

9. Lin, C.H., Liu, E.Z.F., Huang, Y.Y.: Exploring parents' perceptions towards educational robots: gender and socio-economic differences. British J. Educ. Techno. **43**(1) (2012). https://doi.org/10.1111/j.1467-8535.2011.01258.x

10. Llerena-Izquierdo, J.: Adaptation of the curriculum in relation to student learning outcomes in initial programming courses. In: 2023 IEEE World Engineering Education Conference (EDUNINE). pp. 1–6. IEEE (mar 2023). https://doi.org/10.1109/EDUNINE57531.2023.10102894

11. López-Chila, R., Llerena-Izquierdo, J., Sumba-Nacipucha, N., Cueva-Estrada, J.: artificial intelligence in higher education: an analysis of existing bibliometrics. Educ. Sci. **14**(1) (2024). https://doi.org/10.3390/educsci14010047

12. Lopez-Chila, R., Mora-Saltos, N., Cedeño-Tello, A., Llerena-Izquierdo, J.: A learning resource management model for high-enrollment programming courses in engineering. In: 2023 International Conference on Electrical, Communication and Computer Engineering (ICECCE), pp. 1–6 (2023). https://doi.org/10.1109/ICECCE61019.2023.10442311

13. Lytridis, C., Bazinas, C., Papakostas, G.A., Kaburlasos, V.: On measuring engagement level during child-robot interaction in education. In: Merdan, M., Lepuschitz, W., Koppensteiner, G., Balogh, R., Obdržálek, D. (eds.) Robotics in Education, pp. 3–13. Springer International Publishing, Cham (2020). https://doi.org/10.1007/978-3-030-26945-6_1

14. López-Belmonte, J., Segura-Robles, A., Moreno-Guerrero, A.J., Parra-González, M.E.: Robotics in education: A scientific mapping of the literature in web of science. Electronics **10**(3) (2021). https://doi.org/10.3390/electronics10030291

15. Rousouliotis, M., Vasileiou, M., Manos, N., Kavallieratou, E.: El greco platform: a novel python programming learning platform that uses a real robot. Comput. Appl. Eng. Educ. **n/a**(n/a), e22742 (2024). https://doi.org/10.1002/cae.22742

16. Shahmoradi, S., Kothiyal, A., Bruno, B., Dillenbourg, P.: Evaluation of teachers' orchestration tools usage in robotic classrooms. Educ. Inf. Technol. **29**(3), 3219–3256 (2024). https://doi.org/10.1007/s10639-023-11909-z
17. Vasconcelos, L., Gleasman, C., Umutlu, D., Kim, C.: Epistemic Agency in Preservice Teachers' Science Lessons with Robots. J. Sci. Educ. Technol. **33**(3), 400–410 (2024). https://doi.org/10.1007/s10956-024-10092-1
18. de Wit, S., Hermans, F., Specht, M., Aivaloglou, E.: Gender, social interactions and interests of characters illustrated in scratch and python programming books for children. In: Proceedings of the 55th ACM Technical Symposium on Computer Science Education V. 1. p. 262–268. SIGCSE 2024, Association for Computing Machinery, New York, NY, USA (2024). https://doi.org/10.1145/3626252.3630862
19. Yuan, J., Kim, C., Vasconcelos, L., Shin, M.Y., Gleasman, C., Umutlu, D.: Preservice elementary teachers' engineering design during a robotics project. Contemp. Issues Technol. Teacher Educ. **22**(1), 74–104 (March 2022). https://www.learntechlib.org/p/215681
20. Zhang, Y., Zhu, Y.: Effects of educational robotics on the creativity and problem-solving skills of k-12 students: a meta-analysis. Educ. Stud. , 1–19 (2022). https://doi.org/10.1080/03055698.2022.2107873

Pilot Channel Estimation for Improve the Non-Linear Effects on GFDM-MIMO Scheme

Juan Inga[1]([envelope]) [ORCID], Brayan Peñafiel[2] [ORCID], and Andrés Ortega[3] [ORCID]

[1] Ecuador, Grupo de Investigación de Telemática y Telecomunicaciones (GITEL), Universidad Politecnica Salesiana, Guayaquil, Ecuador
jinga@ups.edu.ec
[2] Universidad Tecnológica Empresarial De Guayaquil, Guayaquil, Ecuador
[3] Instituto de Telecomunicaciones Y Aplicaciones Multimedia (iTEAM), Universidad Politecnica De Valencia, España, Valencia, Spain
alortort@upv.edu.es

Abstract. This paper analyzes the performance of a generalized frequency division multiplexing scheme for multiple-input multiple-output configuration (MIMO-GFDM) in the presence of nonlinear power amplification with pilot insertion at the transmitter side to evaluate the saturation effects. For this purpose, two non-ideal transfer functions of high power amplifiers (HPA) are modeled: a Traveling Wave Tube Amplifier (TWTA) and a Solid State Power Amplifier (SSPA). In addition, for nonlinear distortion cancellation, we explore some peak-to-average power ratio (PAPR) reduction techniques, such as selective mapping (SLM), clipping, and clipping and filtering (C&F), to evaluate the best response and adaptability for the MIMO-GFDM scheme. The results show that an optimal power transmitter in the amplifier saturation region with large-scale input back-off (IBO) can mitigate out-of-band (OOB) radiation effects and consequently reduce the error floor in terms of bit error rate (BER) caused by amplifier saturation effects. In addition, we can control some trade-offs between PAPR, OOB emissions, BER performance, and power when the large-scale IBO reaches 4 dB. Finally, multicast effects in the Rayleigh channel can further saturate the amplifier when the transmit power increases, causing a proportional increase of the background noise with the channel variance. Thus, we can determine the maximum transmit power to avoid distortion and error-floor noise in poor channel conditions.

Keywords: HPA · MIMO–GFDM · PAPR · OOB · IBO

1   Introduction

Signal processing for MIMO-OFDM promises a high data rate in mobile communications. However, since the fifth generation (5G) and beyond mobile wireless transmission scenarios, the need to increase spectral efficiency, and maintain low latency and low out-of-band (OOB) power emission has led to the search for alternatives to OFDM. In this context, generalized frequency division multiplexing (GFDM) is one of the candidates for non-orthogonal waveforms [1,2] to achieve the requirements posed by 5G.

© The Author(s) 2025
E. M. Inga Ortega et al. (Eds.): CITIS 2024, LNNS 1331, pp. 12–24, 2025.
https://doi.org/10.1007/978-3-031-87065-1_2

Therefore, the PHY layer is the key to achieving massive connectivity in multiple access networks.

GFDM improves the high Peak-to-Average Power Ratio (PAPR) signals concerning OFDM, a significant advantage, especially for multi-carrier systems. On the other hand, pilot symbols are inserted in the signal transmission to improve the robustness of the system. In this context, [3] shows that the Pilot-aided channel directly affects the PAPR; this can lead to a trade-off that can be complex, especially when the power transmission is involved.

A novel technique is studied in [4] to insert the pilot symbols in the frequency domain for MIMO-GFDM to evaluate least square (LS) and statistical linear minimum mean square error (LMMSE) estimators. Additionally, the impact of pilot design on the signal properties is evaluated with Power Spectral Density (PSD) and PAPR. However, the main consideration for real communications systems is the insertion of High-Power Amplifiers (HPAs). Consequently, some disadvantages arise in the system performance: HPA produces non-linear distortion [5], and OOB emissions are degraded by the interference between adjacent sub-carriers and saturation amplifiers, affecting the spectral efficiency. In this way, the presence of a trade-off between pilot allocation and PAPR reduction is highlighted, because it would help us to improve the spectral efficiency. In addition, the non-linear effects of power amplifiers in GFDM multi-carrier signals produce the spectral regrowth and increase the high Bit Error Rate (BER), there being a trade-off, as is demonstrated in [6]. For this reason, it is necessary to design a robust and low complexity scheme that redefines the design of future physical layer architectures for the great demands of wireless communications, in terms of spectral and power efficiency, to enable the large scale of massively parallel non-linear processing [7].

To avoid such distortion effects, there is the need to use High Power Amplifiers (HPAs) with a large input back-off (IBO) in the transmitted signal. Optimizing the IBO level could limit the working zone of the amplifier to avoid the saturation zone, and thus distort the signal transmission. However, the IBO is inversely proportional to spectral regrowth, a sensitive interference parameter. On the other hand, to mitigate the non-linear effects, GFDM systems [8] could be analyzed for different parameters searching for a trade-off for optimal cost/performance systems. The Clipping technique for PAPR reduction causes severe effects on base-band and band-pass multi-carrier signals. It means that in-band and out-of-band emissions can be traded directly against reductions in the amplitude clipped signal.

Then, our hypothesis for overcoming non-linear distortion in multi-carrier systems is to combine the PAPR techniques with MIMO-GFDM systems to achieve an optimal point in the dynamic range of the amplifier for high power transmission. The PAPR reduction techniques are largely related to high power transmissions and consequently at non-linear effects [9]. However, a new technique called discrete cosine transform (DCT)-based orthogonal time frequency space (OTFS) [10] is promising because does not use a traditional technique for PAPR reduction, and it can reduce approximate 0.5 dB in CCDF of PAPR without compromising the BER.

1.1 Our Contribution

This work extends the investigation carried out in [6,11], the model can obtain an optimal point in the HPA saturation zone, increasing the large scale of input back-off (IBO) to cancel the error floor in the BER performance.

Our paper uses Solid State Power Amplifier (SSPA) and Traveling-Wave Tube Amplifier (TWTA) HPA characterization models, with non-lineal and non-ideal parameters curves operating in the saturation region.

The study applies a Rayleigh fading channel with two opposite scenarios to channel characterization using two criteria sigma noise, this one with solid interference, and another with interference cancellation. These two scenarios emulate the reliability to adapt a pre-distorter between GFDM Modulator and HPA block to know the optimal power transmission to cancel the error floor (BER) produced by non-linear distortion. The pre-distorter could be another scope of study.

The proposed design has been adapted to pilot-aided channel estimation [12], over the MIMO-GFDM block, due to the property of inserting the pilots into the GFDM block using the frequency domain. The number of investigations related to non-linear effects over MIMO-GFDM is limited. Based on the above inclusion and exclusion criteria, only 3 studies could be included in this review, as are shown in Table 1.

Table 1. Summary of related works

Author, Year	PAPR Reduction	Linearization HPA	Pilot-Aided CE	MIMO-OFDM	GFDM/FBMC/UFMC	MIMO-GFDM	Rayleigh Channel	AWGN Channel	Ideal HPA	Non-Ideal HPA
Danneberg, 2015 [13]	-	-	-	-	-	✓	-	✓	✓	-
Ortega, 2016 [6]	✓	-	-	-	✓	-	-	✓	✓	-
Chungbuk, 2016 [11]	-	-	-	✓	-	-	-	✓	-	✓
Aggarwal, 2018 [14]	-	-	-	✓	-	-	✓	-	✓	-
Farzamnia, 2018 [9]	✓	-	-	✓	-	-	-	✓	✓	-
Jayati, 2018 [5]	-	-	-	-	-	✓	-	✓	-	✓
Jayati, 2019 [15]	-	✓	-	-	-	✓	-	✓	-	✓
Kumar, 2021 [16]	✓	✓	-	-	✓	-	-	✓	-	-
Jayati, 2021 [17]	-	-	-	-	-✓	✓	-	✓	-	✓
Anughna, 2022 [18]	-	-	-	-	✓	✓	-	✓	-	-
Present Work, 2024	✓	-	✓	-	-	✓	✓	-	✓	✓

2 Problem Formulation

Figure 1 shows the MIMO-GFDM scheme. A source pseudo-random sequence $b[n] \in \{0,1\}$ is generated, where $n = \{1, 2, \cdots N\}$ represents the element array with length N.

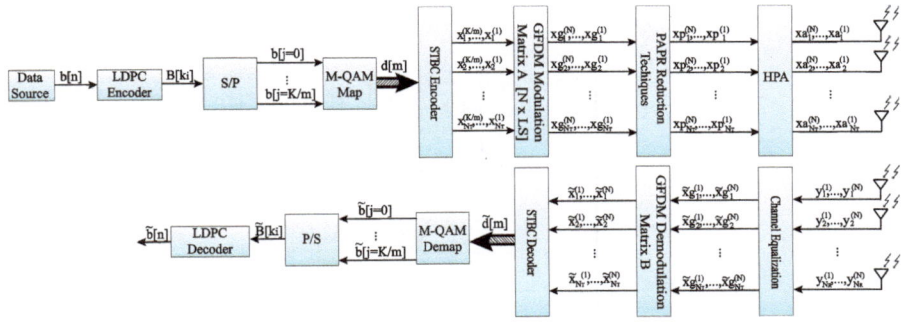

Fig. 1. General MIMO-GFDM scheme.

Next, an LDPC (Low-Density Parity Check) code is applied with a code rate of $R = 1/2$. The coded bits are represented by $b[k] \in \{0,1\}$ of K length, where $k = \{0, \cdots, K\}$. The coded rate has a $R = N/K$ ratio. LDPC has been designed through an irregular Gallager matrix taken from [19].

Therefore, the data is converted from serial to parallel through $b_j [i] \in b [k]$, being $i = [0, \cdots, m-1]$, where $j = \{0, \cdots, K/m\}$ and m is the modulation group M-QAM defined by $m = log_2 M$.

Thus, from the M-QAM modulator comes a sequence of complex symbols $d[j] \in \mathbb{C}^{(K/m) \times 1}$ which will be input to the STBC encoder in order to obtain multiple signals to be transmitted by N_T multiple antennas using the Alamouti scheme MIMO 2×2 and MIMO 4×4. The codeword $\overrightarrow{X}$ has $N_T \times K/m$ symbols and can be defined in matrix form as:

$$\overrightarrow{X} \left(\mathbf{x}_z^{(k/m)} = \begin{pmatrix} x_1^{(1)} & \cdots & x_1^{(k/m)} \\ x_2^{(1)} & \cdots & x_2^{(k/m)} \\ \vdots & \ddots & \vdots \\ x_{N_T}^{(1)} & \cdots & x_{N_T}^{(k/m)} \end{pmatrix} \right) \tag{1}$$

The pilots sequence are added to the codeword in order to obtain the signal modulated by GFDM block $\overrightarrow{X}' = \mathbf{x}_z + \mathbf{x}_{z,z}^{(pilot)}$, represented by N complex symbols to be transmitted by N_T antennas. This signal contains the pilot that is inserted in the time and frequency domain every $\Delta k = m+1$ position for channel estimation [12]. The pilot symbol arrangement is localized in the function of an Identity matrix with dimensions $\mathbf{x}_{z,z}^{(pilot)} = \mathbf{I}_{N_T, N_R}$, as is shown in Fig. 2.

Now, a GFDM modulator is added for each coded STBC, and thus, the MIMO generalized signal is made up. GFDM can increase the matrix dimension concerning the number of symbols to be transmitted. In this way, the GFDM signal has three processes: (i) oversampling, (ii) pulse shaping, and (iii) carrier conversion, defined in Equation (2).

$$xg_z^n = \sum_{l=0}^{L-1} \sum_{s=0}^{S-1} \left(\overrightarrow{X}' \right) g_z^{(sl)} [n] \forall z = \{1, \cdots, N_T\} \atop \forall n = \{1, \cdots, N\} \tag{2}$$

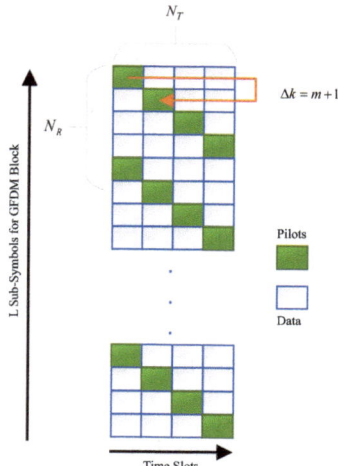

Fig. 2. Lattice pilot arrangement.

where $N = L * S$, is the matrix GFDM dimension distributed in L sub-symbols and S sub-carriers. The oversampling is defined through the pulse shaping filter in this way:

$$g_z^{(sl)}[n] = g[(n - sl) \quad mod(N)] e^{-j2\pi ns/N} \tag{3}$$

The modulation process is described by $\overrightarrow{Xg} = \mathbf{A} \times \overrightarrow{X'}$, where $\mathbf{A}$ is the transmission matrix $N \times N$ and $\overrightarrow{X'}$ is the signal information that contains the pilots symbols. The matrix representation of the output signal from the GFDM modulator block is:

$$\overrightarrow{Xg}\left(\mathbf{xg}_z^n = \begin{pmatrix} xg_1^{(1)} & \cdots & xg_1^{(N)} \\ xg_2^{(1)} & \cdots & xg_2^{(N)} \\ \vdots & \ddots & \vdots \\ xg_{N_T}^{(1)} & \cdots & xg_{N_T}^{(N)} \end{pmatrix} \right) \tag{4}$$

2.1 PAPR Reduction Techniques

To determine the probability level of an amplitude signal that does not exceed a threshold γ_{max}, the Complementary Cumulative Distribution Function (CCDF) is used. Based on the work presented in [6], the best techniques were selected, such as Clipping, Clipping and Filtering, and Selective Mapping (SLM). The output signal for Clipping and Clipping and Filtering is described as follows:

$$\overrightarrow{Xp} = \begin{cases} \sum\limits_{z=1}^{N_T} \sum\limits_{n=1}^{N} xg_z^n & if \ |xg_z^n| \leq \gamma_{max} \\ \sum\limits_{z=1}^{N_T} \sum\limits_{n=1}^{N} \gamma_{max} e^{j\theta[xg_z^n]} & if \ |xg_z^n| > \gamma_{max} \end{cases} \tag{5}$$

where γ_{max} limits so that the signal does not reach its saturation point, and $\theta_{[xg_z^n]}$ is the phase of complex signal $\overrightarrow{Xg}$. The Clipping Ratio (CR) helps the BER performance and it comes defined in the equation by $CR = \dfrac{\gamma_{max}}{\sqrt{\mathbb{E}\left(\overrightarrow{Xg}\right)^2}}$, where $\mathbb{E}$ refers to the expected value of the modulated signal $\overrightarrow{Xg}$ (Table 2).

Table 2. Parameters of Simulation.

Symbol	Values	Description
N	1920	Data bits
K	3840	Coded bits
M	16	length MQAM
L	15	GFDM subsymbols
S	128	GFDM subcarrier
BW [MHz]	500	Bandwidth
CR	2 dB	Clipping Ratio
V	16	Number of sequences SLM
Λ	7 dB	Threshold Saturation
N_T, N_R	1, 2, 4	Number of antennas TX/RX
MP	15	Number of multi-paths
$f_D[Hz]$	100	Doppler Frequency
σ_n^2	[0.1, 1]	Noise Variance

The SLM technique differs from the two previous ones because it takes the signal $\overrightarrow{\mathbf{X}}$ from the STBC encoder to multiply by different sequences V that have different phase values $q \in \{-1, 1\}$. In this way, for every data frame, the coding is multiplied for every sequence phase $q_v^{K/m}$, where $v = 1, \cdots, V$ with a length of K/m. This process is carried out for each N_T antenna. Therefore, the V candidate signals are modulated for each GFDM block, as is described in Eq. (2).

$$xg_{(z,v)}^{(n)} = \sum_{l=0}^{L-1}\sum_{s=0}^{S-1}\left(\vec{x}' q_v^{K/m}\right) g_z^{(sl)}[n]; \forall\, z = \{1, \cdots, N_T\}$$

$$\forall\, n = \{1, \cdots, N\}$$

(6)

When SLM is implemented, the evaluation of computational complexity is necessary, due to the number of IFFT blocks existing in GFDM. The scheme presented in Fig. 3 refers to a conventional SLM whose performance may be compromised in terms of both computational complexity and PAPR at the same time. This issue can be balanced through V sequences. Consequently, if V is increased, the PAPR is reduced, but the computational complexity (BigO) is also increased. Is possible adapt another techniques to offset this problem [20].

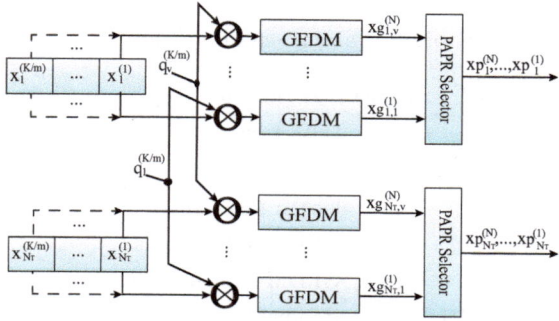

Fig. 3. MIMO-GFDM SLM scheme.

2.2 Pilot Channel Estimation

For pilot detection, we relate the above implication $\overrightarrow{X_a} \subseteq \overrightarrow{X'}$ that contains the pilot component $x_{z,z}^{(pilot)}$. The received signal at pilot-bearing is as follows:

$$\overrightarrow{Y} = \begin{bmatrix} xa_1{}^{(1)} \cdots xa_1{}^{(N)} \\ xa_2{}^{(1)} \cdots xa_2{}^{(N)} \\ \vdots \quad \ddots \quad \vdots \\ xa_{N_T}^{(1)} \cdots xa_{N_T}^{(N)} \end{bmatrix} \times F \begin{bmatrix} h_{1,1} \; h_{1,2} \cdots h_{1,z} \\ h_{2,1} \; h_{2,2} \cdots h_{2,z} \\ \vdots \quad \vdots \quad \ddots \quad \vdots \\ h_{z,1} \; h_{z,2} \cdots h_{z,z} \end{bmatrix} \begin{bmatrix} W_1 \\ W_2 \\ \vdots \\ W_z \end{bmatrix} \tag{7}$$

or

$$\overrightarrow{Y} = \sum_z^{N_T} \sum_z^{N_R} \left(\overrightarrow{X}_a \right) \cdot \overrightarrow{H}_{z,z} + \overrightarrow{W}_z \in \mathbb{C}^{N_T \times N}, z = \{1, \cdots, N_T\}, n = \{1, \cdots, N\} \tag{8}$$

where $\overrightarrow{H}_{z,z} = F \cdot \overrightarrow{h}_{z,z}$. Furthermore, the matrix of channel impulse responses is defined by $\overrightarrow{h}_{z,z} = [h_{1,1}, h_{1,2}, ..., h_{2,1}, h_{2,2}, ..., h_{N_T,N_R}] \in \mathbb{C}^{N_T \times N_R}$ where $N_T = N_R$. The DFT matrix F is introduced at the channel and it contains a deterministic term in the following way:

$$F = \sum_z^{N_T} \sum_z^{N_R} e^{-j \frac{2\pi(z-1)^2}{N_T}} \tag{9}$$

In addition, the interference term is considered a Rayleigh channel model with MP multi-path $\mathbf{W} \sim \mathcal{N}\left(u, \sigma_n^2\right) \in \mathbb{C}^{N_T \times N_R}$.

The multi-path is defined by $\mathbf{W} = \mathbf{w}_I + \mathbf{w}_Q$, where the real and imaginary components are detailed as follows:

$$w_I(nT_s) = \frac{1}{\sqrt{MP}} \sum_{m=1}^{MP} \cos\{2\pi f_D \cos(\varphi n T_s) + \alpha_m^*\} \tag{10}$$

$$w_Q(nT_s) = \frac{1}{\sqrt{MP}} \sum_{m=1}^{MP} \sin\{2\pi f_D \cos(\varphi n T_s) + \beta_m^*\} \tag{11}$$

The number of multi-paths is defined by MP, f_D is the Doppler Frequency, α^* and β^* are uniformly distributed between $[0, 2\pi]$, and finally T_s is the sample period.

To consider the worse effects of channel, two scenarios have been simulated with different values of variance $\sigma^2{}_n = E\left[|\mathbf{W}\mathbf{W}^H|^2\right]$ evenly separated by a decade among them. Thus, the results indicate the effects *i.* worsening of channel *ii.* cancellation of multi-path effects.

However, one can obtain the LS estimate of the channel impulse response by minimizing it with respect to $\mathbf{h}$. In this context, $\widehat{h}_{LSM} = Q_{LSM} \cdot \overrightarrow{Y}$, where Q_{LSM} is calculated in the next way:

$$Q_{LSM} = \left(\left(x_{z,z}^{(pilot)} \cdot F\right)^H \left(x_{z,z}^{(pilot)} \cdot F\right)\right)^{-1} \left(x_{z,z}^{(pilot)} \cdot F\right)^H \tag{12}$$

In this way, the distortion generated by the HPA amplifiers adds the effects of the MIMO channel interference and multi-path fading channel.

3   Analysis of Results

This work analyzed different metrics such as spectral regrowth, out-band emissions, PAPR, BER, applied to the MIMO-GFDM scheme for each HPA model. It determined the optimal IBO value, allowing the system to balance the system to reduce the noise floor caused by the non-linearity of amplifiers.

3.1   PAPR Analysis

The trade-off between PAPR and BER is critical to system performance, because both have the opposite effect. LDPC codes are used to improve the BER, as is shown in Figure 4{a,b. Clipping and Filtering shows exciting results due to the LDPC coded being added, achieving a PAPR < 3 dB when an HPA-SSPA model is used. This result is obtained from Equation (5). In addition, the Clipping technique achieves good performance when the MIMO system is enabled with respect to the SISO system. In Fig. 4b with LDPC coded, MIMO-GFDM 4 × 4 achieves the balancing of the trade-off between PAPR and BER.

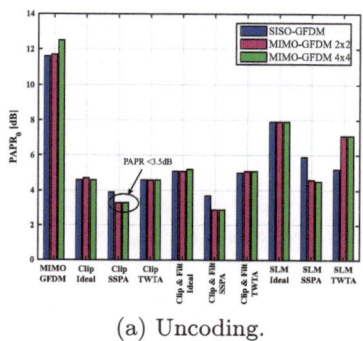

(a) Uncoding.

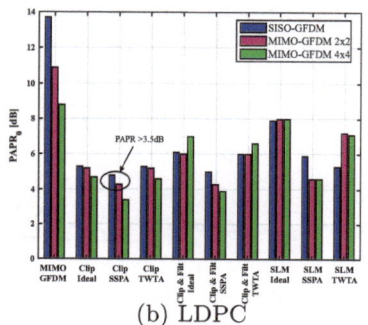

(b) LDPC

Fig. 4. PAPR reduction techniques.

3.2 IBO vs. OBB Analysis

The leakage OBB is evaluated concerning the input back-off IBO. To define the working point on the amplifier, the input back-off (IBO) in dB is defined as $IBO = 10 * log_{10}(P_{max}/P_{in})$, being $P_{max} = (\Lambda)^2/2$, and Λ is the saturation level at the input of the power amplifier. When the OOB is low, the system works in a saturation zone, where the efficiency of different PAPR techniques can be evaluated when subjected to high power levels.

As is shown in Fig. 5a, c, the HPA-SSPA models present a big leakage OOB. The Clipping technique can not support high-level power concerning another technique. For LDPC uncoded, the leakage OOB is not dependent on several antennas for MIMO systems. When IBO is increased, the leakage OOB tends to stabilize in a noise floor of -37 dB. On the other hand, the leakage OOB is increased for all PAPR reduction techniques, except in SLM when the HPA-TWTA model is used.

The spectral regrowth for the Clipping and SLM technique has similar results with both amplifier models, as is shown in Fig. 6a,c respectively. The results show a spectral broadening at -20 dB concerning the GFDM spectrum.

The best performance is the Clipping and Filtering technique using the HPA-TWTA model, since it is closer to the ideal GFDM spectrum. In this case, the spectrum is degraded at -50 dB, as is shown in Fig. 6b.

3.3 Spectral Regrowth Analysis

The spectral regrowth is the Power Spectral Density evaluation concerning the frequency. This parameter is related to spectral efficiency, and it can be susceptible when the non-linear amplifier is considered. This is because the signal transmission is saturated when the power increases, thereby causing bandwidth loss and interference in neighboring bands.

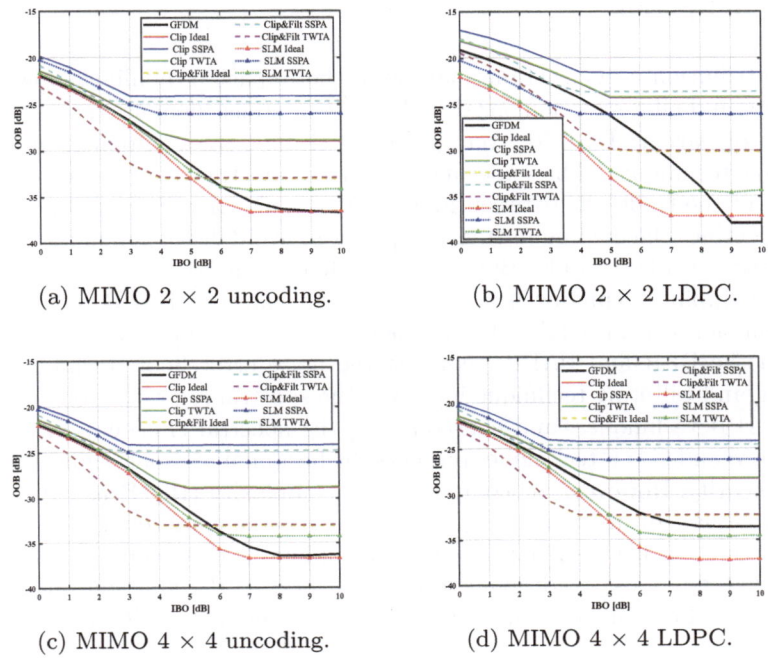

(a) MIMO 2 × 2 uncoding.   (b) MIMO 2 × 2 LDPC.

(c) MIMO 4 × 4 uncoding.   (d) MIMO 4 × 4 LDPC.

Fig. 5. IBO vs. OOB for MIMO-GFDM scheme

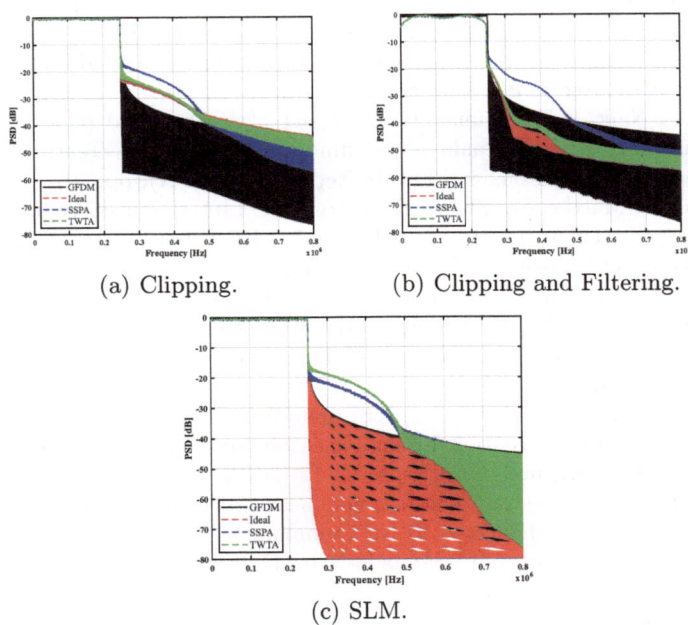

(a) Clipping.   (b) Clipping and Filtering.

(c) SLM.

Fig. 6. PSD of MIMO-GFDM signal.

4   Conclusions

Our contribution is focused on analysis of the impact of nonlinear high-power amplifiers and pilot insertion on GFDM-MIMO systems, demonstrating the effectiveness of various Peak-to-Average Power Ratio (SLM, clipping, and C&F) reduction techniques. The experiment offers interesting insights, especially in relation to the trade-offs between PAPR, OOB emissions, BER and power. The study identifies the maximum transmit power to avoid distortion in poor channel conditions, with large-scale IBO due to amplifier saturation.

The Clipping technique has a higher performance in PAPR reduction and OOB emission when the HPA-TWTA model is used. Clipping achieves a balance between the BER and PAPR, reducing the noise error floor. In addition, the Clipping and Filtering technique improves the spectral regrowth concerning the other PAPR techniques, and consequently, the spectral efficiency. On another hand, SLM and C & F techniques presents the best analysis concerning the effects of nonlinearity. However, the computational complexity (BigO) used by SLM is undeniable. For future works could be considered the traditional techniques for PAPR reduction like as Adaptive SLM Technique with Companding or DCT-OTFS that don't use traditional techniques for PAPR reduction and reduced computational complexity but without compromising the BER performance.

References

1. Liu, Y., et al.: Evolution of noma toward next generation multiple access (ngma) for 6G. IEEE J. Sel. Areas Commun. **40**(4), 1037–1071 (2022)
2. Zhang, X., Wang, Z., Ning, X., Xie, H.: On the performance of gfdm assisted noma schemes. IEEE Access **8**, 88961–88968 (2020)
3. Singh, A., Naik, K.K., Kumar, C.S.: Impact of sc-fdma and pilots on papr and performance of power domain noma-ufmc system. In: 2018 Tenth International Conference on Ubiquitous and Future Networks (ICUFN), pp. 507–511 (2018)
4. Ehsanfar, S., Matthe, M., Zhang, D., Fettweis, G.: Interference-free pilots insertion for mimo-gfdm channel estimation. In: 2017 IEEE Wireless Communications and Networking Conference (WCNC), pp. 1–6 (2017)
5. Jayati, A.E., Suryani, T.: Characteristic of HPA nonlinear distortion effects in MIMO-GFDM systems. In: 9th International Conference on Information and Communication Technology Convergence: ICT Convergence Powered by Smart Intelligence, ICTC 2018, pp. 379–384 (2018)
6. Ortega, A., Fabbri, L., Tralli, V.: Performance evaluation of gfdm over nonlinear channel. In: 2016 International Conference on Information and Communication Technology Convergence (ICTC), pp. 12–17 (2016)
7. Nikitopoulos, K.: Massively parallel, nonlinear processing for 6g: Potential gains and further research challenges. IEEE Commun. Mag. **60**(1), 81–87 (2022)
8. Hilario-Tacuri, A., Fortes, J., Sampaio-Neto, R., Soncco, L., Donaires, D., Borja, J.: Performance evaluation of generalized frequency division multiplexing systems over non-linearities with memory. IEEE Access **7**, 119131–119139 (2019). https://doi.org/10.1109/ACCESS.2019.2936840

9. Farzamnia, A., Min, T.J., Fan, L.C., Moung, E.G., Haldar, M.K.: The non-linearity effect of high power amplitude on ofdm signal and solution to solve by using papr reduction. In: 2018 Fourth International Conference on Advances in Computing, Communication Automation (ICACCA), pp. 1–5 (2018)
10. Kalpage, N.V., Priya, P., Hong, Y.: DCT-based OTFS with reduced PAPR. IEEE Commun. Lett. **28**(1), 158–162 (2024). https://doi.org/10.1109/LCOMM.2023. 3337778
11. An, C., Kim, B., Ryu, H.G.: Design of w-ofdm and nonlinear performance comparison for 5g waveform. In: 2016 International Conference on Information and Communication Technology Convergence (ICTC), pp. 1006–1009 (2016)
12. Ehsanfar, S., Matthé, M., Zhang, D., Fettweis, G.: A study of pilot-aided channel estimation in mimo-gfdm systems. In: WSA 2016; 20th International ITG Workshop on Smart Antennas, pp. 1–8 (2016)
13. Danneberg, M.: Implementation of a 2 by 2 MIMO-GFDM transceiver for robust 5G networks. In: Proceedings of the International Symposium on Wireless Communication Systems 2016-April, pp. 236–240, (2015)
14. Aggarwal, P., Bohara, V.A.: Analytical characterization of dual-band multi-user MIMO-OFDM system with nonlinear transmitter constraints. IEEE Trans. Commun. **66**(10), 4536–4549 (2018)
15. Jayati, A.E., Wirawan, Suryani, T., Endroyono: Nonlinear distortion cancellation using predistorter in MIMO-GFDM systems. Electronics **8**(6), 620 (2019). https://doi.org/10.3390/electronics8060620
16. Kumar, P., Kansal, L., Gaba, G.S., Mounir, M., Sharma, A., Singh, P.K.: Impact of peak to average power ratio reduction techniques on generalized frequency division multiplexing for 5th generation systems. Comput. Electr. Eng. **95**, 107386 (2021). https://doi.org/10.1016/j.compeleceng.2021.107386
17. Jayati, A.E., Destyningtias, B.: The analysis of the high power amplifier distortion on the mimo-gfdm system. In: 2021 IEEE International Conference on Communication, Networks and Satellite (COMNETSAT), pp. 252–257 (2021)
18. Anughna, N., Ramesha, M.: Performance analysis on 5g waveform candidates for mimo technologies. In: 2022 IEEE 2nd Mysore Sub Section International Conference (MysuruCon), pp. 1–8 (2022)
19. Prieto, R., Abril, A., Ortega, A.: Experimental alamouti-stbc using ldpc codes for mimo channels over sdr systems. In: 2017 IEEE 30th Canadian Conference on Electrical and Computer Engineering (CCECE), pp. 1–5, (2017)
20. Teja Sai Vishnu Vardhan, D., Narendra Kumar, A., Vijaya Kumar, P., Chandra Kiran, K., Jyothiraditya, G., Ravi Raja, A.: Fusion of adaptive slm technique with companding for papr reduction in 5g mimo-ofdm system. In: 2023 Second International Conference on Electrical, Electronics, Information and Communication Technologies (ICEEICT), pp. 1–5 (2023)

Random Walks Sampling on the Facebook Network of the Massachusetts Institute of Technology Using Ant Colonies

Rodolfo Bojorque$^{(\boxtimes)}$ (ID), Andrea Plaza (ID), and Pilar Morquecho (ID)

Universidad Politécnica Salesiana, Cuenca 010102, Ecuador
{rbojorque,aplaza,mmorquechoy}@ups.edu.ec

Abstract. This study investigates the effectiveness of using Ant Colony Optimization (ACO) algorithms for random walks sampling in the Facebook network of the Massachusetts Institute of Technology (MIT). Random walks sampling is a crucial technique for network analysis, enabling an understanding of the network's state irrespective of the starting node. By implementing an ACO algorithm, this research demonstrates an efficient method of sampling that ensures all nodes are sampled with uniform probability. The ACO algorithm leverages heuristic methods to significantly reduce the warm-up time required to obtain a sample. Experimental results confirm that the ACO implementation achieves the expected outcomes, demonstrating its efficiency in random sampling by reducing the number of jumps needed. This reduction in warm-up time, along with the uniform sampling capability, positions ACO as a promising alternative to traditional random walk algorithms for network analysis. The findings underscore the potential of bio-inspired algorithms in enhancing network sampling methodologies, offering both theoretical and practical implications for future research in this domain.

Keywords: ant colony · small worlds · node sampling · random walks

1 Introduction

Predicting network behavior based on its current state is a well-studied phenomenon by [12]. Modern complex systems like the Internet, the World Wide Web, financial markets, biological systems, and neurosciences use network structures for communication, functioning, self-organization, and evolution. A challenge in these networks is the unknown global state; we can only know the state of the node where the system resides and possibly information about its neighbors. Therefore, performing random walks on these types of networks is essential. Various studies on random walks, such as Metropolis-Hastings and Centrifugal Walks, provide these possibilities. However, these algorithms require a high warm-up time, consisting of randomly jumping through different nodes of the graph. Our main contribution to this study is to optimize this warm-up

© The Author(s) 2025
E. M. Inga Ortega et al. (Eds.): CITIS 2024, LNNS 1331, pp. 25–34, 2025.
https://doi.org/10.1007/978-3-031-87065-1_3

time using ant colonies by reducing the number of jumps to find the stationary behavior.

Ant Colony Optimization (ACO) is a metaheuristic optimization algorithm inspired by the foraging behavior of ants in nature. The algorithm seeks to find the optimal solution to a given problem by simulating the way ants communicate and collaborate to find the shortest path between their nest and a food source [7].

The core idea behind Ant Colony Optimization is that ants deposit a chemical substance called pheromone on the paths they traverse, and the more ants use a particular path, the stronger the pheromone trail becomes, making it more attractive for other ants to follow [7].

This work presents the results of random walks on a graph using a variation of the ant colony optimization algorithm to obtain random samples. The paper is organized as follows: The Review section defines the fundamental concepts of ant colonies, the Methodology section details the development and design of the experiments, and finally, the Conclusion section outlines the relevant findings and recommendations.

2 Background

2.1 Ant Colony

Ants are social insects that live in colonies and, through mutual collaboration, demonstrate complex behaviors and perform challenging tasks from the perspective of an individual ant. An interesting aspect of ant behavior is their ability to find the shortest paths between their nest and a food source [2]. This behavior has inspired the development of the Ant Colony Optimization (ACO) algorithm, a metaheuristic approach introduced in the early 1990 s by Dorigo and colleagues. ACO mimics the foraging behavior of ants to solve hard combinatorial optimization problems effectively [6].

ACO is utilized to solve optimization problems by exploring a large number of possible solutions. The algorithm operates by simulating the behavior of ants searching for food. Individual ants deposit a chemical substance called pheromone along their paths, creating a trail that other ants can follow. This process allows the colony to find the shortest path between their nest and a food source more efficiently [4]. The pheromone trail acts as a collective memory for the colony, influencing the decision-making process of individual ants [7,15].

The first application of the ACO algorithm was to solve the Traveling Salesman Problem (TSP), a problem that involves finding the shortest possible route that visits a set of cities and returns to the origin city. TSP is a classic example of an NP-hard problem, which makes it a suitable candidate for evaluating the performance of optimization algorithms. The success of ACO in solving TSP demonstrated its potential for addressing other complex optimization problems [5].

2.2   Max Min Ant System

The Max-Min Ant System is a variant of the Ant Colony Optimization algorithm that was developed by H. Hoos and T. Stützle in 1996. This algorithm is part of a family of algorithms known as Ant Colony Optimization, which is inspired by the behavior of real ants [19].The Max-Min Ant System differs from other algorithms within the family in a few key aspects.

First, in the Max-Min Ant System, the algorithm employs multiple ants (m) and iterates over them a maximum number of times (N max). During each iteration, the ants are tasked with finding an optimal solution to a given combinatorial optimization problem, such as the Traveling Salesman Problem.

To do so, the ants construct solutions by moving from one component to another based on a set of rules and heuristics. These rules and heuristics are derived from the foraging behavior of real ants, where they leave pheromone trails to communicate and navigate their environment. In the Max-Min Ant System, the pheromone trails play a crucial role in guiding the ants' search for optimal solutions. Furthermore, the Max-Min Ant System introduces a mechanism to update and manage the pheromone trails more effectively. Instead of updating the pheromone trails after each iteration, the Max-Min Ant System employs a strategy that only updates certain edges with high quality [9,20,21].

2.3   Related Works

Over the years, ACO has been applied to various domains, including the vehicle routing problem [16], job scheduling problem [13], and network routing[11]. Its flexibility and robustness make it a valuable tool for solving diverse optimization problems [1]. For instance, in the vehicle routing problem, ACO helps determine the most efficient routes for a fleet of vehicles to deliver goods to a set of customers [3].

One of the significant advantages of ACO is its ability to find near-optimal solutions in a reasonable amount of time. This characteristic is particularly important for real-world applications where finding the exact optimal solution may be computationally infeasible [20].

Additionally, ACO has been extended to incorporate various strategies to enhance its performance. For example, the Max-Min Ant System (MMAS) introduced by Stützle and Hoos limits the range of pheromone values to prevent premature convergence and improve solution quality [20].

Moreover, ACO has inspired the development of hybrid algorithms that combine it with other optimization techniques. These hybrid approaches leverage the strengths of ACO and other methods to achieve better performance. For example, the Hybrid Ant System combines ACO with local search techniques to refine the solutions obtained by the ants [8].

The success of ACO has also led to its application in dynamic and stochastic environments. Researchers have developed variations of ACO to address problems where the optimization landscape changes over time or where uncertainty is present. These extensions have broadened the applicability of ACO to a wide range of practical problems [10].

In summary, Ant Colony Optimization is a powerful metaheuristic algorithm inspired by the foraging behavior of ants. Its ability to solve complex combinatorial optimization problems efficiently has made it a popular choice in various fields. The ongoing research and development of ACO continue to expand its capabilities and applications, demonstrating its potential as a versatile optimization tool.

3   Methodology

The methodology was divided into three phases:

1. The Metropolis-Hastings random walks algorithm was used as the baseline [18].
2. Experimentation was conducted with our proposed ACO algorithm.
3. Results were compared.

This study used a graph representing the Facebook social network at the Massachusetts Institute of Technology (MIT). The primary objective was to propose a new random walk method based on ant colonies and determine the number of rounds (warm-up) necessary to find the stationary distribution in the graph using random walks and uniform distribution sampling.

The main characteristics of the graph are summarized in Table 1.

Table 1. Graph characteristics obtain with Igraph tool

Characteristic	Detail
Number of nodes	6.402
Number of edges	251.230
Average degree	78.785
Network diameter	8
Average clustering coefficient	0.284
Average path length	2.72

Additionally, the graph is connected and undirected, and its edges are unweighted. However, for this study, the weight of different edges will represent the pheromones left by the ants. The pheromones are initially set to zero, and with each step taken by an ant, this value increases by one. The evaporation rate used is 0.33, as indicated in [14].

A colony of ants is used, where all ants start the walk from the same node, moving randomly through the graph. There are two types of ants: explorers and workers.

Explorers leave a trail of pheromones along the path traveled. Each move evaluates the pheromones on the edges connecting neighboring nodes. In this study, the ants try to avoid paths overloaded with pheromones, making a path with more pheromones less likely to be chosen. This problem adopts a hybrid approach (Hybrid Ant System) where pheromones are used to find different solutions [2].

Pheromones diminish by a third of the increment value each time an ant does not choose a path with pheromones.

Figure 1 illustrates how the explorer ant moves from node 1 to node 3, increasing the pheromone quantity on edge 1–3 based on the number of rounds established for the walk. In contrast, edges 1–2 and 1–4 reduce their pheromone quantity by a third of the increment.

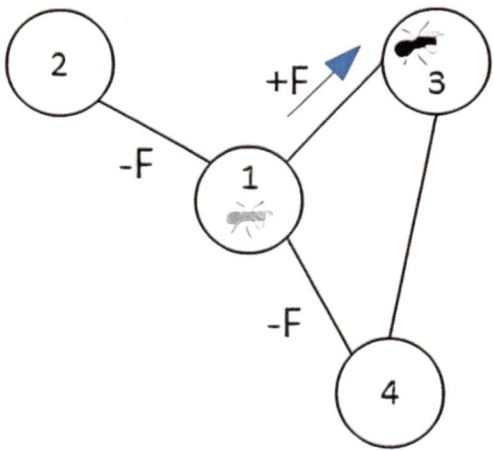

Fig. 1. Pheromones evaporation and assignation.

In this way, less-visited nodes are more likely to be chosen as explorer ants probabilistically prefer new paths, thus revealing the graph's complexity.

A novelty in the algorithm is the avoidance of pheromone paths based on the expectation that explorer ants seek to discover the entire graph for uniform sampling.

Worker ants, on the other hand, follow pheromone trails without altering them and cannot leave their pheromones. Their criterion is inversely proportional to the explorers; paths with pheromones are more likely to be visited.

To determine if the sampling is random, the mean error of the samples is calculated using the formula:

$$error = \frac{\sum_1^n \frac{samples_i}{totalM}}{n} \tag{1}$$

where:

- i represents the node
- n = Total number of nodes in the graph
- $samples_i$ = Total times node i has been sampled
- totalM = Total samples obtained in the entire walk

Random walks were performed starting from various nodes, all showing the same behavior represented in Fig. 2.

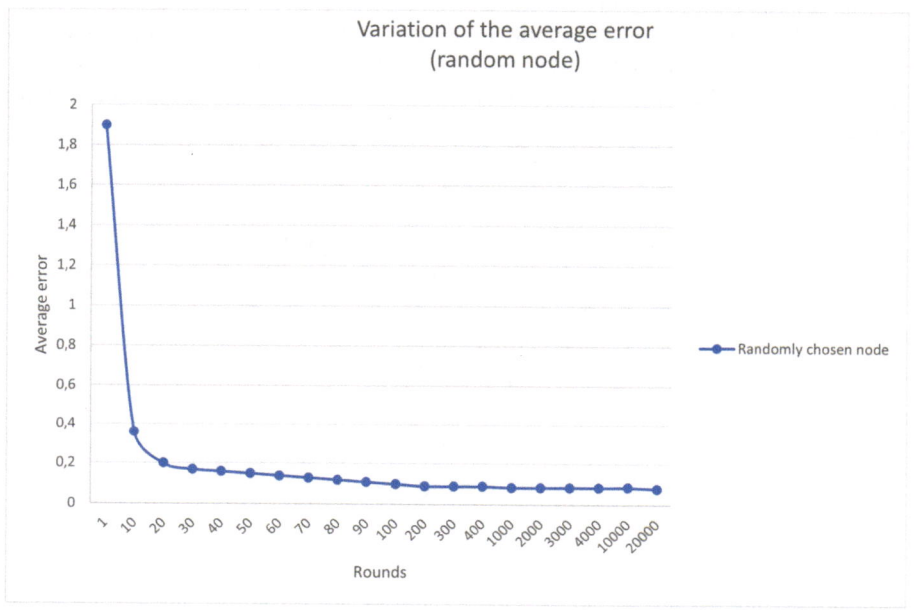

Fig. 2. Variation of the mean error of the random walk initiated by a node taken at random.

By default, 100 jumps are configured to consider a complete ant walk. The number of jumps is significantly reduced compared to other sampling methods like MHRW [18], but the computation required by the ant is much higher due to the need for more calculations to add and remove pheromones from the network.

4   Results

The results demonstrated that ACO significantly reduced the number of jumps needed to achieve uniform sampling compared to traditional methods like MHRW, which require thousands of jumps. Specifically, the ant colony approach achieved similar sampling distributions with only 200 jumps. This reduction is crucial for enhancing the efficiency of network analysis, particularly in large-scale networks where computational resources are a concern.

The analysis of results focuses on two aspects: the first is the sampling distribution, and the second is the results with larger sample sizes.

4.1   Sampling Distribution

Graphing several samples obtained after executing the ant walk consistently shows a uniform distribution represented in Fig. 3 that determines the average number of jumps that the MHRW algorithm (y-axis) must perform to obtain n different samples (x-axis). As can be seen, between 2000 and 2500 jumps provide the maximum number of possible samples [17]. This characteristic is important because social network graphs have nodes with high connectivity and nodes with only one edge.

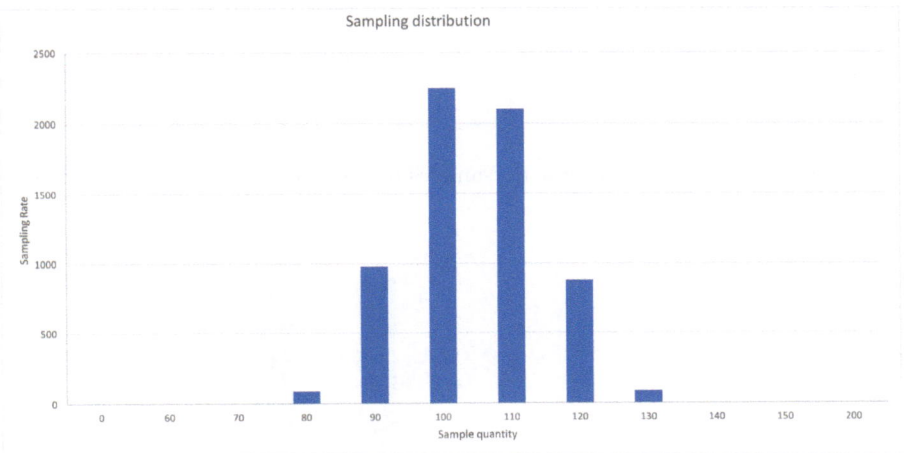

Fig. 3. Sampling Distribution.

In contrast, our approach requires 200 jumps to obtain the same number of samples. However, the estimated time to obtain a sample has increased exponentially because edge weights were used to determine pheromone quantities. Nevertheless, the algorithm reduces the number of jumps required by other sampling techniques.

4.2   Results with Larger Sample Sizes

The algorithm was also tested with a larger sample size, establishing a uniform distribution of 1000 samples per node. The results significantly improved the mean error value, as shown in Fig. 4, comparing the variation of mean errors for walks with 1000 samples per node and 100 samples per node.

The sampling distribution is much better, as observed in Fig. 5, with almost all samples distributed between 1000 and 1100 per node.

A disadvantage of this sampling is the computational cost; obtaining a sample can take around 2 to 15 min depending on the degree of different nodes.

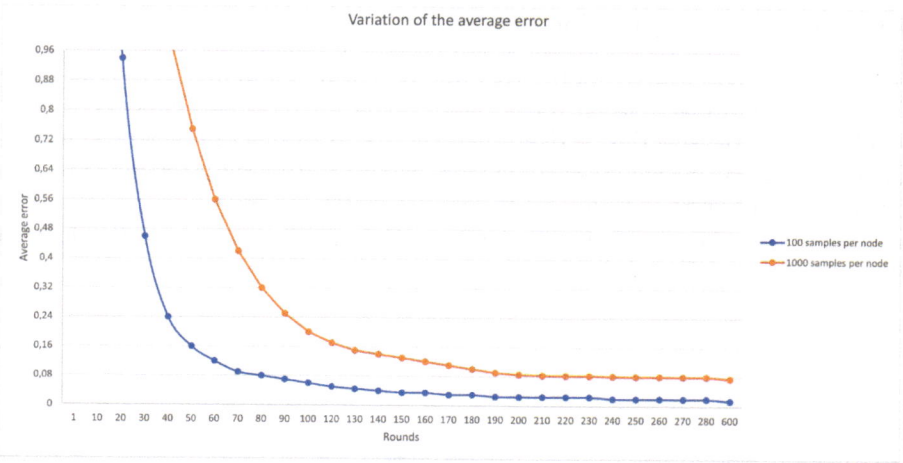

Fig. 4. Variation of mean error for samples of 1000 per node vs. 100 per node.

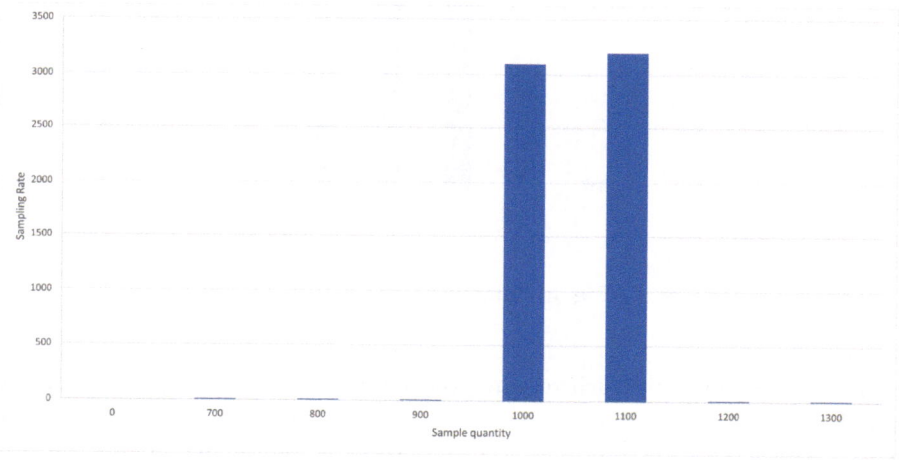

Fig. 5. Sampling Distribution with 1000 samples per node.

5   Conclutions

The study "Random Walks Sampling on the Facebook Network of the Massachusetts Institute of Technology Using Ant Colonies" explores the efficacy of using ant colony optimization (ACO) algorithms for random sampling in complex networks. The primary goal was to address the high warm-up time typically required by conventional random walk algorithms. The study employed an ACO algorithm that leveraged pheromone trails to optimize the process of obtaining uniform samples from the MIT Facebook network graph, which comprised 6,402 nodes and 251,230 edges.

Furthermore, the study highlighted the role of pheromones in guiding the ants' paths. Explorer ants, which leave pheromones on the paths they traverse, tend to avoid heavily pheromone-laden paths, promoting exploration and ensuring a more comprehensive coverage of the network. Worker ants, in contrast, follow these pheromone trails, facilitating a balanced sampling process that combines exploration with exploitation.

However, the computational cost associated with ACO was noted as a significant drawback. The process of calculating and updating pheromone levels for numerous nodes and edges increased the computational burden, particularly in graphs of substantial size. Despite this, the improved sampling efficiency offered by the ACO method provides a compelling trade-off, especially when accurate and uniform sampling is paramount.

In conclusion, the application of ant colonies for random walks in network sampling presents a promising alternative to traditional methods. The ACO algorithm's ability to reduce the warm-up time and number of jumps required for uniform sampling positions it as an effective tool for network analysis. Nevertheless, the computational cost remains a critical consideration, warranting further research into optimizing the algorithm's efficiency. This study underscores the potential of bio-inspired algorithms in solving complex problems in network theory and opens avenues for future exploration in enhancing their scalability and applicability.

References

1. Blum, C., Roli, A., Dorigo, M.: Hybrid metaheuristics: an emerging approach to optimization. In: Lecture Notes in Computer Science, vol. 26, pp. 1–30. Springer (2008)
2. Bonabeau, E., Dorigo, M., Theraulaz, G.: Swarm intelligence: from natural to artificial systems. Oxford University Press (1999)
3. Bullnheimer, B., Hartl, R.F., Strauss, C.: An improved ant system algorithm for the vehicle routing problem. Ann. Oper. Res. **89**, 319–328 (1999)
4. Dorigo, M., Blum, C.: Ant colony optimization theory: a survey. Theoret. Comput. Sci. **344**(2–3), 243–278 (2005)
5. Dorigo, M., Maniezzo, V., Colorni, A.: The ant system: optimization by a colony of cooperating agents. IEEE Transactions on Systems, Man, and Cybernetics, Part B (Cybernetics) **26**(1), 29–41 (1996)
6. Dorigo, M., Stützle, T.: Ant colony optimization. MIT Press (2004)
7. Dorigo, M., Stützle, T.: Ant colony optimization: overview and recent advances. Springer (2019)
8. Gambardella, L.M., Dorigo, M.: Ant-q: A reinforcement learning approach to the traveling salesman problem. In: Proceedings of ML-95, Twelfth International Conference on Machine Learning, pp. 252–260 (1995)
9. Gong, X., Rong, Z., Wang, J., Zhang, K., Yang, S.: A hybrid algorithm based on state-adaptive slime mold model and fractional-order ant system for the travelling salesman problem. Complex Intell. Syst. **9**(4), 3951–3970 (2023)

10. Gutjahr, W.J.: A graph-based ant system and its convergence. Futur. Gener. Comput. Syst. **16**(8), 873–888 (2000)
11. Kooshari, A., Fartash, M., Mihannezhad, P., Chahardoli, M., AkbariTorkestani, J., Nazari, S.: An optimization method in wireless sensor network routing and iot with water strider algorithm and ant colony optimization algorithm. Evol. Intel. **17**(3), 1527–1545 (2024)
12. Lee, C.H., Xu, X., Eun, D.Y.: Beyond random walk and metropolis-hastings samplers: why you should not backtrack for unbiased graph sampling. ACM SIGMETRICS Perform. Eval. Rev. **40**(1), 319–330 (2012)
13. Li, W., Yan, X., Huang, Y.: Cooperative-guided ant colony optimization with knowledge learning for job shop scheduling problem. Tsinghua Sci. Technol. **29**(5), 1283–1299 (2024)
14. Mavrovouniotis, M., Yang, S.: Ant colony optimization with self-adaptive evaporation rate in dynamic environments. In: 2014 IEEE Symposium on Computational Intelligence in Dynamic and Uncertain Environments (CIDUE), pp. 47–54 (2014). https://doi.org/10.1109/CIDUE.2014.7007866
15. Neroni, M.: Ant colony optimization with warm-up. Algorithms **14**(10) (2021). https://doi.org/10.3390/a14100295, https://www.mdpi.com/1999-4893/14/10/295
16. Ren, T., Luo, T., Jia, B., Yang, B., Wang, L., Xing, L.: Improved ant colony optimization for the vehicle routing problem with split pickup and split delivery. Swarm Evol. Comput. **77**, 101228 (2023)
17. Sevilla, A., Mozo, A., Anta, A.F.: Node sampling using random centrifugal walks. J. Comput. Sci. **11**, 34–45 (2015)
18. Stutzbach, D., Rejaie, R., Duffield, N., Sen, S., Willinger, W.: On unbiased sampling for unstructured peer-to-peer networks. In: Proceedings of the 6th ACM SIGCOMM conference on Internet measurement, pp. 27–40 (2006)
19. Stutzle, T., Hoos, H.: Max-min ant system and local search for the traveling salesman problem. In: Proceedings of 1997 IEEE International Conference on Evolutionary Computation (ICEC'97), pp. 309–314. IEEE (1997)
20. Stützle, T., Hoos, H.H.: Max-min ant system. Futur. Gener. Comput. Syst. **16**(8), 889–914 (2000)
21. Zhai, Y., Xu, L., Yang, Y.: Ant colony algorithm research based on pheromone update strategy. In: 2015 7th International Conference on Intelligent Human-Machine Systems and Cybernetics. vol. 1, pp. 38–41. IEEE (2015)

Open Access

Fuzzy Logic in Non-current Assets

María Alexandra Chávez Pullas[1]([⊠]) [iD], María Alejandra Zuñiga Alvarado[1]([⊠]) [iD],
and Miguel Paúl Herrera Estrella[2]([⊠]) [iD]

[1] Universidad Politécnica Salesiana, Guayaquil, Ecuador
{mchavez,mzunigaa}@ups.edu.ec
[2] Asamblea Nacional del Ecuador, Quito, Ecuador

Abstract. According to International Financial Reporting Standards, there can be two non-current asset valuations: The cost model and the fair value model. In this aspect, it is essential to find the causes of the large companies in Ecuador in the accounting policy choice about the non-current assets´ subsequent valuation. In the present investigation, fuzzy logic shows the causality of the model selection. Through the linear regression of fuzzy logic, this investigative work demonstrates that the decisions of the administrations of large companies in Ecuador are based on financial benefits. In this aspect, this research article contributes to the measurement of non-current assets in large companies in Ecuador, through the fuzzy logic method, thus providing a new perspective for analyzing management decisions, and contributing to future scientific research.

Keywords: Non-currents assets · fuzzy logic · IFRS. Subsequent valuation · Companies

1 Introduction

Worldwide, Accounting Standards are issued by two international institutions. The Financial Accounting Standards Board (FASB) issues the GAAP (Generally Accepted Accounting Principles) that are applied in the United States of America and the IFR (International Financial Reporting Standards) that are issued by the IASB (International Accounting Standards Board), which are applied in Europe and Latin America [1]. In Ecuador, IFRS has been applied since 2011, according to Resolution N0. 08.G.DSC.010 of 2008.11.20, R.O. No. 498 of 2008.12.31, published by Superintendencia de Compañías, Valores y Seguros, regulatory institution to whom companies must mandatorily submit their financial information annually.

The information presented in the financial statements should be useful for investors' decision-making therefore, the information presented should be considered reliable i.e. credible [2]. Puerta et al. [3] emphasize that financial analysis involves the study of asset performance as a source of substantial and relevant financial information. Under this scenario, the valuation of non-current assets is determinant in the reading of Ecuadorian entities' financial statements; this measurement according to IFRS is performed at the beginning under the historical cost method, and the subsequent measurement is

© The Author(s) 2025
E. M. Inga Ortega et al. (Eds.): CITIS 2024, LNNS 1331, pp. 35–45, 2025.
https://doi.org/10.1007/978-3-031-87065-1_4

performed at historical cost or fair value [4]. Because of the above, the dilemma of determining the subsequent valuation method arises. In this context, the present re- research aims to establish the determinants of the appropriate subsequent valuation model for non-current assets to reach a more relevant financial reading from an administrative point of view.

Historical cost is defined as the value traditionally established and subjugated to the acquisition price [5]. Smith and Smith [6] argue that the historical cost, from the accounting point of view, is developed by recording assets at amounts that can be verified and thus are reliable. As for fair value, according to International Financial Reporting Standard 13, it is defined as "The amount for which an asset can be exchanged, or a liability canceled, between an interested and duly informed buyer and seller who carry out a free transaction" Gómez et al., 2012, p. 614 [7] Linsmeier [8] establishes comparisons between historical cost and fair value and makes it clear that the neuralgic part of the assets' measurement is the subsequent value, given that the initial value is established at acquisition.

[9] Argue that the comparison between the historical cost model and fair value is relevant, especially in conditions of financial crisis. Tirado [10] states that in times of crisis, accounting information measured at historical cost is not advisable since the figures shown do not represent the financial reality of the entities. For this reason, the subsequent measurement of non-current assets at fair value is a better alternative because the accounting figures are recorded at market values. On the other hand, the research prepared by Palavecinos [11] indicates the existing contrast between advocates of historical cost and fair value. It is concluded that the cost model is useful when the economy is stable and there is no crisis, while the reliability of the fair value is based on the efficiency of the active markets.

Budrionytė and Gaižauskas [12] indicate that the choice between the historical cost model and the fair value model does not have a hard and fast answer since the choice of model is based on situations endemic to each entity such for example short- or long-term objectives. Under this scenario, the choice between the historical cost model and the fair value is strictly a matter of management decisions in the various corporations. Elsiefy and Elgammal [13] stipulate while it is true that International Financial Reporting Standards allow entity managers to choose between historical cost or fair value for the subsequent valuation of non-current assets, the fair value is more statistically related to enterprise value than the historical cost model.

To understand the choice of a model, whether it is a cost or fair value model, academics rely on the agency theory. This theory is responsible for explaining, understanding and pointing out the causes of why different managers make [14]. Agency theory addresses the conflicts of interest that arise between the owners of organizations and those in charge of managing them. In this scenario, incentives must be placed that manage to please the interests of both parties [15]. For Boučková [16] the agency theory is a contract in which one or more people agree with administrative professionals to execute actions that are stipulated in a consensus, with the purpose that the results favor the contracting parties. In this situation, it should be considered that one of the problems presented by the agency theory is the alignment of the objectives of the principals (contractors) and agents (contracted). This situation can lead to results that are not to the linking of the principals. Thus, principals give compensations or bonuses to the agents to favor the objectives set by the principals [17].

Therefore, the objective of this research is to determine the model of non-current assets' subsequent valuation that the various Ecuadorian entities apply in the financial statements, through a fuzzy logic analysis, which allows for defining the causes of the choice of the subsequent valuation model. Previous studies have implemented this fuzzy logic model, such as the one conducted by Mallo et al. [18] in Spain, on the valuation of intangible assets. Another study that argues that fuzzy logic is an advance for accounting sciences is the one published by Muñoz [19]. Castiblanco [20] conducts a study, where he examines with the fuzzy logic model the valuation of cash-generating units. Research by Luna et al. [21] concerning the analysis of financial reports of industries engaged in ceramics uses fuzzy logic to determine the decision-making of companies' management in the sector. Doskočil [22] uses fuzzy logic to determine the status of a preestablished project. Diaz et al. [23] use fuzzy logic to tidily examine the financial statements of the financial sector.

To evaluate the determinants of the non-current assets' subsequent valuation model in the present research, an alternative econometric model called fuzzy logic is established, which attributes to the notions of experts in subjects to be treated, which differs from the traditional models. The methodology explains what refers to the fuzzy logic method, in the following section shows the results, in which the PERT probability distribution is established, which is the most convenient for this research. The last section of the research describes the discussions and conclusions.

2 Methodology

Fuzzy Logic is a method applied for decision-making by various managers [20], allowing decisions made in the field of accounting sciences to be framed in the possibility of an event occurring, under this aspect, the opportunity is established to choose the model of the non-current assets subsequent valuation in Ecuadorian companies using fuzzy logic, through uncertain data that can be obtained at first sight, through the survey technique. In this way, fuzzy logic can measure the perceptions of the stakeholders with the likelihood of different scenarios [24].

In the present work, the Fuzzy-Delphi method is used, which contacts experts on the subject matter to be addressed [25] In this case, the research is directed to interview experts linked to the application of international financial Reporting Standards. Thus, an experton is constructed, which absorbs the expert´s perceptions of non-current assets subsequent valuation in Ecuadorian companies [26].

The selected experts are professionals in the accounting area, with approximately ten years of experience in the valuation of non-current assets and the management of companies in Ecuador. The selected experts are working in different positions in the accounting areas of prestigious auditing firms and important business groups. To obtain data, four areas are proposed: Auditing, Agency Theory, bank loans, and taxes. Each area has questions related to the choice of the historical cost or fair value model for the subsequent measurement of non-current assets in large companies in Ecuador. Table 1 shows the questions.

Table 1. Questions´ categorization

Question number	Description
1	Do you consider that large companies in Ecuador when valuing non-current assets use the historical cost model to avoid problems with auditing firms?
2	Do you consider that when valuing non-current assets with the fair value model the audit firm's fees increase and therefore prefer the historical cost model?
3	Do you believe that the fair value model for valuing non-current assets reliably represents the financial performance of large companies in Ecuador?
4	Do you consider that the objectives of the shareholders are aligned with the objectives of the directors if they apply the fair value model for the subsequent valuation of non-current assets?
5	Do you consider that for large companies in Ecuador the use of the fair value model to value non-current assets is more convenient than the historical cost model to access loans with the national financial system?
6	Do you consider that the municipal tax on assets is a determining factor in deciding between the historical cost model and fair value in large companies in Ecuador?

Source: Authors

The first two questions of the survey are related to the area of auditing. In Ecuador, an external audit report is mandatory for companies with assets of more than $ 500.000. In this regard, the largest companies in Ecuador have been evaluated by an auditing firm.

Therefore, the selection of the questions is based on the causality´s findings presented in the external audit reports. Habbash and Alghamdi [27] state that one of the circumstances in the choice of auditing firm is the fees and the findings that are found. For this reason, the first two questions are linked to situations that are related to external audits, given that the situation of the companies versus the fees and findings of external audits is very relevant.

Questions 3 and 4 are framed within the agency theory, for Barbei and Bauchet [28] the accounting situations are based on the agency theory. It was described in previous paragraphs that the managers are aligned with the objectives of the shareholders; in this aspect, it is necessary to establish whether the choice of the subsequent valuation model of non-current assets in large companies in Ecuador is under the agency theory. The importance of establishing that the company´s objectives are framed in the subsequent valuation of non-current assets is important because agency theory is a determining factor.

Question 5 addresses the area of bank loans since the various entities require their own or third-party financing for their growth. Martínez et. al [29] argue that almost all entities project economic growth and therefore require financing. In this context, financing from the financial system becomes necessary for various entities. Elsiefy and

Elgammal [13] argue that the financial sector is more aligned with market values. Given that the objective of the companies is to have sustained growth over time- agency theory- it is pertinent that the question is asked right after the questions related to agency theory.

Finally, question 6 focuses on the taxation part. According to Zhang et al. [30] fiscal costs play a determining role in the accounting companies´ policies. In Ecuador, there are sectional taxes whose tax base is the assets´ value. The question that relates to the area of taxes is asked in the final part of the survey since Deskins [31] argues that taxes modify business behavior concerning their accounting policies.

The experts´ answers are framed in the possibilities of choices´ occurrence model. Following the fundamental semantic scale of Kaufmann and Gil Aluja [32], it is determined that the answers of the present scientific research have an endecadary scale. The responses are illustrated in the following scale.

Table 2. Endecadary´s scale

Rating	Description
0	False
0,10	Practically false
0,20	Almost false
0,30	Quite false
0,40	More false than true
0,50	Neither false nor true
0,60	More true than false
0,70	Quite true
0,80 0,90 1	Amost true Practically true True

Source: Authors

The answers obtained by the experts are ordered in least probable or minimum values.

Quite probable values or mode and more probable or maximum values. In this way and following García [33] the PERT distribution is used for this research to find the data´s average value provided in the experts´ survey. This distribution necessarily requires the experts to provide three categorized answers as minimum, probable and maximum. The answers are quantified according to Table 2.

In the present research, twelve experts in the accounting area science were interviewed. As described in the previous paragraph, each expert gives three answers. Then the PERT distribution is applied to obtain the mean of answers provided. To establish causality, fuzzy regression is applied using fuzzyreg package with R program.

3   Results

Furthermore, with the data obtained from the survey of the twelve experts in the account-ing area, it proceeds with the probability distribution to obtain the values used in the econometric model. To proceed with PERT distribution, the minimum, most probable, and maximum values must be selected for each question asked. In this way, a table is prepared to illustrate the calculation of the PERT distribution. According to Sanchez and Solarte [34] the PERT distribution is obtained according to the following formula:

$$PERT = (minimum + probable * 4 + maximum)/6$$

The following table shows the results of the PERT distribution obtained from the results about 12 experts in the accounting sciences.

Table 3. PERT´S Distribution

	Question 1	Question 2	Question 3	Question 4	Question 5	Question 6
Minimum	0,1	0	0	0	0	0
Probable	0,1	0,1	0,7	0,7	1	0,4
Maximum	1	1	1	1	1	1
PERT	0,25	0,23	0,63	0,63	0,83	0,43

Source: Authors

The econometric model proposed in this scientific research is circumscribed within the linear regressionmodels in thisaspect is defined as follows where:

$$PERT = \beta 0 + EXP\beta 1 + \varepsilon$$

PERT = The PERT distribution
EXP = The number of questions asked in the survey β0 = The constant
β1 = The regression coefficient ε = Error term
The results obtained using the R program using the fuzzyreg library of the package of the same name are proved in three confidence intervals. Being the central tendency, the lower and upper limit. The R program obtains the result of the central tendency in the following regression.

$$PERT = 0.2311 + 0.0776 * EXP$$

The lower bound of the fuzzy regression is as follows:

$$PERT = 0.1333 + 0.05 * EXP$$

From the upper bound of the regression, it is:

$$PERT = 0.3333 + 0.1 * EXP$$

The root mean square distance between the response and the prediction is 0.16. Therefore, the model explains the determination of the posterior value model of the Ecuadorian companies' non-current assets is 0.84. Likewise, looking at the PERT distribution of each question shows that the most relevant question is question five which is related to financial loans.

The most relevant cause in the choice of non-current assets´ subsequent valuation model is financial loans. Table 3 shows that the PERT distribution is high, giving a probability of 83% in using the fair value model.

In question one the probability that companies use the cost model to avoid an unfavorable audit report is 25%. In this aspect, the choice of audit firm is not relevant for the determination of accounting policy to the non-current assets' subsequent valuation. Besides, question two shows that a determinant for the choice of models for the non-current assets´ subsequent valuation models is the audit firm´s fee. The result yielded a 23% probability of the event occurring. Thus, it is concluded that the choice of an auditing firm to decide the non-current assets subsequent valuation model for non- current assets is not decisive. It is inferred that the large companies in Ecuador, due to their economic capacity, have their accounting personnel updated according to International Financial Reporting Standards and that scenario the implementation of the International Financial Reporting Standards 16 is implemented correctly.

Question three addresses the financial field. In this aspect, the research shows that there is a 63% probability that the fair value model is selected by the entities to perform a better evaluation of their managers. Continuing with the analysis of this question, it is concluded that the obtaining of monetary resources is a determining factor for the choice of the subsequent valuation model for non-current assets, given that the financial ratios improve when calculated based on fair value. It should be noted that the continuous growth of companies is a main objective. There are two types of financing: external and internal. External financing is loans granted by banks. And own financing is the monetary resources of the shareholders. The result of the econometric model is congruent with this situation since it shows that large companies in Ecuador use the fair value model as their accounting policy for the subsequent valuation of non-current assets to access bank loans.

In the next question, the probability that shareholders agree with the fair value model is 63%. In this aspect, the research is consistent with the theory. The shareholders of the different entities seek profitability in their investments. To measure the profitability of their investments, the different shareholders analyze financial indexes, among the most required is the return on assets. This ratio measures the operating performance of the entity, this ratio is the most visualized by the shareholders since it reflects their investment´s profitability.

The penultimate question is limited to the taxes that exist in Ecuador that have non-current assets as their taxable table. The research illustrates a 43% probability of adopting the fair value model for the municipal taxes existing in Ecuador. In this aspect, the robust economic situation of the large companies in Ecuador means that the sectional taxes are not a determining factor in establishing the accounting policy for non-current assets. However, it should be clear that the situation changes when dealing with a small or medium-sized company.

4 Discussions and Conclusions

According to Zúñiga et al. [35] argue that the application of IFRS is to illustrate the essence of economic facts of the various entities and that the valuation of non-current assets is not alien to this situation. That is why the choice of the model from the noncurrent assets´ subsequent valuations is not a trivial matter and, on the contrary, it should be considered essential in the financial analysis of entities.

The valuation of non-current assets is substantial for the financial analysis of various entities. Some financial ratios are used in measuring the performance of managers. Return on assets is one of the most watched financial ratios by shareholders since they need to know. If their investments are profitable. For this reason, the fair value model is established as the most convenient Da Cunha and Mac Hado [36].

In this aspect, Guevara et al. [37] argue that the fair value model is immersed in the subjectivity of accounting science professionals. Therefore, they suggest that this model should be regularized to avoid fraudulent situations. In addition, they argue that in the absence of necessary regulations to avoid fraudulent situations, the historical cost model, which is more conservative, should be selected.

For Budrionyté and Gaizáuskas [12] the main disadvantage of the fair value model is that sometimes there is no market to value the assets and therefore, it is resorted to financial tools that are unreliable, so it suggests in this aspect the historical cost model. In this research, the proposed objective is met, since it is determined that the choice of non-current assets´ subsequent valuation is subject to the objectives of various entities. The econometric model used in this work proves the causality between administrative objectives and accounting policies, thus being congruent with the agency theory. As it is established, the agency theory is framed in the fulfillment of shareholders' interests. In this aspect, the growth of companies supported by external financing is a significant determinant for the choice of the non-current assets´ subsequent valuation. This is shown in Table 3 of question 5, which illustrates that there is an 83% probability of using the fair value model to the detriment of the historical cost model. Thus, the research concludes that the most relevant determinant for adopting the fair value model is financial loans, given that non-current assets are real guarantees that support the different loans requested.

Large companies in Ecuador, due to their size, are oriented to continuous economic growth, which is why their medium-term objective is to obtain sources of financing. For this purpose, the accounting policies of non-current assets´ subsequent valuation help to facilitate the opening of monetary resources of the institutions that belong to the financial system. In the present scientific research, it is established through the proposed econometric model that large companies in Ecuador are in that direction. It is concluded that the existing sectional taxes in Ecuador are less determinant when selecting the model for non-current assets´ subsequent valuation. In other words, large companies prefer to value their non-current assets as fair value, even if they must pay more taxes. It can be inferred that large companies in Ecuador prefer external financing to their financing.

The remains determinants according to the proposed econometric model, are not significant given that the application of fair value leads to additional economic resources for the entities, However, as demonstrated in this econometric model, the loans required for economic growth are a significant determinant in the choice of the fair value model. The present work nourishes future research that requires studying other variables to determine the choice of the non-current assets´ subsequent valuation model through the proposed econometric model or otherwise requires using the fuzzy logic model when it researches: financial, taxing, or administrative areas because fuzzy logic is a very useful alternative for future scientific research.

References

1. Chávez, M., Herrera, M.: Los impuestos diferidos en Ecuador para la reducción de la elusión fiscal. VinculaTégica, pp. 192–201 (2019)
2. Amat, O., Antón, M.: Fiabilidad de las información contable relativa al activo no corriente: Algunos problemas y propuesta de mejora. Boletín de Estudios Económicos LXVII(208), 23–38 (2013)
3. Puerta, F., Vergara, J., Huertas, N.: Análisis financiero: enfoques en su evolución. Criterio Libre 16(28), 85–104 (2018)
4. Argilés, J., Garcia, J., Monllau, T.: Fair value versus historical cost-based valuation for biological assets: Predictability of financial information. Revista de Contabilidad 14(2), 87–113 (2011). https://doi.org/10.1016/S1138-4891(11)70029-2
5. Castellanos, H.: Las acepciones de "valor" en el marco de las Normas Internacionales de Información Financiera. Actualidad Contable FACES, 0(19), 5–18 (2009)
6. Smith, S., Smith, K.: The Journey from Historical Cost Accounting to Fair Value Accounting: the Case of Acquisition Costs. Journal of Business & Accounting, 7(1), 3–10 (2014). Retrieved from http://search.ebscohost.com/login.aspx?direct=true&db=bth&AN=103044587&site=ehost-live&scope=site
7. Gómez, L., Romero, H., Sánchez, W.: Aproximación a una conceptualización del costo. Criterio Libre 17(30), 155–172 (2019). https://doi.org/10.18041/1900-0642/criteriolibre.201 9v17n30.5794
8. Linsmeier, T.J.: A Standard setter's framework for selecting between fair value and historical cost measurement attributes: a basis for discussion of "Does fair value accounting for nonfinancial assets pass the market test?" Review of Accounting Studies 18(3), 776–782 (2013). https://doi.org/10.1007/s11142-013-9238-7
9. Liao, L., Kang, H., Morris, R.: The value relevance of fair value and historical cost measurements during the financial crisis. Accounting and Finance 61(S1), 2069–2107 (2021). https://doi.org/10.1111/acfi.12655
10. Tirado, J.: Innovar Valor contable y la crisis financiera" : las entidades de crédito españolas. Revista Innovar 21(39), 113–122 (2011)
11. Palavecinos, B.: Valor razonable: un modelo de valoración incorporado en las normas internacionales de información financiera. Estudios Gerenciales 27(118), 97–114 (2011). https://doi.org/10.1016/S0123-5923(11)70148-6
12. Budrionytė, R., Gaižauskas, L.: Historical cost vs fair value in forest accounting: the case of Lithuania. Entrepreneurship and Sustainability Issues 6(1), 60–76 (2018). https://doi.org/10.9770/jesi.2018.6.1(5)
13. Elsiefy, E., ElGammal, W.: The effect of using fair value accounting on fundamental analysis: Some evidence from the emerging economies. The Journal of Developing Areas 51(3), 103–121 (2017). https://doi.org/10.1353/jda.2017.0063
14. Ganga, F., Quiroz, J., Maluk, S.: Qué hay de nuevo en la teoría de agencia (TA)? Prisma Social 15, 685–707 (2015)
15. Balcazar. A.: Gobernanza corporativa, una propuesta para el mejoramiento en la gestión administrativa y financiera en el hospital E.S.E nuestra señora del Carmen Tabio-Cundinamarca. Contaduría Universidad de Antioquia, (73), 13–32 (2018). https://doi.org/10.17533/udea.rc.n73a01

16. Boučková, M.: Management Accounting and Agency Theory. Procedia Economics and Finance **25**(15), 5–13 (2015). https://doi.org/10.1016/s2212-5671(15)00707-8
17. Brown, N., Droege, S.: Reflections on the generalization of agency theory: Cross-cultural considerations. Hum. Resour. Manag. Rev. **14**(3), 325–335 (2004). https://doi.org/10.1016/j.hrmr.2004.06.003
18. Mallo, P., Artola, M., Morettini, M., Galante, M., Pascual, M., Busetto, A.: Valuación de activos intangibles con matemática difusa y su adecuación a normas contables españolas e internacionales. Estudios de Economía Aplicada **26**(2), 139–159 (2008)
19. Muñoz, C.: La inteligencia artificial y la contabilidad. Lógica borrosa y representación del conocimiento. Lúmina **15**, 146–173 (2014)
20. Castiblanco, F.: Valor de uso de un activo o unidad generadora de efectivo bajo incertidumbre: el flujo de efectivo esperado mediante metodología borrosa. Cuadernos de Contabilidad **17**(44), 449–465 (2017). https://doi.org/10.11144/javeriana.cc17-44
21. Luna, K., Romero, R., Montes, Y.: Sistemas de información financiero en el sector industrial cerámico de Cuenca- Ecuador. Revista Ibérica de Sistemas y Tecnología de Información, pp. 143–156 (2021)
22. Doskočil, R.: Fuzzy logic: An instrument for the evaluation of project status. Revista de Metodos Cuantitativos Para La Economia y La Empresa **19**(1), 5–23 (2015)
23. Diaz, J., Coba, E., Hidalgo, C., Valencia, E., Bonilla, J.: Conjuntos borrosos aplicado al sector cooperativo del Ecuador. Política y Cultura **47**, 227–253 (2017)
24. Gutiérrez, J.: Aplicación de los conjuntos borrosos a las decisiones de inversión Aplicación de los conjuntos borrosos a las decisiones de inversión. AD-Minister, 0(9), 62–85 (2006)
25. Gil, A., Hernández, M.: The permanence of the client under uncertain estimations. Review, Fuzzy Economic, XVII **I**(2), 45–62 (2013)
26. Barcellos, L., Gil, A.: Algoritmo aplicado en el diálogo con los grupos de interés: un estudio de caso en una empresa del sector de turismo. Revista Del Departamento Académico de Ciencias Administrativas **10**(5), 76–85 (2010)
27. Habbash, M., Alghamdi. S.: Audit quality and earnings management in less developed economies: the case of Saudi Arabia. Journal of Management and Governance, **21**(2), 351–373 (2017). https://doi.org/10.1007/s10997-016-9347-3
28. Barbei, A., Bauchet, A.: Teoría contable positiva: una revisión de sus bases teóricas y la contribución a la teoría general contable. XXXV jornadas universitarias de contabilidad xiii jornadas universitarias de contabilidad de san juan facultad de ciencias económicas y empresariales, pp. 1–9 (2015)
29. Martinez, L.B., Belén Guercio, M., Corzo, L., Vigier Hernán, P.: Determinantes del financiamiento externo de las PyMEs del MERCOSUR. Revista Venezolana de Gerencia **22**(80), 672 (2018). https://doi.org/10.31876/revista.v22i80.23185
30. Zhang, L., Ru, Y., Li, J.: Optimal tax structure and public expenditure composition in a simple model of endogenous growth. Economic Modelling **59**, 352–360 (2016). https://doi.org/10.1016/j.econmod.2016.08.005
31. Deskins, J.A.: Essays on the behavioral effects of tax policy, 135. (2005). http://proquest.umi.com/pqdweb?did=920929211&Fmt=7&clientId=25620&RQT=309&VName=PQD [321(4), 91–102
32. Kaufmann, A., Gil Aluja, J.: Técnicas especiales para la gestión de expertos. Ed Milladoiro (1993)
33. García, J.: El método de subasta como complemento al PERT clásico. Estudios de Economía Aplicada **10**, 71–88 (1998)
34. Sanchez, L., Solarte, L.: Del Project Management Institute-PMBOK ® Guide las especificidades de la qestion de proyectos. Una revision critica Innovar **20**(37), 89–100 (2010)
35. Zuñiga, F., Pacheco, L., Díaz, J.: Convergencia contable: Cambios profundos en la contabilidad chilena. Activo fijo, un caso a considerar Capic Rewiew **7**, 75–82 (2009)

36. da Cunha, T., Machado, C.: Estudio y medición de la correlación entre el valor económico añadido y el valor de mercado agregado en un grupo empresarial cotizado en la bolsa. NYSE Euronext Cuadernos de Contabilidad **12**(31), 455–468 (2011)
37. Guevara-Sanabria, J.-A., Osorio-Ospina, J.-S., Pulgarín-Arias, A.-F.: Medición del valor razonable de los bienes inmuebles en Colombia: un análisis de la actividad de valuación y su coherencia con las NIIF. Contaduría Universidad de Antioquia **75**, 139–161 (2019). https://doi.org/10.17533/udea.rc.n75a06

Hate Speech Detection During the 2023 Chilean Plebiscite Constitutional Reform

Jimmy Paredes[1]([✉]) [iD], Erick Cuenca[1] [iD], Claudio Coloma[2] [iD], and Daniel Grimaldi[2] [iD]

[1] Yachay Tech University, Hacienda San José, Urcuquí 100119, Ecuador
{jimmy.paredes,ecuenca}@yachaytech.edu.ec
[2] Chile 21, San Sebastián 2807 of 416, Las Condes, Santiago de Chile, Chile
{c_coloma,dgrimaldi}@chile21.cl

Abstract. This work presents a classification of hate speech using Natural Language Processing approaches, including collecting and labeling the data, text augmentation using the back-translation technique to address the imbalanced class problem, and data preprocessing. This led to the creation of a model capable of classifying hate in tweets in Spanish from platform X in the context of the 2023 Chilean Constitutional Plebiscite. Results show that approaches based on Convolutional Neural Networks (CNNs) in 1 dimension obtained better results than approaches based on Machine Learning because CNNs can identify patterns and relations between consecutive words, making them more accurate in understanding the context of the tweet. The CNN model achieved an overall accuracy of 86% on the testing dataset, while Machine Learning approaches achieved between 79% and 81% on the testing dataset. It is essential to consider that since the dataset presents an imbalance in the classes, other metrics were also presented, such as precision, F1-score, and recall, where, once again, the best results were obtained using CNN.

Keywords: Hate speech · imbalance · Chilean Constitutional Plebiscite · tweets · CNNs

1 Introduction

In December 2023, Chilean citizens participated in a crucial event that would shape the future of their nation: the 2023 Chilean constitutional plebiscite. This national referendum marked a pivotal moment in Chile's democratic journey as it allowed voters to decide on adopting a freshly crafted constitution meticulously composed by a constitutional convention. During this historic plebiscite, Chileans exercised their democratic rights by casting their votes in favor of or against the proposed new constitution.

Hate speech detection plays a vital role in social media applications, where it is crucial to create a hate-free environment. Hate speech has grown notably on social media in the last decade [18, 21, 24, 25]. In fact, [23] explained that the volume of works related to the investigation of hate speech has grown since 2014; it explains that detecting hate speech in social media is essential. In this work, it is relevant to detect and classify hate in the X

© The Author(s) 2025
E. M. Inga Ortega et al. (Eds.): CITIS 2024, LNNS 1331, pp. 46–55, 2025.
https://doi.org/10.1007/978-3-031-87065-1_5

platform from Chilean users related to the campaign about the plebiscite constitutional 2023.

Some works use Convolutional Neural Networks (CNN) to classify hate speech. It is explained in more detail in [12], where the authors have used CNN in 1 dimension and a Hierarchical Attention-based model to detect hate in tweets, aiming that those models are useful to extract important and complex features from the text efficiently. In [13], the authors explained using a capsule network along with CNN, explaining that this model is vital for retrieving important features to detect hate; their experiments and results were performed on imbalanced and balanced datasets. Another important work is [20], where the authors use some Machine Learning approaches such as Logistic Regression, Support Vector Machines, Naive Bayes, and Random Forest Classifier; they also used CNN in 1 dimension to detect hate speech, the results obtained in this work was that the CNN model performs better results respect to the other models. In [26], the authors proposed a CNN model with GloVe embedding vectors in tweets to detect hate on them, it is explained that CNNs are able to capture the semantics of the sentence.

2 Methodology

In this context, the dataset obtained and the labeling are explained in detail. Then, the preprocessing chosen for this dataset is exposed, and the different techniques we applied to clean the data and obtain the necessary and highest quality data from the tweets. After that, to address the problem related to the imbalanced dataset, it was chosen to apply a text augmentation Back-translation to help the model not overfit, generalize to unseen data, and make the model more robust. So, the model created is explained in detail, including how it was created, the different layers, the results obtained related to the different metrics used, and the hyperparameters chosen for this case.

2.1 Data Collection

The dataset has been collected from the X platform (formerly known as Twitter). It is composed of 15692 Spanish tweets containing at least one of the following hashtags: *#AFavor, #AFavorDeChile, #Apruebo, #ChileEnContra, #ChileVotaInformado, #EnContra, and #Rechazo* or the keyword *Constitución*. Those hashtags are associated with the in favor and against the constitutional campaign between November 22, 2023, and December 18, 2023. For labeling the dataset, we used the Google API Perspective[1], which uses Machine Learning models to calculate the toxicity level in each tweet. To do it, the API calculates scores for different attributes: toxicity, severe toxicity, insult, profanity, identity attack, threat, and sexually explicit. The API returns a probability score for each attribute between 0 and 1, with 0 a lower probability of being a toxic text and one a higher probability of being a toxic text. We used the toxicity score.

Figure 1 shows a 12-interval cumulative frequency histogram where the height of each bar represents the number of tweets in that interval plus the number of tweets in all lower intervals. There are more tweets with lower toxicity levels than tweets with

---

[1] https://perspectiveapi.com/

higher toxicity levels. Based on that, we choose a threshold of 0.4, which means if the toxicity score of any tweet is bigger or equal to 0.4, then the tweet is labeled as "hate"; otherwise, it is labeled as "no-hate" ending up being a binary classification problem.

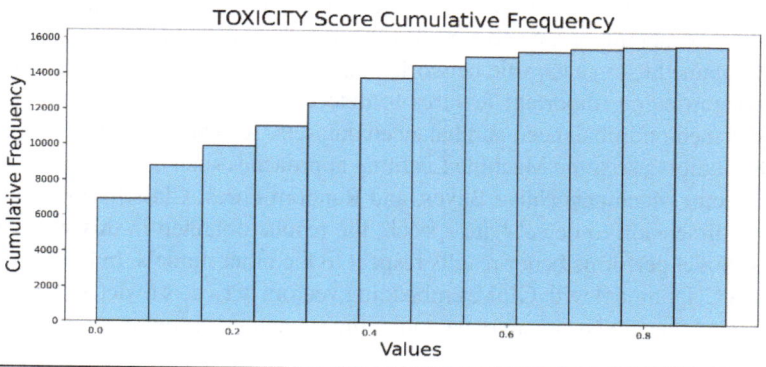

Fig. 1. Cumulative frequency histogram of the toxicity score in the dataset.

The dataset comprises 15,692 labeled samples, with 12,701 samples categorized as "no hate" and 2,991 as "hate". This class distribution reveals an inherent class imbalance, with a larger proportion of samples belonging to the "no hate" than the "hate" class. Class imbalances such as these can pose challenges for machine learning models, as they may struggle to effectively learn from the minority class due to its limited representation. Thus, the class imbalance will be addressed through oversampling to ensure robust performance across both classes.

2.2 Data Preprocessing

Data preprocessing is essential before training any artificial intelligence, machine learning, or deep learning model [9] because depending on the preprocessing steps applied, it can even influence the model learning [10]. The following preprocessing was applied:

- Removing HTML content because this information is irrelevant to the classification task of the work.
- @ and its mentions, as well as # hashtags, were deleted to reduce noise so that the model would focus on its main task. [5]
- Links and URLs to other pages were also removed, which do not provide meaningful information, so removing them will lead to obtaining data with more quality for its purpose [5].
- Stopwords were removed since they do not add significant meaning to the text [28].
- Punctuations, black spaces, numbers, and underscores were decided to be removed to help the model learn substantial patterns and relationships in the text [28].
- The capitalization of the text was preserved to maintain semantic meaning and context [6].

– Emojis and emoticons have not been modified since those sometimes add meaning to the text and could transmit emotions, clarify the intended meaning of the tweet, and reflect the way users naturally interact on the X platform [8].
– Lastly, lemmatization was applied, which consists of converting the words to their root or lemma [19], in other words, to their base form; it is widely used in Natural Language Processing (NLP) [14].

2.3 Text Augmentation

Following the labeling and preprocessing steps, it is necessary to analyze the distribution of the classes in the dataset. Figure 2a illustrates the imbalance in the dataset. The number of samples of "no-hate" (green hue) is 10160, while the number of "hate" (red hue) samples is 2393 on the training dataset. Consequently, to address this problem, a text data augmentation technique must be used.

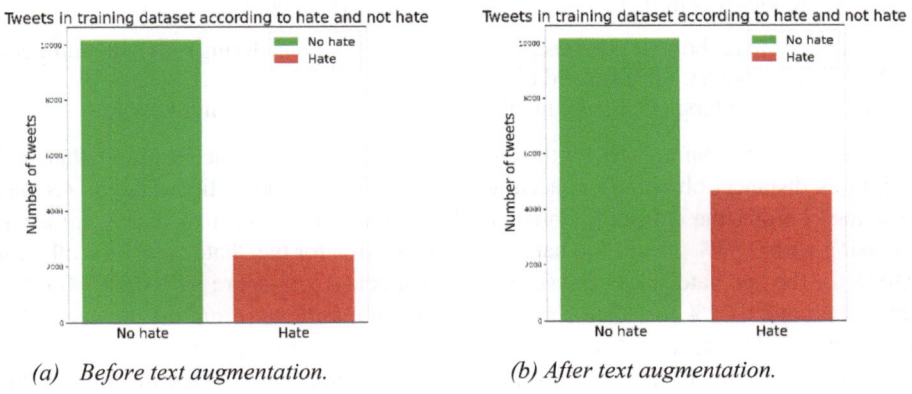

(a)   Before text augmentation.   (b) After text augmentation.

Fig. 2. Number of samples of "no-hate" and "hate" in the training dataset before (a) and after (b) applying text augmentation.

Several strategies exist for text augmentation [1, 27]. Some are generative data augmentation [2, 3], style augmentation [15], back-translation [4, 17], word replacements [11, 16], lexical substitution of word synonyms [7], and others. After an exhaustive analysis, we concluded to use the text augmentation technique back-translation, which consists of translating the text into another language English was chosen in this case since the dataset is in Spanish) and then translating it back to the original language of the data. This procedure used Google Translate services and the Python library *BackTranslation*.

It is vital to notice that before applying the chosen text augmentation technique, the dataset was divided into training, validation, and testing datasets being 80%, 10%, and 10%, respectively, considering the output column to obtain the same percentage of samples of each class on the three datasets because of its imbalance. The dataset division was performed to apply the data augmentation only to the training dataset to prevent data leakage, avoiding the model learning patterns from the testing or validation data. This step makes the model robust enough to correctly classify hate in new tweets. Finally, the number of "hate" samples increased to 4639 (see Fig. 2b). It was chosen not to augment the data again. Doing so could cause the model to overfit when training, as the augmented text could be similar to the original one.

3   Experiments

Before training the model, the words were tokenized. This step consists of converting every word to a token; the tokens can be words, symbols, or even phrases. This procedure is necessary when working with text because how the text is tokenized can even influence the model performance [22]. In this work, machine learning models such as K-Nearest Neighbors (KNN), Decision Tree, Logistic Regression, and deep learning models like Convolutional Neural Networks (CNN) were used to choose the more suitable model for detecting hate in Spanish tweets from the dataset.

3.1   K-Nearest Neighbors (KNN)

The KNN model was trained, and the hyper-parameters tuned were:

– Number of neighbors: It was tuned between 1 and 61, considering only odd numbers.
– Weight parameters: uniform and distance.
– Metrics parameters: euclidean, manhattan, nan_euclidean, and minkowski.

The best hyper-parameters for this model were 59 neighbors, metric Manhattan, and weight as distance, obtaining an accuracy of 0.8183% on the testing dataset. As was explained before, the dataset is imbalanced; for that reason, other metrics were used: precision with 0.83% for the "no hate" class and 0.61% for the "hate class"; recall with 0.98% for the "no hate" class and 0.13% for "hate class"; f1-score with 0.90% for "no hate" class and 0.22% for "hate class". As it can be interpreted from the results of this model, the KNN model tends to classify "no hate" tweets correctly but fails to classify "hate" tweets. Thus, KNN is not a suitable model for hate speech classification in this dataset. The confusion matrix obtained for KNN is shown in Fig. 3.

3.2   Decision Tree

The Decision Tree model was trained, and the hyper-parameters tuned were:

– Depth: It was tuned between 1 and 14.
– Minimum number of samples required to split an internal node: It was tuned between 2 and 24.

After the tuning phase, the best hyper-parameters obtained from this model were 1 for depth and 24 for a minimum number of samples to split; the model obtained an accuracy of 0.8095% on the testing dataset. The results obtained from the other metrics used are precision with 0.81% for the "no hate" class and 0.0% for the "hate class"; recall with 1.0% for the "no hate" class and 0.0% for the "hate class"; f1-score with 0.89% for "no hate" class and 0.0% for "hate class." So, the model cannot classify any tweet as hate; for this reason, it is possible to conclude that this model is not a good choice for classifying hate in this dataset. The confusion matrix obtained for the Decision Tree classifier is presented in Fig. 4, where it is possible to observe the results obtained for this model in the testing dataset.

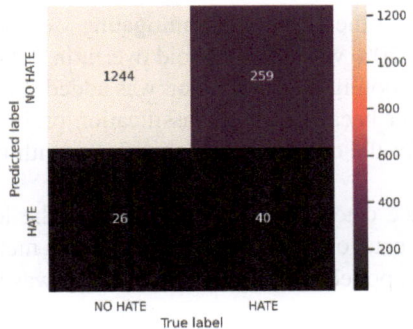
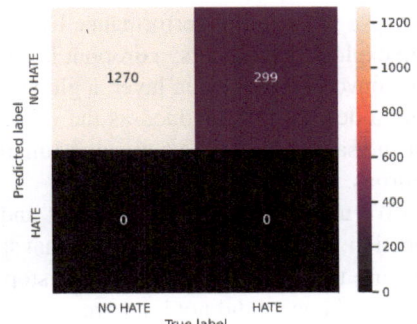

Fig. 3. Confusion matrix of the results obtained using *KNN* in the testing dataset.

Fig. 4. Confusion matrix of the results obtained using *Decision Tree* in the testing dataset.

3.3 Logistic Regression

Logistic Regression was also used with the following hyper-parameters:

– Penalty: 11, 12 and elasticnet.
– Multi class: auto, over and multinomial.
– Saga was used as solver.

Once the tuning phase ended, it returned the following hyper-parameters for this model: 12 for a penalty, auto for multi_class, and the solver used was a saga; with those hyper-parameters, the model obtained an accuracy of 0.7954% on the testing dataset. The results obtained from the other metrics used are precision with 0.82% for the "no hate" class and 0.35% for the "hate class"; recall with 0.96% for the "no hate" class and 0.08% for the "hate class"; f1- score with 0.88% for "no hate" class and 0.13% for "hate class". Analyzing the results obtained from the different metrics, it is concluded that this model is unsuitable for classifying hate speech in this dataset. In the confusion matrix shown in Fig. 5, it is possible to appreciate the performance results of Logistic Regression on the testing dataset.

3.4 Convolutional Neural Networks (CNN)

The CNN model was created using the high-level neural networks API *keras* and *keras tuner* for tuning the model and obtaining the best hyper-parameters; so, based on the tuning and using Bayesian optimization for searching the best configuration for the model, it was built a CNN in 1 dimension because of the text data. The structure of the model is the following: starting from an input layer, it contains 1 Convolutional hidden layer with 250 filters; the size of the convolution window was 6; the convolution window is important because it determines the number of words we want to associate. It used *ReLU* as an activation function and padding 'same' to obtain the same size after each layer and add more convolutional layers in 1 dimension if needed; the model also contains a fully connected hidden layer with 250 neurons and *ReLU* as activation function. After each hidden layer on the model, a batch normalization layer was used to improve the training performance by stabilizing the training and mitigating vanishing and exploding gradients; a dropout layer with 20% was used to avoid overfitting. After the convolutional hidden layer, a global max pooling in 1 dimension was added. In the end, a dense layer is added as the output layer because it is a classification problem; for the same reason, the activation function for the output layer was softmax with two neurons.

We used Adam as an optimizer and space categorical cross-entropy for the loss function; it is important to mention that sparse categorical accuracy was used as a metric because the data is sparse. An early stop was applied to ensure that the model was not overfitted once it did not improve.

After training the model with the hyper-parameters exposed above, it obtains an accuracy of 0.86% on the testing dataset. The results obtained from the other metrics are precision with 0.90% for the "no hate" class and 0.64% for the "hate class"; recall with 0.93% for the "no hate" class and 0.56% for the"hate class"; f1-score with 0.91% for "no hate" class and 0.60% for "hate class". The confusion matrix in Fig. 6 illustrates the results of the CNN model on the testing dataset.

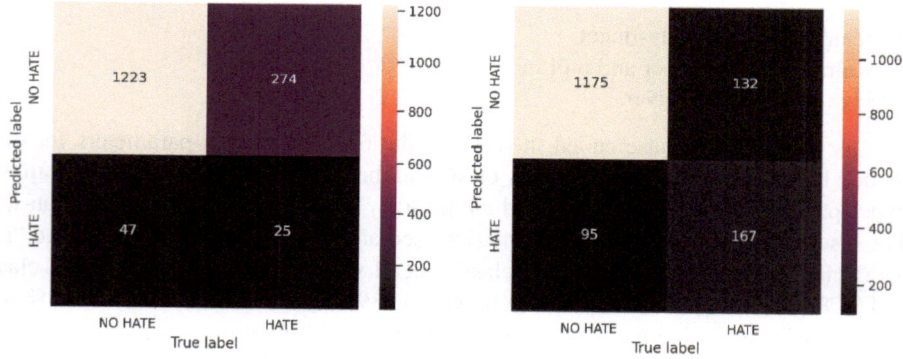

Fig. 5. Confusion matrix of the results obtained using *Logistic regression* in the testing dataset.

Fig. 6. Confusion matrix of the results obtained using *CNN* in the testing dataset.Results and Discussion

4   Results and Discussion

Among the models used, preliminary results depict a higher accuracy level obtained using CNN with an accuracy of 86% in the testing dataset. Considering that the dataset is still unbalanced even after applying the text augmentation technique (back-translation) is essential. Thus, the model must classify both categories (hate and no-hate) correctly. For this reason, and since the other models present a much lower percentage in other metrics, it is possible to observe that the most suitable model for classifying hate speech in the dataset used in this work is CNN in 1 dimension, it is because using CNNs those are capable of understanding patterns in the data; also those CNNs depending on the length of the window will understand the existing relation between consecutive words making them more powerful at the moment of train and test their result.

Several related works have used CNNs, explaining the importance of those models for detecting hate speech by extracting important patterns, understanding complex data, and capturing semantics from the sentences. The models are also useful for extracting features by themselves, making it possible to perform the detection of hate on social media. In [20], the authors compared different Machine Learning models with CNNs and concluded that the best results for hate speech detection were obtained with CNNs in 1 dimension in both metrics accuracy and macro-F1 score. In this work, we analyze the use of CNNs to address this problem over other Machine Learning approaches by comparing the results obtained from the metrics.

5 Conclusions

Our investigation deployed a variety of machine learning algorithms, including K-Nearest Neighbors (KNN), Decision Trees, and Logistic Regression, alongside deep learning approaches such as Convolutional Neural Networks (CNN). The comparative analysis aimed to ascertain the most effective model for identifying hate speech. Preliminary findings highlight CNN's superior efficacy, achieving an impressive accuracy of 86% on the test dataset. This outcome underscores CNN's potential to parse the nuances of hate speech in social media content.

However, the challenge of dataset imbalance persists despite efforts to mitigate it through text augmentation techniques like back-translation. Recognizing the importance of balanced classification across hate and non-hate categories, we extended our evaluation to encompass precision, recall, and the F1-score.

This work anticipates further refinement through hyperparameter tuning for all models examined. This future direction aims to enhance the models' sensitivity and specificity, thereby contributing to developing more robust and effective hate speech detection systems. Future works could explore the use of the models analyzed in this research not only for Spanish datasets but also across datasets from various languages to extend the application of the models to different linguistic environments. Through such advancements, we aspire to provide valuable tools for monitoring and countering hate speech on social media platforms, fostering a safer and more inclusive online environment.

References

1. Bayer, M., Kaufhold, M.A., Reuter, C.: A survey on data augmentation for text classification. ACM Comput. Surv. **55**(7), 1–39 (2022)
2. Cao, R., Lee, R.K.W.: Hategan: Adversarial generative-based data augmentation for hate speech detection. In: Proceedings of the 28th International Conference on Computational Linguistics. pp. 6327–6338 (2020)
3. Casula, C., Tonelli, S.: Generation-based data augmentation for offensive language detection: Is it worth it? In: Proceedings of the 17th Conference of the European Chapter of the Association for Computational Linguistics. pp. 3359–3377 (2023)
4. Cohen, S., Presil, D., Katz, O., Arbili, O., Messica, S., Rokach, L.: Enhancing social network hate detection using back translation and gpt-3 augmentations during training and test-time. Information Fusion **99**, 101887 (2023)

5. Defersha, N., Tune, K.: Detection of hate speech text in afan oromo social media using machine learning approach. Indian J. Sci. Technol. **14**(31), 2567–2578 (2021)

6. Dyer, L., Hughes, A., Shah, D., Can, B.: Comparison of token-and character-level approaches to restoration of spaces, punctuation, and capitalization in various languages. In: Proceedings of the 5th International Conference on Natural Language and Speech Processing (ICNLSP 2022). pp. 168–178 (2022)

7. Emmery, C., Kádár, A., Chrupała, G., Daelemans, W.: Cyberbullying classifiers are sensitive to model-agnostic perturbations. arXiv preprint arXiv:2201.06384 (2022)

8. Farnham, N.: Unlocking the pragmatics of emoji: Evaluation of the integration of pragmatic markers for sarcasm detection (2023)

9. Glazkova, A.: A comparison of text preprocessing techniques for hate and offensive speech detection in twitter. Soc. Netw. Anal. Min. **13**(1), 155 (2023)

10. Jahan, M.S., Oussalah, M.: A systematic review of hate speech automatic detection using natural language processing. Neurocomputing p. 126232 (2023)

11. Jahan, M.S., Oussalah, M., Beddia, D.R., Arhab, N., et al.: A comprehensive study on nlp data augmentation for hate speech detection: Legacy methods, bert, and llms. arXiv preprint arXiv:2404.00303 (2024)

12. Khan, S., et al.: Bichat: Bilstm with deep cnn and hierarchical attention for hate speech detection. Journal of King Saud University-Computer and Information Sciences **34**(7), 4335–4344 (2022)

13. Khan, S., et al.: Hcovbi-caps: hate speech detection using convolutional and bidirectional gated recurrent unit with capsule network. IEEE Access **10**, 7881–7894 (2022)

14. Khyani, D., Siddhartha, B., Niveditha, N., Divya, B.: An interpretation of lemmatization and stemming in natural language processing. Journal of University of Shanghai for Science and Technology **22**(10), 350–357 (2021)

15. MacRae, C.: Noledge: Creating an intelligent search tool for the florida state university computer science departmenbt using fine-tuned transformers and data augmentation (2022)

16. Madukwe, K.J., Gao, X., Xue, B.: Token replacement-based data augmentation methods for hate speech detection. World Wide Web **25**(3), 1129–1150 (2022)

17. Marie, B., Rubino, R., Fujita, A.: Tagged back-translation revisited: Why does it really work? In: Proceedings of the 58th Annual Meeting of the Association for Computational Linguistics. pp. 5990–5997 (2020)

18. Matamoros-Fernández, A., Farkas, J.: Racism, hate speech, and social media: A systematic review and critique. Television & new media **22**(2), 205–224 (2021)

19. Mehta, H., Passi, K.: Social media hate speech detection using explainable artificial intelligence (xai). Algorithms **15**(8), 291 (2022)

20. Ojo, O.E., Hoang, T.T., Gelbukh, A., Calvo, H., Sidorov, G., Adebanji, O.O.: Automatic hate speech detection using cnn model and word embedding. Computación y Sistemas **26**(2) (2022)

21. Pariyani, B., Shah, K., Shah, M., Vyas, T., Degadwala, S.: Hate speech detection in twitter using natural language processing. In: 2021 Third International Conference on Intelligent Communication Technologies and Virtual Mobile Networks (ICICV). pp. 1146–1152. IEEE (2021)

22. Park, K., Lee, J., Jang, S., Jung, D.: An empirical study of tokenization strategies for various korean nlp tasks. arXiv preprint arXiv:2010.02534 (2020)

23. Paz, M.A., Montero-Díaz, J., Moreno-Delgado, A.: Hate speech: A systematized review. SAGE Open **10**(4), 2158244020973022 (2020)

24. Pijal, W., Armijos, A., Llumiquinga, J., Lalvay, S., Allauca, S., Cuenca, E.: Spanish pre-trained catrbeto model for sentiment classification in twitter. In: 2022 Third International Conference on Information Systems and Software Technologies (ICI2ST). pp. 93–98. IEEE (2022)

25. Quelal, A., Brito, J., Lomas, M.S., Camacho, J., Andrade, A., Cuenca, E.: Identifying the political tendency of social bots in twitter using sentiment analysis: A use case of the 2021 ecuadorian general elections. In: Doctoral Symposium on Information and Communication Technologies. pp. 184–196. Springer (2022)
26. Roy, P.K., Tripathy, A.K., Das, T.K., Gao, X.Z.: A framework for hate speech detection using deep convolutional neural network. IEEE Access **8**, 204951–204962 (2020)
27. Shorten, C., Khoshgoftaar, T.M., Furht, B.: Text data augmentation for deep learning. Journal of big Data **8**(1), 101 (2021)
28. Tabassum, A., Patil, R.R.: A survey on text pre-processing & feature extraction techniques in natural language processing. International Research Journal of Engineering and Technology (IRJET) **7**(06), 4864–4867 (2020)

Inverted Classroom Strategy Applied to Students of Higher Basic General Education with Learning Difficulties in Mathematics Considering the Use of Information and Communication Technologies

Marco Rodríguez-Calle[1]([✉]) [iD] and Johnny Jiménez-Contreras[1,2] [iD]

[1] Universidad Politécnica Salesiana, Cuenca, Ecuador
mrodriguezc16@est.ups.ec, jjimenez@ups.edu.ec
[2] Grupo de Investigación en Enseñanza-Aprendizaje de Las Ciencias Para La Ingeniería,
GIEACI, Universidad Politécnica Salesiana, Cuenca, Ecuador

Abstract. In the search for improving the teaching and learning process in the face of the traditional ways of teaching in most educational institutions, the flipped and learning classroom strategy arises. This paper presents a proposal for the creation of virtual classrooms for students with learning disabilities. The methodology used in its first part is historical and descriptive and in its second part the research work uses the cuasi-experimental analytical empirical method with a quantitative and qualitative approach. The objective is to propose an inverted classroom strategy applied to upper basic general education students with learning difficulties in mathematics with the use of information and communication technologies, in a group of 20 students with limited access to the internet in a rural school in Gualaceo, Ecuador. The results obtained through evaluation instruments and surveys show improvement values in academic performance in 52.78% and in the level students acceptance to this new teaching strategy, in 70%.

Keywords: TIC · learning problems · inverted classroom

1 Introduction

Basic education is the most important education that a person receives, because it is here where knowledge is received, and from this, according to their educational level, they can deepen their rational and intellectual sense [1, 2]. In Latin America it has been determined that there are two problems related to education, the first is literacy and the second is mathematical skills, either due to cognitive development or emotional disorders, and if we consider educational institutions in rural areas, these problems increase even more. The countries that reach higher educational levels are those that have established that the best tool for growth is their educational system, even over their natural, economic and technological resources, because they consider that quality education is essential for

© The Author(s) 2025
E. M. Inga Ortega et al. (Eds.): CITIS 2024, LNNS 1331, pp. 56–68, 2025.
https://doi.org/10.1007/978-3-031-87065-1_6

the development of the individual and global progress, since if the individual develops, society will also benefit [3, 3]. Based on this, education becomes the engine that moves the economies of the world, so the more educated countries are, the better the political, social and economic development [5].

Learning mathematics is not only to ensure that students reach an optimal understanding of the material, but also that what they learn helps them in their ability to communicate, engage, represent and solve problems [6, 7]. In the case of mathematics instruction, recommendations have been made, such as that it should involve research by both students and teachers. To this end, teachers must possess the conceptual and pedagogical foundations necessary to be able to implement student-centered practices in the classroom [8, 9]. At present, there is still controversy about the definition of learning disability, however, it is pointed out that learning disabilities are neurodevelopmental disorders with biological basis and cognitive implications. The existence of a specific learning disorder is established, divided into three academic fields: (i). Specific learning disorder with difficulty in reading, (ii). Specific learning disorder with difficulty in writing and (iii). Specific learning disorder with disability in mathematics [10–12].

The advance in information and communication technologies (ICT) undoubtedly become a valuable tool in the educational field for teaching and learning [13]. Thus, ICT improve the opportunities to transmit data, content, knowledge and reading aids, contributes to student learning, as it facilitates access to information through technological devices, here they can find all the information they need, the ease of storing this information, contribute to achieving the objectives of their learning [14].For a student to learn successfully in a traditional classroom, four moments must be considered, namely, (i). Paying attention, (ii). Cognitive processing of the information presented, (iii). Taking notes, and (iv). Reviewing their notes. It is then clear that many students cannot apply these four moments, and therefore the contents presented by the teacher do not reach the students in an adequate way [15]. From this arises the need to seek new methods that help students to improve the learning process in mathematics, having as a valid alternative the inverted classroom model.

For this reason, this research seeks to establish an inverted classroom strategy applied to students in higher general basic education with learning difficulties in Mathematics with the use of information and communication technologies.

1.1 Educational Evolution: Participatory Students and Technological Tools

The current paradigm of learning is based exclusively on the active participation of the student. Thus, learning only happens when students leave their passive role as listeners and become active participants, which is why many researchers and educators have developed new methods that involve students as the center of learning. Nowadays the use of ICT and electronic devices are immersed in all fields of life, and education is no exception, so it is not to be ignored that technology becomes an effective tool to help students learn.

Combining these two elements, i.e., an active student and technology, new approaches and educational models are born, among which we have the inverted classroom [16, 17]. Under the digital environment that we are going through in our times, learning demands that teachers use ICT, in order to provide more accessible knowledge to students. Based on this, it is essential to motivate teachers to improve their teaching processes and to prepare their students for a more virtualized educational system [18]. To achieve better learning results, it is established that students should have a basic knowledge of ICT tools, since the lack of this knowledge will be a factor that limits learning [19].

1.2 Inverted Classroom and Its Applicability in Mathematics Learning

The academic performance of students can be improved and at the same time reduce learning problems in mathematics, proposing the alternative of an inverted classroom (Flipped Learnig), a method that is located within a student-centered framework, in which most of the time dedicated to learning is done through videos, since the idea is to attend classes with a previous knowledge base [20]. The inverted classroom model has the following objectives: to inspire and stimulate students to work actively individually, with their peers and with the teacher during the course [21].

According to the studies reviewed, the advantages of the inverted classroom include the following: complex and abstract theories can be presented in a simpler and easier way if the videos are reviewed several times by the students, this will allow a better understanding of a concept, promotes independent and personalized learning, stimulates discussion and collaboration among students, develops critical thinking and supports problem solving, increases creativity and communication, strengthens self-regulation in learning, and improves academic performance [22].

When talking about mathematics, it is common to hear that it is not exactly one of the favorite subjects for students, and this is because the motivation (meaning by motivation the willingness and desire to actively participate in learning) to learn mathematics is low, which undoubtedly leads to poor performance [23] The traditional way of teaching seems to be coming to an end, and it is at this moment where technology makes its appearance to support the change in the teaching of mathematics in high school [24]. However, not everything is positive, there are also drawbacks under this educational model (Flipped Learning), if the student does not understand the content of the video, it can cause a disconnection of learning at the moment he/she is in the classroom, which would cause an unsuccessful learning, all this would lead to lower motivation to learn and generate frustration in the student [22].

1.3 National and International Exploration of the Inverted Classroom in Mathematics Teaching

At the international level, research indicates that the provision of technological resources in the educational environment was identified as a major challenge for the implementation of flipped classroom and that in general the results obtained in their research indicate that there are promising possibilities for the implementation of the flipped classroom [25]. Likewise, the work of two Spanish authors, who argue that the use of the flipped classroom strategy in the teaching of mathematics has as its main advantage the interaction between teacher and student [26]. At the national level there are few works related to this topic, where they concluded that the contribution of the inverted classroom strategy for learning mathematics is beneficial, because students generate interest in interactive learning, the use of ICT motivates the learning of current students and it is there where the teacher has the opportunity to take advantage of these resources to improve their teaching practice [27, 28].

Although the works carried out at the national level coincide with the subject of mathematics and the age range of the students, it is necessary to emphasize that this research covers an important topic, such as the place in which it is developed, being a rural area, generally lacking the economic and technological conditions found in an urban area, what is attempted is to determine, how this condition can affect the application of the inverted classroom in the students.

2 Methods

First, the descriptive historical method was used to evaluate scientific articles with a global and regional focus; additionally, a bibliometric analysis was performed through VosViewer to identify the countries with the greatest scientific contribution in research on learning disabilities; the universities and research centers that present such research and the most relevant researchers in relation to the number of citations. The database evaluated was Web of Science. The search for these articles used keywords such as: ICT, learning disabilities, flipped and learning and inverted classroom.

Next, the empirical-analytical quasi-experimental method was used with a quantitative and qualitative approach. Among the strategies to be developed in the inverted classroom, we can point out the simplest and most accessible for the students, which consisted of assigning them a work topic so that they could research from home in printed sources such as school texts, books, encyclopedias or other sources and thus acquire the necessary knowledge for the moment of the face-to-face class.

Another strategy, which could be applied and considered ideal, is that in which the topics to be investigated by the students are carried out using technological resources such as the internet, computers, cell phones, etc., through a platform of the educational institution, however, the context in which this work was carried out is a public school located in a rural area, in which most of the students' families are in a rural area. However, the context of the school where this work was carried out is a public school located in a rural area, where most of the students' families are made up of parents who have a basic level of education and their knowledge in computer issues is scarce, which makes it difficult for them to be a support for their children in digital education issues, so it was not possible to apply this strategy in its entirety. Thus, after an analysis of the advantages and disadvantages of what would be the most viable strategy to improve mathematics performance in students with learning disabilities, at the time of applying the flipped classroom (flipped learning) and considering that the proposal of this research involved the use of information and communication technologies, it was decided to carry it out using technological devices, most of which were cell phones, because almost all students did not have computers at home, It was decided to carry it out using technological devices that were mostly cell phones, because almost all students did not have computers at home, in addition to the fact that students had internet signal at home and to solve the lack of a digital platform of the institution, it was decided to use a free and open access platform for students, such as Google's Classoroom; It was possible to upload all kinds of useful and necessary information for the teaching of the subject, such as explanatory videos of mathematical topics, PDF documents of the theoretical part that complemented the teaching, interactive sheets to solve online exercises and games where students interacted while learning by playing; it should be noted that these resources were previously selected from free and open access websites such as You Tube and Liveworksheets and other materials such as texts and games were developed by the teacher.

The research was directed to a group of high school students of the Basic Education School Dr. Mariano Cueva belonging to the rural parish Daniel Córdova of Gualaceo, Azuay-Ecuador; during the school year 2022–2023, the same that was applied in the fourth partial with a duration of 7 weeks.

In this study, 8th, 9th and 10th grade students participated freely and voluntarily, taking into account that there is only one parallel per year, the number of participants was limited (N = 44), and due to the context of the study, it was relevant to obtain information from all students, so it was finally decided to work with a convenience sample.

The students were between 12 and 15 years of age. For the selection of the sample, each year of elementary school was considered: eighth, ninth and tenth. In each classroom, the students were classified by academic performance on a scale of high, medium and low; noting that in this last scale were the students with learning difficulties in mathematics, for this a range of grades was determined, where all students who had grades lower than 7.00 over 10.00, according to the report card of the third partial, were part of this group, in total there were 20 students, distributed as shown in Table 1. This group was called group "A" while the other 24 students, who also participated in this research, were called group "B".

Table 1. Distribution of students according to their academic performance.

Academic Performance	Qualification	Learning Difficulty	Years of High School			Total
			8th	9th	10th	
High	Between 9 and 10	NO	4	3	1	8
Medium	Between 7 and 8.99	NO	5	4	7	16
Under	Less than 7	YES	8	6	6	20
Totals			17	13	14	44

The topics covered for each year of basic education were divided into blocks, and during the fourth partial period, the time in which the study was conducted, two blocks were assigned for each year, as detailed in Table 2 below.

The research is of an applicative type since it consisted of creating a virtual classroom in the Classroom application, for each year of high school, thus, 3 classrooms were formed, one for each year of study, 8th, 9th and 10th year, maintaining the basic structure of the inverted classroom; before class, during class and after class.

Table 2. Topics by years of basic distributed in blocks applied in the fourth partial.

Year of Basic	Blocks
8°	Laws of Logic and Functions; Statistics and Probability
9°	Geometry and Measurement; Statistics and Probability
10°	Trigonometric Ratios; Statistics and Probability

The following is a description of the flipped classroom process, which was developed for the application of this research:

1. Before classes: in this phase, students reviewed the contents created by the teacher, which consisted of resources such as: video tutorials (freely available on YouTube) on each topic, digital tools for practice and readings where they could solve any doubts that the previous activities could not clarify. As additional material, in each classroom, interactive games were included in one of the weeks, which, in addition to reinforcing the contents of the subject, motivated the students to learn by playing. In this way, the teacher identified the areas in which the students had difficulties and was able to plan the face-to-face class. At this stage, 53 videos with a total duration of 6 h and 11 min and 34 interactive activities from the Liveworksheets platform were used.
2. During the class: during this phase, the teacher summarized the material previously studied by the students and clarified the doubts that could be observed in the previous phase, the objective was to make the students remember and understand, which allowed them to apply and analyze concepts with the help of the teacher using an approach that involved the resolution of a practical case. Most of the time was used to solve problems, tasks and activities, both individually and in groups, and the students were given the corresponding feedback they required. Worksheets were developed for this phase.
3. After class: in this phase, content knowledge was reinforced with homework that students took home and solved autonomously. Here the teacher was able to provide explanations and additional resources, motivating students to share their knowledge with their classmates.

It is important to mention that the Inverted Classroom process with its three phases mentioned above was replicated during each of the 7 weeks that the study lasted, at the end of each week, an individual evaluation was made to each member of both groups"A" and "B", through questionnaires, with a 10-point scale, in order to collect the necessary information for subsequent quantitative analysis.

Academic performance was evaluated in two parts, the first one by means of formative activities where inputs and resources available on the network were used, such as Liveworksheet, which was intended for students to put into practice what was observed in the video tutorials and take any doubts that arose to class, to be solved together with the teacher and their classmates. Meanwhile, for the summative evaluation, inputs were prepared for the weekly tests, which were carried out individually and in the classroom.

All evaluations were taken in both groups, group "A" and group "B", at the same time and date.

Throughout this research, the connectivity of the students to study in the classroom was done only in their homes, since the school does not have an internet service, which limited the development of interactive activities within the institution. Most of the students used a cell phone as a connection device and few used a computer, because in rural areas, it is not common for families to have this type of equipment, mainly because of its high cost.

Regarding the perception and impact caused by the application of the Inverted Classroom strategy in students with learning problems in mathematics who participated in this research, and in order to obtain information for subsequent qualitative analysis, a 12-question questionnaire was designed, using the Likert scale, which was applied in person in each classroom.

3   Results

In order to evaluate the effectiveness of the application of the inverted classroom process, the academic performance of the students was analyzed. It is worth noting that initially there were 20 students in the low performance group, with grades below 7 points. This group represented 45.46% of the total number of students.

Table 3. Comparison of academic performance between groups, before the Inverted Classroom and after the Inverted Classroom.

Performance	Number of students before the Inverted Classroom	Percentage of students before Inverted Classroom	Number of students after Inverted Classroom	Percentage of students after Inverted Classroom
High	8	18,18%	14	31,82%
Medium	16	36,36%	21	47,73%
Under	20	45,46%	9	20,45%
Total	44	100,00%	44	100,00%

After the application of the inverted classroom, according to Table 3, it can be observed that this percentage is 20.45%, which represents an improvement of 25.01%. It is necessary to point out that as a result of this improvement, the other categories also showed changes in their composition, thus, in the medium level whose grades range from 7.00 to 8.99, it went from being with an initial percentage of 36.36% to 47.73%, that is, there was an increase of 11.37%, and in the high level whose grades are between 9.00 and 10.00; this started with 18.18% and ended with 31.82%, which represents an improvement of 13.64%.

Figure 1 shows the changes in the number of students by basic years who improved their performance in the category: Low performance.

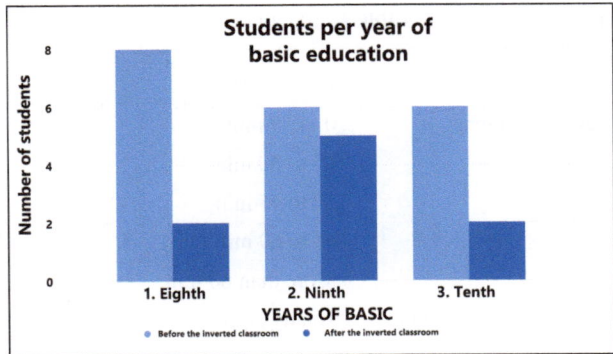

Fig. 1. Comparison of academic performance by years of basic education, before and after the application of the inverted classroom.

Figure 1 clearly shows that the eighth year was the grade where the best results were obtained, reaching an improvement percentage of 75%, followed by the tenth year, where the improvement percentage reached 66.67%, leaving the ninth year as the grade where the improvement percentages were the lowest, reaching only 16.67%.

Regarding the answers given to the questionnaire that was formulated to the 20 students with learning problems in mathematics, the results were as follows:

Table 4. Survey Analysis

N°	Variables	Dimensions	Results
1	Device used	Tablet	5%
		Cell phone	95%
2	Expectations of the inverted classroom	Videos help me understand the content	30%
		Learning mathematics in a better way	35%
		Improve my grades	20%
		Interactive activities complete the class objective	15%
3	Watching videos helps to understand content	Totally Agree	20%
		Agree	55%
		Undecided	20%
		In Disagreement	5%
4	The practical exercises helped	Totally Agree	20%
		Agree	55%
		Undecided	15%
		In Disagreement	10%

(continued)

Table 4. (*continued*)

N°	Variables	Dimensions	Results
5	Time to review the contents	0 to 15 min	10%
		15 to 30 min	40%
		30 to 45 min	30%
		45 to 60 min	10%
		More than 60 min	10%
6	I learn through classroom activities	Totally Agree	30%
		Agree	45%
		Undecided	20%
		In Disagreement	5%
7	Improved my mathematical understanding	Totally Agree	25%
		Agree	50%
		Undecided	20%
		In Disagreement	5%
8	Improved responsibility and independence	Totally Agree	10%
		Agree	70%
		Undecided	15%
		In Disagreement	5%
9	Increased my motivation	Totally Agree	30%
		Agree	50%
		Undecided	20%
10	Which model is best for learning	Inverted classroom model	70%
		Traditional classroom model	30%
11	The model would be useful in other subjects	Totally Agree	40%
		Agree	50%
		Undecided	5%
		Strongly Disagree	5%
12	What feeling did this experience generate	Trust	45%
		Satisfaction	25%
		Enthusiasm	10%
		Safety	10%
		Anxiety	5%
		Frustration	5%

According to the results shown in Table 4, in the survey it was determined that 95% of students used the cell phone for classes; 35% considered that the greatest expectation when starting with this strategy was that they could learn mathematics in a better way; 45% of the students dedicated between 15 to 30 min to see the contents in the classroom; 50% of students indicated that they improved their understanding in mathematics; 70% indicated that they were more responsible and independent in their studies; 70% considered that this inverted classroom model is better for their learning; finally, 90% of students indicated positive feelings such as confidence, satisfaction and security at the end of this study experience.

4 Discussion

Although this article has limitations with respect to the methodology used, it shows similarities with research conducted in Cuenca-Ecuador [28], where it was established that the Inverted Classroom strategy favors the teaching/learning process, since it provokes active and practical learning in students, so that they are the ones who generate their knowledge. Similarly, it was agreed that the Inverted Classroom favors the learning of mathematics, since it is an area of study, where systematic processes and logical reasoning are involved, it contributes to the students learning autonomously and at their own pace, from their own home.

In relation to academic performance, it was determined, as in the study conducted in Guayaquil -Ecuador [27], that the application of the Inverted Classroom, improved the average of the partial with respect to the period in which this strategy was not applied, evidencing the feasibility and effectiveness of the proposal.

5 Conclusions

Finally, after the results obtained and in accordance with the objectives proposed at the beginning of this work, the following conclusions can be drawn:

It was possible to establish the inverted classroom strategy for students of higher basic general education with learning difficulties in Mathematics with the use of ICT.

The application of the inverted classroom strategy based on ICT, provoked a different and positive reaction in the students, regarding the way of learning inside and outside the classroom.

It was possible to demonstrate that both qualitatively and quantitatively, the improvement of mathematical concepts, according to the results obtained and the assessments given by the students.

Being a new proposal in the institution, which involved the use of technology, it generated in them the necessary interest to know and use it, which can serve as an incentive for teachers to apply it in other subjects, as stated by the students themselves.

It should be considered that internet access for some students may be a limiting factor for the application of the flipped classroom. It is recommended that the sample of students be expanded for future research.

References

1. Cardona-Reyes, H., Ortiz-Esparza, M.A., Munoz-Arteaga, J.: Use of Learning Paths Through a Digital Ecosystem to Support Children with Learning Problems in Basic Math. Revista Iberoamericana de Tecnologias del Aprendizaje **17**(1), 79–88 (2022). https://doi.org/10.1109/RITA.2022.3149784

2. Visvizi, A., Daniela, L., Chen, C.W.: Beyond the ICT- and sustainability hypes: A case for quality education. Comput Human Behav **107**, 2018–2020 (2020). https://doi.org/10.1016/j.chb.2020.106304

3. Ruiz-Jiménez, M.C., Martínez-Jiménez, R., Licerán-Gutiérrez, A., García-Martí, E.: Students' attitude: Key to understanding the improvement of their academic RESULTS in a flipped classroom environment. International Journal of Management Education, 20(2), (2022). https://doi.org/10.1016/j.ijme.2022.100635

4. English, J.L., Keinonen, T., Havu-nuutinen, S., Sormunen, K.: Education sciences A Study of Finnish Teaching Practices : How to Optimise Student Learning and How to Teach Problem Solving. Educ Sci (Basel), (2022). https://doi.org/10.3390/educsci12110821

5. Chen, L., Iuculano, T., Mistry, P., Nicholas, J., Zhang, Y., Menon, V.: Linear and nonlinear profiles of weak behavioral and neural differentiation between numerical operations in children with math learning difficulties. Neuropsychologia **160**(July), 107977 (2021). https://doi.org/10.1016/j.neuropsychologia.2021.107977

6. Tinungki, G.M.: Education sciences Team-Assisted Individualization Type of the Cooperative Learning Model for Improving Mathematical Problem Solving , Communication , and Self-Proficiency : Evidence from Operations Research Teaching. Educ Sci (Basel), (2022). https://doi.org/10.3390/educsci12110825

7. Purba, S.W.D., Hwang, W.Y., Pao, S.C.: Effect of ubiquitous physics app on learning achievements in authentic contexts. Proceedings - 2019 12th International Conference on Ubi-Media Computing, **13**(3), 273–278 (2019). https://doi.org/10.1109/Ubi-Media.2019.00060

8. Serrano Corkin, D., Coleman, S.L., Ekmekci, A.: "Navigating the Challenges of Student-Centered Mathematics Teaching in an Urban Context. Urban Review **51**(3), 370–403 (2019). https://doi.org/10.1007/s11256-018-0485-6

9. Wright, G.W., Park, S.: The effects of flipped classrooms on K-16 students' science and math achievement: a systematic review. Stud. Sci. Educ. **58**(1), 95–136 (2022). https://doi.org/10.1080/03057267.2021.1933354

10. Bishara, S., Kaplan, S.: Inhibitory Control, Self-Efficacy, and Mathematics Achievements in Students with Learning Disabilities. Intl J Disabil Dev Educ **69**(3), 868–887 (2022). https://doi.org/10.1080/1034912X.2021.1925878

11. Nja, C.O., Orim, R.E., Neji, H.A., Ukwetang, J.O., Uwe, U.E., Ideba, M.A.: Students' attitude and academic achievement in a flipped classroom. Heliyon **8**(1), e08792 (2022). https://doi.org/10.1016/j.heliyon.2022.e08792

12. Förster, M., Maur, A., Weiser, C., Winkel, K.: Pre-class video watching fosters achievement and knowledge retention in a flipped classroom. Comput. Educ. **179**(March), 2022 (2021). https://doi.org/10.1016/j.compedu.2021.104399

13. Sarimanah, E., Soeharto, S., Dewi, F.I., Efendi, R.: Investigating the relationship between students' reading performance, attitudes toward ICT, and economic ability. Heliyon **8**(6), e09794 (2022). https://doi.org/10.1016/j.heliyon.2022.e09794

14. Trinh Thi Phuong, T., Nguyen Danh, N., Tuyet Thi Le, T., Nguyen Phuong, T., Nguyen Thi Thanh, T., Le Minh, C.: Research on the application of ICT in Mathematics education: Bibliometric analysis of scientific bibliography from the Scopus database. Cogent Education, **9**(1), (2022). https://doi.org/10.1080/2331186X.2022.2084956

15. Feudel, F., Fehlinger, L.: Using a lecture-oriented flipped classroom in a proof-oriented advanced mathematics course. Int. J. Math. Educ. Sci. Technol. (2021). https://doi.org/10.1080/0020739X.2021.1949057

16. Tutal, Ö., Yazar, T.: Flipped classroom improves academic achievement, learning retention and attitude towards course: a meta-analysis. Asia Pac. Educ. Rev. **22**(4), 655–673 (2021). https://doi.org/10.1007/s12564-021-09706-9

17. Islam, A.Y.M.A., Mok, M.M.C., Gu, X., Spector, J., Hai-Leng, C.: ICT in Higher Education: An Exploration of Practices in Malaysian Universities. IEEE Access **7**, 16892–16908 (2019). https://doi.org/10.1109/ACCESS.2019.2895879

18. Hu, J., Hu, J.: Teachers' Frequency of ICT Use in Providing Sustainable Opportunity to Learn: Mediation Analysis Using a Reading Database. Sustainability *(Switzerland)*, **14**(23), (2022). https://doi.org/10.3390/su142315998

19. Slater, E.V., Barwood, D., Cordery, Z.: Pre-service teachers' use of ICT to collaborate to complete assessment tasks. The Australian Educational Researcher, (0123456789), (2022). https://doi.org/10.1007/s13384-022-00580-x

20. Fredriksen, H.: Investigating the affordances of a flipped mathematics classroom from an activity theoretical perspective. Teaching Mathematics and its Applications **40**(2), 83–98 (2021). https://doi.org/10.1093/teamat/hraa011

21. Ölmefors, O., Scheffel, J.: High school student perspectives on flipped classroom learning. Pedagog. Cult. Soc. **00**(00), 1–18 (2021). https://doi.org/10.1080/14681366.2021.1948444

22. Silverajah, V.S.G., Wong, S.L., Govindaraj, A., Khambari, M.N.M., Rahmat, R.W.B.O.K., Deni, A.R.M.: A Systematic Review of Self-Regulated Learning in Flipped Classrooms: Key Findings, Measurement Methods, and Potential Directions. IEEE Access **10**, 20270–20294 (2022). https://doi.org/10.1109/ACCESS.2022.3143857

23. Li, C.T., Hou, H.T., Li, M.C., Kuo, C.C.: Comparison of Mini-Game-Based Flipped Classroom and Video-Based Flipped Classroom: An Analysis of Learning Performance, Flow and Concentration on Discussion. Asia-Pacific Education Researcher **31**(3), 321–332 (2022). https://doi.org/10.1007/s40299-021-00573-x

24. Bailey, J.: Learning to Teach Mathematics Through Problem Solving. N. Z. J. Educ. Stud. **57**(2), 407–423 (2022). https://doi.org/10.1007/s40841-022-00249-0

25. Youhasan, P., Henning, M.A., Chen, Y., Lyndon, M.P.: Developing and evaluating an educational web-based tool for health professions education: the Flipped Classroom Navigator. BMC Med Educ, **22**(1), (2022). https://doi.org/10.1186/s12909-022-03647-6

26. Seco Izquierdo, A., Pérez, P.: Mathematics with Flipped Classroom in a primary school classroom. Universidad de Cantabria (2017)

27. Guerrero Salazar, C., Prieto López, Y., Noroña Medina, J.: La aplicación del aula invertida como propuesta metodológica en el aprendizaje de matemática. Espíritu Emprendedor TES **2**(1), 1–12 (2018). https://doi.org/10.33970/eetes.v2.n1.2018.33

28. Pañi, T., Tacuri, A.: Aprendizaje de la Matemática mediante la aplicación del Aula Invertida (2019)

Multi-Method Spectral Predictive Models with FTIR-ATR in the Simultaneous Quantification of Ethanol and Legal Methanol Limits in Ecuadorian Clear Spirits

Wilson Cajilima$^{(\boxtimes)}$ and Pablo Arévalo

Politécnica Salesiana University, Cuenca, Ecuador
wcajilima@est.ups.edu.ec, parevalo@ups.edu.ec

Abstract. In this research, the feasibility of employing multiple machine learning algorithms with FTIR-ATR spectroscopy for the simultaneous quantification of ethanol and legal limits of methanol in distilled and artisanal beverages characteristic of Ecuador is investigated, based on spectral matrix similarity.

Initially, spectra acquired in the range of 4000 to 400 cm^{-1} underwent spectral preprocessing including baseline correction, smoothing, normalization, first and second derivative, and their combinations. Forty-eight distinct treatments were used to construct models employing Principal Component Regression (PCR) and Partial Least Squares 2 (PLS2). The treatment yielding superior metrics was employed for constructing an Artificial Neural Network combined with Principal Component Analysis (PCA-ANN) and Recursive Feature Elimination (RFE-ANN) utilizing a Decision Tree Regressor as a variable selector. Based on confidence intervals and hypothesis testing of statistics such as root mean squared error of prediction (RMSEP) the PCR and PLS2 models exhibited superior performance. PCR achieved detection and quantification limits of 0.25% and 0.7%, respectively. In commercial beverages, predictions were compared with results obtained via gas chromatography, where again PCR and PLS2 demonstrated the finest metrics, showcasing speed and high cost-effectiveness, rendering them viables alternatives for preliminary quality control analysis.

Keywords: FTIR-ATR spectroscopy · Preprocessing · Predictive models

1 Introduction

The WHO ranks Ecuador 11th in alcohol-related mortality in Latin America [1], with the country holding the second position in 2011. Furthermore, the National Institute of Statistics and Census (INEC) indicates that over 900 thousand Ecuadorians consume alcohol [2].

In the festivities and clandestine retail points of the country, the sale of artisanal alcohol such as the so-called 'guanchaca' or 'punta' is common. These beverages, while inexpensive and possessing potent intoxicating effects, lack sanitary registration. They

E. M. Inga Ortega et al. (Eds.): CITIS 2024, LNNS 1331, pp. 69–80, 2025.
https://doi.org/10.1007/978-3-031-87065-1_7

have been implicated in cases of mass intoxication due to adulteration with methanol [3, 4], a substance used by illegal producers as a cheaper alternative to ethanol production owing to its molecular similarity. However, methanol follows a metabolic pathway that produces toxic metabolites, leading to severe metabolic acidosis and even permanent blindness [5]. The Ecuadorian Technical Standard (INEN) [6] employs gas chromatography for the control of these substances; nevertheless, factors such as analysis time or high cost per sample pose significant limitations. Moreover, the reliability of informational labels on commercial ethyl products, such as the claimed ethanol percentage, is essential to ensure that consumers receive products that meet their specifications and quality standards. Several chemometric approaches have been proposed in the chemical analysis of contaminant substances using NIR, MIR, and UV spectroscopy [7–12]. In regression tasks, several authors [13–17] concur that the utilization of spectral treatments enhances predictive capacity, alongside the selection of optimal spectral ranges [12, 18]. Previous research has aimed to quantify ethanol, using Raman [19] or FTIR spectroscopy [15] and to analyze ethanol and methanol in fruit brandies [20]. In distilled beverages, these analytes have also been investigated using Raman spectroscopy [21], and there have been endeavors in controlling various congeners with the ATR technique [16]. However, model validation has relied on adulterated commercial samples, and different models have been constructed for each specific substance.

In contrast, this study focuses on simultaneous prediction and, for the first time according to the author's knowledge, delves into the quantification of legal limits using FTIR-ATR spectroscopy.

2 Methods

2.1 Instrumentation

A Nicolet iS10 FTIR Spectrometer equipped with a Mid-infrared Ever-Glo source, a DTGS detector, KBr beamsplitter, and single-bounce ATR accessory was used to collect spectra and visualized via OMNIC (v9.2) software. A pre-experimental phase was implemented to establish instrumental optimal parameters that minimize sources of variability and demonstrate interday and intraday repeatability.

2.2 Reagents and Standards

HPLC-grade ethanol and methanol, Milli-Q water (Millipore system) and 14 clear spirits collected from illegal artisan alcohol sales points in southern Quito and northern Cuenca during characteristic festivities of these cities.

An experimental mixture design with constraints was implemented denoted by:

$$0 \leq a_i \leq x_i \leq b_i \leq 1 \tag{1}$$

where a_i represents the lower constraint for component x_i of the mixture, and b_i denotes the upper constraint. Working in volume/volume percentage, the following limits were considered for the analytes: $30 \leq$ ethanol ≤ 50 and $0 \leq$ methanol ≤ 2.

Then, standards were prepared in Eppendorf tubes with a final volume of 1 ml. For methanol, serial dilutions were used from a 5% stock solution (methanol - water).

Table 1. Optimal Operating Parameters

	Variable	Value
FTIR	Spectral resolution	$4\,cm^{-1}$
	Data point spacing	$0.4821\,cm^{-1}$
	Number of scans	32
	Spectral range	$4000\,\text{-}400\,cm^{-1}$
	Nitrogen gas pressure	68.95 kPa
SAMPLE	Sample Volume	55 μl
	Wait time between samples	1 min
	Vortex time	5 s

Table 2. Range of concentrations obtained by the experimental design.

Level	Ethanol (% v/v)	Methanol (% v/v)
1	30	0.00 - 0.05 - 0.10 - 0.30 - 0.50 - 1.00 - 1.50 - 2.00
2	32.5	
3	35	
4	37.5	
5	40	
6	42.5	
7	45	
8	47.5	
9	50	

Note. Seventy-two mixtures obtained, each with three replicates, totaled 216 samples

2.3 Spectral Quantification of Ethanol and Methanol

The spectra were analyzed based on the location of characteristic bands of the target analyses and the significant spectral overlap of the various components in the samples.

Preprocessing Methods. To address the complex shape of the baseline, a polynomial (degree $= 2, 3$) and rubberband function were tested. To optimize the algorithms' performance, Zeroone normalization was applied. For enhanced peak resolution, smoothing with the Savitzky-Golay filter was used with order "p" and a small filter length "n" suitable for the low-concentration samples under study, using R software (version 4.3.1) and the IR package.

The first and second derivatives were also applied to reveal the subtle spectral changes due to low concentrations; treatments are summarized in Table 3.

Table 3. Different spectral treatments

Treatment	Baseline Correction	Normalization	Smoothing	Derivative
T1	2	Zeroone	$p = 2; n = 5$	1st Der
T2	3	Zeroone	$p = 2; n = 5$	1st Der
T3	Rubberband	Zeroone	$p = 2; n = 5$	1st Der
T4	2	Zeroone	$p = 3; n = 5$	1st Der
T5	3	Zeroone	$p = 3; n = 5$	1st Der
T6	Rubberband	Zeroone	$p = 3; n = 5$	1st Der

Note. The 2nd derivative implemented just like the 1st derivative totaled 12 treatments.

2.4 Predictive Models

Principal Component Regression (PCR) and Partial least Squares 2 (PLS2). Using the 'pls' package in R, PCR and PLS2 were implemented. The preprocessed spectra were randomly split into training and test sets (80:20 ratio) and subjected to principal component analysis (PCA) for preliminary examination. The models were constructed using six components, K-fold cross-validation (K=8), and different fitting algorithms: singular value decomposition for PCR due to its numerical stability, and for PLS2, wide kernel pls to handle the "wide" datasets of the study, kernel pls to capture non-linear relationship and orthogonal scores pls to remove orthogonal variation (noise) from predictors.

The limit of detection and quantification were also calculated, as well as the metrics Standard Error of Prediction (SEP), Root Mean Squared Error of Prediction (RMSEP), and Coefficient of Determination (R2). The optimal spectral treatment was selected through bootstrapping (1000 iterations) of the RMSEP for each model with its respective algorithm, followed by ANOVA ($\alpha = 0.05$) alongside Tukey's HSD.

Artificial Neural Network combined with PCA (PCA-ANN) and Recursive Feature Elimination (RFE-ANN). After applying PCA to the treatment with the best prediction metrics in PCR and PLS2 the first three components were used as input for the development of an ANN with two hidden layers, the ReLu activation function, and mean squared error as the loss function. For the RFE-ANN model, the same best treatment was used, recursive feature elimination was implemented as variable selection method using a Decision Tree regressor as the selector and Lasso and Ridge regularization. Both models were constructed in Python (3.11.8).

Validation of the Models. A subset comprising five beverages including commercial brands and artisanal products was analyzed using gas chromatography to determine the alcoholic content and methanol levels, aiming to assess accuracy. Finally, prediction metrics were obtained, and the model with the best performance was chosen using the same approach for selecting the optimal treatment as mentioned earlier.

3 Results and Discussion

3.1 Optimal Spectral Range

The spectra of the complex mixtures used exhibit significant overlap in the diagnostic region, as well as a strong absorption by the solvent (Milli-Q water) in this region, leading to its exclusion, a common practice in chemometrics [19–21].

Conversely, the fingerprint region shows partial overlap between the bands of the pure spectra. Thus, the discrimination of the range was based on the location of characteristic bands of the target analytes (Fig. 1).

For ethanol, the bands with peaks at 1085, 1044, and 876 cm^{-1} correspond to CH_3 rocking, C-O stretching, and C-C stretching, respectively [22, 23]. For methanol, the band with peak at 1017 cm^{-1} corresponds to C-O stretching [24].

Consequently, the optimal spectral range selected was: 1250 – 800 cm^{-1}.

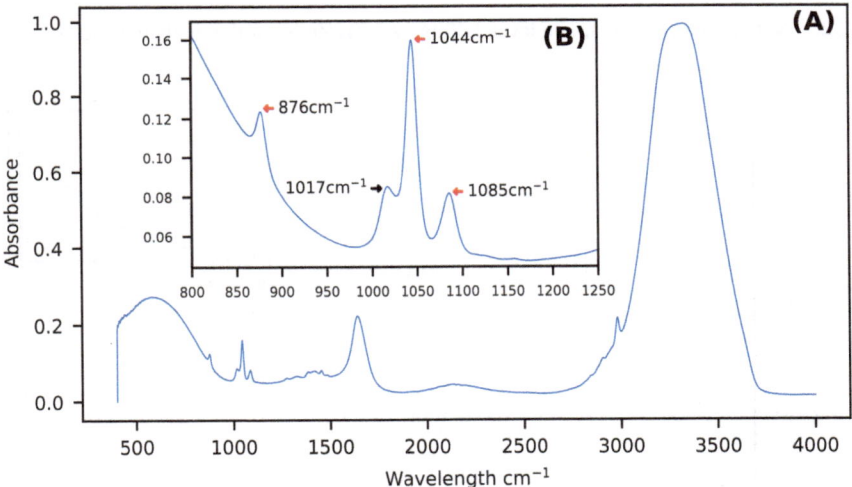

Fig. 1. Spectrum of a 30% ethanol and 5% methanol mixture (v/v). (A) Full range (4000–400 cm^{-1}), (B) Optimal range (1250–800 cm^{-1}) with bands of ethanol (red) and methanol (black).

3.2 Spectral Treatments

Due to the low concentrations used for methanol (0–2%), only slight perturbations were observed in its spectral band (1017 cm^{-1}) as shown in Fig. 2.

This led to the use of the first and second derivative. In Fig. 3 (B), those very slight 'perturbations' of methanol are now revealed as prominent waves. It should be emphasized that the maxima and minima point on the first derivative curve are the points of maximum rate of change and not the maxima and minima of the original peaks since these have a value of zero [25].

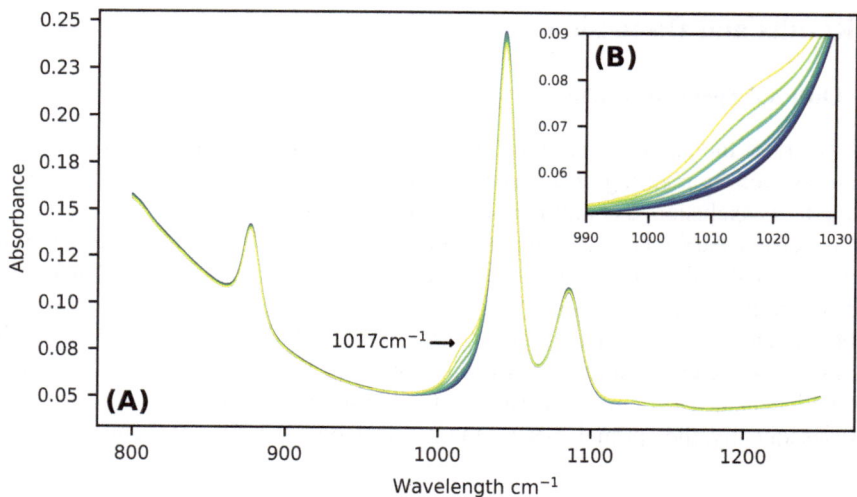

Fig. 2. (A) Optimal range (sample level 3, Table 2), (B) Zoom of the methanol spectral band.

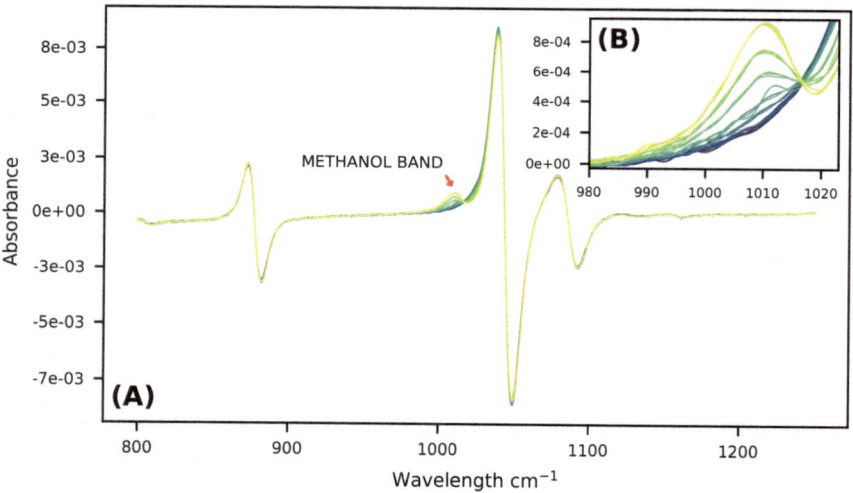

Fig. 3. First spectral derivative (1st Der) (sample level 3). (A) Optimal range and methanol band, (B) Zoom of the methanol spectral band.

3.3 Principal Component Regression (PCR) and Partial Least Squares (PLS2)

Based on the analysis of the root mean squared error of Cross-Validation (RMSECV) plot, for PCR the optimal number of components selected was three, which explain 99.9% of the variance in the data. Additionally, the PCA of the model was inspected.

In Fig. 4, it is observed that PC1 separates the training samples according to the ethanol concentration gradient, while PC2 separates them according to methanol.

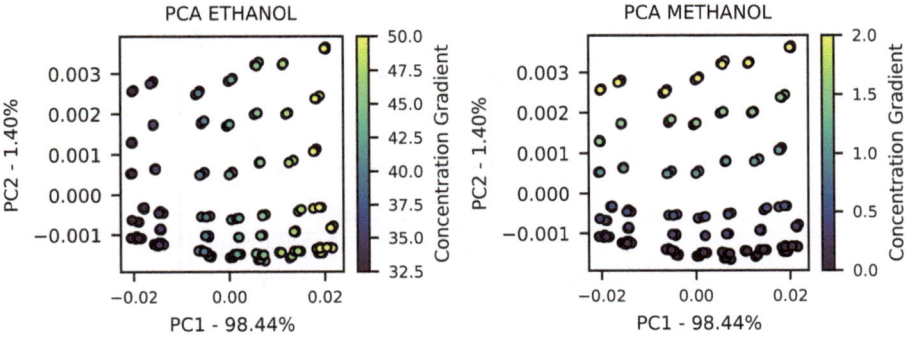

Fig. 4. PCA plot for the two target variables.

In the loading plot (Fig. 5.) the first component exhibits a characteristic pattern of the first derivative of ethanol, while the second component does so with methanol.

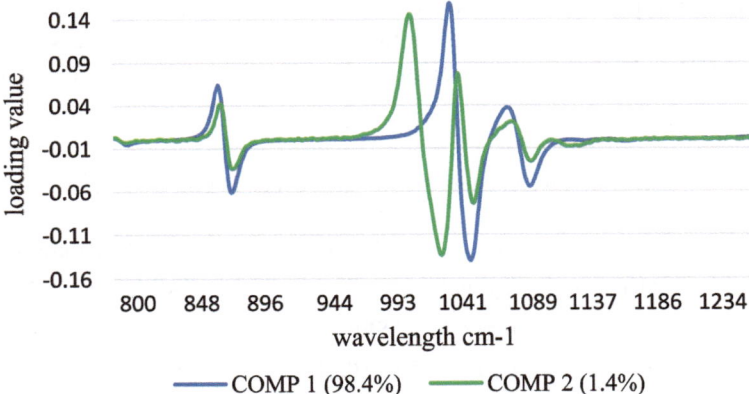

Fig. 5. Loading plot of the two principal components of PCR

For PLS2, the T6 treatment with the orthogonal scores pls algorithm showed the best metrics which is assumed to be due to its ability to remove noise from the predictors, interpreted as orthogonal variation. The optimal number of latent variables was three with 99.9% explained variance, and a similar loading plot like PCR was obtained; limits of detection and quantification are summarized below.

Limit of Detection (LOD) and Limit of Quantification (LOQ). In Fig. 6 (the methanol band in the 1st Der - Fig. 3 (B)) six linear relationships are observed; however, eight different concentrations were used for methanol. Graphically, at a concentration of 0.1%, the absorbance values obtained can no longer be distinguished from 0% methanol, this slight singularity being the first indication of the detection limit.

The LOD defined as "the lowest amount that can be distinguished from the absence of that substance within a specified confidence limit", and LOQ as "the lowest analyte

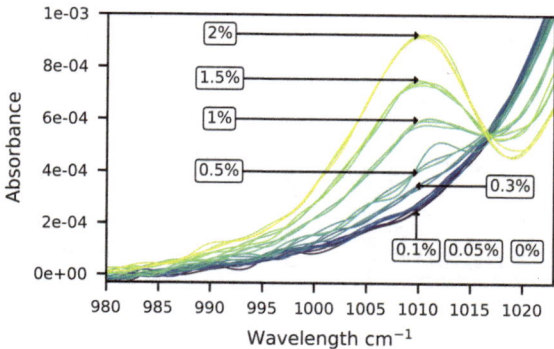

Fig. 6. Absorbance signals as a function of methanol concentrations (sample level 3, 1ˢᵗ Der).

concentration that can be quantitatively detected with a stated accuracy and precision" [26]. Using a "pseudo univariate" approach, LOD and LOQ were determined as an extension of a univariate method to a multivariate calibration [27–29], considering the probabilities of Type I and Type II errors ($\alpha = 0.01$), the Standard Error of Prediction (SEP) and the model sensitivity (calibration curve of predicted vs measured values) (Table 4). LOD and LOQ are shown below and are like those found in the literature [16, 21].

Table 4. Prediction metrics for the PCR and PLS2 models.

Parameters	PCR - Treatment 6		Parameters	PLS2 - Treatment 6	
	Methanol	Ethanol		Methanol	Ethanol
SEP	0.07	0.517	SEP	0.08	0.49
RMSEP	0.079	0.511	RMSEP	0.07	0.48
R^2	0.989	0.981	R^2	0.986	0.98
Components	3		Latent Variables	3	
L.O.D	0.25%	1.74%	L.O.D	0.29%	1.63%
L.O.Q	0.7%	5.28%	L.O.Q	0.88%	4.95%

3.4   Artificial Neural Network Combined with PCA (PCA-ANN) and Recursive Feature Elimination (RFE-ANN)

No significant differences were observed between the prediction metrics of PCA-ANN and PCR, PLS2 (Table 5), as documented by other authors [30, 31]. However, the potential loss of nonlinear relationships between the original variables when forming the PC's [32] considerably limits the true capability of the ANN, which is why RFE-ANN was implemented.

For RFE-ANN, different parameters of the Decision Tree Regressor such as max depth or min samples leaf were optimized; however, no improvement in predictions was found (Table 5), on the contrary, they worsened.

Table 5. Prediction metrics for the PCA-ANN and RFE-ANN models.

Parameters	PCA-ANN		Parameters	RFE-ANN	
	Methanol	Ethanol		Methanol	Ethanol
SEP	0.08	0.76	SEP	0.23	1.49
RMSEP	0.082	0.9	RMSEP	0.28	1.65
R^2	0.98	0.93	R^2	0.82	0.80
S.D	0.08	0.76	S.D	0.231	1.49

3.5 Validation of the Models

All models were evaluated in commercial clear spirits with and without sugar, as well as clear artisanal spirits sold during Ecuadorian festivities. As expected, small fluctuations appeared across all beverage spectra, particularly in those with high sugar content, even near the characteristic wavenumbers of the analytes under study, likely due to extra components (flavorings, sugars, congeners, etc.) present even in clear beverages with triple distillation (e.g., Iceland Vodka).

The most accurate predictions were obtained from clear spirits without sugar, presumably due to their high similarity to the training standards. The predictions were less accurate with spectra from a complex matrix (e.g., Cristal Roja) whose components were not included in the training set, suggesting that exact simulation of such matrices beyond clear, sugar-free spirits is still unachievable (Table 6).

However, outstanding approximations to those obtained by gas chromatography were achieved.

Table 6. Prediction metrics in commercial beverages compared to gas chromatography.

Parameters	ETHANOL				Parameters	METHANOL			
	M1	M2	M3	M4		M1	M2	M3	M4
RMSEP	3.19	3.03	3.22	4.2	RMSEP	0.34	0.37	0.39	0.41
R^2	0.957	0.95	0.91	0.47	R^2	0.68	0.67	0.26	0.11
S.D	1.29	1.35	1.65	4.40	S.D	0.12	0.11	0.64	0.20

Note. M1, M2, M3, M4, correspond to the PCR, PLS2, PCA-ANN, and RFE-ANN models, respectively

No significant difference was found between the RMSEPs of the PCR and PLS2 models; nevertheless, a significant difference was found when comparing them with

PCA-ANN and RFE-ANN. Thus, the lower RMSEPs of PCR and PLS2 allowed us to conclude their superior performance, highlighting their repeatability of predictions.

For methanol, the legal limit stipulated by NTE-INEN 1837 is 10 mg/100 cm3 (0.012% v/v), and although its $R^2 = 0.68$ (between 0.5–0.7) demonstrates good separation of samples into high, medium, and low concentrations, the calibration can only be used for screening purposes [33, 34] due to its limit of detection (0.25% v/v).

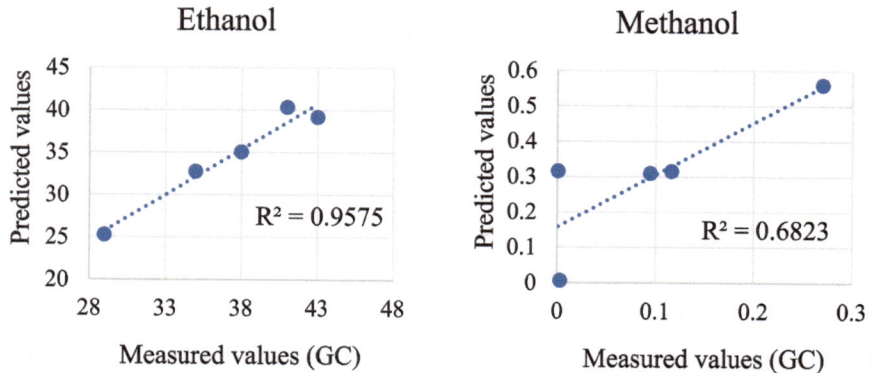

Fig. 7. Calibration curve for ethanol and methanol in commercial beverages by the PCR model.

On the other hand, regarding ethanol, the obtained $R^2 = 0.95$ with GC indicates excellent quantitative information, even in complex matrices such as distilled beverages (Fig. 7).

Based on the obtained metrics, the sale of artisanal alcohol such as "punta" or "guan-chaca" on Ecuadorian streets can be rapidly analyzed using the developed models; a 1 ml sample is sufficient, and predictions, obtained in seconds after spectra collection, aid efficient on-site assessment, ensuring safe and high-quality commercial ethyl products.

4 Conclusion

The PCR and PLS2 models in conjunction with the sixth spectral treatment using the first derivative, provided the best prediction metrics. The second derivative did not improve the predictions. Furthermore, for the characteristic beverages from the main cities of Ecuador, excellent quantitative information for ethanol was obtained, even in complex matrices, making them a reliable quantitative tool. On the other hand, the limit of quantification for methanol restricts the prediction of legal limits; however, its limit of detection allows it to be considered an effective preliminary analysis alternative in quality control.

References

1. Chrystoja, B.R., Monteiro, M.G., Owe, G., Gawryszewski, V.P., Rehm, J., Shield, K.: Mortality in the Americas from 2013–2015. Addiction **116**(10), 2685–2696 (2015)

2. INEC: Encuesta Nacional de Ingresos y Gastos en lugares Urbanos y Rurales - Ecuador. Available: https://www.ecuadorencifras.gob.ec/mas-de-900-mil-ecuatorianos-consumen-alc ohol/
3. El Diario: Muertes por alcohol adulterado, Empresa Periodística el Diario (2024)
4. La Hora: Intoxicaciones por metanol, Empresa Periodística La Hora, (2024)
5. Ashurst, J.V., Nappe, T.M.: Methanol Toxicity (2023)
6. INEN: Servicio Ecuatoriano de Normalización. Available: http://apps.normalizacion.gob.ec/ descarga/
7. Vinciguerra, L., Marcelo, M., Motta, T., Meneghini, L., Bergold, A., Ferrão, M.: Chemometric Tools and Ftir-ATR Spectroscopy. Quim Nova (2019)
8. Yaman, H.: A rapid method for detection adulteration in goat milk by using vibrational spectroscopy. J. Food Sci. Technol. **57**(8), 3091–3098 (2020)
9. Liu, P., Wang, J., Li, Q., Gao, J., Tan, X., Bian, X.: Rapid identification and quantification of Panax notoginseng with its adulterants by near infrared spectroscopy. Spectrochim Acta A Mol Biomol Spectrosc **206**, 23–30 (2019)
10. Filoda, P.F., et al.: Fast Methodology for Identification of Olive Oil Adulterated with a Mix of Different Vegetable Oils. Food Anal **12**(1), 293–304 (2019)
11. de Freitas, A.G., et al.: FTIR spectroscopy with chemometrics for determination of tylosin residues in milk. J Sci Food Agric **101**(5), 1854–1860 (2021)
12. Sahlan, M., et al.: Identification and classification of honey's authenticity by attenuated total reflectance Fourier-transform infrared spectroscopy and chemometric method. Vet World **12**(8), 1304–1310 (2019)
13. Devianti, S., et al.: Rapid and Simultaneous Detection of Hazardous Heavy Metals Contamination in Agricultural Soil Using Infrared Reflectance Spectroscopy. IOP Conf Ser Mater Sci Eng **506**, 012008 (2019)
14. Du, L., Lu, W., et al.: Authenticating Raw from Reconstituted Milk Using Fourier Transform Infrared Spectroscopy and Chemometrics. J Food Qual **2019**, 1–6 (2019)
15. Zamani Mazdeh, F., et al.: Rapid Determination of Ethanol in Non-Alcoholic Malt Beverage by ATR-FT-IR Spectroscopy and Headspace Gas Chromatography Confirmation. Journal of Agricultural Science and Technology **25**(2), 391–401 (2023)
16. Anjos, O., Santos, A.J.A., Estevinho, L M., Caldeira, I.: FTIR–ATR spectroscopy applied to quality control of grape-derived spirits. Food Chem, pp. 28–35 (2016)
17. Riahi, S., et al.: A new technique for spectrophotometric determination of Pseudoephedrine and Guaifenesin in syrup and synthetic mixture, Drug Test Anal, pp. 319–324 (2011)
18. Assis, C., Pereira, H.V., Amador, V.S., Augusti, R., de Oliveira, L.S., Sena, M.M.: Combining mid infrared spectroscopy and paper spray mass spectrometry in a model to predict the composition of coffee blends. Food Chem. **281**, 71–77 (2019)
19. Picard, A., Daniel, I., Montagnac, G., Oger, P.: In situ monitoring by quantitative Raman spectroscopy of alcoholic fermentation by Saccharomyces cerevisiae under high pressure. Extremophiles **11**(3), 445–452 (2007)
20. Coldea, T.E., Socaciu, C., Fetea, F., Ranga, F., Pop, R.M., Florea, M.: Rapid Quantitative Analysis of Ethanol and Prediction of Methanol Content in Traditional Fruit Brandies from Romania, using FTIR Spectroscopy and Chemometrics. Not Bot Horti Agrobot Cluj Napoca **41**(1), 143 (2013)
21. Boyaci, I.H., Genis, H.E., Guven, B., Tamer, U., Alper, N.: A novel method for quantification of ethanol and methanol in distilled alcoholic beverages using Raman spectroscopy. J. Raman Spectrosc. **43**(8), 1171–1176 (2012)
22. Klein, D.: "Espectroscopia Infrarroja," in Química Orgánica, Editorial Médica Panamericana, pp. 671–690 (2013)
23. Brown, D.: Advanced Organic Chemistry: The infrared spectrum of ethanol.

24. Virtual Planet Laboratory: "Methanol (CH3OH) IR."
25. Thermo Electron Corporation: "Using OMNIC Algorithms." Available: https://knowledge1.thermofisher.com/
26. Jannetto, P.J.: Therapeutic drug monitoring using mass spectrometry. In: Mass Spectrometry for the Clinical Laboratory, pp. 165–179. Elsevier (2017). https://doi.org/10.1016/B978-0-12-800871-3.00008-0
27. Allegrini, F., Olivieri, A.C.: IUPAC-Consistent Approach to the Limit of Detection in Partial Least-Squares Calibration. Anal. Chem. **86**(15), 7858–7866 (2014)
28. Ortiz, M.C., et al.: Capability of detection of an analytical method evaluating false positive and false negative (ISO 11843) with partial least squares. Chemom. Intell. Lab. Syst. **69**(1–2), 21–33 (2003)
29. Ortiz, M.C., Sarabia, L.A., Sánchez, M.S.: Tutorial on evaluation of type I and type II errors in chemical analyses: From the analytical detection to authentication of products and process control. Anal. Chim. Acta **674**(2), 123–142 (2010)
30. Peguero Gutiérrez, A.: La espectroscopia NIR en la determinación de propiedades físicas y composición química de intermedios de producción y productos acabados. Universitat Autònoma de Barcelona, Barcelona (2017)
31. Loele, G., De Luca, M., Dinç, E., Oliverio, F., Ragno, G.: Artificial Neural Network Combined with Principal Component Analysis for Resolution of Complex Pharmaceutical Formulations. Chem Pharm Bull (Tokyo) **59**(1), 35–40 (2011)
32. Davies, T., Fearn, T.: Back to basics: the principles of principal component analysis, Spectroscopy Europe, pp. 16–20 (2004)
33. Ozili, P.K.: The Acceptable R-Square in Empirical Modelling for Social Science Research. SSRN Electronic Journal (2022)
34. Sapra, R.L.: Using R2 with caution. Curr Med Res Pract **4**(3), 130–134 (2014)

Analysis of Students' Emotions in Real-Time During Class Sessions Through an Emotion Recognition System

Miguel-Ángel Quiroz-Martínez(✉) 📵, Sergio Díaz-Fernández📵,
Kevin Aguirre-Sánchez📵, and Mónica-Daniela Gómez-Ríos📵

Computer Science Department, Universidad Politécnica Salesiana, Cuenca, Ecuador
{mquiroz,mgomezr}@ups.edu.ec, {sdiazf,kaguirres2}@est.ups.edu.ec

Abstract. Recognizing and responding to students' emotional states during classes is crucial for optimizing learning outcomes, yet it remains challenging for educators. This study presents the development and implementation of a real-time emotion recognition system using convolutional neural networks (CNNs) to analyze students' facial expressions during in-person classes. Trained on the FER2013 dataset, the system classifies seven distinct emotions with 85% accuracy. An experiment with 20 university students aged 18–25 compared emotional responses across three teaching methodologies: collective, group, and experiential. Experiential teaching elicited the most positive emotions, with 50% of expressions classified as happiness, while collective teaching generated more negative responses. Statistical analysis revealed a significant positive correlation between happiness and academic performance ($r = 0.65$, $p < 0.01$) and a negative correlation between fear and performance ($r = -0.54$, $p < 0.05$). The system provides educators with quantitative emotional data, enabling real-time adaptation of teaching strategies and retrospective analysis of class dynamics. This research contributes to AI-enhanced education, offering insights into creating more responsive and student-centered learning environments while addressing privacy and data protection considerations throughout the study.

Keywords: Artificial intelligence · Classrooms · Emotion recognition · Teaching methodologies

1 Introduction

Analyzing student emotions during class sessions has emerged as a critical factor in optimizing educational outcomes and enhancing the learning experience. Maintaining student engagement in today's diverse and technologically advanced classrooms has become increasingly challenging due to various internal and external distractions. Monitoring and understanding students' emotional states in real-time is crucial for improving the effectiveness of teaching and learning processes [1].

Recent advancements in AI, particularly in computer vision and deep learning, have opened new avenues for automated emotion recognition in educational settings. These

© The Author(s) 2025
E. M. Inga Ortega et al. (Eds.): CITIS 2024, LNNS 1331, pp. 81–92, 2025.
https://doi.org/10.1007/978-3-031-87065-1_8

technologies offer the potential to provide educators with valuable insights into their students' emotional engagement, allowing for more responsive and adaptive teaching methodologies.

Several recent studies have explored the application of emotion recognition systems in educational contexts. [2] evaluated a camera-based emotion recognition system to measure student engagement and participation during classes [3].

Li et al. proposed using an emotion recognition system as a tool for real-time monitoring of student emotions, enabling educators to adjust their teaching approaches to meet individual student needs [4]. Chen et al. further emphasized the potential of these systems to improve student performance and emotional well-being [5].

Building upon this foundation, our research introduces an AI-based emotion recognition system utilizing convolutional neural networks (CNNs) to analyze student emotions in real-time during in-person class sessions. We selected CNNs for their proven efficacy in image classification tasks, particularly in facial expression recognition [6].

In this article, we introduce the use of an AI-based emotion recognition system utilizing a convolutional neural network to analyze student emotions in real-time during class sessions. A webcam is used to capture the students' facial expressions, and an AI algorithm to identify their emotions. The results of our research are explored, and we discuss how this technology can be used to enhance the interaction between teachers and students and provide a more personalized and focused learning experience tailored to the needs of each student group.

2 Literature Review

2.1 Cognitive Emotions and Learning Environments

Emotions exert a significant influence on the learning process, exhibiting a direct correlation with memory formation and emotional stimuli originating in the hippocampus and amygdala. Benavidez and Flores emphasize the necessity for implementing learning strategies that enhance student cognition through effective emotion management by educators [7]. The integration of emotion recognition systems in educational environments has emerged as a prominent area of research. [2] developed a camera-based emotion recognition system to quantify classroom engagement, demonstrating the potential of these technologies in providing objective insights into student participation [2].

Liu et al. investigated real-time emotion detection in computer-assisted language learning, illustrating how these systems can provide valuable feedback to both students and educators [8]. Chen et al. conducted a comprehensive review of promising techniques for emotion recognition in online learning, highlighting the potential of multimodal approaches that integrate facial expression analysis with other physiological signals [2].

The present research extends these previous studies by implementing a real-time emotion recognition system directly into the physical classroom environment. In contrast to earlier approaches that focused on post-hoc analysis or were limited to online settings, this system provides immediate feedback to educators during the class session, facilitating real-time adjustments to pedagogical methodologies.

Zhang et al. explored the application of deep learning models for continuous emotion recognition in educational contexts, demonstrating the potential for more nuanced and dynamic emotional analysis throughout a learning session [9]. This study builds upon their work by implementing a similar continuous recognition approach, with a specific focus on real-time application in physical classroom settings.

By synthesizing insights from cognitive neuroscience with advanced artificial intelligence technologies, this study offers a novel perspective on the practical implementation of emotional awareness in quotidian classroom environments. This interdisciplinary approach, which integrates educational psychology, neuroscience, and computer vision, represents a significant advancement in the field of emotion-aware education.

2.2 Facial Emotion Recognition (FER) Systems

Facial recognition technologies have significantly advanced, evolving beyond mere identification to analyze expressions, behaviors, and moods. Andrejevic and Selwyn [10] highlight the implementation of these technologies in educational contexts for various purposes, including security enhancement, attendance monitoring, and emotion detection via video devices [11].

The application of FER systems in education has been explored in several studies. Krithika and Lakshmi [12] developed a system to monitor student attention levels in e-learning environments using continuous facial expression analysis. Their approach demonstrated the potential of FER in providing real-time feedback on student engagement.

[13] proposed a deep learning-based facial expression recognition system for intelligent tutoring systems. Their model achieved high accuracy in recognizing seven basic emotions, showcasing the potential of AI in personalizing educational experiences. Deep Learning, particularly through convolutional neural networks (CNNs), [14] has emerged as a powerful tool in emotion recognition. Jaiswal et al. [6] describe the typical process involving face detection, feature extraction, and emotion classification. These models have shown remarkable accuracy in capturing subtle facial cues indicative of emotional states [5, 14–16].

Advancements in computer vision libraries have facilitated the implementation of FER systems in real-time educational settings. Eltenahy [17] discusses the use of OpenCV for real-time video and image processing in facial emotion recognition applications within educational contexts. Recent developments, such as the Streamlit framework described by Yalamarthi et al. [18], have addressed challenges in implementing these systems in web browsers. This open-source framework allows for the development of interactive applications utilizing artificial intelligence and data science, integrating seamlessly with libraries such as TensorFlow, Keras, and Numpy.

The novelty of our approach lies in the real-time application of FER in physical classroom environments, combining advanced AI technologies with insights from cognitive neuroscience. Unlike previous studies focused on e-learning or post-hoc analysis, our system provides immediate emotional feedback during in-person class sessions. This enables educators to make real-time adjustments to their teaching strategies, potentially revolutionizing the way student emotions are understood and addressed in the learning process [19].

3 Methodology

This research employed a mixed-methods approach, combining quantitative data from the emotion recognition system with qualitative insights from participant feedback. The study was designed as a quasi-experimental investigation with repeated measures. A sample of 20 university students (12 females, 8 males) aged 18–25 (M = 21.3, SD = 1.8) was recruited through purposive sampling. Inclusion criteria were: Full-time enrollment in an undergraduate program; No prior experience with emotion recognition systems; Willingness to participate in all three experimental sessions.

The study consisted of three experimental conditions, each corresponding to a different teaching methodology: Collective: Traditional lecture-style teaching; Group: Collaborative learning in small groups; Experiential: Hands-on, activity-based learning.

Each condition was implemented in a 50-min class session, with a one-week interval between sessions to minimize carryover effects.

To obtain the data, the following was carried out: Continuous data collection throughout each session; to assess students' self-reported emotional states and engagement levels; a short quiz at the end of each session to measure immediate learning outcomes; conducted with a subset of participants (n = 5) after the completion of all sessions.

The study examines the impact of the independent variable, teaching methodology (collective, group, experiential), on dependent variables including emotional states (as detected by the system), self-reported engagement, and academic performance scores. To minimize confounding factors, counterbalancing was used to randomize the order of teaching methodologies across participants, consistent content was maintained across sessions, and all sessions were conducted in the same classroom with controlled lighting and seating arrangements.

The primary data source for the study was the FER2013 dataset [20], containing 35,887 grayscale images of facial expressions from diverse subjects, which enhances model generalization. The dataset was split into 28,709 images for training, 3,589 for validation, and 3,589 for testing. Data augmentation techniques, including random rotations ($\pm10°$), horizontal flips, and slight zoom variations ($\pm10\%$), were applied to improve model robustness. All images were standardized to a resolution of 48x48 pixels and normalized to values between 0 and 1.

A CNN was designed using TensorFlow 2.10.0 and Keras [21] to classify facial emotions with a focus on computational efficiency and accuracy. The architecture included four convolutional layers with 32, 64, 128, and 256 filters, respectively, each followed by batch normalization and max pooling. The max pooling layers used a 2x2 window to reduce spatial dimensions. Two fully connected layers with 512 and 256 units followed, integrating high-level features for emotion classification. The output layer had 7 units for different emotions, using ReLU activation for all layers except the output layer, which used softmax activation to generate a probability distribution.

The model was trained using the Adam optimizer with an initial learning rate of 0.001, adjusted by a scheduler reducing the rate by 0.1 when validation loss plateaued for 5 epochs. Training ran for up to 100 epochs with a batch size of 64. Early stopping with a patience of 15 epochs was employed to prevent overfitting, and dropout layers (rate 0.5) were used after each fully connected layer. Key performance metrics, including training and validation accuracy, loss functions, and the confusion matrix, were monitored throughout the training process for real-time assessment and early detection of issues.

The trained convolutional neural network (CNN) was integrated into a real-time video processing pipeline for continuous emotion recognition in classrooms. Face detection used OpenCV's Haar Cascade classifier [17], ensuring accurate detection across various poses and lighting conditions. Detected faces were preprocessed to 48x48 pixels and converted to grayscale, matching the CNN's input requirements. The preprocessed images were then classified by the CNN. A graphical user interface, developed with Streamlit [18], provided real-time emotion visualization, displaying the dominant emotion for each detected face on the video feed. Post-session analysis tools allowed educators to review emotion trends and statistics after each class session (Fig. 1).

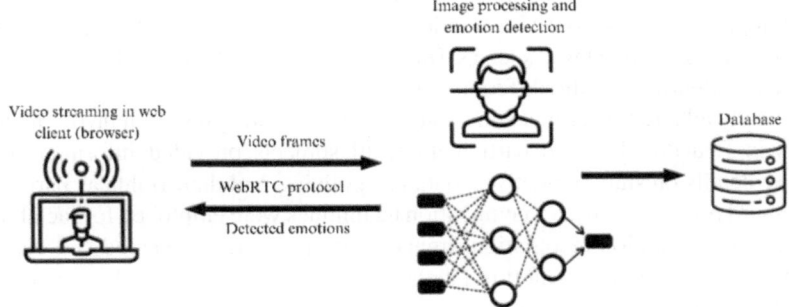

Fig. 1. Design of the general emotion recognition process in the system prototype.

The emotion recognition system's performance was assessed using various evaluation metrics. Overall accuracy was calculated as the ratio of correct to total predictions across all emotions. Precision, recall, and F1-score were computed for each emotion class. A confusion matrix was generated to visualize performance and identify misclassification patterns. The ROC-AUC was calculated for each class, measuring the model's discrimination ability, with values closer to 1 indicating better performance. Cohen's Kappa coefficient was also computed to assess agreement between the model's predictions and human annotations, accounting for chance agreement and providing a more nuanced measure of reliability.

Various validation techniques ensured the reliability and generalization of the emotion recognition system. K-fold cross-validation (k = 5) assessed performance consistency across different data subsets. Hold-out validation used a separate test set (20% of the dataset) not involved in training or tuning, offering an unbiased performance estimate. Real-world testing through pilot tests in classroom settings validated the system's performance under practical conditions, considering factors like lighting, student movement, and occlusions. Ablation studies, which involved systematically removing or altering components of the model, helped identify the most critical elements of the architecture.

The experimental phase involved three 50-min class sessions, each using a distinct teaching methodology: collective (lecture-style), group (collaborative learning), and experiential (activity-based learning). These methods were chosen to explore their varying impacts on student emotions. The emotion recognition system monitored facial expressions continuously during each session, sampling data at 5 frames per second to balance temporal resolution, computational efficiency, and data storage requirements.

After collecting data from the emotion recognition system, analytical procedures were employed to derive insights. Emotion percentages were calculated for each teaching methodology to characterize the emotional landscape associated with each approach. Pearson's correlation coefficient was used to explore relationships between detected emotions and academic performance, with significance set at $p < 0.05$ and corrections for multiple comparisons applied as needed. Time series analysis, including moving averages and autocorrelation techniques, was conducted to uncover patterns in emotional states throughout class sessions. To compare the effectiveness of different teaching methodologies in eliciting positive emotions, a one-way Analysis of Variance (ANOVA) was performed. Post-hoc tests, such as Tukey's Honest Significant Difference test, were employed to identify specific differences between methodologies.

The study adhered strictly to ethical guidelines governing research involving human subjects in Ecuador. Prior to participation, all subjects provided informed consent, including details on study objectives, data use policies, and their rights as participants. To safeguard privacy, robust anonymization techniques were employed for facial images captured during emotion recognition. Images were promptly processed and discarded, retaining only anonymized emotion labels for analysis. This protocol ensured that no personally identifiable information was stored beyond each class session, maintaining confidentiality throughout the study.

4 Results

The following results were obtained from a study conducted at the Salesian Polytechnic University in Ecuador. The experiment involved 20 university students aged between 18 and 25 who participated voluntarily in three different class sessions, each employing a distinct teaching methodology.

4.1 Emotion Recognition Accuracy

The emotion recognition system achieved an overall accuracy of 85% in classifying the seven distinct emotions.

The system demonstrated high precision and recall across all emotion categories, with happiness showing the highest F1-score (0.93) and fear the lowest (0.81). This suggests that the model is particularly adept at recognizing expressions of happiness, while expressions of fear may be more challenging to detect accurately (Table 1).

4.2 Emotion Distribution Across Teaching Methodologies

The experiential teaching methodology elicited the highest percentage of positive emotions, with 50% of students experiencing happiness, compared to 41.75% in group

Table 1. The table details the performance metrics for each emotion category.

Emotion	Precision	Recall	F1-score
Happiness	0.92	0.94	0.93
Sadness	0.86	0.83	0.84
Anger	0.88	0.85	0.86
Surprise	0.89	0.91	0.90
Fear	0.83	0.80	0.81
Disgust	0.85	0.82	0.83
Neutral	0.88	0.90	0.89

teaching and 25% in collective teaching. Interestingly, the experiential method also resulted in the highest percentage of fear (25%), which may be attributed to the novel and challenging nature of hands-on activities.

Collective teaching showed the most even distribution of emotions, suggesting that this traditional method may not strongly evoke any particular emotional response. Group teaching, on the other hand, showed a marked increase in happiness compared to collective teaching, possibly due to the social interaction component (Fig. 2).

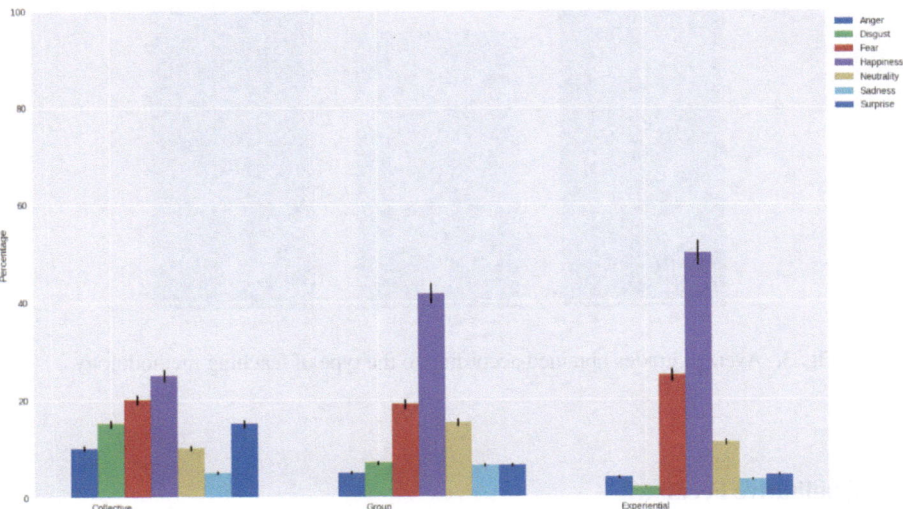

Fig. 2. Emotion Distribution Across Teaching Methodologies with 95% CI

4.3 Correlation Between Emotions and Academic Performance

A significant positive correlation was found between happiness and academic performance ($r = 0.65$; $p < 0.01$). This suggests that students who experienced more happiness during class sessions tended to perform better academically. Conversely, a negative correlation was observed between fear and academic performance ($r = -0.54$; $p < 0.05$), indicating that higher levels of fear were associated with lower academic performance.

These findings highlight the potential importance of fostering positive emotional states in the classroom to enhance learning outcomes. However, it's important to note that correlation does not imply causation and further research is needed to establish causal relationships.

4.4 Impact of Teaching Methodologies on Academic Performance

The experiential teaching methodology resulted in the highest average grade (9.1), followed by collective teaching (7.5) and group teaching (7.0). This suggests that hands-on, experiential learning may be more effective in promoting academic performance than traditional lecture-style (collective) or group-based approaches.

However, it's worth noting that the difference between collective and group teaching methodologies was relatively small. This could indicate that while group work promotes positive emotions, it may not necessarily translate directly into improved academic performance (Fig. 3).

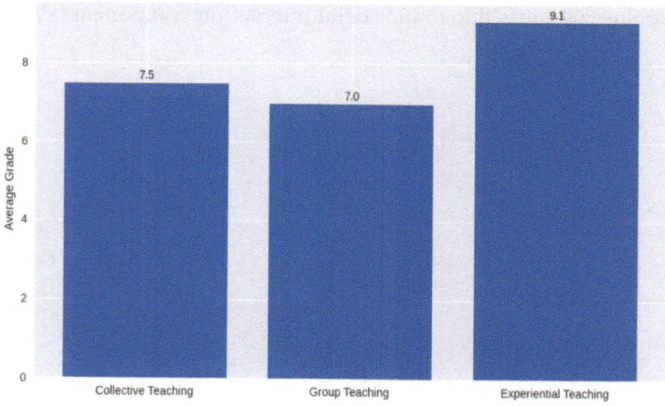

Fig. 3. Average grades obtained according to the type of teaching methodology.

4.5 Qualitative Feedback

Semi-structured interviews conducted with participants revealed several key themes: Perceived Usefulness: 80% of teachers reported that the system provided valuable insights for real-time adjustment of teaching strategies.

- Privacy Concerns: Initially, 40% of students expressed privacy concerns, which were largely alleviated after detailed explanations of the anonymization process.
- Engagement Impact: 73% of students reported increased self-awareness of their emotional states during classes, leading to improved self-regulation and participation.

• Technical Limitations: Two teachers noted occasional difficulties in emotion detection under certain conditions, such as when students wore face masks or in poor lighting.

These results suggest that the emotion recognition system has the potential to enhance teaching practices, although privacy concerns and technical limitations should be addressed in future iterations.

5 Discussion

The present study demonstrates a significant relationship between teaching methodologies, student emotions, and academic performance. The experiential teaching methodology elicited the highest proportion of positive emotions, with 50% of students exhibiting happiness. This observation aligns with the findings of Chen et al., who reported that interactive and hands-on learning experiences tend to evoke more positive emotional responses in educational settings.

In contrast, the collective teaching methodology was associated with a higher prevalence of negative emotions, with 10% of students displaying anger and 15% expressing disgust. This outcome is consistent with the research conducted by Moreno et al., which suggests that traditional lecture-style teaching may be less engaging for a substantial portion of the student population [7].

The analysis revealed a significant positive correlation between happiness and academic performance ($r = 0.65$; $p < 0.01$), corroborating the findings of Liu et al. regarding the contribution of positive emotional states to enhanced learning outcomes [1]. Conversely, a negative correlation was observed between fear and academic performance ($r = -0.54$; $p < 0.05$), which is congruent with the research of Araya-Pizarro and Espinoza Pastén on the detrimental effects of negative emotions on cognitive processes [22].

5.1 Limitations and Future Directions

While this study provides valuable insights into the interplay between emotions, teaching methodologies, and academic performance, several limitations must be acknowledged. Firstly, the sample size of 20 students is relatively small, potentially limiting the generalizability of the findings. Future research should aim to include larger and more diverse student populations to enhance the external validity of the results.

Secondly, although the emotion recognition system achieved an accuracy of 85%, which is promising, there remains room for improvement. Further refinement of the convolutional neural network model and the incorporation of a more extensive training dataset could potentially enhance the system's performance and reliability.

Thirdly, the current study focused exclusively on facial expressions for emotion recognition. Future investigations could benefit from incorporating additional modalities, such as voice analysis or physiological measures, to provide a more comprehensive and nuanced assessment of emotional states in educational contexts.

Lastly, ethical considerations regarding privacy and data protection warrant ongoing attention. Future implementations of emotion recognition systems in educational settings should explore robust methods for data anonymization and ensure that informed consent is obtained and respected throughout the research process.

In conclusion, this study contributes to the growing body of literature on emotion-aware education and highlights the potential benefits of adapting teaching methodologies to optimize emotional engagement and academic performance. However, further research is necessary to address the identified limitations and explore the long-term implications of emotion recognition technologies in educational environments.

6 Conclusion

The real-time analysis of students' emotional states during classroom sessions, facilitated by a webcam-based emotion recognition system and convolutional neural network, offers a valuable tool for continuous assessment throughout the learning process. Findings from implementing three teaching methodologies—collective, group, and experiential—highlight that the experiential approach leads to the highest levels of student happiness. This emphasizes the importance of integrating practical and meaningful activities into the pedagogical framework, enhancing both student satisfaction and emotional well-being.

Furthermore, leveraging a convolutional neural network for emotion detection enhances the accuracy and reliability of these assessments, proving highly beneficial for educators. This technological approach empowers teachers to effectively tailor and adapt their instructional methods, creating a positive and conducive learning environment. This integration of advanced technology and pedagogical innovation not only improves academic outcomes but also plays a crucial role in the holistic development of students.

Acknowledgments. This work has been supported by the GIIAR research group of the Universidad Politécnica Salesiana.

References

1. Bahreini, K., Nadolski, R., Westera, W.: Towards real-time speech emotion recognition for affective e-learning. Educ Inf Technol (Dordr) **21**, 1367–1386 (2016). https://doi.org/10.1007/S10639-015-9388-2/TABLES/6
2. Pabba, C., Kumar, P.: An intelligent system for monitoring students' engagement in large classroom teaching through facial expression recognition. Expert. Syst. **39**, e12839 (2022). https://doi.org/10.1111/exsy.12839
3. Andujar, A., Medina-López, C.: Exploring New Ways of eTandem and Telecollaboration Through the WebRTC Protocol: Students' Engagement and Perceptions. International Journal of Emerging Technologies in Learning (iJET) **14**, 200 (2019). https://doi.org/10.3991/ijet.v14i05.9612
4. Yin, Z., Zhao, M., Wang, Y., et al.: Recognition of emotions using multimodal physiological signals and an ensemble deep learning model. Comput. Methods Programs Biomed. **140**, 93–110 (2017). https://doi.org/10.1016/J.CMPB.2016.12.005
5. Miguel Ángel Quiroz Martínez, M., Ginnette Andreina Granda Villon, S., Davis Israel Maldonado Cevallos, S., Yelandi Leyva Vázquez, M.: Análisis comparativo para seleccionar una herramienta de reconocimiento de emociones aplicando mapas de decisión difusos y TOPSIS. Dilemas contemporáneos: Educación, Política y Valores (2020). https://doi.org/10.46377/DILEMAS.V8I1.2441

6. Jaiswal, A., Krishnama Raju, A., Deb, S.: Facial Emotion Detection Using Deep Learning. In: 2020 International Conference for Emerging Technology (INCET), pp. 1–5. IEEE (2020)
7. Benavidez, V.V., Flores, P.R.: La importancia de las emociones para la neurodidáctica. Wimb Lu **14**, 25–53 (2019). https://doi.org/10.15517/wl.v14i1.35935
8. Hassouneh, A., Mutawa, A.M., Murugappan, M.: Development of a Real-Time Emotion Recognition System Using Facial Expressions and EEG based on machine learning and deep neural network methods. Inform Med Unlocked **20**, 100372 (2020). https://doi.org/10.1016/J.IMU.2020.100372
9. Jain, P., Murali, M., Ali, A.: Face Emotion Detection Using Deep Learning. Proceedings of the 5th International Conference on I-SMAC (IoT in Social, Mobile, Analytics and Cloud), I-SMAC 2021, pp. 517–522 (2021). https://doi.org/10.1109/I-SMAC52330.2021.9641053
10. Andrejevic, M., Selwyn, N.: Facial recognition technology in schools: critical questions and concerns. Learn. Media Technol. **45**, 115–128 (2020). https://doi.org/10.1080/17439884.2020.1686014
11. Paredes-Velasco, M., Rios, M.G., Velázquez-Iturbide, A.: Analysis of the Emotions Experienced by Learning Greedy Algorithms with Augmented Reality (2020)
12. Krithika, L.: Science LG-PC, 2016 undefined Student emotion recognition system (SERS) for e-learning improvement based on learner concentration metric. ElsevierLB Krithika, LP GGProcedia Computer Science, Elsevier (2016)
13. Alruwais, N.M., Zakariah, M.: Student Recognition and Activity Monitoring in E-Classes Using Deep Learning in Higher Education. IEEE Access **12**, 66110–66128 (2024). https://doi.org/10.1109/ACCESS.2024.3354981
14. Martínez, Q., Ángel, M.I., Hernández, G., et al.: Análisis comparativo para seleccionar una herramienta de reconocimiento de emociones aplicando el modelo AHP. Revista UNIANDES Episteme **6**(3), 453–463 (2019)
15. Martinez, M.A.Q., Bermello, E.X.M., Leon, W.P.C., Briones, L.: A Hybrid Approach Entropy - TOPSIS for the Selection of Machine Learning Classifiers for Software Defect Prediction. Intelligent Human Systems Integration (IHSI 2022): Integrating People and Intelligent Systems 22: (2022). https://doi.org/10.54941/AHFE1001088
16. Martinez, M.A.Q., Rugel, D.T.L., Alcivar, C.J.E., Vazquez, M.Y.L.: A Framework for Selecting Classification Models in the Intruder Detection System Using TOPSIS. In: Ahram, T., Taiar, R., Langlois, K., Choplin, A. (eds.) Human Interaction, Emerging Technologies and Future Applications III: Proceedings of the 3rd International Conference on Human Interaction and Emerging Technologies: Future Applications (IHIET 2020), August 27-29, 2020, Paris, France, pp. 173–179. Springer International Publishing, Cham (2021). https://doi.org/10.1007/978-3-030-55307-4_27
17. Eltenahy, S.A.M.: Facial Recognition and Emotional Expressions Over Video Conferencing Based on Web Real Time Communication and Artificial Intelligence, pp. 29–37 (2021). https://doi.org/10.1007/978-981-33-6129-4_3
18. Yalamarthi, R.N.S.V., Shaik, S., Singh, D., Rakhra, M.: Real-Time Face Mask Detection Using Streamlit, TensorFlow, Keras and Open-CV. IEEE International Conference on Data Science and Information System, ICDSIS **2022**, 10 (2022). https://doi.org/10.1109/ICDSIS55133.2022.9915817
19. Gómez-Rios, M.D., Paredes-Velasco, M., Hernández-Beleño, R.D., Fuentes-Pinargote, J.A.: Analysis of emotions in the use of augmented reality technologies in education: A systematic review. Comput. Appl. Eng. Educ. **31**, 216–234 (2023). https://doi.org/10.1002/CAE.22593
20. Goodfellow, I.J., Erhan, D., Carrier, P.L., et al.: Challenges in Representation Learning: A Report on Three Machine Learning Contests. Lecture Notes in Computer Science (including subseries Lecture Notes in Artificial Intelligence and Lecture Notes in Bioinformatics) 8228(LNCS), 117–124 (2013). https://doi.org/10.1007/978-3-642-42051-1_16

21. Chollet, F., Chollet, F.: Keras: The Python Deep Learning library. ascl ascl:1806.022 (2018)
22. Araya-Pizarro, S.C., Espinoza Pastén, L.: Aportes desde las neurociencias para la comprensión de los procesos de aprendizaje en los contextos educativos. Propósitos y Representaciones **8** (2020). https://doi.org/10.20511/pyr2020.v8n1.312

Optimization of a Rectangular Warehouse Design Using Heuristics Techniques

Guillermo O. Pizarro-Vasquez[(✉)] [iD]

Universidad Politecnica Salesiana, Cuenca, Ecuador
gpizarro@ups.edu.ec

Abstract. This research aims to determine the number of cross aisles that minimizes distance traveled. To achieve this, a rectangular warehouse with various transverse aisles was defined, and simulations were conducted with 0 to 90 transverse aisles using heuristic algorithms (FCFS, G01, G02, G03, RANDOM, and SOP) for order batching, and S-Shape and L-Gap for order picking. The simulations considered a list of 20 to 100 orders or carts with a capacity of 30 to 75 order items. As a result, ANOVA analysis demonstrates the significance of order batching and picking algorithms, number of orders and cart capacity. The order batching algorithms that provide minimum distances in these experiments are G03 and RANDOM, and for order picking, LGAP algorithm is most effective. The number of cross aisles in configuration presented in this research that yields the minimum distances is 26; however, this is not conclusive to ensure that minimum distances will always be achieved with this number of transverse aisles.

Keywords: Optimization · Heuristic · Warehouse design

1 Introduction

Warehousing operations are a critical component of the supply chain, playing a vital role in ensuring that products are stored, handled, and distributed efficiently. Among the various operations within a warehouse, order picking stands out as the most labor-intensive and costly process. This task involves retrieving products from their storage locations to fulfill customer orders, and it can significantly impact a warehouse's overall efficiency and profitability. According to [3], order picking can account for up to 50% of a warehouse's operational costs. This is corroborated by other studies; [11] and [2] suggest that these costs can be even higher, reaching up to 60% or 65%, respectively.

1.1 Warehouse Design

The design of a warehouse refers to the layout of the aisles within the warehouse. Some of the most well-known designs are: rectangular (Fig. 1) [1,9], V-shaped (Fig. 2) [4] or parabolic- shaped [15], among others [7,8].

© The Author(s) 2025
E. M. Inga Ortega et al. (Eds.): CITIS 2024, LNNS 1331, pp. 93–104, 2025.
https://doi.org/10.1007/978-3-031-87065-1_9

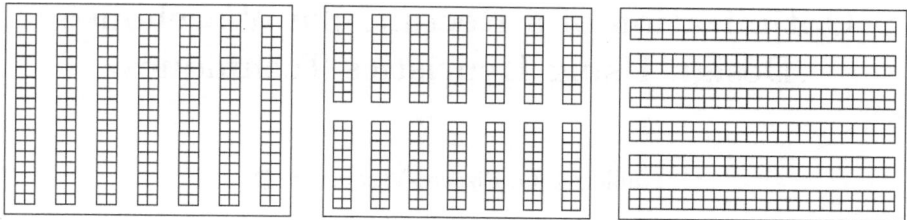

Fig. 1. Rectangular warehouses design.

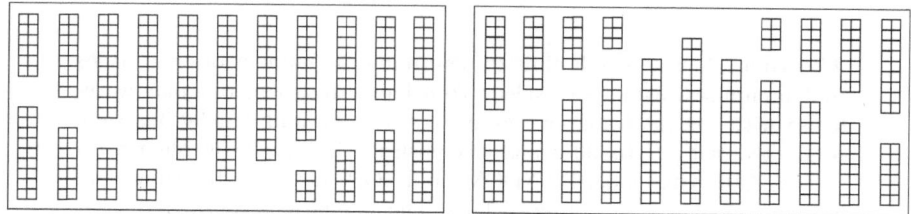

Fig. 2. V-shaped warehouses design.

Specifically, in this research work we are going to use a rectangular design with one or more transversal aisle, in addition to the aisles at the front and back.

The research question of this study is: Does the number of transversal aisles influence the total distance required to pick orders in a rectangular warehouse?

This research work is organized as follows: introduction section, describes the design of the rectangular warehouse and different designs currently in use; methods section, details the model formulation, formal definition, explains order batching algorithms and order picking algorithms; results section, mentions sources of the instances used for experiments, warehouse technical specifications, and presents an exploratory data analysis (EDA); discussion section, presents the main findings; and finally, respective conclusions are provided.

2   Methods

To optimize the number of transversal aisles in a rectangular warehouse, it is necessary to optimize the total travel distance for picking orders in the warehouse. To achieve this, it is first necessary to propose an order batching strategy, then proceed with the order picking and then we are going to get a metric (distance) to know if the number of cross aisles is the correct.

2.1   Model Formulation

Let $\mathbf{P}$ be the set of products in a warehouse, and let $\mathbf{L}$ be the set of locations in the warehouse, where each location $L \in \mathbf{L}$ contains a subset $P_L \subset \mathbf{P}$.

Let $\mathbf{O}$ be the set of orders placed by customers, and each order $o \in \mathbf{O}$ contains $P_o \subseteq \mathbf{P}$.

Consider a graph (Fig. 3) as the representation of a warehouse, including a vertex for each location (white vertices).

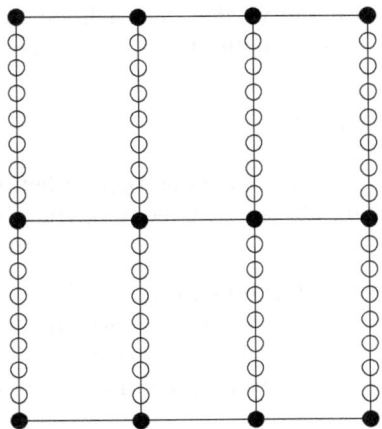

Fig. 3. Warehouse Graph.

For each order $o \in \mathbf{O}$, there is a subset of locations $L_o \subseteq \mathbf{L}$ that contains all the products in P_o (it is possible that some products of the same order are in the same location).

Let $L(\mathbf{O}) = \bigcup_{o \in O} L_o$ be a set of locations containing all the products that need to be picked for all orders.

Let $d_{lm} \geq 0$ be the distance between locations $l, m \in \mathbf{L}$, which is symmetric, i.e., $d_{lm} = d_{ml}$.

Often, $L(\mathbf{O}) \neq \mathbf{L}$ because there are locations without products assigned in customer orders; in these cases, the graph representing the warehouse is reduced to only the locations with requested orders $\mathbf{L} \setminus L(\mathbf{O})$.

Let b_o be the capacity of the picking cart, meaning the number of items that can be placed in the picking cart. It is assumed that products from various orders can be placed in one picking cart; however, it is not possible to place products from one order in different picking carts.

Let $\mathbf{s}$ be the starting point from where the pickers begin their route.

2.2   Formal Definition

Here is the formal definition of the order batching and picking problem in a rectangular warehouse.

Let $\mathbf{D}$ be a directed and connected graph, $\mathbf{D} = (\mathbf{V}, \mathbf{A})$, where $\mathbf{V}$ is the set of vertices given by the union of:

- **s**, the starting point;
- **V(O)**, the set of vertices for each location where $L \in L(\mathbf{O})$;
- **V**$_A$, the set of artificial locations (black vertices), located at the corners between aisles and cross-aisles; in addition, they do not contain products to be picked.

Let $\mathbf{V}_o$ be the set containing vertices for each location where $L \in L(\mathbf{O})$. Let $\mathbf{V(O)} = \bigcup_{o \in O} V_o$; additionally, $V = \{s\} \bigcup \mathbf{V(O)} \bigcup \mathbf{V}_A$.

2.3   Order Batching Problem

For the order grouping, the following constructive heuristic algorithms will be used: Random, FCFS, SOP, Greedy 1, Greedy 2, and Greedy 3.

Random. Create batches with random orders. Given a list of orders, select an order at random and assign it to a batch. Continue adding orders to the same batch randomly until the batch reaches its capacity. Repeat this process to create additional batches until all the initial orders have been assigned to a batch.

First Come First Served (FCFS). Create batches with sequential orders. Given a list of orders, select the first order and place it in a batch. Then select the second order (following the order in the list) and continue this process until the batch reaches its capacity. After that, create another batch and repeat the process until all the initial orders have been assigned to batches.

Strick Order Picking (SOP). Create batches with a single order. Given a list of orders, select one order and create a batch. Then take another order and create a new batch. Repeat this process until all the orders have been assigned to batches.

Greedy 1. Given a list of orders, sort it in descending order by the number of items. Starting with the first order, create batches up to their capacity in sequence. Repeat this process until all orders have been assigned to batches.

Greedy 2. Given a list of orders, sort it in ascending order by the number of items. Starting with the first order, create batches up to their capacity in sequence. Repeat this process until all orders have been assigned to batches.

Greedy 3. Given a list of orders, create batches up to their capacity with orders that have items located close to each other in the warehouse. Repeat this process until all orders have been assigned to batches.

2.4 Picker Routing Problem

After creating the order batches, the next step is to proceed with order picking. For this, the S-Shape and Largest Gap algorithms were chosen.

S-Shape. Starting from the depot, proceed to the last block and begin walking in a zigzag pattern, as shown in Fig. 4. This behavior is repeated with the block closest to the depot and continues in the same way.

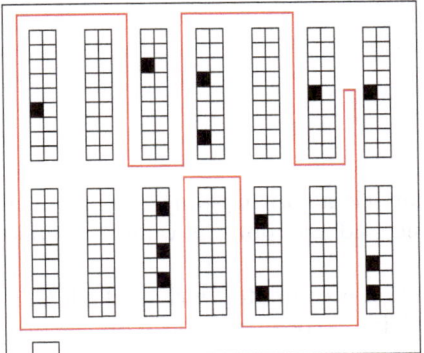

Fig. 4. S-Shape Algorithm.

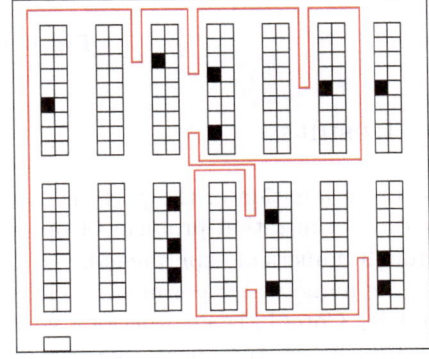

Fig. 5. Largest Gap Algorithm.

Largest Gap. Begin at the depot, go along to the last block in the warehouse, then algorithm identifies the largest gap between items to be picked in each aisle. A gap is defined as the distance between two adjacent items. Enter the first aisle and proceed along the aisle until you reach the largest gap, at the largest gap, cross over to the next aisle, continuing to pick items, This pattern continues, ensuring that the picker crosses aisles at the points where the gaps between items are the largest, minimizing the travel distance. The process repeats for all aisles until all items have been picked, as you see in Fig. 5.

2.5 Multiple Cross Aisles

After the order batching and picker routing algorithms are implemented, the simulation of grouping and picking is performed starting with no aisle (Fig. 6), then one cross aisles (Fig. 7), two aisles (Fig. 8) and continuing up to 90 cross aisles.

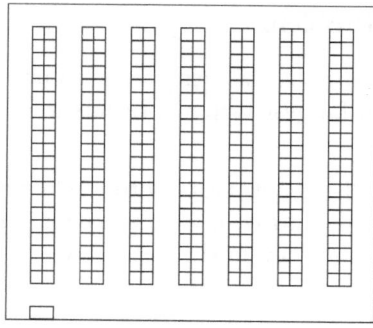

Fig. 6. One block.

3   Results

For the evaluation of the performance of heuristic algorithms and their impact on the optimization problem of order picking and grouping, it would be done through numerical experiments.

The numerical experiments were executed on a computer with an 11th Gen Intel(R) Core(TM) i5 processor with 2.42 GHz, 4 cores, 8 logical processors, 16GB of RAM, running Windows 11 Operating System.

The algorithms were implemented in Java, using Java(TM) SE Development Kit 20.0.1, and developed in Visual Studio Code with the Extension Pack for Java.

The source code for the multi-aisle warehouse configuration in this research is based on the Perl source code programmed from the following scientific articles: [13,14], and [12].

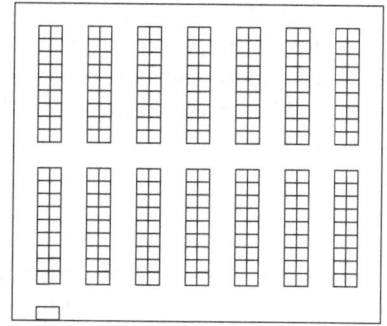

Fig. 7. Two blocks.

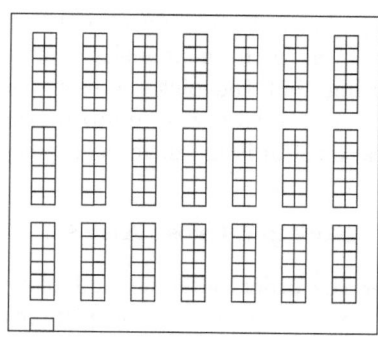

Fig. 8. Three blocks.

3.1 Instances

The instances used in the experimentation are partially based on [5,6], as these authors used the instances in a single-block warehouse (Fig. 6); however, for this research work, similar to [10], the instances were configured to operate in two or more blocks warehouse.

Below is the detailed configuration of the warehouse and the orders:

Warehouse. The warehouse design can have several cross aisles; however, we will describe it in detail for two blocks.

The warehouse consisting of two blocks (Fig. 7) has 10 aisles and a total of 900 locations (45 locations on each side of the aisle; that is, 90 locations per picking aisle).

The dimensions of the warehouse have the following specifications: a **location** is 1.5 unit length (UL) wide and 1 UL long, the **cross aisle** is 1 UL wide and 1 UL long, the width of an **aisle** is 2 ULs (the distance between two opposite locations in a picking aisle). If a picker moves from one aisle to another, they would travel 5 ULs; and if they traverse an entire picking aisle (from the rear cross aisle to the front aisle), it would be 47 ULs. And if there are more cross-aisles, the respective changes are made according to the configuration detailed in this research work.

The assignment policy in the warehouse, according to the instances used, is randomly located with a uniform distribution.

Orders. The order sizes are: 20, 30, 40, 50, 60, 70, 80, 90, and 100; the capacity of the cart for order pickup has the following values: 30, 45, 60, and 75. The number of products per order ranges from {5, ..., 25} and varies with a uniform distribution.

3.2 Exploratory Data Analysis

Applying ANOVA can help understand how the different factors (obp_algorithm, prp_algorithm, capacity_device, and num_orders) affect the variable num_cross_aisles, with the aim of minimizing the traveled distance (distance). In this case, the model has been applied as follows:

```
distance ~ obp_algorithm * prp_algorithm * capacity_device *
           num_orders + num_cross_aisles
```

In ANOVA analysis, all main factors and their interactions have a significant effect on the traveled distance, with p-values $< 2e\text{-}16$, indicating that these effects are unlikely to be due to chance.

Regarding heuristic algorithms for order batching (obp_algorithm), when considering minimal travel distance and number of orders, the algorithms that provide minimal distances are: **G03** and **RANDOM** (Fig. 9), and result is the

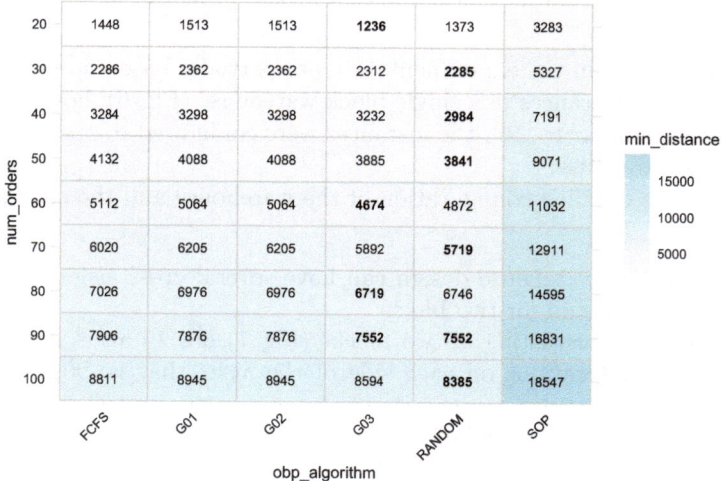

Fig. 9. OBP Algorithms vs. minimal distance vs. number of orders.

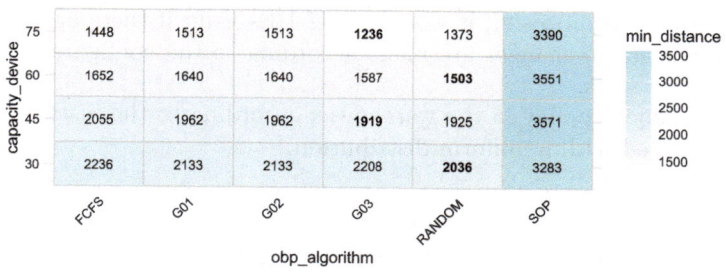

Fig. 10. OBP Algorithms vs. minimal distance vs. capacity device.

same when considering minimal distance and cart capacity (Fig. 10). Heuristic algorithms for picking routing (prp_algorithm), considering minimal travel distance and number of orders is: **LGAP** (Fig. 11), and result is the same when considering minimal distance and cart capacity (Fig. 12).

The following are the data for order batching (Tables 1 and 2) and picking (Tables 3 and 4) algorithms selected for having shortest distances and respective number of cross aisles used in the simulation.

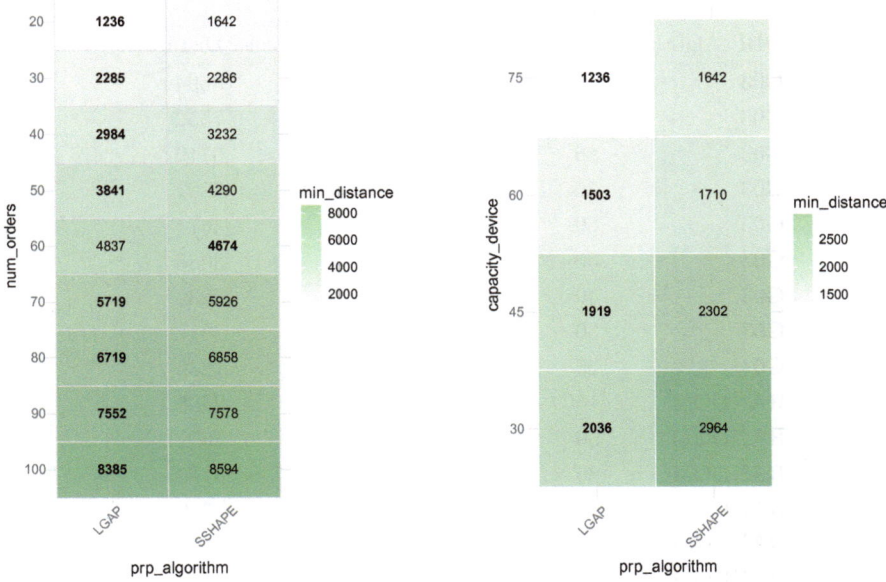

Fig. 11. PRP Algorithms vs. minimal distance vs. number of orders.

Fig. 12. PRP Algorithms vs. minimal distance vs. capacity device.

Table 1. OBP Algorithms vs. capacity device

OBP Algorithms	Capacity device	Num. of cross aisles	Distances
G03	30	10	2208
G03	45	15	1919
G03	60	12	1587
G03	75	26	1236
RANDOM	30	8	2036
RANDOM	45	18	1925
RANDOM	60	18	1503
RANDOM	75	9	1373

Table 2. PRP Algorithms vs. capacity device

PRP Algorithms	Capacity device	Num. of cross aisles	Distances
LGAP	30	8	2036
LGAP	45	15	1919
LGAP	60	18	1503
LGAP	75	26	1236

Table 3. OBP Algorithms vs. num. of orders

OBP Algorithms	Num. of orders	Num.of cross aisles	Distances
G03	100	3	8594
G03	90	13	7552
G03	80	21	6719
G03	70	15	5892
G03	60	5	4674
G03	50	26	3885
G03	40	3	3232
G03	30	3	2312
G03	20	26	1236
RANDOM	100	13	8385
RANDOM	90	24	7552
RANDOM	80	24	6746
RANDOM	70	24	5719
RANDOM	60	3	4872
RANDOM	50	15	3841
RANDOM	40	16	2984
RANDOM	30	13	2285
RANDOM	20	9	1373

Table 4. PRP Algorithms vs. num. of orders

PRP Algorithms	Num. of orders	Num.of cross aisles	Distances
LGAP	100	13	8385
LGAP	90	13	7552
LGAP	80	21	6719
LGAP	70	24	5719
LGAP	60	10	4837
LGAP	50	15	3841
LGAP	40	16	2984
LGAP	30	13	2285
LGAP	20	26	1236

4   Discussion

According to ANOVA analysis, there is a relationship between order batching and picking algorithms, number of orders, cart capacity, number of cross aisles, and distance traveled. Indeed, it is observed that distance and cart capacity are inversely proportional, while distance and number of orders are proportional.

However, in general, it can be observed that minimum distances are obtained between 3 and 26 transverse aisles.

5   Conclusions

There is no direct and linear relationship between the number of transverse aisles and the minimum distance.

A greater number of transverse aisles does not always guarantee a lower minimum distance; however, for number of cross aisles equal to 26, some of the lowest minimum distances are observed.

References

1. Al-Shboul, M.A.: Design and control order picking route of a retailer warehouse using simulation to increase labour productivity. Acta Logistica **10**, 121–133 (2023). https://doi.org/10.22306/al.v10i1.367
2. Coyle, J.J., Bardi, E.J., Langley, C.J.: The Management of Business Logistics: A Supply Chain Perspective. South-Western, 7th edn. (2003)
3. Frazelle, E.H.: World-Class Warehousing and Material Handling. McGraw-Hill Education (2016)
4. Gue, K.R., Ivanovic, G., Meller, R.D.: A unit-load warehouse with multiple pickup and deposit points and non-traditional aisles. Trans. Res. Part E: Logist. Trans. Rev. **48**(4), 795–806 (2012). https://doi.org/10.1016/j.tre.2012.01.002
5. Henn, S., Koch, S., Doerner, K.F., Strauss, C., Wäscher, G.: Metaheuristics for the order batching problem in manual order picking systems. Bus. Res. **3**(1), 82–105 (2010). https://doi.org/10.1007/BF03342717
6. Henn, S., Wäscher, G.: Tabu search heuristics for the order batching problem in manual order picking systems. Eur. J. Oper. Res. **222**(3), 484–494 (2012). https://doi.org/10.1016/j.ejor.2012.05.049
7. Ozturkoglu, O.: Investigating the robustness of aisles in a non-traditional unit-load warehouse design: leverage. In: 2015 IEEE Congress on Evolutionary Computation (CEC), pp. 2230–2236 (May 2015). https://doi.org/10.1109/CEC.2015.7257160
8. Ozturkoglu, O., K.R., G., Meller, R.: A constructive aisle design model for unit-load warehouses with multiple pickup and deposit points. Euro. J. Operat. Res. **236**(1), 382 – 394 (2014). https://doi.org/10.1016/j.ejor.2013.12.023
9. Pohl, L.M., Meller, R.D., Gue, K.R.: An analysis of dual-command operations in common warehouse designs. Trans. Res. Part E: Logist. Trans. Rev. **45**(3), 367–379 (2009). https://doi.org/10.1016/j.tre.2008.09.010
10. Scholz, A., Wäscher, G.: Order Batching and Picker Routing in manual order picking systems: the benefits of integrated routing. CEJOR **25**(2), 491–520 (2017). https://doi.org/10.1007/s10100-017-0467-x
11. Tompkins, J.A., White, J.A., Bozer, Y.A., Tanchoco, J.: Facilities Planning. John Wiley & Sons Ltd., UK (2010)
12. Valle, C.A., Beasley, J.E.: Order batching for picker routing using a distance approximation. arXiv preprint arXiv:1808.00499 (2018)
13. Valle, C.A., Beasley, J.E., da Cunha, A.S.: Modelling and solving the joint order batching and picker routing problem in inventories. In: Cerulli, R., Fujishige, S., Mahjoub, A.R. (eds.) ISCO 2016. LNCS, vol. 9849, pp. 81–97. Springer, Cham (2016). https://doi.org/10.1007/978-3-319-45587-7_8

14. Valle, C.A., Beasley, J.E., da Cunha, A.S.: Optimally solving the joint order batching and picker routing problem. Eur. J. Oper. Res. **262**(3), 817–834 (2017). https://doi.org/10.1016/j.ejor.2017.03.069

15. Zhang, Z.Y., Liang, Y., Hou, Y.P., Wang, Q.: Designing a warehouse internal layout using a parabolic aisles based method. Adv. Product. Eng. Manag. **16**, 223–239 (2021). https://doi.org/10.14743/APEM2021.2.396

An Approach to Bibliometric Analysis of Biosignal Acquisition Systems to Determine Neurological and Physiological Aspects of Driving Stress

Génesis-Gabriela Macansela-Tapia⬤, Jean Márquez⬤, Esteban Ordóñez[✉]⬤, and Juan Inga⬤

Grupo de Investigación en Telecomunicaciones GITEL, Universidad Politécnica Salesiana, Cuenca, Ecuador
eordonez@ups.edu.ec

Abstract. It is of paramount importance to determine the level of stress that drivers face while driving and processing a large amount of information coming from their vehicle, other vehicles and the surrounding environment. This is essential to improve road safety, driver performance and driver well-being. In addition, it enables the design of new advanced driving assistance systems that are coupled to the driving profile of the vehicle user, which would reduce the costs associated with accidents, promoting the development of intelligent traffic systems. In this context, the present study employs a bibliometric analysis of biosignal acquisition systems to examine the neurological and physiological aspects of driving stress. Scientific publications related to the acquisition of biosignals, including brain activity, heart rate, eye movement, muscle electrical activity, and skin conductance, are investigated. After these signals are processed, it is possible to determine the person stress level. A bibliometric analysis enables the identification of research trends, the most influential authors, the most relevant journals, and the countries engaged in research on the subject. Furthermore, the most commonly utilized systems and equipment for biosignal acquisition are identified. This approach provides a comprehensive overview of the current state of research, which allows us to direct our future research and highlights the importance of technology in improving road safety and the driver well-being.

Keywords: Bibliometric Analysis · Biosignals Acquisition Systems · Stress in drivers

1 Introduction

It is well documented that driver stress and emotional state contribute significantly to the occurrence of traffic accidents. Indeed, it has been estimated that

E. M. Inga Ortega et al. (Eds.): CITIS 2024, LNNS 1331, pp. 105–115, 2025.
https://doi.org/10.1007/978-3-031-87065-1_10

30% of these accidents are related to driver stress [1]. Furthermore, studies have shown that driving-induced stress can vary in severity, ranging from mild discomfort to significant impairment of driver performance [2]. The relationship between stress and driver performance is a critical topic in road safety research.

Elevated stress levels have been demonstrated to impair driver attention, increase reaction time, and negatively impact decision-making in emergency situations. This critical factor significantly affects driver performance. In this context, biosignal acquisition systems have been fundamental in monitoring and analyzing the neurological and physiological aspects of stress. The most commonly used biosignals are electroencephalography (EEG), electrocardiography (ECG), electromyography (EMG), electrooculography (EOG), and skin conductance (SC) [3].

Biosignal acquisition is a fundamental process in biometrics and health sciences, where physiological and neurological data are collected through various sensors. EEG signals are captured using electrodes placed on the scalp, allowing the monitoring of the brain electrical activity to assess states such as sleepiness and alertness [4]. Meanwhile, ECG signals measure the electrical activity of the heart through electrodes placed on the skin, providing crucial information about heart rate variability and arrhythmia detection [5]. EMG signals record the electrical activity produced by skeletal muscles, which is essential for studying muscle tension and fatigue [6]. SC signals measure changes in skin electrical conductance in response to emotional stress [7]. These biometric data are subjected to advanced signal processing techniques and machine learning algorithms, which permit the identification of intricate patterns associated with various physiological and emotional states [8]. The analysis of these signals not only enhances the understanding of physiological and neurological states but also has practical applications in health monitoring and the development of adaptive technologies for various fields [8].

Biosignal processing is a crucial technique for detecting stress levels and emotional states in a person. The acquisition of the biosignals indicated above, undergo various signal processing and data analysis techniques to extract relevant features. Machine learning algorithms, such as neural networks and support vector machines, are widely used to classify and predict emotional states from these features [9]. In the case of EEG, specific frequency bands (alpha, beta, gamma) are analyzed to assess attention and mental alertness [10]. For example, an increase in beta wave activity (13–30 Hz) is frequently associated with heightened alertness and stress. This is because these waves reflect increased cortical activity related to attention and intense cognitive processing [11]. In addition, heart rate variability derived from ECG is a crucial tool for identifying stress responses, as changes in heart rate are sensitive indicators of emotional state [12]. Moreover, EMG is employed to assess muscle and nerve activity, thereby enabling the identification of indications of tension and physical stress. Indeed, elevated muscle activity levels have been demonstrated to be associated with stress states [13]. On the other hand, EOG monitors eye movements and blinking, which can increase significantly under stress and fatigue. This provides additional indicators of emotional state [14]. SC is also utilized as a dependable indicator of emotional distress, reflecting the activity of the autonomic nervous system [15]. The skin conductance response (SCR), which is a component of SC, increases in response to emotional stimuli due to the activation of sweat glands controlled by the sympathetic nervous system. This increase in skin conductance is a direct indicator of emotional arousal and stress, as it is associated with the activation of the autonomic nervous system in stressful or threatening situations [16]. These techniques permit the construction of systems that are capable of monitoring emotional states in real-time and providing feedback for emotional regulation, as exemplified by biofeedback applications [17]. The integration of multiple signals and the utilization of sophisticated analysis methodologies permit a more precise and individualized assessment of stress and emotional states, thereby enhancing the capacity of systems to respond to individual requirements [18].

Biosignal acquisition systems, when combined with advanced data analysis techniques and machine learning algorithms, can reveal complex patterns and provide a deep understanding of emotional and physiological responses to stress. This offers an approximate reading of the emotional state of a person [9] [12] [17]. In this context, biosignal acquisition and processing systems can be utilized to determine the level of stress experienced by vehicle drivers when they are driving and receiving a large amount of information from the vehicle itself, other vehicles around them, and the surrounding environment. The detection and management of high levels of stress in drivers can enhance their safety. This information can be utilized to develop intelligent driving assistants that adapt to the driving profile of the user, filtering information based on predefined priorities and rules. This opens new possibilities for the development of safer and smarter transportation systems [3] [19]. Therefore, this document aims to carry out a bibliometric analysis of the main approaches that present biosignal acquisition systems aimed at determining the neurological and physiological aspects of driving stress.

This paper is structured as follows: Section 2 presents systems approaches aimed at acquiring biosignals in drivers to determine their stress levels. While Sect. 3 describes the methodological process that was followed to establish the most prominent scientific articles in the field of study of this work. The results and discussion are presented in Section IV. Finally, Sect. 5 states the conclusions and future work.

2 Biosignal Acquisition for Detecting Driving Stress

The acquisition of multiple biosignals can enhance the prediction of stress in vehicle drivers, providing a comprehensive assessment of their emotional and alert states by detecting conditions of anxiety and fatigue. This data can be used to develop next-generation Advanced Driver Assistance Systems (ADAS) based on artificial intelligence aim to improve the driving experience and contribute to reducing stress levels [20][21][22]. For instance, the use of EEG in ADAS enhances the detection of drowsiness and alertness states, thereby enhancing the effectiveness of forward collision warning systems [23] [19]. Similarly, the continuous monitoring of parameters such as ECG and SC can also provide crucial information about the emotional state and stress level of the driver [24]. A number of studies have demonstrated that the combination of EEG with other biosignals, such as heart rate variability, can provide a more accurate detection of driver fatigue [25].

An illustration of the manner in which biosignal acquisition systems bolster road safety is DeCaDrive, which is a driver assistance system (ADAS) that employs a multitude of sensors, including an electrocardiogram (ECG), to gather physiological data. The data collected from the various sensors is utilized to monitor the driver's condition and intentions, classify the level of drowsiness, and issue warnings when dangerous levels are detected, thereby preventing traffic accidents [26].

The integration of these technologies in both conventional and smart vehicles would not only enhance safety but also optimize the driving experience by reducing the cognitive load and stress associated with prolonged driving [27]. Such systems can facilitate the provision of early alerts, thereby assisting in the prevention of accidents that may be related to driver stress [28]. For example, through a haptic feedback system, as illustrated in Fig. 1, it is possible to apply small shear forces to the skin on the driver hand, providing tactile signals about the vehicle acceleration and direction. Drivers can easily perceive and distinguish these signals, allowing them to receive notifications of intentions in semi-autonomous vehicles. This improves navigation by alerting the driver to potential collisions and lane changes [29].

This bibliometric review aims to analyze trends and advancements in the research of biosignal acquisition systems aimed at the analysis and monitoring of neurological and physiological signals for the detection of driver stress, with the objective of safeguarding their safety and well-being. We believe that systems with these characteristics can provide a solid foundation for the development of future intelligent and adaptive transportation systems [30].

Fig. 1. Steering wheel skin stretch haptic feedback system [29].

3 Methodology

3.1 The Database and Search Criteria

In order to conduct the bibliometric review focusing on the search for articles related to biosignal acquisition systems for stress detection in individuals, the PRISMA methodology was employed [31]. The SCOPUS database was selected for article search due to its status as one of the largest academic databases worldwide, covering a wide range of academic and scientific disciplines. This choice is based on the database ability to provide comprehensive coverage of relevant literature, ensuring that important studies in the field of bio-signal acquisition for stress detection are not overlooked.

Initially, the search equation (SE1) "biosignal AND acquisition AND system" was used in all fields, resulting in 2.282 documents from 2005 to 2024, distributed across 27 areas of knowledge and considering 152 secondary keywords. Most documents are published, with some indicated as "in press." The top five contributing countries are, in descending order, the United States, China, India, Germany, and Italy. Most articles are published primarily in English, followed by Chinese. However, most articles are concentrated in the fields of engineering and computer science.

To address this concentration, a second search (SE2) was conducted using the same SE1 equation but only searching in the fields of "article title, abstract, keywords". This resulted in a total of 258 articles found, distributed across 25 areas of knowledge (maintaining the trend of concentration in engineering and computer science). Additionally, 148 secondary keywords were identified. In this case, the top five contributing countries are led by Germany, followed by the United States, India, Italy, and Portugal. As in the SE1 search, the most used language is English, followed by Chinese. In both SE1 and SE2 searches, conference articles, journal articles, and book chapters were considered, with most documents being published.

In the third stage of the search (SE3), two additional terms were added to SE1: "stress" and "driving." This was done to identify articles focused on bio-signal acquisition systems that characterize the neurological and physiological aspects necessary to determine stress levels. These aspects are not limited to the general population but are also relevant to drivers exposed to real driving situations. Consequently, the following search equation was used for the third stage (SE3): "biosignal AND acquisition AND system AND stress OR drivers." The "OR" function was used to introduce the final search term "drivers," as articles on bio-signal acquisition systems intended to determine a person stress level are developed in a laboratory or other controlled environments, these approaches could be adapted (with necessary considerations) to determine stress levels in drivers. Using the "OR" operator instead of the "AND" operator allowed for obtaining a sufficient number of articles, as using the "AND" operator produced only one article specifically focused on bio-signal acquisition in drivers. The SE3 search equation yielded 14 documents.

4   Results and Discussion

4.1   Analysis of Publications in Thematic Areas

The 14 documents identified using the PRISMA methodology were published between 2005 and 2023. The articles are primarily in the fields of engineering and computer science and are written in English. However, in contrast to the previous searches (SE1 and SE2), Taiwan leads the list this time, followed by India, the United States, and the United Kingdom. There are 12 conference papers (one of which is a review) and 2 journal articles. The initial document was presented in 2005, and since that time, approximately one article has been published annually, with the exception of the years 2011, 2013 and 2015, during which two articles were published per year. Additionally, it is observed that the article [32] published in 2008 is the most cited, with 169 citations to date. Furthermore, it is observed that the article [33] is the one identified when using the last operator in SE3 as "AND." This demonstrates that changing the operator "AND" to "OR" in the last term resulted in a more productive search.

Figure 2 illustrates the thematic areas of the 14 articles, indicating that the two areas with the highest concentration of publications are Engineering and Computer Science.

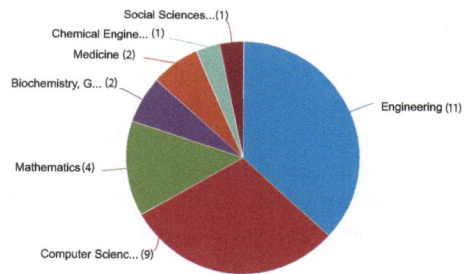

Fig. 2. Documents published in 7 subject areas.

Figure 3 depicts the geographical distribution of the publications, with Taiwan emerging as the most prolific country, followed by India and the United States.

4.2   Cultural Contexts

The findings regarding biosignal acquisition systems and their ability to detect driver stress may vary depending on the cultural context. For example, a driver with a specific cultural background may not react in an appropriate manner in a foreign environment due to a lack of understanding of the local driving culture. The use of systems that can monitor stress levels in such situations would enable drivers to be advised to make better decisions based on the specific characteristics of their environment [34].

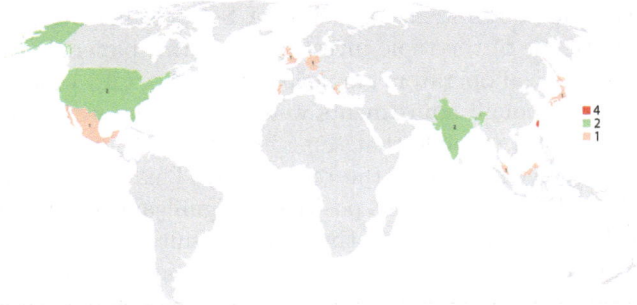

Fig. 3. Publications by country.

4.3   Ethical Aspects

The continuous monitoring of a driver's stress using biosignal acquisition systems presents several ethical considerations that must be addressed to ensure respect for the privacy and autonomy of individuals. Given that monitoring systems collect a large amount of sensitive biometric data, including EEG, ECG, EMG, EOG, and SC, it is of the utmost importance to ensure that this information is stored and processed securely. Techniques such as encryption, obfuscation, and differential privacy are essential for safeguarding the integrity of bio-information and the secure storage of personal data, thereby minimizing the risk of data misuse [35].

4.4   Comparison with Previous Studies

Previous studies have demonstrated the effectiveness of stress detection in drivers using biosignals. In [3], it is confirmed that it is possible to detect high levels of stress in drivers through ECG analysis in real driving conditions, suggesting applications in the improvement of road safety. On the other hand, [25] demonstrated that the combination of ECG and EEG improves the accuracy of stress and fatigue detection, highlighting the effectiveness of a multimodal approach. Furthermore, in [13] the authors found that EMG is useful in assessing muscle and nerve activity, identifying indications of physical tension and stress, and demonstrating that elevated levels of muscle activity are associated with stress states. Similarly, in [14] it is confirmed that EOG monitors eye movements and blinking, activities that can increase significantly when a person is under stress, providing additional indicators of emotional state. Finally, [15] showed that the level of skin conductance is a reliable indicator of emotional distress, reflecting the activity of the autonomic nervous system.

4.5   Limitations of the Study

The use of the SCOPUS database, although one of the largest and most comprehensive, may limit the scope of the bibliometric review. Other relevant databases, such as IEEE Xplore, PubMed, and Web of Science, may contain additional publications pertinent to the topic investigated in this paper. Furthermore, the search terms and logical operators employed ("AND", "OR") may influence the results obtained. Although adjustments were made to improve the relevance of the articles found, it is possible that some important studies were omitted due to the specificity or generality of the terms used. This may also contribute to publication bias, where studies with positive or significant results are more likely to be published than those with negative or null results, leading to a skewed representation of the effectiveness and usefulness of biosignal acquisition systems in driver stress detection. Although studies from several countries were identified, most publications come from a limited number of regions (Taiwan, India, United States). This may introduce geographic and cultural biases, as stress responses and driving conditions can vary significantly across different cultural contexts.

5   Conclusions

This paper presents an approach to bibliometric analysis of biosignal acquisition systems for the purpose of determining the neurological and physiological aspects of driving stress. The process of conducting the literature review was framed in accordance with the PRISMA methodology, which consisted of three stages. In the initial stage of the search, a total of 2.282 articles were obtained through a search in all fields of the SCOPUS database. Documents were found in research areas other than engineering and computer science, which were the areas of greatest interest to us. In a second stage, the search was refined by focusing on the specific search fields of "titles, abstracts, and keywords", the results of this second stage were 258 articles. Finally, in the third and final stage, a total of 14 relevant articles were obtained, with the majority concentrated in the areas of our interest (engineering and computer science). It was found that the 14 articles were distributed across 10 countries, with Taiwan at the top with four articles, followed by India and the United States with two articles each, and the rest of the countries with one article each.

The investigation of systems that acquire, monitor, and process biosignals from drivers in real-world environments can provide valuable insights into the factors influencing driving behavior. These findings will facilitate the development of ADAS and virtual driving assistants aimed at supporting more efficient intelligent transportation systems, which will adapt to the driver's profile and integrate internal vehicle data with external environmental information.

The 14 articles identified were determined to be reliable sources of information based on the search criteria. This allows for the analysis of these articles with confidence, particularly those that are more representative and have a higher number of citations. This provides an opportunity to continue advancing our future work aimed at developing a biosignal acquisition and processing system to determine the level of stress to which a driver is exposed in real driving environments.

Acknowledgements. Thanks for the support of the *"RED PARA ACELERAR LA TRANSICIÓN DE PYMES A INDUSTRIA 4.0 CON TECNOLOGÍA DE BAJO COSTO" (REDTPI4.0)"*.

References

1. World Health Organization (WHO), "Road Traffic Injuries," December 2023
2. Pêcher, C., Lemercier, C., Cellier, J.-M.: The influence of emotions on driving behavior. Traffic Psychol. Inter. Perspect., 145–158 (2011)
3. Healey, J.A., Picard, R.W.: Detecting stress during real-world driving tasks using physiological sensors. IEEE Trans. Intell. Transp. Syst. **6**(2), 156–166 (2005)
4. Vidal, J.J.: Toward direct brain-computer communication. Annu. Rev. Biophys. Bioeng. **2**(1), 157–180 (1973)
5. Sahoo, J.P.: Analysis of ECG signal for Detection of Cardiac Arrhythmias. PhD thesis (2011)
6. Merletti, R., Farina, D.: Surface electromyography: physiology, engineering, and applications. John Wiley & Sons (2016)
7. Allen, J.: Photoplethysmography and its application in clinical physiological measurement. Physiol. Meas. **28**(3), R1 (2007)
8. Heikenfeld, J., et al.: Wearable sensors: modalities, challenges, and prospects. Lab Chip **18**(2), 217–248 (2018)
9. Picard, R.W., Vyzas, E., Healey, J.: Toward machine emotional intelligence: analysis of affective physiological state. IEEE Trans. Pattern Anal. Mach. Intell. **23**(10), 1175–1191 (2001)
10. Klimesch, W.: Eeg alpha and theta oscillations reflect cognitive and memory performance: a review and analysis. Brain Res. Rev. **29**(2–3), 169–195 (1999)
11. Sterman, M.B., Mann, C.A.: Beta EEG (13–30 Hz) and mental performance. Elsevier, Amsterdam (2000)
12. Shaffer, F., Ginsberg, J.P.: An overview of heart rate variability metrics and norms. Front. Public Health **5**, 290215 (2017)
13. Taelman, J., Vandeput, S., Spaepen, A., Van Huffel, S.: Influence of mental stress on heart rate and heart rate variability. In: 4th European Conference of the International Federation for Medical and Biological Engineering, pp. 1366–1369. Springer (2008). https://doi.org/10.1007/978-3-540-89208-3_324
14. He, J., Wu, Z., Sun, L.: Monitoring of eye movements for human-computer interaction. J. Biomed. Sci. Eng. **4**(1), 17 (2011)
15. Critchley, H.D.: Electrodermal responses: what happens in the brain. Neuroscientist **8**(2), 132–142 (2002)
16. W. Boucsein, *Electrodermal Activity*. Springer Science & Business Media, 2012
17. Lehrer, P.M., Gevirtz, R.: Heart rate variability biofeedback: how and why does it work? Front. Psychol. **5**, 104242 (2014)
18. Lisetti, C., Nasoz, F., LeRouge, C., Ozyer, O., Alvarez, K.: Developing multimodal intelligent affective interfaces for tele-home health care. Int. J. Hum Comput Stud. **59**(1–2), 245–255 (2003)
19. Wascher, E., Alyan, E., Karthaus, M., Getzmann, S., Arnau, S., Reiser, J.E.: Tracking drivers' minds: continuous evaluation of mental load and cognitive processing in a realistic driving simulator scenario by means of the eeg. Heliyon **9**(7) (2023)

20. Li, Y., Huang, Z.: Basics and applications of ai in adas and autonomous vehicles. In"Advanced Driver Assistance Systems and Autonomous Vehicles: From Fundamentals to Applications, pp. 17–48 (2022)
21. Shruti, S., Shravya, M., Mohan, S., Suhail, Y., Radha, R.: Ai-based solutions for adas. In: 2022 6th International Conference on Intelligent Computing and Control Systems (ICICCS), pp. 1009–1012. IEEE (2022)
22. Niermann, D., Lüdtke, A.: Predicting vehicle passenger stress based on sensory measurements. In: Arai, K., Kapoor, S., Bhatia, R. (eds.) IntelliSys 2020. AISC, vol. 1252, pp. 303–314. Springer, Cham (2021). https://doi.org/10.1007/978-3-030-55190-2_23
23. Kulkarni, A.M., Nandi, A.V., Nissimagoudar, P.: Driver state analysis for adas using eeg signals. In: 2019 2nd International Conference on Signal Processing and Communication (ICSPC), pp. 26–30. IEEE (2019)
24. Dong, Y., Hu, Z., Uchimura, K., Murayama, N.: Driver inattention monitoring system for intelligent vehicles: a review. IEEE Trans. Intell. Transp. Syst. **12**(2), 596–614 (2010)
25. Schmidt, P., Reiss, A., Duerichen, R., Laerhoven, K.V.: Multimodal biosignal-based detection of driver's stress and fatigue. J. Ambient. Intell. Humaniz. Comput. **10**(5), 1835–1849 (2019)
26. Li, L.: A Multi-Sensor Intelligent Assistance System for Driver Status Monitoring and Intention Prediction. Technische Universität Kaiserslautern (2017)
27. Lisetti, C.L., Nasoz, F.: A smart affective environment to regulate driving stress. IEEE Trans. Affect. Comput. **9**(3), 297–311 (2018)
28. Mateos-García, N., Gil-González, A.-B., Luis-Reboredo, A., Pérez-Lancho, B.: Driver stress detection from physiological signals by virtual reality simulator. Electronics **12**(10), 2179 (2023)
29. Ploch, C.J., Bae, J.H., Ju, W., Cutkosky, M.: Haptic skin stretch on a steering wheel for displaying preview information in autonomous cars. In: 2016 IEEE/RSJ International Conference on Intelligent Robots and Systems (IROS), pp. 60–65. IEEE (2016)
30. Alarcón, J.A., Guerrero, L.: Machine learning techniques for driver drowsiness detection based on bio-signals. In: Proceedings of the International Conference on Machine Learning and Computing, pp. 23–28 (2018)
31. Page, M.J., et al.: The prisma: statement: an updated guideline for reporting systematic reviews,". BMJ **372**, 2021 (2020)
32. Lin, C.-T., et al.: Development of wireless brain computer interface with embedded multitask scheduling and its application on real-time driver's drowsiness detection and warning. IEEE Trans. Biomed. Eng. **55**(5), 1582–1591 (2008)
33. Sengupta, J., Baviskar, N., Shukla, S.: Biosignal acquisition system for stress monitoring. In: Das, V.V., Chaba, Y. (eds.) Mobile Communication and Power Engineering. CCIS, vol. 296, pp. 451–458. Springer, Heidelberg (2013). https://doi.org/10.1007/978-3-642-35864-7_69
34. Linkov, V., Zámečník, P.: Cultural differences-induced mistakes in driving behaviour: an opportunity to improve traffic policy and infrastructure. In: Mistakes, Errors and Failures across Cultures: Navigating Potentials, pp. 605–619 (2020)
35. D. Martens, *Data science ethics: Concepts, techniques, and cautionary tales.* Oxford University Press, 2022

JointCare: An Integrated Platform for EMG Signal Analysis in Knee Osteoarthritis Rehabilitation

Tatiana Dolores Cárdenas-Guaraca[1]([⊠]) (ID), Danilo Andrés Molina-Vidal[2] (ID), and Vladimir Espartaco Robles-Bykbaev[1] (ID)

[1] Universidad Politécnica Salesiana, Grupo de Investigación de Inteligencia Artificial – GIIATA, Cuenca, Ecuador
tcardenas@est.ups.edu.ec
[2] Biomedical Engineering Program, Alberto Luiz Coimbra Institute for Graduate Studies and Research in Engineering (Coppe), Federal University of Rio de Janeiro, Rio de Janeiro, Brazil

Abstract. Knee osteoarthritis represents a medical care challenge, requiring tools that improve the accuracy of monitoring during rehabilitation. The tools found present limitations for a complete analysis. In this perspective, we developed a platform to analyze electromyography (EMG) signals stored on file or captured online by an acquisition module. To evaluate the platform, an EMG acquisition of four leg muscles of a healthy subject was performed. Also, EMG signals from the database were used to visualize and calculate typical values used for EMG rehabilitation analysis. The interface allowed online capture and processing of 4 EMG channels. Using the database signals, we applied the envelope, histogram, Fourier transform, rectified, gain, and mean frequency. The values found are in the range found in the literature. The results suggest the user interface can be used for monitoring knee osteoarthritis rehabilitation or other muscle diseases requiring EMG-biofeedback.

Keywords: EMG · osteoarthritis · rehabilitation · EMG-biofeedback

1 Introduction

Knee osteoarthritis is a chronic condition that affects millions of people worldwide [1–3], Globally, it is estimated that 365 million people are affected by osteoarthritis [4]. The disease is more prevalent in women, affecting 2,693 out of every 100,000, compared to 1,770 out of every 100,000 men [5], receiving constant attention from the medical and scientific community [6, 7]. Rehabilitation is crucial for this disease, as it helps improve joint function, reduce pain, and maintain mobility [8]. However, challenges persist in the effectiveness of treatments due to the lack of specialized and comprehensive tools. Research, such as that described in [9, 10], describes the development of systems that capture electromyography (EMG) signals and visualize them through a graphical interface to obtain feedback (EMG-biofeedback) for rehabilitation. In [11], it is highlighted that EMG, along with electrical stimulation, are considered promising alternatives for limb rehabilitation.

© The Author(s) 2025
E. M. Inga Ortega et al. (Eds.): CITIS 2024, LNNS 1331, pp. 116–125, 2025.
https://doi.org/10.1007/978-3-031-87065-1_11

On the other hand, EMG signals have also been used to detect knee osteoarthritis [9, 12]. Innovation in this field has generated significant developments, creating graphical interfaces in various areas [13].

These graphical user interfaces have a fundamental role in the analysis and processing of biomedical signals. Techniques and architectures for efficiently acquiring and processing EMG signals are constantly evolving [14].

Sengchuai et al. [9] developed a graphical interface using LabVIEW to adapt the analysis to the characteristics of EMG signals. The system provides typical EMG features and instantaneous range of motion measurements for biofeedback.

Raeissadat et al. [15] found a significant reduction in knee pain in the group that received EMG-biofeedback compared to the control group. Therefore, having interfaces to improve the treatment and rehabilitation of conditions such as knee osteoarthritis is relevant.

These and other studies that use graphical interfaces for rehabilitation feedback form a basis that highlights that this area is still developing to improve the effectiveness of rehabilitation. These software applications show parameters individually, such as signal envelope, frequency analysis (Fourier transform and mean frequency), histograms, or simply the filtered EMG signal for analysis [16]. The utility of these features extends to clinical, research, and educational areas. On the other hand, not all equipment allows the connection of more than one EMG channel, which limits comprehensive analysis [17].

The ability to visualize multiple EMG channels and their main features/parameters is essential for thoroughly understanding muscle activity in the knee joint. This would enable researchers and physicians to understand how osteoarthritis affects the muscles and allow for more effective strategies for diagnosis and treatment [18].

The aim of this work is to develop a comprehensive platform for processing and analyzing multiple EMG signals in the rehabilitation of knee osteoarthritis.

2   Materials and Methods

The general structure of the proposal (Fig. 1) includes a four-channel electromyography (EMG) signal capture module and a user interface developed in Matlab.

2.1   Device to EMG Signals Acquisition

The EMG capture module uses the Olimex Shield EMG-EKG board [19]. It has already been used in protocols that require the capture of EMG signals [20, 21]. For this work, four Shield EMG-EKG boards were coupled to capture EMG signals.

Each board has a high-voltage protection circuit at its input and an instrumentation amplifier responsible for converting the analog differential signal (EMG biopotential). Next, filters minimize unwanted signals during EMG capture. The total gain depends on a precision potentiometer included on the board.

The filtered signal is connected to an analog-to-digital converter (ADC) of the Atmega328 microcontroller (UC). The UC sends the data from each channel (separated by commas, e.g., "0.015, −0.986, 0.555") to the user interface using the serial communication protocol.

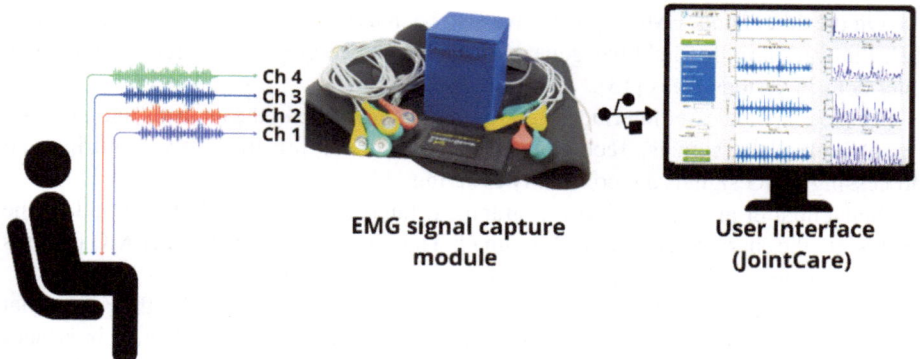

Fig. 1. General Structure.

On the other hand, adding sensors (e.g. accelerometers), digital inputs/outputs, and more EMG capture channels such as commercial clinical equipment compromises the processing capability of the device. Also, the filters included in the board and the 10-bit ADC limit the comprehensive analysis of EMG signals.

2.2 User Interface (UI)

The user interface (UI), JointCare, provides access to a series of tools and the most used functions for EMG signal analysis. JointCare allows offline analysis with recorded signals and online interaction with the EMG capture module.

In each mode, the user can configure and visualize up to four EMG signals and access processing/calculation tools such as envelope, Fourier transform, histogram, rectification, gain adjustment, and mean frequency calculation for each channel. These options have been used in other studies to analyze EMG signals in knee osteoarthritis rehabilitation [9, 13, 17, 21].

The tools are applied to all channels simultaneously. This limits the independent analysis of each parameter in each signal.

2.3 User Interface Evaluation

For the online mode, a real-time capture was performed with a 22-year-old person with no history of knee problems. Eight electrodes (four bipolar channels) were connected to the following muscles (all on the right side): biceps femoris, vastus medialis, rectus femoris, and semitendinosus [22, 23].

Online. I) The area for electrode placement was cleaned with 70% alcohol to minimize the impedance between the electrode and the skin, ii) Disposable Ag/AgCl electrodes, 10 mm in diameter, were placed according to the recommendations of the SENIAM platform [24], iii) The sampling frequency was then set to 1 kHz, and acquisition began, iv) The volunteer was instructed to perform leg muscle contractions with intervals between 1 and 2 s for 25 s, v) Finally, the signals were saved and analyzed.

Offline. For the offline mode, a database was used that includes EMG signals from healthy individuals and individuals with knee problems. This database corresponds to a study conducted in Colombia to analyze muscle behavior associated with the knee, including walking, leg extension from a seated position, and leg flexion [25]. The signals were visualized, and all processing options available in the interface were applied. Additionally, analysis of variance (ANOVA) was performed. The significance level was $\alpha = 5\%$.

3   Results and Discussions

Figure 2 shows the setup of the device for acquiring signals from the 4 muscles. For this purpose, 4 Olimex ECG-EMG acquisition modules were used (Fig. 2a), along with an Atmega328 microcontroller module (Fig. 2b). Disposable Ag/AgCl electrodes were utilized (Fig. 2c). A USB cable was employed for power supply and data transmission (Fig. 2d).

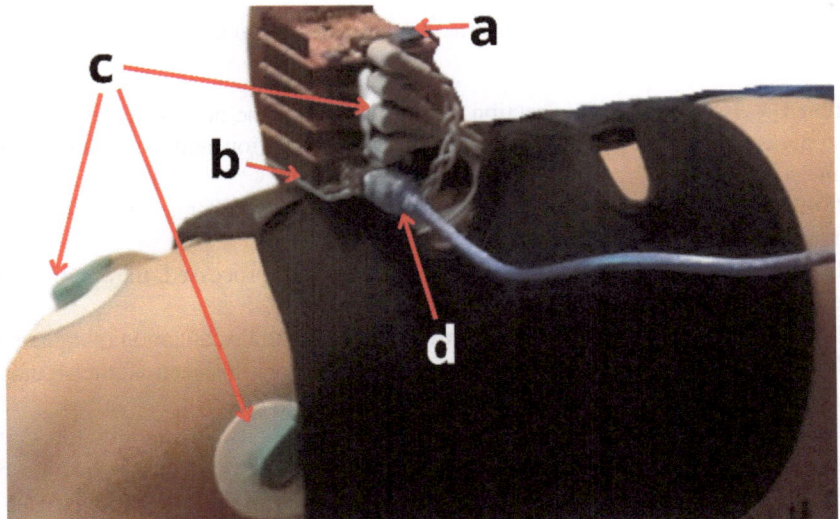

Fig. 2. Signal acquisition module: a) Shield EMG-EKG modules b) Microcontroller based on Atmega328 c) Signal acquisition cables and electrodes d) Data and power cable for the modules.

3.1   JointCare

The user interface was devolvement in Matlab. Figure 3 shows the interface in sections.

Section A – Configuration: For the online mode, the user can select the acquisition time. The sampling frequency (fs) is configured for online and offline modes.

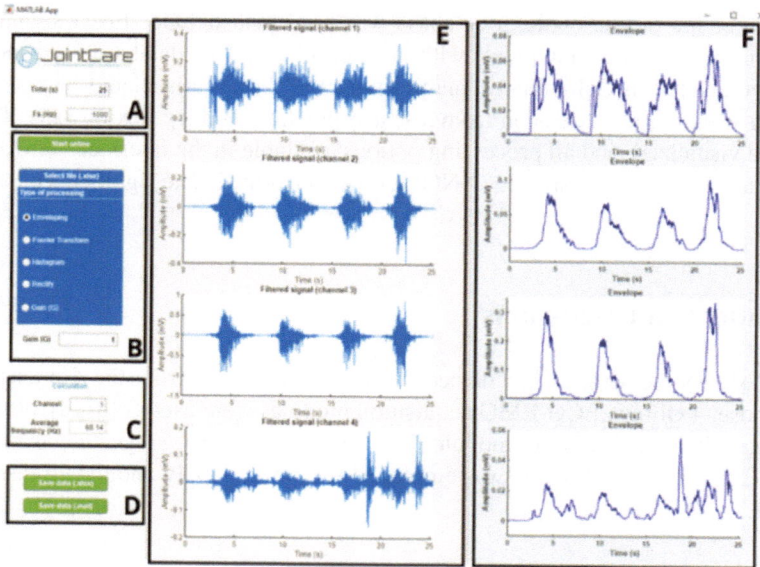

Fig. 3. User Interface (JointCare) in sections.

Section B – Processing: This section allows us to start the acquisition or load a file and apply envelope, Fourier transform, histogram, rectification, and amplification for all channels.

Section C – Average frequency: It allows the user to select the channel and automatically calculate the average frequency.

Section D – Save data: Once the signals are captured or processed, they can be saved in an *xlsx* or *mat* file.

Section E – Temporal Signal Visualization: It allows the visualization of signals sent by the acquisition module (online) or loaded from a file. In this section, the signals are displayed only in the time domain.

Section F – Processed Signal Visualization: The signals, histograms, or Fourier transforms are displayed, depending on the type of processing selected in Section B.

3.2 Online Acquisition

Figure 4 shows the EMG signals from the online capture configured to calculate the Fourier transform. In the visualization section, 17 muscle contractions were counted during the 25-s acquisition. The mean frequency of each repetition was calculated for the biceps femoris muscle (channel 1), vastus medialis muscle (channel 2), rectus femoris muscle (channel 3), and semitendinosus muscle (channel 4) were 80.09 Hz ($\pm$16.30 Hz), 90.17 Hz ($\pm$13.67 Hz), 71.67 Hz ($\pm$6.33 Hz), and 108.27 Hz ($\pm$7.08 Hz), respectively. The semitendinosus muscle presented a statistical difference ($p < 0.05$) from the other muscles. There was no statistical difference between the biceps femoris muscle and the vastus medialis muscle ($p = 0.09$) and neither between the biceps femoris muscle and the rectus femoris muscle ($p = 0.23$).

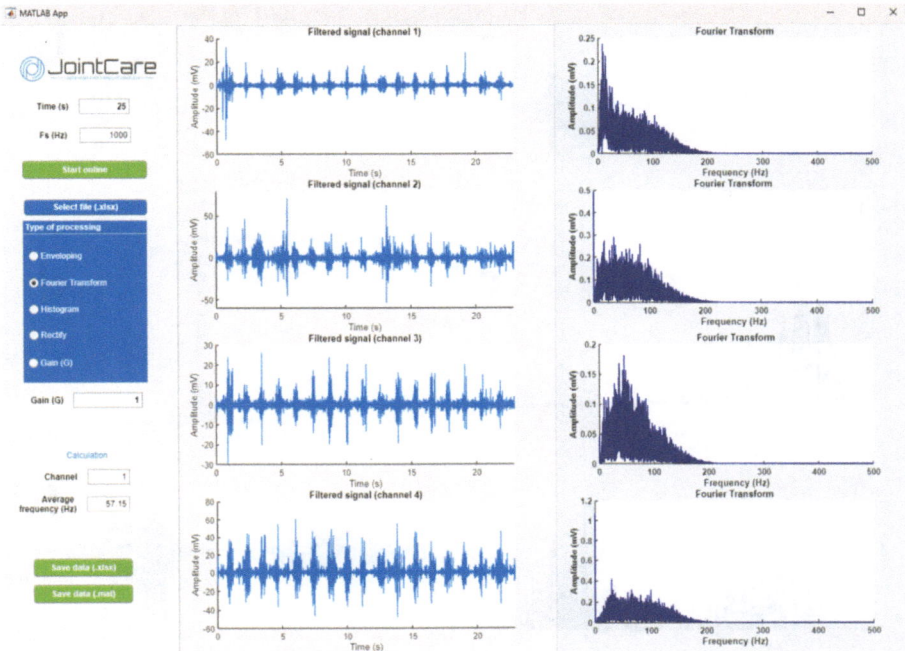

Fig. 4. UI with results of online acquisitions for 4 EMG signals.

3.3 Offline Mode – Database

From the database used in [25], a random file was selected to visualize and process in the interface. Figure 5 shows the results of the EMG signal from channel 4 along with each of the processing options.

In Fig. 5a, b, e, f, two 3-s muscle contractions can be observed. To visualize the signal in microvolts, it was amplified 1000 times (Fig. 5f). The morphology of the histogram (Fig. 5d) corresponds to what is found in the literature [26].

On the other hand, most of the signal energy from channel 4 is around 80 Hz (Fig. 5c). The mean frequency of the channel confirms this result. The frequencies for channels 1, 2, 3, and 4 were 65.72 Hz, 84.76 Hz, 55.08 Hz, and 78.09 Hz, respectively.

Based on the results of the offline and online mode, this study contributes to the existing literature by providing an integrated tool for the analysis of multiple EMG signal parameters, specifically focused on the rehabilitation of knee osteoarthritis.

4 Conclusions

The EMG signal acquisition module, based on the Olimex EMG-EKG Shield board, allowed the capture of up to four bipolar channels, unlike other works that have only one acquisition channel [17]. Additionally, the interface is compatible with any other module that sends data using the protocol established in the methodology of this work (Sect. 2.1).

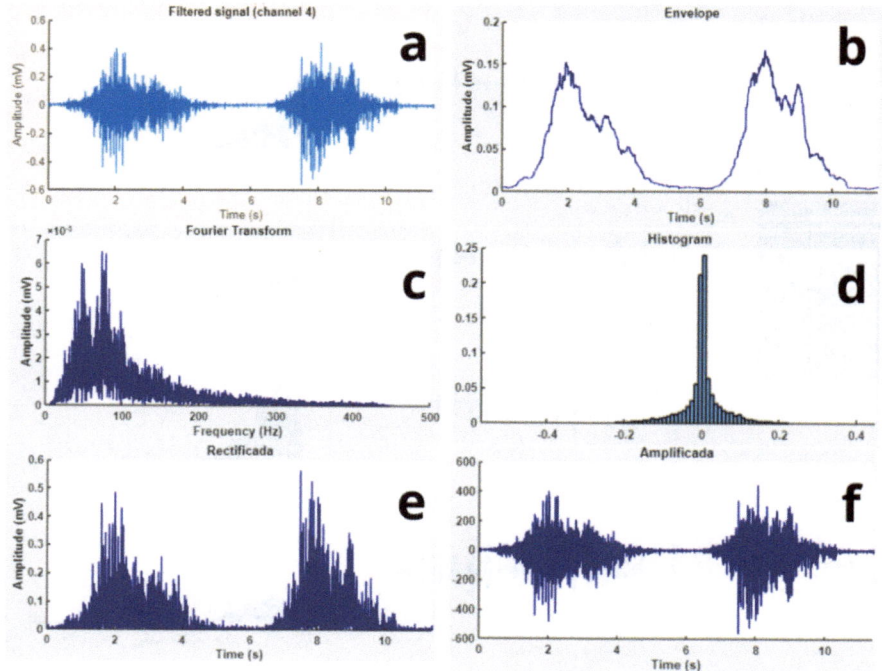

Fig. 5. Results of EMG signal of channel 4. a) Temporal EMG signal, b) Envelope, c) Fourier transform, d) Histogram, e) Rectified signal, and f) EMG signal amplified 1000 times.

The user interface (JointCare), developed in Matlab, constitutes a significant advancement in monitoring and rehabilitating knee osteoarthritis, overcoming current limitations for effective treatment. JointCare enables online analysis as well as analysis of stored data. Our platform offers typical tools for electromyographic signal analysis such as envelope application, Fourier transform, histograms, signal rectification, and amplification, as well as calculation of mean frequency. In tests conducted, the interface allowed efficient acquisition and processing of EMG signals, showing mean frequencies of 80.09 Hz ($\pm$16.30 Hz), 90.17 Hz ($\pm$13.67 Hz), 71.67 Hz ($\pm$6.33 Hz), and 108.27 Hz ($\pm$7.08 Hz) for the biceps femoris, vastus medialis, rectus femoris, and semitendinosus muscles, respectively. These values are within the frequency range found in the literature [27, 28].

The ability to visualize multiple channels of EMG signals is essential for understanding muscular activity in the knee joint, allowing for an understanding of how osteoarthritis affects muscles and developing more effective treatment strategies, in addition to providing feedback on rehabilitation [29]. The data obtained are consistent with those reviewed in the literature, supporting the validity and reliability of the system.

EMG signals are useful for rehabilitation [30] and the results of our proposed graphical interface can be used as a tool in applications requiring EMG biofeedback, such as rehabilitation and monitoring of knee osteoarthritis.

On the other hand, more specific analyses that require analyzing specific bands of frequency, energy, duration of muscle activity, among others, are not contemplated in the interface and can be implemented in future work. In addition, implement an online statistical analysis and perform more experimental tests in healthy people to have a control group and in patients with knee osteoarthritis.

Acknowledgments. The authors would like to express their gratitude to GIIATA and PEB/LAPIS for their support.

References

1. Sánchez Lozano, J., Martínez Pizarro, S.: Ozonoterapia para reducir el dolor en la osteoartritis de rodilla. Revista Cubana de Reumatología **26**, 1310 (2024)
2. López, C., Cedeño, M., Moscoso, G.: Efectos De La Hidroterapia En Pacientes Con Osteoartritis De Rodilla. Revista Científica Arbitrada Multidisciplinaria PENTACIENCIAS. **6**, 237–249 (2024)
3. Vea-Huerta, M., Salazar-López, J., Flores-Bautista, P.: Signo de digito-presión en pacientes con osteoartritis de rodilla. Acta Ortop Mex. **38**, 101–104 (2024). https://doi.org/10.35366/115079
4. Jahn, J., Ehlen, Q.T., Huang, C.Y.: Finding the goldilocks zone of mechanical loading: a comprehensive review of mechanical loading in the prevention and treatment of knee osteoarthritis. Bioengineering **11** (2023). https://doi.org/10.3390/bioengineering11020110
5. Costa, A.P., et al.: Correlation between muscle strength and functional improvement after a neuromuscular electrical strengthening associated with undenatured type II collagen in knee osteoarthritis. Compr. Clin. Med. **3** (2021). https://doi.org/10.1007/s42399-021-00830-6
6. Clark, P., et al.: Analysis of musculoskeletal disorders-associated disability in Mexico from 1990 to 2021. Gac Med. Mex. **159**, 517–526 (2023). https://doi.org/10.24875/GMM.23000394
7. Laterza, J.B.: Prevalencia y tratamiento de la artrosis de miembros inferiores en deportistas profesionales retirados (2022)
8. Mayoral Rojals, V.: Epidemiology, clinical impact and therapeutic objectives in osteoarthritis. Revista de la Sociedad Espanola del Dolor **28**, 4–10 (2021). https://doi.org/10.20986/resed.2021.3874/2020
9. Sengchuai, K., et al.: Development of a real-time knee extension monitoring and rehabilitation system: range of motion and surface EMG measurement and evaluation. Healthcare (Switzerland). **10** (2022). https://doi.org/10.3390/healthcare10122544
10. Jarque-Bou, N.J., Gracia-Ibáñez, V., Roda-Sales, A., Bayarri-Porcar, V., Sancho-Bru, J.L., Vergara, M.: Toward early and objective hand osteoarthritis detection by using EMG during grasps. Sensors **23** (2023). https://doi.org/10.3390/s23052413
11. Sarhan, S.M., Al-Faiz, M.Z., Takhakh, A.M.: A review on EMG/EEG based control scheme of upper limb rehabilitation robots for stroke patients. Heliyon **9**, 16 (2023). https://doi.org/10.1016/j.heliyon.2023.e18308
12. Pilkar, R., Momeni, K., Ramanujam, A., Ravi, M., Garbarini, E., Forrest, G.F.: Use of surface EMG in clinical rehabilitation of individuals with SCI: barriers and future considerations. Front Neurol. **11** (2020). https://doi.org/10.3389/fneur.2020.578559
13. Tavares, M.C., Pizzetta, A.B., Costa, M.H., Pinheiro, M.M.C.: Microcontroller-based acquisition system for evoked otoacoustic emissions: protocol and methodology. Biomed. Signal Process. Control **87** (2024). https://doi.org/10.1016/j.bspc.2023.105453

14. Al-Ayyad, M., Owida, H.A., De Fazio, R., Al-Naami, B., Visconti, P.: Electromyography monitoring systems in rehabilitation: a review of clinical applications, wearable devices and signal acquisition methodologies. Electronics (Switzerland) **12**, 1520 (2023). https://doi.org/10.3390/electronics12071520
15. Raeissadat, S.A., et al.: The efficacy of electromyographic biofeedback on pain, function, and maximal thickness of vastus medialis oblique muscle in patients with knee osteoarthritis: a randomized clinical trial. J. Pain Res. **11**, 2781–2789 (2018). https://doi.org/10.2147/JPR.S169613
16. Rodríguez Jouvencel, M.: Fatiga muscular: Cuestiones previas principios de electromiografía de superficie (2020)
17. Zhou, Y., et al.: EMG signal processing for hand motion pattern recognition using machine learning algorithms. Arch. Orthop. **1** (2020). https://doi.org/10.33696/Orthopaedics.1.005
18. Castillo Morillo, R.: Calidad de vida en pacientes adultos mayores con osteoartritis de rodilla (2018)
19. OLIMEX SHIELD: SHIELD-EKG-EMG bio-feedback shied USER'S MANUAL (2012)
20. Priya, E., Savithri, C.N., Nirmal Raja, K.L., Varunapriyan, K.: A machine learning approach to control a Prosthetic arm via signals from residual limb – a boon for amputees. In: 2023 14th International Conference on Computing Communication and Networking Technologies (ICCCNT), pp. 1–6 (2023)
21. Pedrosa, S.F.: Sistema EMG para controlo MIDI (2023)
22. Sandoval, C., González, S., Becerra, L., Salido, R.: Caracterización y Clasificación Automática de Señales Electromiográficas Registradas durante la Marcha en Sujetos con Lesión en Miembro Inferior. In: Memorias del Congreso Nacional de Ingeniería Biomédica, pp. 53–60 (2020)
23. Gonzales, L.: Evaluación Funcional de la Rodilla Mediante Cinemática y Electromiografía en Sujetos con Lesión de Ligamento Cruzado Anterior (2023)
24. Investigación Biomédica (BIOMED II) de la Unión Europea: Recommendations for sensor locations in hip or upper leg muscles – Seniam
25. Sanchez, O., Sotelo, J., Gonzales, M., Hernandez, G.: EMG dataset in lower limb data set. UCI Mach. Learn. Repository **2** (2014)
26. Oo, T., Phukpattaranont, P.: Signal-to-noise ratio estimation in electromyography signals contaminated with electrocardiography signals. Fluctuat. Noise Lett. **19** (2020). https://doi.org/10.1142/S0219477520500273
27. Kim, H., Lee, J., Kim, J.: Electromyography-signal-based muscle fatigue assessment for knee rehabilitation monitoring systems. Biomed. Eng. Lett. **8**, 345–353 (2018). https://doi.org/10.1007/s13534-018-0078-z
28. Murphy, J., Hodson-Tole, E., Vigotsky, A.D., Potvin, J.R., Fisher, J.P., Steele, J.: Surface electromyographic frequency characteristics of the quadriceps differ between continuous high- and low-torque isometric knee extension to momentary failure. J. Electromyogr. Kinesiol. **72** (2023). https://doi.org/10.1016/j.jelekin.2023.102810
29. Domingo, F.J.: Rehabilitation Process Using Electromyography and Biofeedback (2021)
30. Lita, A., et al.: Elbow angle estimation for medical rehabilitation device based on EMG sensor with ARIMAX method. In: AIP Conference Proceedings. American Institute of Physics Inc. (2022)

Mobility

Determination of the Real Percentages of Fuel Expansion in the Automotive Sector Influenced by Ambient Temperatures in a City 50 m. Above Sea Level, Using Thermographic Tools

Pablo Renato Fierro[1,2]([⊠]) [iD], Leonel Martínez[1], and Cristian Guevara[1]

[1] Smart Mobility Research Group of Salesian Polytechnic University
(GMOVINT, UPS), Guayaquil, Ecuador
pfierro@ups.edu.ec

[2] University Institute of Automobile Research, Technical University of Madrid
(INSIA, UPM), Madrid, Spain
r.fierro@alumnos.upm.es

Abstract. In this article, a study is carried out in a city of 50 m. Above sea level, on the volumetric expansion of 87 and 92-octane gasoline and diesel fuel (named: Ecopais, Super, and Diesel) inside the tanks. The aim is to determine the volume variations due to temperature changes. The justification is based on the lack of specific information on the subject. The proposed methodology includes a mathematical model supported by thermographic analysis experiments to visualize the dilatation of fuels. A volume variation of up to 0.55% was evidenced in the fuel tanks as a result of ambient temperature fluctuations. However, the use of thermographic tools as a measurement method does not facilitate the detection of this volumetric variation in the tanks; it depends on the volume and camera capacities. Using Pearson's index shows that the correlation is direct, and the influence of temperature affects the perception of volume.

Keywords: Expansion fuel · Infrared thermography · Temperature

1 Introduction

The study delves into fuel volumetric expansion, potentially altering liquid volume with temperature changes [11,18] affecting tank fuel level measurement accuracy and logistic efficiency. Understanding these thermal dynamics [6,10] is pivotal for precise fuel distribution and economic viability. This research is prompted by Ecuador's escalating fuel demand, with transportation consuming almost half (48.9%) of all energy in 2021, dominated by diesel at 51.6% and

Salesian Polytechnic University.

E. M. Inga Ortega et al. (Eds.): CITIS 2024, LNNS 1331, pp. 129–139, 2025.
https://doi.org/10.1007/978-3-031-87065-1_12

gasoline at 47.3%, while heavy-duty land transport utilized around 47% and light-duty 21% [2].

Studies highlight temperature's impact on fuel storage volume, emphasizing the importance of understanding expansion coefficients. One study found that at 30.0°C, a tank of biodiesel increases by approximately 83 liters per 1,000 liters [19], while another, from Bogota, observed a 354-liter fuel volume increase in 36900 liters of gasoline when heated [7]. Both stress comprehending temperature effects on fuel handling to ensure precise logistic calculations. Guayaquil, Ecuador, experienced a significant temperature rise, reaching 36.1°C, signaling local thermal trend shifts [12]. Volumetric expansion occurs due to heightened molecular kinetic energy, influenced by fuel quality [3]. Thermal expansion coefficients are vital in liquids, reflecting fluid expansion rates [4, 16], impacting company finances through temperature-induced fuel losses during transport [8].

Infrared thermography, an unconventional yet effective diagnostic technique, aids automotive industry material and component visualization by detecting thermal anomalies indicating defects or wear not visible through standard methods [5, 14, 17].

A planned comparative study on hydrocarbon volume expansion at various temperatures aims to establish mathematical relationships, using statistical methods to identify patterns and relationships between variables, providing a deeper understanding of temperature's influence on fuel volumetric behavior, crucial in the automotive industry.

2 Materials and Methods

2.1 Materials

Thermographic Camera. The FOTRIC 346A, 346A [9] is used, this equipment measures the infrared energy of objects, showing a color scale that provides the thermographic image. This is related to the emissivity, determined by comparing the thermal radiation emitted by a surface or object with the temperature difference between them.

Controlled Environment. Simulating a controlled environment to take thermographic images with temperature variations required careful planning: 1. choice de environment, 2. temperature control (heater), 3. choice of environment generation of temperature variations, 4. temperature measurement, 5. emissivity control, 6. data logging and 7. data analysis, in addition the use of specific tools. By following these steps, It was able to effectively simulate a controlled environment for thermographic imaging, allowing me to study objects or devices under specific conditions and without relying on external temperature fluctuations.

Sectorization. Consideration of the most popular automobiles in Ecuador during the year 2023 [1], as well as the evaluation of the types of vehicles and the materials used in the fuel tanks, is fundamental to carrying out a comprehensive and contextualized study in the automotive field.

In summary, the inclusion of the most sold vehicles, vehicle types, and fuel tank materials in an automotive study offers a complete and up-to-date perspective of the Ecuadorian market. This not only improves the quality and relevance of the research, but also provides valuable information for consumers, manufacturers, and regulators, contributing to the sustainable and safe development of the automotive sector in Ecuador.

Technical Specifications. In recent years, an increase in the use of fuel tanks made of high-density polyethylene (HDPE) has been observed due to its potential benefits such as weight reduction, flexibility in design, and lower costs [13]. Therefore, a tank is produced from this material, in addition to the more common one, which is steel.

2.2   Methods

Experimental Design. The description of the experimental setup includes the specification, as well as the order and procedures necessary to ensure the achievement of the objectives of this experiment. As for the data acquisition process, the following aspects will be considered, as detailed in the methodology shown in Fig. 1. As can be seen, this is divided into 2 phases: experimental and analytical.

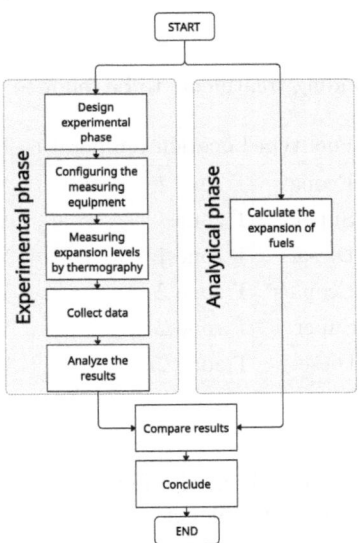

Fig. 1. Flow chart of the application of the research methodology.

Experimental Phase. In the experimental stage, one or more variables are deliberately manipulated to evaluate their impact on another variable of interest. This experimental process consists of 5 phases, as illustrated in Fig. 1. First, it is necessary to define the relevant parameters or variables for the study, in this case, the temperature and the type of fuel. Secondly, the calibration of the equipment is carried out, where the emissivity is determined according to the material and the distance required for data collection [9]. Thirdly, we proceed with the development of the experiment; once the parameters are established, data collection begins. Fourthly, all the data obtained during the execution of the experiment is collected for subsequent analysis.

Subsequently, it is required to determine the volumetric expansion of the fuels Ecopaís (Fuel composed of 5% ethanol and 95% pre-blended gasoline.), Súper (Complex mixture of 200 to 300 hydrocarbons such as butane, light naphtha, heavy naphtha, treated naphtha, etc.), and Diésel (Distributed in normal kerosenes, mono-naphthenic, mono-aromatic, and mono-aromatic-naphthenic, which cover 80% of the total composition.) [15] in a tank car, considering the variation of ambient temperatures in Guayaquil city.

Table 1 shows the number of treatments to be followed in this experimental phase, where each of the data described in the table should be recorded: T zone 1: ambient temperature corresponding to the tank car supply location, measured in °C. T zone 2: ambient temperature corresponding to the tank truck unloading site. Vo zone 1: Initial volume corresponding to the tank truck supply site, measured in L and gal. Vf zone 2: Final volume corresponding to the unloading site of the tank truck.

Table 1. Corresponding treatments to be followed for data collection.

Treatments	Fuel type	Location temperature	Fuel volume
1	Ecopaís	T zone 1	Vo zone 1
2	Super	T zone 1	Vo zone 1
3	Diesel	T zone 1	Vo zone 1
4	Ecopaís	T zone 2	Vf zone 2
5	Super	T zone 2	Vf zone 2
6	Diesel	T zone 2	Vf zone 2

Area Selection. The test area is in the facilities of the Clyan service station located in Durán, Juan León Mera Street, passage C. J1. The tank vehicle necessary to carry out the study is located at this site.

Evaluation of Meteorological Conditions. Meteorological conditions are a crucial factor when considering thermal imaging. For this reason, meteorological and environmental parameters are defined according to the guidelines set out in ISO 18434-1 [20]. This standard ensures that operational and environmental conditions during data acquisition are uniform and consistent.

Location

The focus of the study is on the tank of the vehicle in charge of transporting fuel to the service station.

Equipment preparation

The equipment calibration procedure is detailed in Fig. 2:

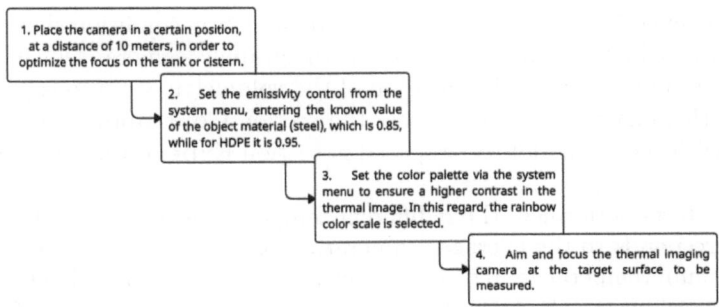

Fig. 2. Political Equipment calibration procedure.

2.3   Analytical Phase

In this stage, the fuel volume variation in response to changes in ambient temperature is determined and compared with the results obtained in the experimental phase to contrast data through a mathematical approach.

Since liquids are contained in vessels that also expand with temperature, it is necessary to consider the expansion of both the liquid and the vessel. The amount of liquid displaced is not equal to the actual expansion of the liquid, but is smaller and is called apparent expansion. Therefore, the following equation is established: [11].

$$\triangle V = V_0(\beta_{liq} - \beta_{rec})\triangle T \tag{1}$$

where

$$\beta = coefficient\ of\ cubic\ dilatation\ determined\ in\ 1/^{\circ}C\ or\ ^{\circ}C^{-1}$$
$$V_0 = initial\ volume\ determined\ in\ cubic\ meters,\ liters\ or\ gallons\ (m^3,\ L,\ gal).$$
$$\triangle T = Variation\ of\ the\ initial\ temperature\ measured\ in\ degrees\ Celsius\ (^{\circ}C).$$
$$\triangle V = Volumetric\ variation$$

Data Processing. Microsoft Excel software is used to record the data and the data is processed in Minitab software to develop a statistical method that facilitates the appreciation of the trend, the calculated percentages, and the correlation between temperature variation and volume variation.

3   Results

3.1   First Experimental Phase

The initial experimental phase was carried out with conditioned results due to restrictions in the capacity of the thermographic camera, which hindered obtaining a more accurate visualization of the volumetric variations present. To carry out this phase, the steps proposed in section **Controlled environment**, together with the methodology proposed in section **Experimental phase**, were followed.

For the tests performed, three different temperatures were considered, 30°C, which corresponds to the average temperature of the city of Guayaquil, 36.1°C (the maximum temperature recorded in the city of Guayaquil during the last few years) [12], belongs to the highest temperature recorded in the city of Guayaquil and 40°C, to observe the behavior of the volumetric expansion of the fuels at temperatures higher than those recorded in the city of Guayaquil, as shown since Fig. 3.

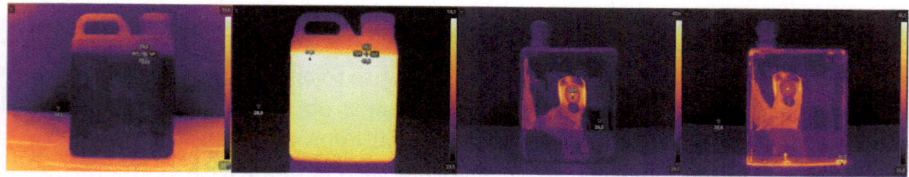

Fig. 3. (a)HDPE low at temperature (b)HDPE high at temperature (c)Steel low at temperature (d)Steel high at temperature.

Analytical Phase. This article presents the results of the analytical stage and the subsequent data analysis, to establish the percentage variation of the volume in the reservoir. The most sold vehicles in Ecuador in 2023, according to the AEADE [1] are taken into account, as shown in Table 2.

Figures 4a and 4b, show a visual representation of various graphs organized by categories, as specified in each one of them. For the generation of these graphs, it was crucial to know the coefficients of volumetric expansion of both the fluid and the material that contains it. Likewise, the initial volumes and temperature variations during the process were taken into account, using the equation detailed in the mathematical approach section.

Table 2. Most sold vehicles during the year 2023 in Ecuador [1].

	Vehicle	Tank capacity	Material	Fuel
1	Chevrolet D-Max	76L	Steel	Diesel
2	Kia Soluto	43 L	Steel	Ecopaís/Super
3	Kia Sonet	45L	Steel	Ecopaís /Super
4	Shineray G01	54L	HDPE	Ecopaís/Super
5	Jetour x70	55L	HDPE	Ecopaís/Super

Figure 4 shows a graph that has been organized according to the "cars" category. In this graph, the volumetric variation of fuels from model to model is evident, where there are models with a greater degree of inclination about the increase in temperature, affecting the perception of the fuel level.

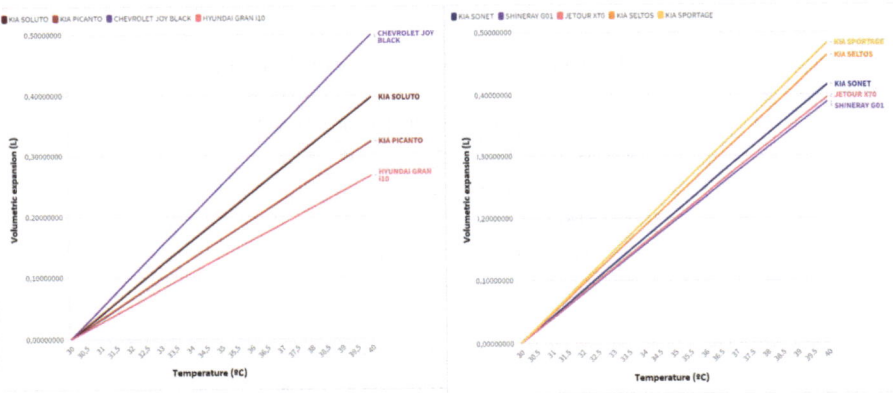

Fig. 4. Fuel dilation of the best-selling Cars (a) and SUVs (b)

Figure 4 shows a graph categorized according to the classification of motor vehicles. This graph allows visualizing the volumetric variation of fuels about the increase in temperature. The straight lines are narrower from model to model, it can be seen in which the volumetric expansion is greater.

Table 3 shows the percentage results of the volumetric variation obtained in the analytical phase, as well as the correlations using Pearson's index, in which for all the models a direct correlation is established as they are equal to 1 and rate of change that allows observing the growth of the volume as a function of temperature shown in the different graphs.

Second Experimental Phase. This part aimed to delve deeper into the effects of environmental temperature fluctuations on volumetric changes in large fuel storage containers. For the Ecopaís fuel container, Fig. 5c illustrates a noticeable temperature decrease near the fuel surface, while for the Super fuel container,

Table 3. Percentage results of volume variations of the most sold vehicles in Ecuador.

Item	Vehicle	%Δ v vs. Δ t 0.1°C	%Δ v vs. Δ t 6.1°C	%Δ v vs. Δ t 10 °C	Exchange rate
1	C. D-Max	0,00856	0,52185	0,8555	0,06502
2	Kia Soluto	0.00925	0.56455	0.9255	0.03980
3	Kia Sonet	0,00925	0,56455	0,9255	0.04165
4	Shineray G01	0,0072	0,4392	0,72	0,03888
5	Jetour x70	0,0072	0,4392	0,72	0,03960

Fig. 5c provides dimensions and parameters. Variations in surface temperature degrees beyond the Diesel fuel surface boundary are observed, influenced by volatile components (Fig. 5d). Analyzing data from this phase (Fig. 5), an Excel spreadsheet was used to correlate volumetric expansion with temperature variations of 0.1°C, up to the maximum recorded temperature of 33°C. Beyond the boundary of the fuel surface, there is a variation in surface temperature degrees, influenced by the volatile components present in that area.

Table 4. Thermographic parameters of different fuels

Measures		Test 1	Test 2	Test 3	Test 4
Ar1	Max	76,6	77,1	76,6	76,7
EI1	Max	71,6	73,8	71,6	71,5
	Min	55,6	73,7	55,6	71,5
	Average	64,1	73,7	64,1	71,5
EI2	Max	55,9	64,9	55,9	61,5
	Min	55,5	64,8	55,5	61,0
	Average	55,7	64,8	55,7	61,3
EI3	Max	51,8	61,0	51,8	57,8
	Min	43,2	46,5	43,2	47,6
	Average	47,6	54,0	47,6	53,4
EI4	Max	43,3	44,4	43,3	50,0
	Min	43,2	44,4	43,2	46,3
	Average	43,3	44,4	43,3	47,5

Parameters: Emissivity & 0.85 Temp.
Refl. & 33°C

Temperature, fuel tank material, and the coefficient of volumetric expansion of fuels are variables and factors that play a crucial role in determining the physical properties of hydrocarbons used as fuels. Several variables are explored that directly impact fundamental characteristics inherent in hydrocarbons used specifically for fuel purposes. This analysis encompasses a comprehensive understanding of how these variables and factors affect the physical properties of hydrocarbons in the context of their application as energy sources.

Table 5 shows the estimated change in fuel volume, regardless of the type of vehicle and tank material, for Ecopaís and Super fuels with a temperature vari-

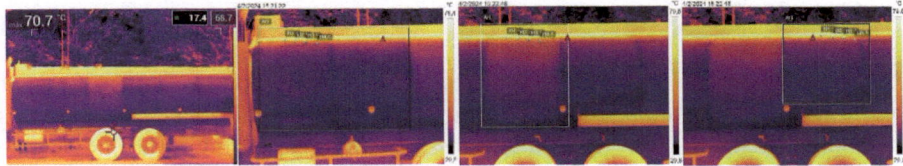

Fig. 5. Thermographic image averages of (a) Generic tank (b) Ecopais 2000 gal. (c) super 1000 gal. (d) diesel 1000 gal. capacity.

ation of 0.1°C in Guayaquil city. The volumetric expansion is 0.0092%, while for a temperature variation of up to 10°C, a volumetric expansion of 0.92% is obtained. For diesel, at a temperature variation of 0.1°C, a volumetric expansion of 0.0085% is achieved, while for a temperature variation of up to 10°C, a volumetric expansion of 0.85% is reached.

Table 5. Percentage estimate of fuel volume change.

Fuel's type	Δt (° C)	$\Delta v (L)$	%
Ecopais/Super Fuel	0.1	0.00397965	0.0092
	6.1	0.24275865	0.56
	10	0.397965	0.92
Diesel Fuel	0.1	0.00650185	0.0085
	6.1	0.3966098	0.52
	10	0.65018	0.85

4 Discussion

The study demonstrates temperature's impact on fuel expansion and validates it through thermographic images and mathematical analysis. Results reveal a substantial change in fuel volume, confirmed by direct measurements and calculations. From EP Petroecuador Pascuales to Clyan service station, the tanker's fuel, initially at 27 °C, rose to 33 °C, aligning with the literature on temperature and volume correlation [7].

These findings enhance fuel supply and transport efficiency and safety, indicating volumetric expansions of 0.0072% to 0.0092% for 0.1 °C temperature shifts, and 0.44% to 0.56% for up to 6.1 °C shifts. For 10 °C variations, expansions range from 0.72% to 0.92%. Notably, a 33 °C temperature yielded volume increases of 21.88 gal (0.55%) for Ecopaís, 5.08 gal (0.51%) for diesel and 5.47 gal (0.55%) for Super fuel in the tank truck's compartments.

5   Conclusions

The study examines fuel volumetric expansion, crucial for tank measurement accuracy. At lower volumes, changes are minor, often under 0.92% for Ecopaís and Super fuels, and 0.85% for diesel, possibly imperceptible via thermographic cameras. However, with larger volumes like 6,000 gallons, changes reach 0.55% per consignment, significant for service station management.

Calculations reveal volumetric expansion under 0.92% for volumes below 76 L. These findings are confirmed by thermographic analysis, although larger volumes show noticeable variations, up to 0.55%, confirmed by manual measurements.

When the Pearson index was applied, a value of 1 was obtained, which indicates a direct correlation. In other words, it suggests that there is a directly proportional relationship between them, where the increase in one variable is reflected in a proportional increase in the other variable, where the existing rate of change ranges from 2.6% to 6.5%.

Thermography negatively impacts tank measurement accuracy. In small volumes, variations are minimal and may go unnoticed by both thermographic equipment and human perception.

References

1. AEADE: Boletín de ventas para prensa: ventas de vehículos – octubre 2023 (2023). https://www.aeade.net/boletines-de-prensa-venta-de-vehiculos/
2. Attfield, P.V., Bell, P.J.L., Grobler, A.S.: Reducing carbon intensity of food and fuel production whilst lowering land-use impacts of biofuels. Fermentation **9**, 633 (2023). https://doi.org/10.3390/fermentation9070633
3. Camilo, A., et al.: DilataciÓn tÉrmica de liquidos
4. Chakraborty, N.: Influence of thermal expansion on fluid dynamics of turbulent premixed combustion and its modelling implications (2021). https://doi.org/10.1007/s10494-020-00237-8
5. Cinthia, E.D.A., Aurora, F.D.: Termografía infrarroja pasiva aplicada a sistemas de detecciÓn de fallas: Una revisiÓn, pp. 28–40 (2021). https://revistas.uaq.mx/index.php/ciencia/article/view/536/597
6. Deng, T., et al.: Modeling of solid oxide fuel cell sintering stress and deformation. Int. J. Mech. Sci. **265**, 108895 (2024). https://doi.org/10.1016/j.ijmecsci.2023.108895
7. Escudero, F.R.M.: afecta la expansión volumétrica de los fluidos, los balances en el transporte de crudo por carrotanques?, http://www.semana.com/economia/
8. Flores, S.R.D., Asesor, C.P., Pino, M.C.E.T.: Mermas de combustible no acreditadas para la determinación del impuesto a la renta en los grifos-arequipa, 2020 (2022)
9. FOTRIC: Cámara térmica de alta temperatura industrial fotric 346a, https://www.fotric.com/product-page/fotric-346a
10. Fu, P., Yan, M., Zeng, M., Wang, Q.: Sintering process simulation of a solid oxide fuel cell anode and its predicted thermophysical properties. Appl. Therm. Eng. **125**, 209–219 (2017). https://doi.org/10.1016/j.applthermaleng.2017.06.061

11. Godino, O., Grijioni, L., Marini, S., Oliva, A.: Temperatura y Dilatación
12. de Meteorología e Hidrología, I.N.: Instituto nacional de meteorología e hidrología, https://www.inamhi.gob.ec/#
13. Josué, C.E.R., Isaías, P.B.A.: Desarrollo de un método de inspección utilizando termografía como herramienta para mantenimiento predictivo de la batería de alto voltaje de ni-mh de vehículo hibrido tipo sedán (2023)
14. López, A., et al.: Influence of dynamic analysis by infrared thermography (IRT) and non-destructive testing (NDT) on alternative predictive diagnosis of automotive brake rotors in light vehicles
15. Mayorga, C.M.G.: Ficha tÉcnica de productos quÍmicos (2012). https://aplicaciones2.ecuadorencifras.gob.ec/SIN/co_quimico.php?id=33310.01.04
16. Nayagam, V., Dietrich, D.L., Williams, F.A.: Unsteady droplet combustion with fuel thermal expansion. Combust. Flame **195**, 216–219 (2018). https://doi.org/10.1016/j.combustflame.2018.01.035
17. Oscar, A., et al.: Infrared thermographic dynamic analysis and non-destructive testing (NDT) for camshaft diagnostics sciencedirect infrared thermographic dynamic analysis and non-destructive testing (NDT) for camshaft diagnostics. www.sciencedirect.com
18. Saha, A., Deshmukh, A., Grenga, T., Pitsch, H.: Dimensional analysis of vapor bubble growth considering bubble – bubble interactions in flash boiling micro-droplets of highly volatile liquid electrofuels. Int. J. Multiph. Flow **165**, 104479 (2023). https://doi.org/10.1016/j.ijmultiphaseflow.2023.104479
19. Silva, T.A.D., Santos, D.Q., Lima, A.P.D., Neto, W.B.: Volumetric property for tankage of biodiesel from residual oil. Revista Virtual de Quimica **5** (2013). https://doi.org/10.5935/1984-6835.20130058
20. for Standardization, I.I.O.: Condition monitoring and diagnostics of machines – thermography – art 1: General procedures (2008). https://www.iso.org/standard/41648.html

Influence of the Acceleration Vector on Measured and Estimated Pollutant Emissions of Hybrid Vehicles, Focusing on Traction and Power Operating Parameters

Jackson Steeven Vidal Suarez[1,2,3]($\boxtimes$) (iD), Walter Josué Semiglia Pineda[1,2,3] (iD), and Néstor Diego Rivera Campoverde[1,2,3] (iD)

[1] Universidad Politécnica Salesiana, Cuenca, Ecuador
jvidals3@est.ups.edu.ec
[2] Grupo de Investigación en Ingeniería del Transporte (GIIT), Cuenca, Ecuador
[3] Vehicle Technology Society IEEE (VTS), Cuenca, Ecuador

Abstract. In the city of Cuenca – Ecuador (2550 m.a.s.l), a significant introduction of HEVs in the automotive fleet is observed. These vehicles are adopted due to measures aimed at reducing fuel consumption and harmful emissions. Due to high-power cold start phenomena and deficiencies in catalytic converters, CO emissions show high increases compared to conventional vehicles at sea level. This necessitates the present analysis, which is based on data acquisition on "RDE" routes using data logger equipment to measure operational, positioning, and emission parameters. This analysis leverages machine learning skills linked to ANNs and Random Forest, aiming to estimate the pollutants of hybrids and demonstrate the influence of acceleration on each gas and calculated operational traction parameter (ICE-EM Powers, Aerodynamic Traction Force, SOC, etc.) in high-altitude cities. The results indicate a strong relationship of the longitudinal acceleration vector on CO_2 and NO_x gases, twice as high as CO and HC, corroborating related studies mentioning that emissions reduction principles are most closely related to acceleration in the present investigation.

Keywords: HEVs · data-logger · ANNs · Random Forest · SOC · ICE · EM

1 Introduction

Currently, environmental issues related to global warming are directly linked to transportation modes relying on fossil fuels [1], making the electrification of vehicles in hybrid and electric lines necessary [2]. However, it is widely known that implementing electric vehicles (EVs) becomes a critical challenge due to factors such as limited range, infrastructure, and overload on electrical grids. As a solution, the introduction of hybrid electric vehicles (HEVs) is considered, as they overcome the mentioned disadvantages [3, 4]. These vehicles typically consist of an internal combustion engine (ICE) and one or two electric motors (EM1–EM2), providing traction under different conditions depending on the operation stage monitored by the hybrid control [5]. While it is true that

E. M. Inga Ortega et al. (Eds.): CITIS 2024, LNNS 1331, pp. 140–150, 2025.
https://doi.org/10.1007/978-3-031-87065-1_13

hybrids are designed and integrated into the vehicle fleet to reduce pollution levels, they also experience instantaneous periods during which they generate higher emission rates, as occurs due to the high-power cold start phenomenon [6]. Other factors influencing the emission performance of HEVs include engine calibration issues and poor operating conditions of catalytic converters [7]. Thus, there arises the need to study which vehicle operating parameters directly influence pollutant emissions.

Ehrenberger et al. [8] mention in an analytical project involving three hybrid vehicles that the state of charge of the battery pack (SOC) influences CO emissions; while there is greater displacement due to electric traction, carbon monoxide emissions decrease. Suarez et al. [9] study the influence of driver acceleration on carbon dioxide (CO_2) and energy consumption, resulting in a 5% divergence in gas emission rates. Costagliola et al. [10] conducted an analysis titled "Impact of road grade on real driving emissions from two Euro 5 diesel vehicles," concluding that to determine the influence of a parameter on emissions, there must initially be a linear and symmetrical correlation between both parts. Zhai et al. [11] demonstrate in their study that HEVs emit higher levels of CO and PN compared to conventional vehicles, by 183.6% and 15.8% respectively. Rivera, Molina et al. [12], describe the importance of certain engine operating parameters on CO, CO_2, HC, and NO_x gases using machine learning techniques associated with the Random Forest method. They also adhere to the guidelines of EURO 6 standards for conducting "RDE" data acquisition routes. Rivera et al. [13], in an analysis of the energy consumption of an HEV, highlight the importance of including the powers of the different hybrid traction sources as study variables, referring to the internal combustion and electric motors (ICE – EM).

The purpose of the research is to understand the influence of longitudinal acceleration on the pollutant emissions of a matrix of HEVs (Sedan, Crossover, SUV) with data acquired on "Real Driving Emissions" routes. Applying machine learning strategies such as neural networks to estimate (with excellent coefficients of determination – R2) and Random Forest to determine the degree of importance of variables with each other. A graphical analysis with variables from the energy traction branch (MCI – ME powers, Battery pack discharge rate, SOC, aerodynamic – traction forces) reveals the behavior of the acceleration vector at certain time intervals, referring to the operating stages of a hybrid vehicle.

2   Material and Methods

Data acquisition is carried out on RDE routes using measurement equipment and data loggers to obtain operational, positioning, and emission parameters. New variables related to traction and energy are calculated. The influence of acceleration on harmful gases and calculated variables is analyzed using the RF algorithm. Using ANNs, factors and gas indices such as CO, CO_2, HC, and NO_x are estimated, followed by graphical and descriptive analyses (Fig. 1).

2.1   Experimental Data Acquisition

In this experimental stage, the PID signals (Data Identification Parameters) are obtained, describing their units and nomenclature as follows (Table 1):

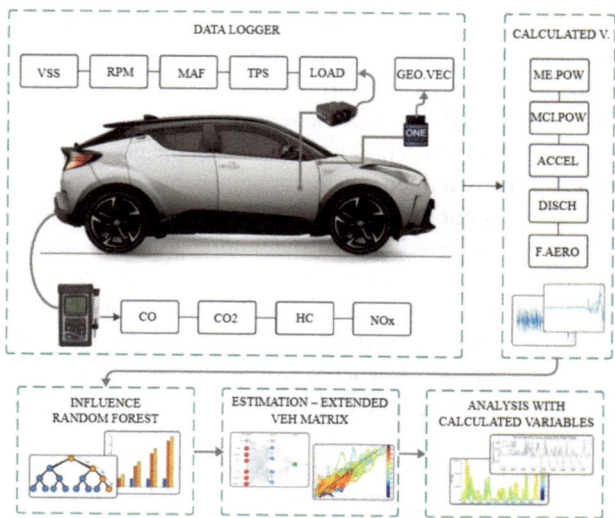

Fig. 1. The stages of the study are shown in an orderly fashion along with the statistical methods, from data acquisition to the analysis of results.

Table 1. Variables of study

Numbering	Nomenclature	Description	Units
1st	Alt	Altitude	[m.a.s.l]
2nd	Lat	Latitude	[m]
3rd	Long	Longitude	[m]
4th	VSS	Vehicle speed	[km/h]
5th	RPM	Engine revolutions	[RPM]
6th	TPS	Accelerator pedal opening	[%]
7th	MAF	Air mass flow	[g/s]
8th	LOAD	Vehicle load	[%]
9th	VOLT	Battery pack voltage	[V]
10th	CURR	Battery pack amperage	[A]
12th	TOR	Percent torque	[%]
13th	FRV	Fuel consumption rate	[g/s]
14th	CO	Carbon monoxide	[%]
15th	CO_2	Carbon dioxide	[%]
16th	HC	Hydrocarbons	[ppm]
17th	NO_x	Nitrous oxides	[ppm]

The positioning parameters (1–3) along with time were obtained using a data logger device called Freematics ONE+. Similarly, using an OBDLink MX, the vehicle operating parameters linked to the engine and high-voltage batteries (4–13) were acquired. Finally, a portable gas analyzer KANE 7 measures gases at the exhaust outlet (15–17).

2.2 PID's Signal Processing

Each of the study variables undergoes a validation stage, segmented by filtering to smooth out the PID signals and eliminate noise [14]. Then, equal sizes between variables and emission indexes are equalized. For the execution of the pollutant gas estimation process as an extension strategy for HEVs, it is necessary to adapt the normalization method to the maximum in the input and output vectors related to the gases that are more complex to estimate accurately and effectively: Hydrocarbons (HC) and Nitrogen Oxides (NO_x). The speeds and distances of the different routes follow the guidelines set forth by the "EURO 6" standard, with equal urban, rural, and highway routes (Fig. 2).

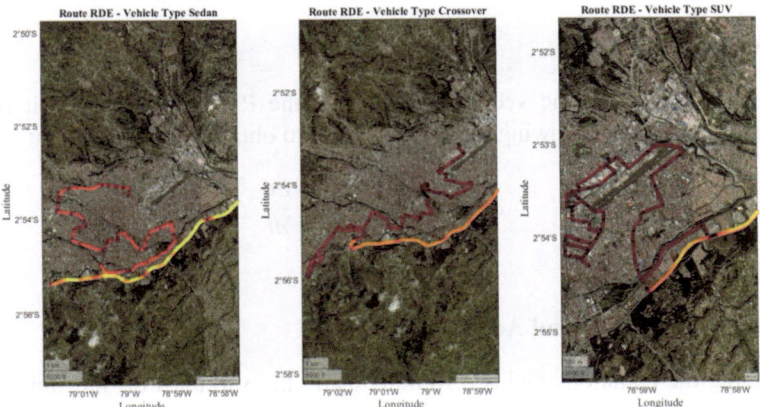

Fig. 2. The three different RDE routes traveled by the HEVs are depicted based on speed, indicated by varying shades of color.

2.3 Emission Factors

Using Eq. (1), the mass flow rate of the exhaust is calculated, and then Eq. (2) is applied to convert concentrations from dry basis to wet basis using the correction factor declared by the European Union [15].

$$\dot{m}_{es} = \dot{m}_{air} + \dot{m}_{comb} \tag{1}$$

$$C_{wet,j} = k_{w,j} C_{dry,j} \tag{2}$$

With Eq. (3), the mass of the pollutant is calculated in grams, and finally, Eq. (4) is used to obtain the emission factor with respect to a distance or travel:

$$m_{gas} = \mu_{gas} * c_{gas} * \dot{m}_{es} * 10^{-3} \tag{3}$$

$$F_{gas} = \frac{m_{gas}}{s} \tag{4}$$

2.4 Traction Force

This force is directly linked to the vehicle mass (m), acceleration (a_x), aerodynamic force, rolling resistance (F_{res}), slope resistance (R_p), and braking force (F_{brk}), which is null as the vehicle remains in motion.

$$F_t = ma_x + F_{res} + R_p + F_{brk} \tag{5}$$

2.5 Acceleration

With the velocity and time vectors available in the PID signal matrix, it becomes necessary to apply the following equation in order to obtain the acceleration:

$$a_x = \lim_{t \to 0} \left(\frac{\Delta v}{\Delta t} \right) = \frac{dv}{dt} \tag{6}$$

2.6 Rolling Resistance and Aerodynamic Force

Synthesizing the equations yields a formula using the variables of: the weight of the vehicle (mg), the coefficient of rolling resistance (C_T), air density (p), frontal area of the test vehicle (A_f), aerodynamic coefficient (C_x) and the instantaneous speed (V_i).

$$F_{res} = (mg * C_T) + \left(\frac{1}{2} * p * C_x * A_f * V_i^2 \right) \tag{7}$$

2.7 Slope Resistance

Considering that the analysis is conducted in a high-altitude city, it is essential to calculate the resistance exerted on the slopes of the different areas of the route.

$$R_p = mg * \sin \left(\frac{Alt_{i+1} - Alt_i}{S_{i+1} - S_i} \right) \tag{8}$$

2.8 Internal Combustion Engine (ICE) and Electric Motor (EM) Powers

Once the torques of the different traction motors are obtained, it becomes straightforward to calculate the powers they require using the following equation:

$$P_{ICE} = T_{ICE} * RPM_{ICE} * \left(\frac{\pi}{30}\right) \tag{9}$$

$$P_{EM} = T_{EM} * RPM_{EM} * \left(\frac{\pi}{30}\right) \tag{10}$$

2.9 Structure and Training Characteristics of ANNs

Table 2. Neural networks dedicated to estimation.

ANN	Description	Hidden Layers	Number of Neurons	R^2
1st	Carbon monoxide (CO)	2	20–20	0.973
2nd	Carbon dioxide (CO_2)	2	20–20	0.972
3rd	Hydrocarbons (HC)	2	25 – 25	0.832
4th	Nitrous oxides (NO_x)	2	25–25	0.937

It becomes necessary to work with two hidden layers in the ANNs training for better performance, also in hydrocarbon gases and nitrous oxides the number of neurons is increased in order to achieve a remarkable determination coefficient (Table 2).

3 Results and Discussion

3.1 Quantitative Indices of Parameter Importance in Acceleration Vector and Pollutant Emissions

In the initial stages, the degrees of influence of the longitudinal acceleration vector on each of the harmful gases are known, concerning the different types of HEVs. The influence of operating parameters on vehicle longitudinal acceleration is then analyzed (Fig. 3).

3.2 Estimation of Pollutant Gases and Emission Factors

Carrying out the process to determine the different emission factors for the various pollutant gases, the following statistics were obtained (Fig. 4, Table 3):

The emission factors of each of the test vehicles are in accordance with the EU standards, and the prediction has a very tight margin of error as the values between the SUV and the estimated one are very similar. The CO_2 factor of the Sedan Type is lower than the rest due to the driving conditions to which the vehicle was subjected in the data acquisition (lower speeds and longitudinal accelerations).

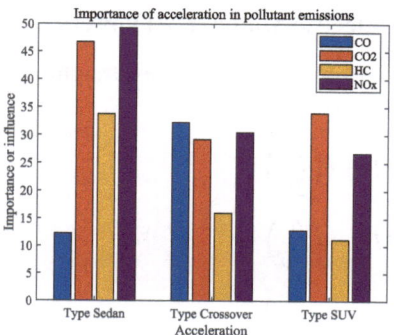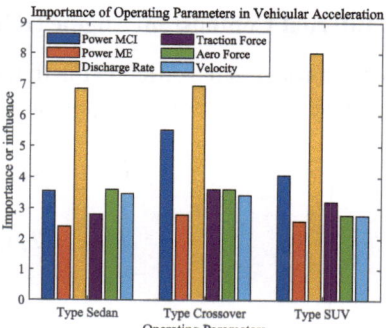

Fig. 3. A) Influence of acceleration on gases. B) Influence of parameters on vehicle acceleration.

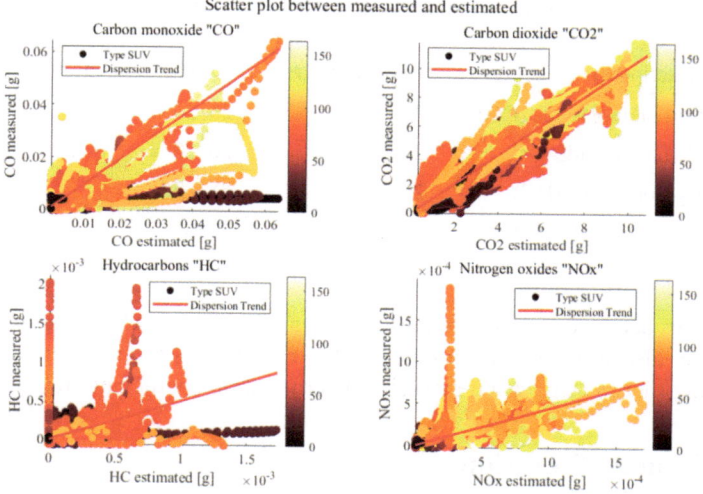

Fig. 4. Scatter plots between actual and estimated indices of each gas, as a function of speed.

Table 3. Emission factors of measured and estimated SUV pollutant gases.

Pollutant gas	Sedan [g/km]	Cross [g/km]	SUV [g/km]	Estimated [g/km]
Carbon monoxide (CO)	0.0735	0.9143	0.1868	0.1899
Carbon dioxide (CO_2)	43.1812	90.2931	92.8487	92.8700
Hydrocarbons (HC)	0.0039	0.0051	0.0033	0.0040
Nitrous oxides (NO_x)	0.0059	0.0151	0.0073	0.0031

3.3   Analysis Between Instantaneous Gas Samples and Vehicle Acceleration

At the time instant (A – 2825 s), significantly high peaks of pollutant gases exceeding the mean values are observed, along with high accelerations (2 m/s^2). During the period

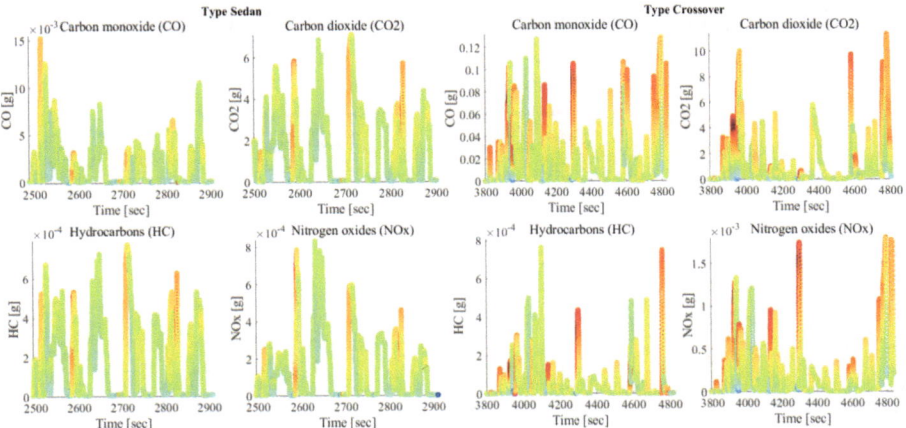

Fig. 5. Instant samples of polluting gases vs. time, depending on the acceleration of the vehicle for the units (1) "Sedan Type" and (2) "Crossover Type".

(B – 2650 s), emission peaks are higher, while this occurs during zero or relatively low accelerations. In the Crossover Type Vehicle, the case is repeated, with most emission peaks being associated with high and extreme accelerations (2 m/s^2). However, there are cases such as at instant (C – 4080 s) where it is graphically noted that the gas peaks are not due to high acceleration (the ICE is not engaged in the traction of the HEV, indicating a probable stage of charging the HV batteries with the ICE) (Fig. 5).

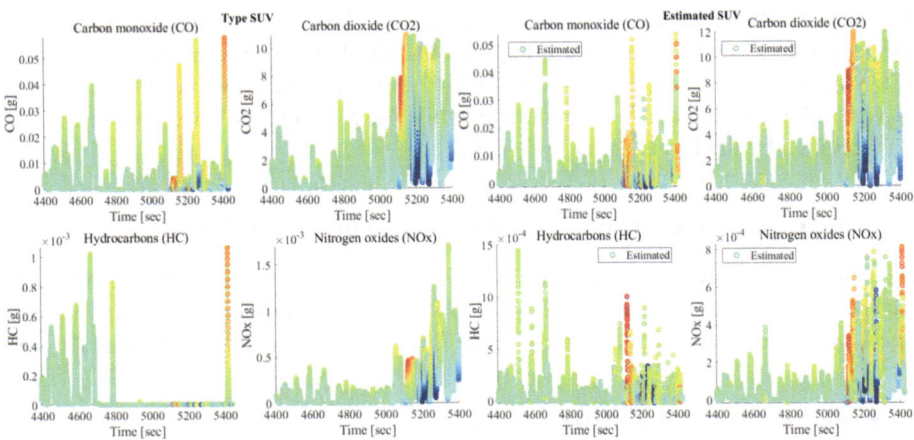

Fig. 6. Instant samples of polluting gases vs. time, depending on the acceleration of the vehicle for the units (3) "SUV Type" and (4) "SUV estimated Type".

At instant (D – 5150 s), for the measured SUV vehicle, low emissions (except for CO_2) are recorded, related to high acceleration; indicating a stage where the ICE is disabled and the traction is generated by the EM. Similarly, the indices and peaks of the predicted SUV are similar, showing a good degree of accuracy in the estimation (Fig. 6).

3.4 Relationship Between Cumulative Emissions and SOC Discharge

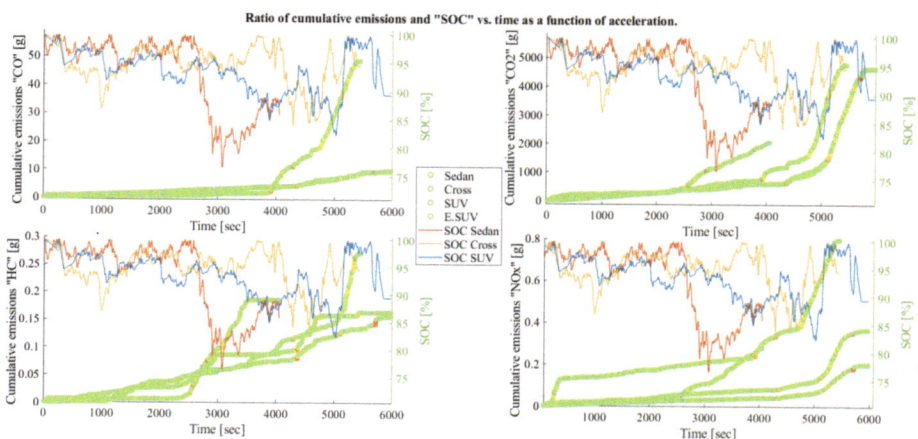

Fig. 7. Cumulative emissions vs. SOC for each of the vehicles as a function of acceleration.

Points of inflection of the cumulative emissions are shown, which are directly related to the respective peak drop in the battery pack discharge (SOC). The color intensity at these points increases according to the acceleration emitted by the study vehicle (since there are four curves representing the three measured HEVs and the estimated one). The HC and NO_x graphs follow the same analysis patterns. An example shown at instant (E – 4450 s): the acceleration is severe (>2 m/s^2) in the curve of the Crossover vehicle and its estimate, and at the same time, the SOC discharge shows a vertical parabola; justifying an acceleration propelled by the EM and ICE. The growth of the cumulative curve is related to the combustion engine, while the SOC parabola is generated by the work emitted by the EM (Fig. 7).

3.5 Powers Demanded by the Different Traction Motors (ICE-EM)

The behavior of vehicle traction powers is analyzed in order to understand when the ICE and the EM provide significant amounts of kW, as well as to identify the time instants when there are records of negative power (EM) related to regenerative braking (Fig. 8).

4 Conclusions and Future Work

The estimation of pollutant gases for the SUV category of HEVs achieved remarkable coefficients of determination in the training of the different artificial neural networks: 0.973, 0.972, 0.832, 0.937; showing precise and effective indices with slight dispersion between the measured and estimated emissions. Longitudinal acceleration shows a strong influence on CO_2 and NO_x gases, doubling the importance degree compared to CO and HC. The Crossover vehicle is the only one in the matrix that levels similar degrees in its pollutant gases. Regarding the operational parameters of traction, the

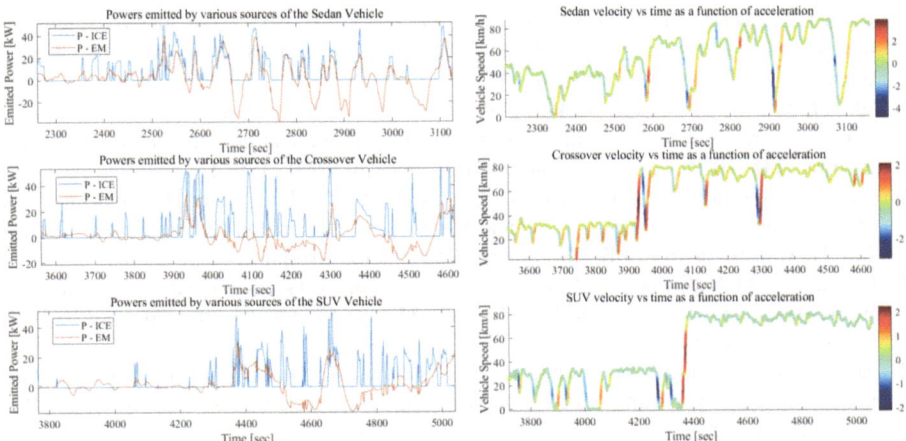

Fig. 8. Left illustrations are traction powers vs. time, while right illustrations are speeds vs. time.

battery pack discharge rate is the most influential factor in acceleration, followed by the power generated by the internal combustion engine. This is because a higher battery pack discharge rate can compromise its ability to provide traction to the vehicle, requiring the combustion engine to operate simultaneously to maintain tire traction and charge the batteries through a generator motor (EM1 or EM2). Finally, future studies are needed to address the influence of acceleration on emissions by temporally segmenting the different operating stages of the vehicle.

References

1. Valladolid, J.D., Patino, D., Gruosso, G., Correa-Flórez, C.A., Vuelvas, J., Espinoza, F.: A novel energy-efficiency optimization approach based on driving patterns styles and experimental tests for electric vehicles. Electronics (Switzerland) **10**(10) (2021). https://doi.org/10.3390/electronics10101199
2. Skouras, T.A., Gkonis, P.K., Ilias, C.N., Trakadas, P.T., Tsampasis, E.G., Zahariadis, T.V.: Electrical vehicles: current state of the art, future challenges, and perspectives. Clean Technol. **2**(1) (2020). https://doi.org/10.3390/cleantechnol2010001
3. Solanke, T.U., Khatua, P.K., Ramachandaramurthy, V.K., Yong, J.Y., Tan, K.M.: Control and management of a multilevel electric vehicles infrastructure integrated with distributed resources: a comprehensive review. Renewable Sustainable Energy Rev. **144** (2021). https://doi.org/10.1016/j.rser.2021.111020
4. Ehsani, M., Singh, K.V., Bansal, H.O., Mehrjardi, R.T.: State of the art and trends in electric and hybrid electric vehicles. Proc. IEEE **109**(6), 967–984 (2021). https://doi.org/10.1109/JPROC.2021.3072788
5. Liu, W.: Introduction to Hybrid Vehicle System Modeling and Control. John Wiley & Sons, New Jersey (2013)
6. Jiang, Y., Song, G., Wu, Y., Lu, H., Zhai, Z., Yu, L.: Impacts of cold starts and hybrid electric vehicles on on-road vehicle emissions. Transp. Res. D: Transp. Environ. **126** (2024). https://doi.org/10.1016/j.trd.2023.104011

7. Bagheri, S., Huang, Y., Walker, P.D., Zhou, J.L., Surawski, N.C.: Strategies for improving the emission performance of hybrid electric vehicles. Sci. Total Environ. **771** (2021). https://doi.org/10.1016/j.scitotenv.2020.144901

8. Ehrenberger, S.I., Konrad, M., Philipps, F.: Pollutant emissions analysis of three plug-in hybrid electric vehicles using different modes of operation and driving conditions. Atmos. Environ. **234** (2020). https://doi.org/10.1016/j.atmosenv.2020.117612

9. Suarez, J., Makridis, M., Anesiadou, A., Komnos, D., Ciuffo, B., Fontaras, G.: Benchmarking the driver acceleration impact on vehicle energy consumption and CO_2 emissions. Transp. Res. D: Transp. Environ. **107** (2022). https://doi.org/10.1016/j.trd.2022.103282

10. Costagliola, M.A., Costabile, M., Prati, M.V.: Impact of road grade on real driving emissions from two Euro 5 diesel vehicles. Appl. Energy **231**, 586–593 (2018). https://doi.org/10.1016/j.apenergy.2018.09.108

11. Zhai, Z., Xu, J., Zhang, M., Wang, A., Hatzopoulou, M.: Quantifying start emissions and impact of reducing cold and warm starts for gasoline and hybrid vehicles. Atmos. Pollut. Res. **14**(1) (2023). https://doi.org/10.1016/j.apr.2022.101646

12. Rivera-Campoverde, N.D., Arenas-Ramírez, B., Muñoz Sanz, J.L., Jiménez, E.: GPS data and machine learning tools, a practical and cost-effective combination for estimating light vehicle emissions. Sensors **24**(7), 2304 (2024). https://doi.org/10.3390/s24072304

13. Rivera, N., Bermeo, K., Juca, J., Ortuño, J.: Analysis of the energy consumption of a series-parallel hybrid electric vehicle. In: ECTM 2023 – 2023 IEEE 7th Ecuador Technical Chapters Meeting, Institute of Electrical and Electronics Engineers Inc. (2023). https://doi.org/10.1109/ETCM58927.2023.10309048

14. Schafer, R.W.: What is a savitzky-golay filter? IEEE Signal Process. Mag. **28**(4), 111–117 (2011). https://doi.org/10.1109/MSP.2011.941097

15. European Commission and Council of the European Union Commission Regulation (EU). 2016/427 of 10 March 2016 amending Regulation (EC) No 692/2008 as regards emissions from light passenger and commercial vehicles (Euro 6) (Text with EEA relevance). Off. J. Eur. Union **82**, 1–98 (2016)

Comparison of Environmental Impact Between Hybrid Electric Vehicles: Analysis of Pollutant Emissions in Real Driving Emissions Test

Walter Josué Semiglia Pineda[1,2](✉) [iD], Jackson Vidal Suarez[1,2] [iD],
and Néstor Rivera Campoverde[1,2] [iD]

[1] Universidad Politécnica Salesiana, Cuenca, Ecuador
waltersemiglia24@gmail.com
[2] Grupo de Investigación en Ingeniería del Transporte – GIIT, Cuenca, Ecuador

Abstract. This article conducts a comparative analysis of emission factors from three vehicle types (sedan, crossover, SUV) during a Real Driving Emissions (RDE) test, adhering to Euro 6 norms for accuracy. Hybrid electric vehicles (HEVs), combining internal combustion engines with electric propulsion, are examined in real-time driving conditions in a high-altitude city. Data collection involves PID signals, pollutant gas measurements, and GPS data, analyzed using engineering software and statistical methods. The study identifies that aggressive driving behaviors and acceleration phases notably impact CO_2 emissions, particularly on motorways. These findings underscore the role of driving style in influencing environmental emissions, highlighting the potential of HEVs to reduce carbon footprint across urban and highway scenarios.

Keywords: HEV · Emission factor · PEMS · RDE

1 Introduction

Cities at high altitudes, such as Cuenca, Ecuador, located at 2,560 m above sea level, face unique challenges regarding vehicle emissions. The reduced oxygen density and lower atmospheric pressure at these altitudes decrease combustion efficiency in vehicle engines, leading to increased emissions of pollutants such as carbon monoxide (CO) and hydrocarbons (HC). Additionally, lower temperatures can delay engine warm-up, thereby increasing fuel consumption and emissions during the startup phase.

In current times, it is crucial to properly assess and quantify the emissions from automotive vehicles to understand their impact on urban air quality and population exposure to urban traffic [1, 2] This approach becomes increasingly relevant as the demand for clean urban environments rises [2] According to data from the European Union [3], the transportation sector contributes approximately a quarter of the total carbon dioxide (CO_2) emissions, with road transport accounting for 71.7% of these emissions.

This underscores the importance of addressing vehicle emissions to reduce the carbon footprint of transportation. In response to this issue, vehicles with hybrid technology have

© The Author(s) 2025
E. M. Inga Ortega et al. (Eds.): CITIS 2024, LNNS 1331, pp. 151–160, 2025.
https://doi.org/10.1007/978-3-031-87065-1_14

emerged, which combine internal combustion engines with electric propulsion systems [4] These vehicles present significant potential for reducing pollutant emissions and improving fuel consumption efficiency, making them a promising option for addressing the environmental challenges associated with urban and long-distance transportation [4, 5]. The hybrid architecture of these vehicles allows harnessing the advantages of both propulsion systems: internal combustion engines provide autonomy and power for long-distance trips, while electric systems reduce emissions and improve fuel efficiency, especially in urban driving [6].

In urban traffic conditions characterized by frequent stops and starts, hybrid electric vehicles can utilize their electric motor, significantly reducing emissions. Furthermore, the ability to operate in electric mode reduces dependence on the internal combustion engine, which is beneficial in high-altitude settings where combustion efficiency is lower. The integration of advanced technologies, such as regenerative braking and intelligent engine management, optimizes the performance and efficiency of hybrid vehicles [6, 7], this results in a significant reduction in emissions of pollutants, thus contributing to improving urban air quality and mitigating the environmental impact of transportation [7]. The objective of this study is to analyze the environmental impact caused by hybrid vehicles during real driving emissions in a high-altitude city.

This article presents a real-time comparative analysis of emission factors among three types of vehicles (sedan, crossover and sport utility vehicle) by obtaining parameter identifier signals of characteristic driving variables such as TPS, MAF, RPM, and VSS from an OBD device. Additionally, GPS data is utilized, and real-time data on pollutant gases (CO_2, CO, HC, and NOx) was collected.

2 Material and Methods

According to [8], direct methods for obtaining emission factors of a vehicle in real operating situations, such as on-road measurement, are established. To obtain emission factors during real-time driving, the onboard measurement method is used, which involves using a portable emissions measurement system device to record emissions while the vehicle is in motion. This method considers relevant variables such as speed, load, and mass airflow [8].

Taking into account important aspects such as vehicle selection, route, among others, the following steps are proposed to obtain accurate and reliable results for each type of vehicle:

- Selection of test vehicle types.
- Real-time data acquisition of driving and emissions.
- Estimation of emission factors.
- Data processing and presentation of results.

2.1 Test Vehicle Types

The selection of vehicle types for conducting road condition tests is a crucial step in evaluating the performance and efficiency of vehicles in different driving scenarios.

In this context, The study encompasses three Toyota hybrid vehicles: the Corolla, CH-R, and Corolla Cross. Each vehicle is equipped with a 1.8 L DOHC 16-valve inline-4 engine featuring VVT-i technology, paired with an Electronically Controlled Continuously Variable Transmission (eCVT).

All vehicles have undergone the manufacturer-recommended maintenance regimen, which includes regular oil changes, tire rotations, brake inspections, and hybrid system evaluations, ensuring optimal performance and reliability throughout the measurement campaign.

Additionally, each of these models successfully reached their estimated sales targets in their respective years of release, reflecting their market acceptance and consumer demand. The characteristics of the vehicles are presented in Table 1.

Table 1. Characteristics of the selected vehicles

Vehicle	Type	Displacement [cc]	Maximum torque [Nm]	Weight [Kg]
1	Sedan	1798	142	1388
2	Crossover	1798	142	1860
3	SUV	1798	142	1850

2.2 Real-Time Data Acquisition of Driving and Emissions

2.2.1 Portable Emissions Measurement System (PEMS)

A Kane Autoplus 4-2 gas analyzer is used, which operates using an infrared system to measure CO_2 [%], CO [%], and HC [ppm], and an electrochemical cell to measure O2 [%] and NOx [ppm]. The measurement system provides a representation of the data in digital or graphical form.

Additionally, it stores real-time data on a PC in CSV format.

2.2.2 Data Logger

Vehicle operating parameters are obtained through a Bluetooth OBD adapter called OBDLink MX+, which is compatible with all legislated OBD-II protocols and also supports SW-CAN and MS-CAN protocols.

Additionally, GPS information is captured using the Freematics ONE+ data logger at a frequency of 15.15 Hz, and it is stored on a micro SD card. The device is equipped with a U-blox Neo-6M GPS module offering 2.5-m accuracy and acquisition times of 27 s for cold starts and 1 s for warm starts.

The main operating parameters used to develop the study are shown in Table 2.

The data obtained from the data logger is saved in CSV format, generating a file for each route test. After obtaining the necessary data, the files are vectorized to obtain a data matrix.

The Savitzky-Golay filtering algorithm is applied to each variable with the purpose of smoothing the measured data in the signal processing [9]. The PEMS and data logger

Table 2. Operating parameters

Numbering	Nomenclature	Description	Units
1st	Lat	Latitude	[m]
2nd	Alt	Altitude	[m.a.s.l]
3rd	Long	Longitude	[m]
4th	VSS	Vehicle speed	[km/h]
5th	RPM	Engine revolutions	[RPM]
6th	TPS	Accelerator pedal opening	[%]
7th	MAF	Air mass flow	[g/s]
8th	ECT	Coolant temperature	[°C]
9th	FR	Fuel consumption rate	[g/s]

measurement equipment have different sampling frequencies; therefore, the vector size equalization method proposed by [10] is utilized.

2.2.3 Test Vehicle Routes

Euro 6 regulations, which set stringent limits on vehicle emissions, aim to reduce the environmental impact of automotive pollutants. With the objective of studying the emissions behavior of hybrid vehicles in accordance with the Real Driving Emissions (RDE) test, the city of Cuenca, Ecuador is selected for data collection. The RDE test cycle adheres to Euro 6 guidelines and includes diverse driving conditions: the urban sector (downtown and its surroundings), rural sector, and motorway (Panamericana Cuenca-Azogues). Figure 1 illustrates the RDE routes and the vehicle's speed throughout the entire test trip [11].

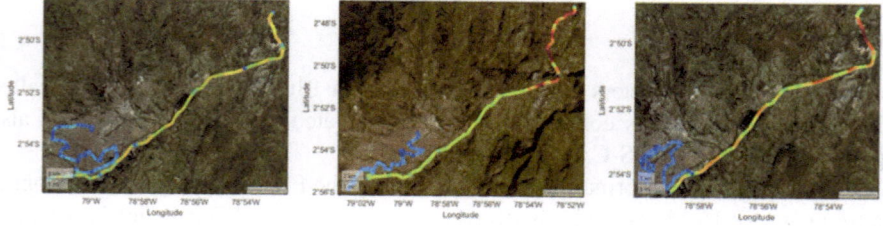

Fig. 1. The three different RDE routes traveled by the selected vehicles for the study: sedan, crossover, suv, respectively.

During the test, the environmental conditions included a temperature of 18 °C, with no precipitation or strong winds. The test was conducted with the windows closed, the air conditioning deactivated, and under minimal traffic conditions. The vehicle was fueled with 92-octane gasoline, as recommended by the manufacturer.

2.3 Emission Factor Estimation

To establish the emission factors of the selected vehicles during an RDE test, the procedure described in [11] is employed. First, the exhaust mass flow is estimated using the Eq. (1).

$$\dot{m}_{ex} = \dot{m}_{in} + \dot{m}_f \tag{1}$$

$\dot{m}_{in}$ represents the mass flow of air and $\dot{m}_f$ The fuel flow from the parameters obtained from the OBDlink Mx data logger. Pollutant emissions are measured on a dry basis, so they must be corrected to a wet basis using the following Eqs. (2) and (3):

$$C_{wet,j} = k_{w,j} C_{dry,j} \tag{2}$$

$$k_w = \frac{1.008}{1 + 0.005\alpha(C_{CO_2} + C_{CO})} \tag{3}$$

$C_{wet,i}$ represents the concentration on a wet basis of each pollutant gas. j, $C_{dry,j}$ the concentration is on a dry basis of the pollutant, k_w it is a correction factor from dry to wet basis, α It is the molar ratio of hydrogen, and $C_{CO_2} + C_{CO}$ They are the concentrations on a dry basis of CO_2 and CO. The instantaneous mass emissions of each pollutant are obtained using the Eq. (4) (Fig. 2):

$$\dot{m}_{j,i} = c_{j,i} \mu_{j,i} \dot{m}_{ex,i} 10^{-3} \tag{4}$$

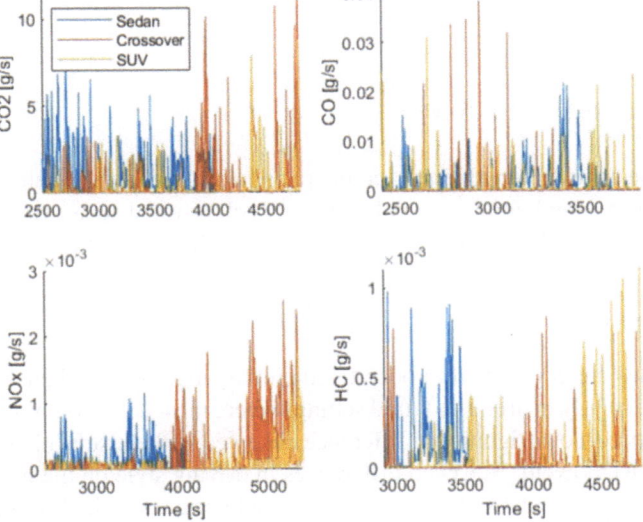

Fig. 2. Exhaust emissions [g/s]

The emission of each pollutant gas m_j (g) During the RDE test cycle, it equals the summation of its instantaneous emission over time. The emission of each pollutant is represented by (5):

$$m_j = \sum_{i=1}^{n} \dot{m}_{j,i} \Delta t \tag{5}$$

The emission factors F_j for each pollutant gas (g/km) during the RDE test cycle were determined by the Eq. (6):

$$F_j = \frac{m_j}{s} \tag{6}$$

The emission factors were calculated in accordance with the methodology outlined in [11, 12].

Resulting in the average emission factors of the test vehicles during the RDE test, in Table 3.

Table 3. Average emission factors

Pollutant gas	Sedan (g/km)	Crossover (g/km)	SUV (g/km)
CO_2	43.21	90.27	92.85
CO	0.0735	0.91	0.187
NOx	0.0059	0.0151	0.0073
HC	0.0039	0.0051	0.0033

3   Results and Discussion

To perform the environmental impact comparison between the study vehicles, the emission factors corresponding to each pollutant for each traveled sector (urban, rural, and motorway) during the RDE test are determined for each selected vehicle type (Fig. 3).

3.1   Emissions CO_2

Hybrid vehicles are engineered to operate efficiently and mitigate their environmental footprint, particularly in urban and rural settings where most daily commutes occur [13]. Therefore, there is not a significant difference observed in the results obtained for these sectors of the RDE cycle. However, differences in driving style on the highway lead to a significant variation in CO_2 emissions among the hybrid test vehicles. Although these vehicles share similar characteristics, the more aggressive driving style on the motorway, typical of larger and heavier vehicles such as crossovers and SUVs, can result in increased fuel consumption and, consequently, CO_2 emissions.

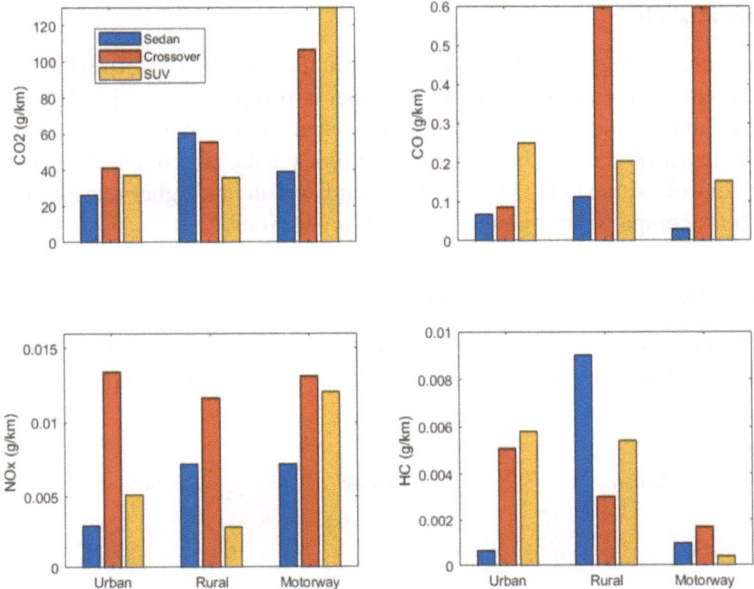

Fig. 3. Emission factors for each vehicle during the RDE test

3.2 Emissions CO

During the RDE test, it is observed that the CO emissions from the three types of vehicles comply with the Euro 6 emission standards in all evaluated sectors. However, a significant difference is noted in the Crossover vehicle compared to the other two types of vehicles, with considerably higher emissions in the rural and motorway sectors. These differences are attributed to various influencing factors, such as specific test conditions in these sectors for the Crossover vehicle, which may have impacted combustion efficiency and the air-fuel ratio. It's possible that driving conditions in rural areas and motorways may have imposed an additional load on the Crossover vehicle's engine, which could negatively impact its performance in terms of CO emissions.

3.3 Emissions NOx

During the RDE cycle, vehicles comply with Euro 6 emissions regulations, but Crossover vehicles exhibit the highest NOx emissions compared to other vehicle types. This increase is attributed to a combination of factors, including higher combustion engine load, driving dynamics during the RDE cycle, and thermal engine ignition to charge the high-voltage battery in hybrid vehicles. Additionally, further validation efforts are required to ensure the accuracy of NOx emission estimates, considering the complex interplay of variables inherent in hybrid vehicle operation [14].

3.4 Emissions HC

A notable increase in unburned hydrocarbon emissions is observed from the sedan vehicle in rural environments compared to the other two types of vehicles. It is noted that this increase may be attributed to lower traffic density and more uniform driving speeds, characteristics of rural areas, factors that can influence the effectiveness of the vehicle's emission control system [14, 15] and consequently result in a higher release of unburned hydrocarbons compared to urban or motorway conditions.

3.5 The Influence of Acceleration on CO_2 Emissions

To observe the increase in CO_2 emissions, the motorway sector of each vehicle is analyzed during its RDE test, where aggressive driving behavior is exhibited due to the high speeds reached.

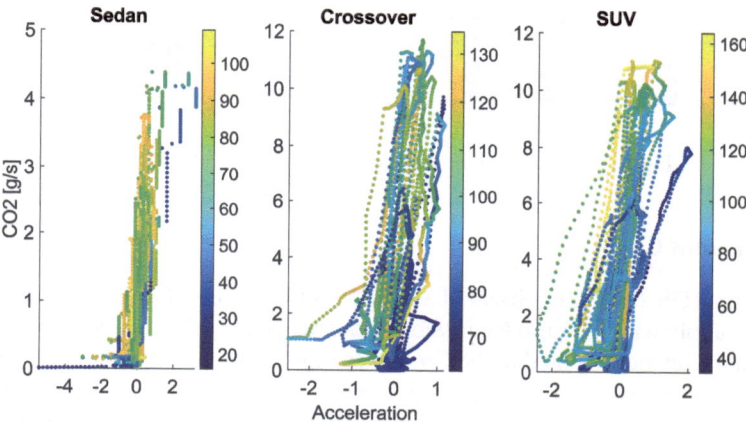

Fig. 4. Acceleration vs CO_2 pollutant emissions in the motorway sector during RDE test.

As shown in Fig. 4, during an RDE cycle, the acceleration phase has a direct influence on the increase of CO_2 emissions in hybrid vehicles on motorways. During acceleration, the combined use of the internal combustion engine and the electric motor results in a temporary increase in emissions, especially in situations requiring additional power or in a sporty driving style.

4 Conclusions and Future Work

Hybrid vehicles are recognized as effective solutions for reducing emissions of pollutants like CO_2, CO, NOx, and HC, thereby enhancing air quality and mitigating climate change impacts. By integrating internal combustion engines with electric motors, hybrids achieve higher energy efficiency and extended driving range, making them adaptable to diverse driving conditions and user requirements. However, the environmental performance of hybrid vehicles can vary significantly based on factors such as driving style and the operational environment.

The most aggressive driving styles and the acceleration phase has a direct influence on the increase of CO_2 emissions especially on motorways.

Specific test conditions, such as engine load and driving dynamics, can significantly influence NOx emissions, with crossover vehicles showing the highest emissions in this regard.

In conclusion, it is imperative to undertake emission studies on hybrid vehicles under real-road conditions to analyze emissions generated by sectors and assess primary influencing factors on various pollutant gases.

References

1. Johnson, T.: Vehicular emissions in review. SAE Int. J. Engines **9**(2), 1258–1275 (2016). https://doi.org/10.2307/26284895
2. Special RepoRt 17 Traffic-Related Air Pollution: A Critical Review of the Literature on Emissions, Exposure, and Health Effects A Special Report of the HEI Panel on the Health Effects of Traffic-Related Air Pollution (2010). www.healtheffects.org
3. European Parliament. CO_2 emissions from cars: facts and figures (infographic)
4. Lebeau, P., De Cauwer, C., Van Mierlo, J., Macharis, C., Verbeke, W., Coosemans, T.: Conventional, hybrid, or electric vehicles: which technology for an urban distribution centre? Sci. World J. (2015). https://doi.org/10.1155/2015/302867
5. Hannan, M.A., Azidin, F.A., Mohamed, A.: Hybrid electric vehicles and their challenges: a review. Renew. Sustain. Energy Rev. **29**, 135–150 (2014). https://doi.org/10.1016/j.rser.2013.08.097
6. Sabri, M.F.M., Danapalasingam, K.A., Rahmat, M.F.: A review on hybrid electric vehicles architecture and energy management strategies. Renew. Sustain. Energy Rev. **53**, 1433–1442 (2016). https://doi.org/10.1016/j.rser.2015.09.036
7. Introduction_to_hybrid_electric_vehicles_State_of_art
8. Gallus, J., Kirchner, U., Vogt, R., Benter, T.: Impact of driving style and road grade on gaseous exhaust emissions of passenger vehicles measured by a Portable Emission Measurement System (PEMS). Transp. Res. D: Transp. Environ. **52**, 215–226 (2017). https://doi.org/10.1016/j.trd.2017.03.011
9. Schafer, R.W.: What is a savitzky-golay filter? IEEE Signal Process. Mag. **28**(4), 111–117 (2011). https://doi.org/10.1109/MSP.2011.941097
10. Rivera-Campoverde, N.D., Arenas-Ramírez, B., Luis Muñoz Sanz, J., Jiménez, E.: A practical and cost-effective approach for estimating light vehicle emissions using GPS with machine learning. Sensors **24** (2024)
11. REGLAMENTO (UE) 2016/ 427 DE LA COMISIÓN – de 10 de marzo de 2016 - por el que se modifica el Reglamento (CE) n.o 692/2008 en lo que concierne a las emisiones procedentes de turismos y vehículos comerciales ligeros (Euro 6)
12. euro 6
13. Prati, M.V., Costagliola, M.A., Giuzio, R., Corsetti, C., Beatrice, C.: Emissions and energy consumption of a plug-in hybrid passenger car in Real Driving Emission (RDE) test. Transport. Eng. **4** (20210. https://doi.org/10.1016/j.treng.2021.100069
14. Vidal, J., Semiglia, W.: I UNIVERSIDAD POLITÉCNICA SALESIANA SEDE CUENCA CARRERA DE INGENIERÍA AUTOMOTRIZ (2024)
15. Wang, Y., et al.: Fuel consumption and emission performance from light-duty conventional/hybrid-electric vehicles over different cycles and real driving tests. Fuel **278** (2020). https://doi.org/10.1016/j.fuel.2020.118340

Application of Topological Design to Brake Discs of Light Vehicles to Improve Heat Dissipation

Pablo Renato Fierro[1,2](✉)(iD), Jerys Macías Molina[1], and Josué Macías Molina[1]

[1] Smart Mobility Research Group of Salesian Polytechnic University
(GMOVINT, UPS), Cuenca, Ecuador
[2] University Institute of Automobile Research, Technical University of Madrid
(INSIA, UPM), Madrid, Spain
pfierro@ups.edu.ec, r.fierro@alumnos.upm.es

Abstract. Topological optimization is a process that allows the design of optimal structural elements, maximizing or minimizing specific characteristics without altering their structural composition. In the case of brake discs, a thorough load analysis was performed to find a balance between applied stresses, safety factors, and mass reduction, without loss of mechanical properties. The discs that passed this analysis were subjected to computational fluid dynamics analysis to determine which disc would most efficiently dissipate the heat generated during operation, with the option of redesigning models. The results showed that the disc with 5 mm perforations reduced the maximum temperature by 8 °C. It is important to remember that applying topology in brake disc design must be an iterative process, performing successive evaluations to determine the optimal shape, thickness, or arrangement of the proposed geometric changes. This ensures that the results are valid and justifiable from a technical and functional point of view.

Keywords: Topology · CFD analysis · Structural thermal analysis · Heat dissipation · Convection coefficient · Brake disc · Temperature

1 Introduction

The FARS (Fatality Analysis Reporting System) is a national census that provides annual data on fatal injuries in motor vehicle crashes in the United States. Between 2011 and 2014, 21% of accidents were due to failures in airbags, seat belts, and brake discs, with only a 0.3% reduction by 2021 [1]. The ANT reported 21,230 accidents related to steering and braking system failures [2]. Vehicle braking distribution typically sees 70% on the front axle and 40% on the rear, leading to front axle discs reaching temperatures up to 900°C [3,4]. This overheating causes cracks, a common symptom of thermal fatigue in brake discs, due to stress-compression cycles that deform the material [5,6]. European standards focus on engine emissions, ignoring that 16% to 55% of non-exhaust emissions

Salesian Polytechnic University.

E. M. Inga Ortega et al. (Eds.): CITIS 2024, LNNS 1331, pp. 161–173, 2025.
https://doi.org/10.1007/978-3-031-87065-1_15

come from brakes, with 80% to 98% suspended in the air, exceeding PM limits in urban areas [7,10]. Topology optimization can mitigate these issues by reducing temperature gradients and mass under thermal conditions [11,13], achievable by drilling holes in the disc surface to reduce cooling time, increase heat dissipation, and decrease maximum working temperature, avoiding thermal fatigue or loss of mechanical properties (fading) Perforations in heat sinks decrease temperature and increase heat transfer, acting as inverted heat sinks, allowing internal convection, improving heat dissipation and stress distribution on the blade surface, and increasing heat transfer for radial and straight blades [15,18]. A methodology is proposed to optimize light vehicle brake discs, aiming to reduce critical temperature and improve temperature distribution through material characterization, mechanical analysis, and computational simulation software, validated using CFD with high accuracy.

2   Materials and Methods

Disc brake systems operate under Pascal's law, increasing the output force of the caliper piston through a hydraulic circuit, In this way, the force produced in the system is transmitted to the brake pad and distributed over the surface in contact with the disc so that the pads can slow or stop the wheels from turning. With the pressure with which the pads press the disc and the fact that the brake pad material has a high coefficient of friction, a transformation from kinetic to heat energy will occur. This energy is the heat flow that is induced internally in the disc and raises its temperature, These are the 2 main boundary conditions: the stress on the disc and the working temperature.

2.1   Brake Disc Load Calculations

The analytical study and the theory behind the calculation of heat flow and the forces arising in braking are discussed in this section. The expression for the heat flow in a brake disc is obtained using two approaches: a general energy-based solution analysis and a specific equation obtained by Reimpel.

Braking Loads. Several loads are generated as a result of vehicle braking. These loads depend on the geometry of the disc and the pads.

In Eq. (1), determine the force on the brake pad when the brake pedal is pressed to slow the vehicle or bring it to a complete stop. The force acting on the pad comes from hydraulic pressure and is "theoretically" evenly distributed over the contact surface.

$$F_h = \frac{\pi(d^2)}{4} .P_h \qquad (1)$$

where:

d is the diameter of the hydraulic cylinder (m).

P_h is the hydraulic pressure the brake system exerts (Pa).

Equation (2), calculates the pressure on the brake disc generated by the pad force and the pad surface. This pressure acts on the disc surface varies depending on the pad size.

$$P = \frac{F_h}{S} \tag{2}$$

where:

F_h is the force present in the pellet (N).

S is the area of the pad that is in contact with the disk (m^2)

In equation (3), the braking moment that occurs due to the opposition produced by the brake to the movement of the wheels is calculated, it is a negative torque that brakes the vehicle

$$M = \frac{1}{3} \cdot \pi \cdot d^2 \cdot \mu_{zap} \cdot \left(\frac{r_e^2 + r_e \cdot r_i + r_e^2}{r_e + r_i} \right) \cdot P_h \tag{3}$$

where:

d is the diameter of the hydraulic cylinder (m).

μ_{zap} friction coefficient of the pad.

P_h is the hydraulic pressure the brake system exerts (Pa).

r_e outer radius of the pellet (m).

r_i inner radius of the pellet (m).

Heat Flow. It can be assumed that all lost kinetic energy variation is converted into thermal energy during braking, and is determined by Eq. 4:

$$\Delta KE = \frac{1}{2} \, m(v_f^2 - v_i^2) \tag{4}$$

Also, in Eq. 5 the velocities are related to the total distance and deceleration.

$$(v_f^2 - v_i^2) = 2ad \tag{5}$$

where a is the deceleration and d is the distance traveled, replacing Eq. 4 we obtain Eq. 6.

$$\Delta KE = mad \tag{6}$$

According to Jazar, [?], the total thermal energy is a function of the change in KE, Eq. 7.

$$Q_{rueda} = \frac{1}{2} \, f \Delta KE \tag{7}$$

where f is the dynamic load distribution factor. Therefore, the heat flux $q = \frac{Q_{rueda}}{f \Delta t}$ y Δt is the operating time of the wheel. The initial heat flux through the rotor face can also be obtained by the expression given by Reimpel, Eq. 8.

$$Q_0 = \frac{(1 - \phi)}{2} \, \frac{mgv_0 z}{2A_d \varepsilon_p} \tag{8}$$

where $z = \frac{a}{g}$ is the braking effectiveness, ϕ is the distribution of braking forces between the front and rear axles, A_d is the area of the disc surface that is swept by a brake pad, v_0 is the initial speed of the vehicle, ε_p is the load factor distributed over the disc surface, m is the mass of the vehicle, and $g = 9.81\frac{m}{s^2}$ is the acceleration of gravity. To find the time it takes for the braking system to stop the vehicle completely, it is necessary to know the deceleration of the car. The regulation n 13-H establishes that under emergency braking conditions the minimum deceleration in modern cars should be of $6.49\frac{m}{s^2}$ for conventional vehicles, the measured deceleration may be up to $7.98\frac{m}{s^2}$. With these first data, the braking time and the braking distance can be calculated and the resulting values are shown in Fig. 1.

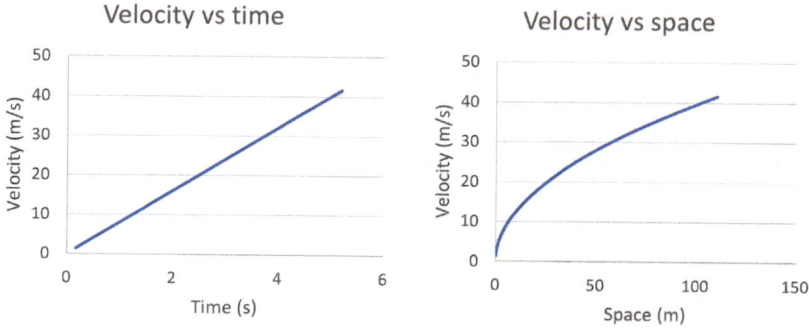

Fig. 1. (a) Theoretical stopping time at different speeds. (b) Theoretical stopping distance at different speeds.

2.2   Calculation of the Convection Coefficient

To determine the convection between the airflow and the brake disk theoretically, an average temperature between the brake disk and the air must be determined, so the values obtained will be used to determine the Prant number, the Reynolds number, and the Nusselt correlations, which are based on thermodynamic tables for different temperatures and pressures.

With the Reynolds, Prant, and Nusselt values, convection can be estimated theoretically at the periphery of the entire disk and also internally. The results obtained are presented in Table 1

With the convection calculated for 4 different speeds, a trend function can be approximated to estimate the convection for all vehicle speeds. Obtaining the convection for each velocity point, we observe a directly proportional behavior concerning the velocity, with which it can be deduced that the velocity is the most influential variable for convection, which in turn suggests a greater heat transfer.

Table 1. Theoretical convection coefficient for different velocities.

Speed	Disc area	Convection coefficient $(W/m^2 - °C)$	Total convection $(W/m^2 - °C)$
40km/h	Peripheral disc	41,45	118,35
	Bell periphery	26,78	
	Friction surface	26,16	
	Rotation	23,74	
	Ducts	0,20	
80km/h	Peripheral disc	70,27	187,56
	Bell periphery	45,41	
	Friction surface	37,15	
	Rotation	33,72	
	Ducts	0,99	
120km/h	Peripheral disc	90,94	237,30
	Bell periphery	58,76	
	Friction surface	44,95	
	Rotation	40,80	
	Ducts	1,82	
150km/h	Peripheral disc	102,85	269,47
	Bell periphery	66,46	
	Friction surface	50,07	
	Rotation	45,45	
	Ducts	4,63	

2.3   Physical, Mechanical, and Thermal Properties

The determination of the brake disc properties is based on the disc manufacturing data sheet, Fremax BD7500 is a gray cast iron disc. The material properties were provided by the technical standard ASTM 48 specifications from grade 35 to 40 or EN-GJL-250 for cast irons, Table 2.

Table 2. ASTM A48 Grade 40 Material Properties.

Physical properties	
Density (Kg/m^3)	7200
Mechanical properties	
Brinell hardness	138 to 225
ultimate tensile stress (MPa)	275
ultimate compressive stress (MPa)	1034
Modulus of elasticity (GPa)	120
Thermal properties	
Thermal conductivity (W/m.K)	46
Specific heat (J/kg-K)	490
Thermal Expansion (um/m.K)	11
Melting temperature (F)	2050 to 2120

2.4   Mesh Convergence for Structural Thermal Analysis

For the mesh analysis, being a structural static module, the mesh convergence will be taken into account, for this type of analysis it is advisable to establish loads and refine the mesh in areas of higher stress, ensuring the convergence of results by finding the optimal size [31. The difference between the standard mesh and the refined mesh is shown in Fig. 2. In very complex geometries, singularities may appear in the software, the most common is the infinite stress, problem. This problem prevents the mesh from converging to a possible value, increasing the resulting stress value obtained whenever the mesh is refined. As the nodes and elements are increased, the stress value increases, so it is necessary to correct the geometry before performing the refinement.

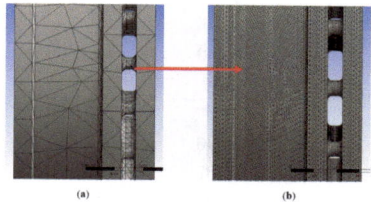

Fig. 2. (a) standard mesh generated by the software. (b) general mesh refinement.

It can be solved by eliminating the edges where stresses exist in the simulation, in this way, the value will converge. The convergence tool uses the stress generated by applying the Von Mises criterion to refine the mesh automatically until the minimum convergence value is found, thus obtaining the number of nodes and elements with which the mesh converges for our study, Table 3. This is a starting parameter because a convergence of 0.50% is an automatic value that seeks the minimum dispersion of results, to save more computational resources, the mesh is re-meshed manually because it allows greater control of the element size per zone and focusing the mesh on the accumulation of stresses, having as a stopping parameter up to 5% difference. Although the new mesh begins to converge, it takes more than 1.5 million nodes and does not reach the value of 59 MPa which is the minimum convergence according to ANSYS, and to achieve this convergence only 1.1 million nodes are needed. The new mesh after the whole process reveals a much smaller size in the blades of the disk, Because of its size and how it receives the load, it is mandatory to perform the refinement in this area, while those that do not receive direct load will be left with a much larger element size, usually 3–5 mm, The resulting stress value is 59.1 MPa which is very close to 59.7 MPa. This new mesh has 811.379 nodes and 497.064 elements.

Boundary Conditions, Thermal-Structural Analysis. The values of these were previously estimated for a vehicle traveling at a speed of 150 km/h and how the loads and supports are placed for the structural analysis is shown in Fig. 3

Table 3. Automatic Ansys Convergence.

Equivalent stress	Error	Nodes	Elements
52,53		39983	21424
59,42	12,312	256781	159967
59,76	0,5685	1112863	752207

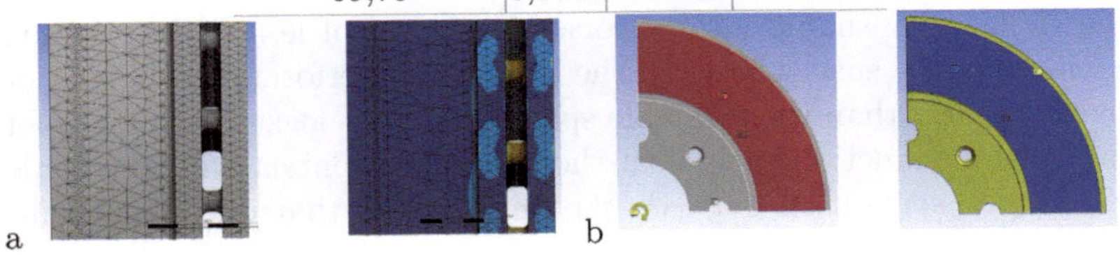

a   b

Fig. 3. (a)Final and refined mesh of the element. (b)Boundary conditions for structural analysis and thermal analysis.

as well as being based on previous studies. For the thermal analysis, the heat flux generated on the disk surface and convection coefficient on all surfaces of the disk periphery, bell, rotary, and ducts previously calculated at the indicated speed are placed as shown in Fig. 3. Table 4. shows the boundary conditions applied to study of brake disc loads and where they should be applied.

Table 4. Boundary conditions of the structural thermal analysis.

Border	Border condition	Parameters
Fixed support	Disc anchor (bolts)	On the disc fastening bolts
Pressure	Brake pad pressure	4.7 MPa
Rotational speed	Rotational speed of the disc	142 rad/s
Convection	Brake disc convection	260 W/m.C
Heat flow	Heat flow in the brake disc	22,419 W/m^2

2.5   Topological Optimization Method

In a shape and size distribution problem, aspects of the structural design problem are addressed, where the objective is to find the optimal thickness of a linearly elastic plate balancing between minimum or maximum distribution and extreme working loads.

Cross-Sectional Shape of the Hole. In the mechanics of materials, there are essential criteria for manufacturing parts. If a part will be subjected to cyclic loads, it is crucial to minimize stress concentrates, rounder or circular shapes contribute to mitigate this effect.

Number of Boreholes. The criterion to be applied for the number of holes is the compliance topology, which states that any change made to the part must ensure compliance without risk of failure.

Location of Boreholes. The perforations should not be radial to avoid friction between them, they should also be located in areas of less stress accumulation which will be in the spaces between the blades, the perforations should not have a diameter greater than the available space leaving at least 2 to 3 mm. per side, this has a double function to prevent the holes have contact with the blades and establish a safe separation between the perforations that does not accumulate stresses.

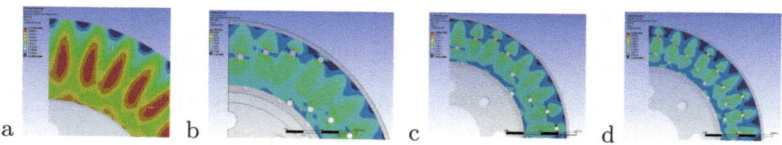

Fig. 4. (a) Stress distribution on the disk surface. (b) Stress distribution model 1. (c) Stress distribution model 2. (d) Stress distribution model 3.

2.6   CFD Analysis with ANSYS FLUENT

CFD uses numerical methods and computational algorithms to analyze and solve problems related to the behavior of fluids, such as gases and liquids. It uses mathematical models to simulate and study fluid dynamic phenomena, such as air flow around an object or the dispersion of substances in a fluid. The tool used for the analysis is ANSYS FLUENT, which allows simulations of rotating machinery, heat exchanges, and turbines, among others.

Boundary Conditions for CFD Simulation. The boundary conditions estimated for the CFD analysis are based on previous studies in which a symmetric brake disk section and a wind speed estimated at the disk rotational speed 142 rad/s, were established. Table 5 shows boundary condition values.

The flow enclosure (enclosure), air flow inlet, and outlet were arranged similarly according to the study mentioned above and are shown in Fig. 5 (a).

Mesh Analysis for CFD Simulation. For the case of meshing the part in computational fluid dynamics, it is advisable to use a fine unstructured mesh near the geometry and a coarser structure where the fluid is to be represented [?,?]. In this way more reliable results will be obtained when the fluid approaches the surfaces in contact with the airflow, the final mesh used in the CFD analysis is shown in Fig. 5 (b)

Table 5. Boundary conditions of the CFD analysis..

Border	Border condition	Parameters
Entrance	Inlet flow velocity	Vehicle speed 150 km/h
Output	Outlet pressure	Atmospheric pressure
Domain Border	Symmetry	Symmetry
Disc surface	Wall	360 K

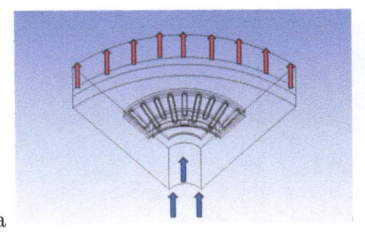 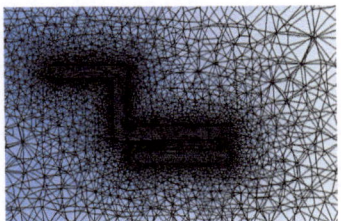

a   b

Fig. 5. (a) Encapsulation for CFD analysis. (b) Mesh used in the CFD analysis.

3   Results

After iterating the process of loading the disc with different drilling diameters, the values of maximum stress on the load application surface, safety factor on the load application surface, maximum temperature in degrees Kelvin at which the disc reaches, and the resulting mass in kilograms after having applied the topology optimization are compiled, the values obtained are shown in Table 6, Fig. 6 (a).

Table 6. Table of topological optimization results.

Brake disc	Temperature[°C]	E. Stress [MPa].	R. mass [g]	SF (x10)	Rt. Mass [kg] (x10)
Original Model	81	21,95	0	73	41,6
Model 1	70.28	60,61	62,52	42	41
Model 2	71.16	66,88	42,34	43	41,2
Model 3	71.15	44,25	47,63	40,9	41,1

E. Stress=Equivalent Stress, R. mass=Reduced mass, SF=Safety factor, Rt Mass=Resulting Mass

Once the new topologies were verified without risk of high structural failure due to holes, their heat dissipation was evaluated. First, a thermal transient analysis is developed, which will allow the observation of the cooling behavior in a controlled manner, applying a heat flux for 10 s and then convection for 50 s, and the cooling is observed only geometrically, since the convention is not changed despite being larger, this to check if only the geometry itself dissipates heat better, as shown in Fig. 6. The heat flow from the CFD analysis of the optimized disks shows an increase in heat dissipation, the holes show a wind flow through the disk at a lower temperature than the disk becoming cooling zones, and around the whole temperature contour a better heat dissipation is shown, since the higher temperature shedding regions are larger in the modified disks, resulting in a better distribution of thermal stresses around the disk as shown in Figure 7.

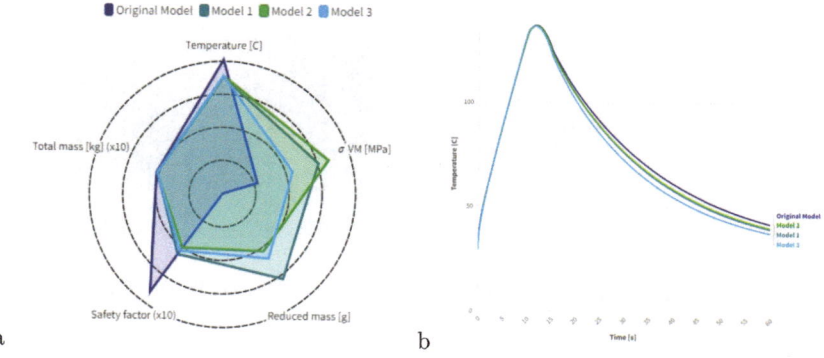

Fig. 6. (a) Comparison of brake disc characteristics (b) Thermal transient cooling values of the models.

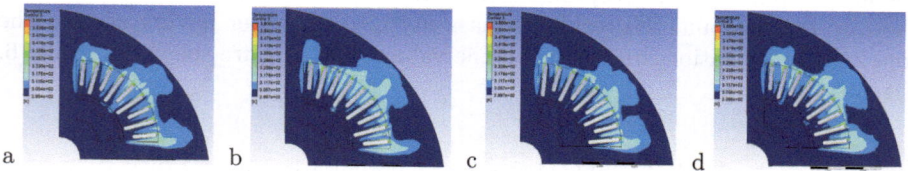

Fig. 7. (a) Temperature contour of the original disk. (b) Temperature contour of model 1. (c) Temperature contour of model 2. (d) Temperature contour of model 3.

4   Discussion

The results revealed that the three models analyzed experienced an increase in stress in the brake pad application zone. However, an improvement in the distribution of these stresses is observed in comparison with the original disc. Model 2 shows the largest increase in stress, reaching 66 MPa. This increase is attributed to the arrangement of its holes, which are close to each other, generating areas of high-stress concentration. On the other hand, model 3 shows a stress of 44 MPa due to the smaller diameter of its holes (3 mm). In contrast, model 1 presents an intermediate value, with a stress of 60 MPa, despite having a larger diameter (4,5 mm). This is because the arrangement of its holes avoids excessive stress concentration. In the CFD analysis, all three models were subjected to the same temperature of 360 K. The temperature contours obtained for each disk indicate a higher heat dissipation than the original disk. However, it is model 1 that shows a larger and more uniform heat dissipation surface area. This is attributed to the size of its perforations, which allow a greater flow of surrounding air through the disk, thus favoring better cooling, as shown in Figure 7 (b).

5   Conclusions

ASTM standards define material properties crucial for brake disc simulation, ensuring they meet regulatory and manufacturer specifications. Thermal and mechanical properties are especially important, influencing component behavior and load resistance. Using ASTM A 48 grade 40 or EN-GJL-250 material was specific to this study. In topology-based design improvements, shape, quantity, and location are key considerations. Circular cross-sections are preferred to reduce stress concentration compared to edged geometries. Element quantity should balance safety factor and grip surface. Element location should avoid stress concentration and maintain vane integrity, limiting holes per abutment to two.

CFD simulation is vital for analyzing thermal behavior in brake discs and visualizing heat dissipation and temperature fluctuations at maximum vehicle speed. Encapsulation or flow re circulation analysis models airflow interaction with the disc, improving simulation accuracy. Symmetric segmentation optimizes computational efficiency for detailed simulations.

Structural and functional integrity are paramount in brake disc design, ensuring safety and functionality. Any improvements should prioritize these aspects, as failure can lead to severe road safety consequences. Once these fundamentals are ensured, other areas for enhancement can be explored.

References

1. N. Highway Traffic Safety Administration and U. Department of Transportation, TRAFFIC SAFETY FACTS Crash - Stats Early Estimate of Motor Vehicle Traffic Fatalities in 2022 (2022)
2. "Accident rate viewer - Statistics - Agencia Nacional de Tránsito del Ecuador - ANT.". Accessed: Nov. 26, 2023. https://www.ant.gob.ec/visor-de-siniestralidad-estadisticas/
3. García-León, R.A., Echavez Díaz, R.D., Flórez Solano, E.: Análisis termodinámico de un disco de freno automotriz con pilares de ventilación tipo NACA 66-209. INGE CUC 14(2), 9–18 (2018). https://doi.org/10.17981/ingecuc.14.2.2018.01
4. Talati, F., Jalalifar, S.: Investigation of heat transfer phenomena in a ventilated disk brake rotor with straight radial rounded vanes. J. Appl. Sci. 8(20), 3583–3592 (2008). https://doi.org/10.3923/jas.2008.3583.3592
5. Gao, C.H., Lin, X.Z.: Transient temperature field analysis of a brake in a non-axisymmetric three-dimensional model. J. Materials Process. Technol. 129(1–3), 513–517 (2002). https://doi.org/10.1016/S0924-0136(02)00622-2

6. Gao, C.H., Huang, J.M., Lin, X.Z., Tang, X.S.: Stress analysis of thermal fatigue fracture of brake disks based on thermomechanical coupling. J. Tribol. **129**(3), 536–543 (2006). https://doi.org/10.1115/1.2736437

7. Grigoratos, T., Martini, G.: Brake wear particle emissions: a review. Environ. Sci. Pollut. Res. **22**(4), 2491–2504 (2015). https://doi.org/10.1007/s11356-014-3696-8

8. Perricone, G., et al.: A concept for reducing PM10 emissions for car brakes by 50%. Wear **396–397**, 135–145 (2018). https://doi.org/10.1016/J.WEAR.2017.06.018

9. Hak, C., Larssen, S., Randall, S., Guerreiro, C., Denby, B., Horálek, J.: Traffic and Air Quality Contribution of Traffic to Urban Air Quality in European Cities (2010). http://air-climate.eionet.europa.eu/

10. Saha, D., Sharma, D., Satapathy, B.K.: Challenges pertaining to particulate matter emission of toxic for-mulations and prospects on using green ingredients for sustainable eco-friendly automotive brake compo-sites. Sustain. Mater. Technol. **37**, e00680 (2023). https://doi.org/10.1016/J.SUSMAT.2023.E00680

11. Huang, Y., Tian, X., Wu, L., Zia, A.A., Liu, T., Li, D.: Progressive concurrent topological optimization with variable fiber orientation and content for 3D printed continuous fiber reinforced polymer composites. Compos. B Eng. **255**, 110602 (2023). https://doi.org/10.1016/J.COMPOSITESB.2023.110602

12. Zheng, Y., Luo, Z., Wang, Y., Li, Z., Qu, J., Zhang, C.: Optimized high thermal insulation by the topological design of hierarchical structures. Int. J. Heat Mass Transf. **186**, 122448 (2022). https://doi.org/10.1016/J.IJHEATMASSTRANSFER.2021.122448

13. Zhang, K., Li, Y., Chang, S.M., Hu, L., Wang, X., Yu, M.: Hydraulic and thermal performance enhancement for the cold plate using topology optimization. Appl. Therm. Eng. **236**, 121829 (2024). https://doi.org/10.1016/J.APPLTHERMALENG.2023.121829

14. Zhu, Z., Bao, J., Yin, Y., Chen, G.: Frictional catastrophe behaviors and mechanisms of brake shoe for mine hoisters during repetitive emergency brakings. Indust. Lubric. Tribol. **65**(4), 245–250 (2013). https://doi.org/10.1108/00368791311331220

15. Hui, Y., et al.: Fading behavior and wear mechanisms of C/C-SiC brake disc during cyclic braking. Wear **526–527**, 204930 (2023). https://doi.org/10.1016/J.WEAR.2023.204930

16. Chin, S.-B., Foo, J.-J., Lai, Y.-L., Yong, T.K.-K.: Forced convective heat transfer enhancement with perforated pin fins. Heat Mass Transfer **49**(10), 1447–1458 (2013). https://doi.org/10.1007/s00231-013-1186-z

17. Lee Jin, M.: Three-Dimensional Analysis on Thermal Stresses in A Perforated Disc Brake of Vehicle. UNI-VERSITI TEKNOLOGI PETRONAS, Perak Darul Ridzuan (2010)

18. Yan, H., Feng, S., Lu, T., Xie, G.: Experimental and numerical study of turbulent flow and enhanced heat transfer by cross-drilled holes in a pin-finned brake disc. Int. J. Thermal Sci. **118**, 355–366 (2017). https://doi.org/10.1016/j.ijthermalsci.2017.04.024

A Model for Electric Vehicle Battery Aging: A Vehicle-to-Grid Perspective

Liliana González Garzón[1]([✉]) [iD], Cesar Diaz-Londono[2] [iD], José Vuelvas[1] [iD],
and Giambattista Gruosso[2] [iD]

[1] Pontificia Universidad Javeriana, Bogotá, Colombia
{gonzalezliliana,vuelvasj}@javeriana.edu.co
[2] Politecnico di Milano, Milan, Italy
{cesar.diaz,giambattista.gruosso}@polimi.it
http://www.springer.com/gp/computer-science/lncs

Abstract. This study presents a comprehensive battery model for electric vehicles (EVs), designed to enhance the understanding and optimization of energy management systems. The model integrates the effects of calendar and cyclic aging, as well as the degradation impacts from Vehicle-to-Grid (V2G) interactions, essential for maintaining battery health and operational efficiency. Through comparative analysis with existing models, new insights into battery behavior under V2G conditions are highlighted. The findings are poised to significantly contribute to the development of sustainable mobility solutions by promoting enhanced battery longevity and performance. Key areas for future research are identified, and potential enhancements are proposed that could advance both practical applications and theoretical understanding in EV energy management. This model serves as a valuable tool for researchers focused on optimizing battery systems, enabling more effective energy strategies for electric vehicles. The study highlights the crucial role of advanced battery models in achieving efficient and sustainable operations of electric vehicles.

Keywords: Battery modeling · Electric vehicles · Battery aging · Cycle aging · Calendar aging · Battery degradation

1 Introduction

The battery is the central component of electric vehicles (EVs), serving as the primary storage unit for the electrical energy required to power the motor [1]. This storage capacity is critical, determining the vehicle's range and directly influencing its overall performance and operational efficiency [2]. Additionally, the battery gives EVs their sustainable character, crucial for mitigating the environmental impact of road transport [3–5]. Recent reports highlight the rapid adoption of EVs, driven by advancements in battery technology and strong policy support [6].

E. M. Inga Ortega et al. (Eds.): CITIS 2024, LNNS 1331, pp. 174–184, 2025.
https://doi.org/10.1007/978-3-031-87065-1_16

Vehicle-to-Grid (V2G) technology further augments EV functionalities, allowing them to draw energy from and feed it back to the grid. This bidirectional flow enhances smart charging benefits [5, 7]. While V2G offers advantages like grid stabilization and emergency power, it also poses challenges in battery degradation. These challenges must be addressed in battery modeling to balance energy provision and battery longevity.

Developing advanced battery technologies is crucial; however, understanding existing battery behavior to optimize use and extend lifespan is equally important [2]. A significant challenge in battery modeling is incorporating both cyclic and calendar aging, which coexist and affect battery performance [8, 9]. Many models do not integrate both aging types and often overlook V2G-induced degradation, frequently focusing on a single battery type, limiting simulations of energy management systems or specific EV strategies.

This paper introduces a comprehensive battery model for EVs, addressing key challenges in energy management. Designed as a realistic tool for researchers, the model incorporates advanced features for performance analysis, including V2G impacts. We compare this model against existing ones, identifying their strengths and limitations, and highlighting areas for future research. The goal is to deepen the understanding of battery behavior, enhancing both theoretical and practical knowledge in the field.

2 Methods

The approach used in this work consists of a systematic review of the literature specialized in battery models for electric vehicles. This methodology is developed in six well-defined stages that allow for an exhaustive and structured evaluation of existing studies (Fig. 1).

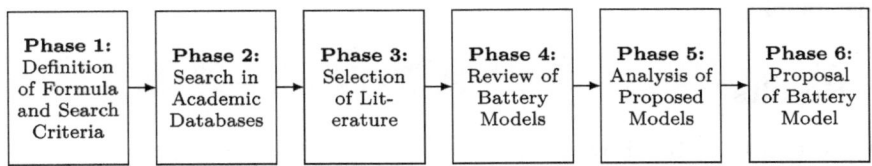

Fig. 1. Block Diagram of the Systematic Review Methodology for Electric Vehicle Battery Models

The first stage involves the Definition of the Formula and Search Criteria. In this initial step, specific parameters and keywords are established to guide the search across various academic databases. An example of the search formula used is: ("Battery Aging" OR "Battery Degradation") AND ("Electric Vehicle*") OR ("V2G" OR "Vehicle to grid") . This formula ensures that only studies specifically addressing the impacts of battery aging and degradation in electric vehicles are collected.

Subsequently, the Search in Academic Databases takes place, specifically in IEEE Xplore and Science Direct. Using the previously defined criteria, an exhaustive search is conducted on these recognized platforms to ensure a broad coverage of relevant literature.

The third stage is the Selection of Literature, is crucial in the systematic review to ensure a comprehensive analysis of battery models for electric vehicles. This meticulous selection was carried out in April 2024, using a specific filter during the database search to include only documents published between 2012 and 2024.

The search yielded 12 unique and detailed references across academic journals, papers, archives, and conference proceedings, focusing on innovative battery models for electric vehicles. This selective process, emphasizing clarity and novelty, prevents redundancy and enriches our understanding of diverse modeling strategies and approaches within the field.

The fourth phase involves Review of Battery Models. Each model is thoroughly documented and key aspects are assessed, including the types of aging, battery types, and the sources of variables from datasheets or suppliers. We also evaluate the models' structures—linear, nonlinear, or MILP—to check their incorporation of V2G functionalities and integration into optimization strategies. This careful examination aids in clearly differentiating and comparing the various battery models.

The penultimate stage consists of the Analysis of Proposed Models, where different models are compared and contrasted to validate which aspects and/or variables are relevant to consider in formulating a battery model for an electric vehicle.

Finally, in the Battery Model Proposal, a battery model for electric vehicles will be defined that integrates both cyclic and calendar aging, as well as additional aging due to the discharge and charge cycles generated by Vehicle-to-Grid (V2G) applications. This model will be explained in detail to ensure complete understanding.

3   Results

3.1   Literature Review

Battery aging refers to the irreversible processes of physical and electrochemical degradation of a battery, typically classified into two types: calendar aging and cyclic aging [9]. Calendar aging refers to the gradual deterioration of battery capacity over time, even in the absence of active use, due to factors such as temperature, exposure to electrolyte, and undesirable chemical reactions that affect energy storage capacity. On the other hand, cyclic aging of batteries occurs due to repeated charging and discharging throughout the battery's lifecycle, which involves physical changes, specifically decomposition in the anode and cathode, and also includes losses of active lithium material due to mechanical stresses in the battery materials [10].

Electric vehicles, particularly domestic models, spend most of their time parked [11], suggesting a greater influence of calendar aging; however, it is important to highlight that both types of aging coexist and interact in the real application of electric vehicles. The introduction of Vehicle-to-Grid (V2G) technologies further complicates this interaction, as the cyclic use of batteries for energy feedback into the grid can accelerate aging processes, necessitating careful consideration in battery models.

Nevertheless, there are few models in the scientific literature that address the overlap of these two types of aging along with the additional stresses imposed by V2G applications.

In this section of the document, we conduct a detailed review of the existing literature regarding the type of battery model used, including considerations such as the model's approach to specific coefficients related to battery type, cyclic and calendar aging, and the impact of V2G on aging (see Table 1).

Table 1. Comparison of the electric vehicle battery models proposed in works related to the proposed work

Reference	Model Type	Type of Aging	Battery	Temp.	Coeff.	V2G
[11]	Nonlinear	Cycle and Calendar	Lithium-ion	Yes	Yes	Yes
[12]	Nonlinear	Cycle (Considers SoC and DoD)	Lithium-ion	No	No	No
[13]	MILP	Cycle and focus on Calendar	-	Yes	No	No
[14]	Linear	Cycle	-	Yes	No	Yes
[15]	Linear	Cycle	-	No	No	No
[16]	Nonlinear	Cycle and Calendar	Lithium-ion	Yes	No	No
[17]	Nonlinear	Cycle and Calendar	Lithium-ion	Yes	No	No
[18]	Nonlinear	Cycle and Calendar	Lithium-ion	Yes	No	No
[19]	-	Real cases	Lithium-ion	Yes	No	No
[20]	Nonlinear	Cycle and Calendar	Lithium-ion	Yes	No	Yes
[21]	Nonlinear	Cycle and Calendar	Lithium-ion	Yes	No	Yes
[22]	Nonlinear	Cycle and Calendar	Lithium-ion	Yes	No	No
Model proposal	Nonlinear	Cycle and Calendar	All types	Yes	Yes	Yes

In reviewing the literature on battery models for electric vehicles, we identified diverse approaches, ranging from linear to nonlinear models, aimed at characterizing battery aging ([11,13,16–18,20,21]). Some models focus on calendar and cyclic aging ([11,13,16]), while others incorporate charge optimization strategies or battery management systems that consider aging ([12,21]). Further, several models introduce aging-related penalties to reflect cost or performance degradation over time ([11–13,15,21]). However, limitations arise from highly specific parameters, like temperature and datasheet-derived coefficients, which constrain their broader applicability.

Models such as [11,13], and [16] offer a comprehensive approach by addressing both cyclic and calendar aging, making them particularly valuable for applications involving a mix of frequent usage and periods of inactivity, such as storage systems or bidirectional interactions. Their holistic approach provides a robust framework for analyzing degradation across various use cases.

Conversely, models like [12,15] are primarily focused on cyclic aging under uncertain operating conditions and optimizing charging for Home Energy Management Systems (HEMS), respectively. While model [12] evaluates the impact of efficiency and costs on cyclic aging, the proposal [15] explores the efficient management of household energy. However, both models are somewhat limited by their lack of detailed analysis on calendar aging, which is an inevitable aspect of all batteries.

The model [18] incorporate advanced simulation that integrate multiple physical disciplines, such as thermodynamics and mechanics. For instance, the work [18] employs a multi-physics approach to delve into the nuances of battery aging. These models excel in accurately representing battery responses under various stress conditions, though they primarily do not focus on bidirectional operations.

The approach [14] capitalizes on the energy storage capabilities of electric vehicles for vehicle-to-building (V2B) applications, promoting smart energy management strategies that can significantly enhance the integration of renewable energy sources and improve building energy efficiency. However, it primarily focuses on cyclic aging and does not provide an in-depth analysis of calendar aging.

The [17] model integrates both cyclic and calendar aging with a special emphasis on thermal management. It leverages MATLAB-Simulink for sophisticated simulations, ensuring that the effects of battery aging are well-characterized under various operational conditions.

Model [22] introduces two innovative nonlinear mathematical models for lithium-ion batteries in electric vehicles. The first model effectively estimates the state of charge (SOC) based on temperature data, offering enhanced precision under dynamic conditions. The second model goes beyond the traditional Thevenin model by replacing its RC network with a battery transfer function that takes temperature and current into account. This model has demonstrated improved performance over empirical models based on the Nernst equation, as validated by simulation results using the mean square error (MSE) index.

Finally, [20,21] address the bidirectional interaction between electric vehicles and the power grid. These models evaluate various applications like V2G, V2H, and V2B to stabilize the grid and optimize charging management in V2G scenarios, highlighting the necessity for models that integrate both cyclic and calendar aging to optimize battery longevity and mitigate degradation effectively.

3.2   Model Proposal

In this section, the proposed model for evaluating battery aging in electric vehicles is detailed, incorporating considerations of cyclic and calendar aging, as well as additional were induced by Vehicle-to-Grid (V2G) interactions. This model builds upon the principles set forth in previously mentioned models, such as Model [11] and Model [21], and is characterized by its comprehensive approach that addresses the multifaceted mechanisms of degradation subjected to batteries under conventional usage conditions and dynamic bidirectional charging operations. This section will elaborate on each type of aging, presenting the formulas utilized and meticulously defining the proposed model, thereby ensuring it addresses the specific needs and challenges associated with contemporary battery applications in electric vehicles.

Calendar Aging. Calendar aging focuses on the degradation that occurs over time, regardless of the number of charge and discharge cycles. This type of aging is particularly sensitive to temperature and the duration of inactivity.

$$C_{\text{cal}} = A \cdot e^{-\frac{E_a}{R \cdot T}} \cdot t \tag{1}$$

- A (Pre-exponential factor): Reflects the battery's sensitivity to non-cyclic aging. This value is specific to each battery type and composition
- E_a (Activation energy): Indicates the energy required for the chemical reactions that lead to battery aging. Appropriate E_a values for different battery types under operational conditions can be found in the table below.
- R (Universal gas constant): A constant value used in thermodynamic equations, R= 8.3145 the universal gas constant in [J/mol/°C]
- T (Absolute temperature in Kelvin): Directly affects the rate at which the chemical reactions of degradation occur, $T(K)= T(C\bar{o})+273.15$ in [°C]
- t (Time in years): The duration of the battery's inactivity or storage period.

Table 2. Energies of activation for various battery types under different charge/discharge conditions. **CC/20**: Charge/discharge completed in 20 h. **CC/2**: Charge/discharge completed in 2 h. **C1C**: Charge/discharge completed in 1 h, indicating a high-intensity rate. Information taken from [23]

Battery Type	Ea (kJ/mol)	Conditions
Li(NiMnCo)O2 based 18650 lithium-ion batteries	58.0	CC/20
LiNi$_{1/3}$Mn$_{1/3}$Co$_{1/3}$O$_2$	43.60	CC/20
Li(NiMnCo)O2 based 18650 lithium-ion batteries	≈55	CC/20
Lithium-ion LiFePO4 / Carbon	≈69.6	C1C
LiNi$_{1/3}$Mn$_{1/3}$Co$_{1/3}$O$_2$	>28	C1C
Lithium ion (Graphite/NMC + Spinel Mn oxide)	24.5	CC/2
LiFePO4 / Carbon	≈86	C1C
Lithium Iron Phosphate (LiFePO$_4$)	20.6	C1C

Cyclic Aging. This type of aging is caused by charge and discharge cycles, affecting the battery's capacity and internal resistance (Table 2).

$$C_{\text{cyc}} = k_{\text{cyc}} \cdot N \tag{2}$$

- k_{cyc} (Cyclic degradation constant): Re

$$C_{\text{cyc}} = k_{\text{cyc}} \cdot N \tag{3}$$

th of discharge and usage conditions
- N (Total number of charge and discharge cycles): The number of complete cycles the battery has undergone.

Additional Degradation Due to V2G. Additional degradation due to V2G addresses the wear caused by charge and discharge operations in V2G contexts, where the battery is used not only for driving but also for grid energy management. The additional degradation due to V2G operations is treated as a distinct component and not implicitly included in the cyclic aging model. This distinction arises from the unique stress factors associated with V2G interactions that extend beyond the conventional charge and discharge cycles of regular vehicle use. V2G operations often involve more frequent and variable depth of discharge (DoD) cycles, which can accelerate battery degradation in ways not accounted for by typical cyclic aging patterns. Additionally, V2G requires bidirectional energy flows, introducing additional thermal and electrical stress that impacts the battery's internal resistance and capacity. These factors necessitate a separate consideration to accurately capture the degradation mechanisms specific to V2G operations. By isolating V2G-induced degradation, the model proposal can more precisely predict the impacts of these operations on battery longevity, thereby enabling more effective optimization and management of battery health in both standard and V2G applications.

$$C_{\text{V2G}} = \delta \cdot (DOD_{\text{V2G}})^{\alpha} \cdot N_{\text{V2G}} \tag{4}$$

- δ (V2G degradation coefficient): Reflects the severity of V2G usage on degradation, based on empirical studies or simulations.
- DOD_{V2G} (Depth of Discharge for V2G cycles): The level to which the battery is discharged during a V2G cycle.
- N_{V2G} (Number of V2G charge/discharge cycles): The number of times the battery has been used in V2G applications.
- α (Exponent): Modifies the impact of the depth of discharge on degradation, adjustable based on empirical data

Proposed Model for Evaluating Battery Aging in Electric Vehicles

$$C_{\text{total}} = C_{\text{cal}} + C_{\text{cyc}} + C_{\text{V2G}} \tag{5}$$

where:

- C_{cal} is calendar aging
- C_{cyc} is cycling aging
- C_{V2G} is the additional degradation due to V2G

3.3 Model Simulation Results

The degradation model considers three main factors contributing to battery aging: calendar aging (C_{cal}), cyclic aging (C_{cyc}), and additional degradation due to Vehicle-to-Grid (V2G) interactions (C_{V2G}). Each component is distinctly influenced by parameters such as the activation energy (E_a), the depth of discharge (DOD), and the number of cycles (N), tailored to V2G applications.

The graphical representation of the simulation depicts the projected degradation over a span of five years for various types of batteries. These results were derived from empirical data and theoretical calculations, employing specific E_a, DOD_{V2G}, and N_{V2G} values for different battery chemistries. As illustrated, batteries with higher E_a and DOD_{V2G} exhibit slower rates of degradation, underscoring their potential resilience and suitability for V2G applications. This graph emphasizes the differential impact of operational strategies in V2G on battery longevity, reinforcing the need for targeted management approaches for diverse battery technologies (Fig. 2).

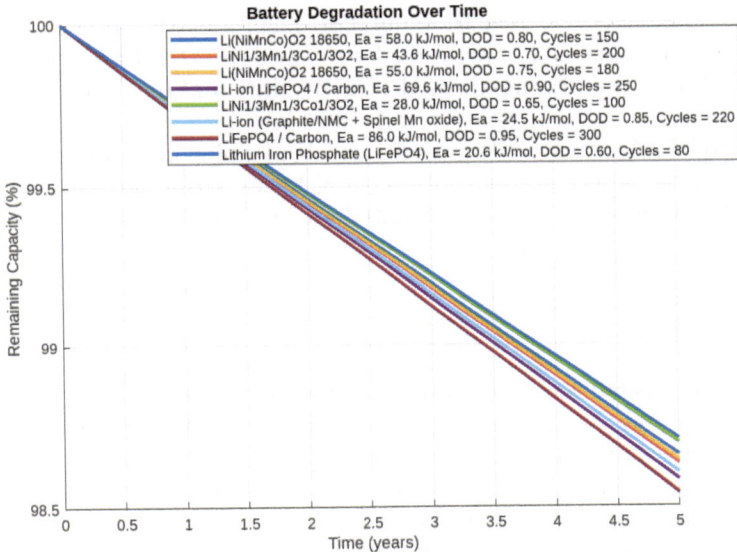

Fig. 2. Simulated battery degradation over five years illustrating the impact of various operational parameters on different battery chemistries.

To reproduce the experiment and its methodology, the following constant values were used based on experimental data and literature: the universal gas constant $R = 8.3145$ J/mol/K, the temperature $T = 298.15$ K ($25°C$), the cyclic degradation coefficient $k_{cyc} = 0.0002$ per cycle, and the activation energy E_a specific to each battery type as per previous studies. The V2G degradation coefficient $\delta = 0.00015$ was derived from empirical studies. This information was crucial for conducting the simulation.

3.4   Applications of the Model Proposal

This model provides a robust framework for predicting battery life, enabling engineers and fleet managers to optimize maintenance and operational strategies effectively. By incorporating real-world conditions and V2G interactions, the model not only enhances the accuracy of degradation predictions but also supports strategic planning for the deployment and management of battery systems in electric vehicles. It is particularly valuable for those involved in the development and integration of electric mobility and renewable energy solutions, ensuring that battery health is maintained under varied operational stresses.

This proposed model, drawing upon the foundational principles from models [11, 21], offers a comprehensive assessment of battery aging in electric vehicles, integrating both calendar and cyclic aging processes along with additional degradation induced by V2G applications. The unified equation presented provides engineers and fleet managers with a framework that allows for the prediction and management of battery life more effectively under various operational and environmental conditions.

This model not only delivers precise estimations of degradation under normal operations and V2G scenarios but also facilitates the implementation of optimized maintenance and operational strategies to extend the durability and efficiency of batteries in contexts of electric mobility and renewable energy integration.

This model is ideal for analysis and strategic planning in the development and management of battery systems in electric vehicles, ensuring that both typical usage and V2G interactions are accounted for in a cohesive and comprehensive manner.

4   Discussion

This paper presented a computationally efficient model that incorporates calendar aging, cyclic aging, and degradation from Vehicle-to-Grid (V2G) interactions, making it versatile for simulations across various battery types. This adaptability is especially beneficial for Energy Management Systems (EMS) needing quick and dynamic analyses.

The model's approach allowed for rapid scenario testing on different electric vehicle (EV) platforms without extensive resources. By including specific degradation mechanisms linked to V2G operations, it facilitates understanding EV batteries' dual role as transportation power sources and grid energy storage units.

Future work will refine degradation coefficients to improve the model's accuracy and expand it to encompass a wider range of battery chemistries and configurations, enhancing its utility in predicting battery life across various EVs and storage solutions. Assessing the economic impacts of battery degradation within V2G systems is also planned, focusing on the cost-effectiveness of deploying different battery technologies in grid-connected vehicles. This will provide deeper insights into their financial viability and environmental benefits.

The model offers valuable insights for stakeholders aiming to develop effective EV deployment strategies within energy systems, promoting sustainable energy management practices. Its simplicity and flexibility make it an ideal tool for strategic planning and optimization in the rapidly evolving fields of electric mobility and renewable energy integration.

References

1. U. D. of Energy, Alternative fuels data center: How do all-electric cars work? 2020
2. I. R. E. Agency, Watts to wheels: Why ev-battery innovation is key to sparking a renewable revolution. 8 (2023)
3. I. E. Agency, Global ev outlook 2023: Catching up with climate ambitions (2023)
4. I. E. Agency, Electrification - energy system (2023)
5. I. E. Agency, Electric vehicles (2023)
6. I. E. Agency, Global ev outlook 2024 moving towards increased affordability (2024)
7. I. E. A. IEA, Global ev outlook 2022 securing supplies for an electric future (2022)
8. Tremblay, O., Dessaint, L.-A.: Experimental validation of a battery dynamic model for EV applications. World Electr. Veh. J. **3**(2), 289–298 (2009). https://doi.org/10.3390/wevj3020289
9. Redondo-Iglesias, E., Venet, P., Pelissier, S.: Calendar and cycling ageing combination of batteries in electric vehicles. Microelectron. Reliab. **88–90**, 1212–1215 (2018). https://doi.org/10.1016/j.microrel.2018.06.113
10. McBrayer, J.D., Harrison, K.L., Allcorn, E., Minteer, S.D.: Chemical contributions to silicon anode calendar aging are dominant over mechanical contributions. Front. Batteries Electrochem. **2** (2023). https://doi.org/10.3389/fbael.2023.1308127
11. Lee, C.F., Bjurek, K., Hagman, V., Li, Y., Zou, C.: Vehicle-to-grid optimization considering battery aging. IFAC-PapersOnLine **56**(2), 6624–6629 (2023). https://doi.org/10.1016/j.ifacol.2023.10.362
12. Jaworski, J., Zheng, N., Preindl, M., Xu, B.: Vehicle-to-grid fleet service provision considering nonlinear battery behaviors 1 2023
13. Majewski, K., Seydenschwanz, M., Gottschalk, C., Weiland, S.: Linear approximation of calendar battery aging costs for milp-based power dispatch optimization. IEEE PES Innovative Smart Grid Technologies Conference Europe (2023)
14. Nazari, S., Borrelli, F., Stefanopoulou, A.: Electric Vehicles for Smart Buildings: A Survey on Applications, Energy Management Methods, and Battery Degradation. Proc. IEEE **109**(6), 1128–1144 (2021). https://doi.org/10.1109/JPROC.2020.3038585
15. Abdelaal, G., Gilany, M.I., Elshahed, M., Sharaf, H.M., El'Gharably, A.: Integration of electric vehicles in home energy management considering urgent charging and battery degradation. IEEE Access **9**, 47713–47730 (2021)
16. Guenther, C., Schott, B., Hennings, W., Waldowski, P., Danzer, M.A.: Model-based investigation of electric vehicle battery aging by means of vehicle-to-grid scenario simulations. J. Power Sources **239**, 604–610 (2013). https://doi.org/10.1016/j.jpowsour.2013.02.041
17. Sarasketa-Zabala, E., Martinez-Laserna, E., Berecibar, M., Gandiaga, I., Rodriguez-Martinez, L.M., Villarreal, I.: Realistic lifetime prediction approach for Li-ion batteries. Appl. Energy **162**, 839–852 (2016). https://doi.org/10.1016/j.apenergy.2015.10.115
18. Wang, Y., et al.: Parameter sensitivity analysis of a multi-physics coupling aging model of lithium-ion batteries. Electrochim. Acta **477**, 143811 (2024). https://doi.org/10.1016/j.electacta.2024.143811
19. Bamdezh, M.A., Molaeimanesh, G.R.: Aging behavior of an electric vehicle battery system considering real drive conditions. Energy Convers. Manage. **304**, 118213 (2024). https://doi.org/10.1016/j.enconman.2024.118213
20. Adegbohun, F., von Jouanne, A., Agamloh, E., Yokochi, A.: A review of bidirectional charging grid support applications and battery degradation considerations. Energies **17**(6), 1320 (2024). https://doi.org/10.3390/en17061320

21. Schwenk, K., Meisenbacher, S., Briegel, B., Harr, T., Hagenmeyer, V., Mikut, R.: Integrating battery aging in the optimization for bidirectional charging of electric vehicles. 11 (2021)
22. Valladolid, J.D., Patiño, D., Ortiz, J.P., Minchala, I., Gruosso, G.: Proposal for modeling electric vehicle battery using experimental data and considering temperature effects. IEEE Milan PowerTech. PowerTech **2019**(6), 2019 (2019)
23. Werner, D., Paarmann, S., Wetzel, T.: Calendar aging of li-ion cells—experimental investigation and empirical correlation. Batteries **7**(2), 28 (2021). https://doi.org/10.3390/batteries7020028

Developing a Methodology to Reduce Fuel Consumption and Classify Driving Styles for a Fleet of Vehicles

Mauro Batallas and Paúl Molina[✉][iD]

Grupo de Investigación en Ingeniería Del Transporte (GIIT),
Universidad Politécnica Salesiana, Quito 170146, Ecuador
mbatallas@est.ups.edu.ec, pmolinac1@ups.edu.ec

Abstract. Driving patterns have a strong impact on fuel consumption and pollutant emission rates. Optimizing these factors can lead to more efficient and environmentally friendly driving. This study aims to categorize drivers into normal, timid, and aggressive types in order to estimate the differences in instantaneous fuel consumption. The data was obtained using a data acquisition device connected to the OBD port to extract the data identifier parameters of four vehicles monitored for 15 consecutive days, which belong to a service vehicle fleet. Through the implementation of a regression tree, we can estimate fuel consumption using the most important predictors. This method yields an accuracy of 99.7%, recall of 99.9%, and F1-score of 99.8%. Mathematical optimization is used to obtain the speed of the vehicle, accelerator pedal position, and engine speed that minimize fuel consumption for each gear. This study demonstrates that employing a normal driving style can lead to an average fuel flow of 3,015 $\cdot 10^{-4}$ liters per second, which is significantly different from the fuel flow of 1,832 $\cdot 10^{-3}$ liters per second associated with an aggressive driving style.

Keywords: VSP · Fuel efficiency · Fuel consumption · Driving behavior

1 Introduction

Global air quality has deteriorated due to the emission of particulate matter and gases resulting from industrialization and urbanization. This global trend, driven by transport growth and increased consumption of fossil fuels, represents a significant health problem [12]. In a 2016 study conducted by the World Health Organization (WHO), it was found that more than a quarter of cities in Latin America exceed the pollution limits considered harmful by two or three times. In Ecuador, cities such as Quito have 18 μ g/m^3, Santo Domingo μ g/m^3, and Milagro 32 μ g/m^3 of PM, which is above the national limit of 15 μ g/m^3. In Ecuador, many vehicles lack modern technology and run on low-quality fuels with high

Supported by Universidad Politécnica Salesiana.

© The Author(s) 2025
E. M. Inga Ortega et al. (Eds.): CITIS 2024, LNNS 1331, pp. 185–194, 2025.
https://doi.org/10.1007/978-3-031-87065-1_17

levels of sulfur, which can reach up to 500 ppm, while the international standard is 10 ppm according to regulations such as Euro 6 [9]. When operating a vehicle, the behavior of the driver affects the driving style, which is mainly determined by acceleration and abrupt speed changes. These variations in driving styles can significantly impact fuel consumption and the environment by affecting pollutant emissions. Therefore, it is important to recognize these driving styles and assess pollutant emissions through actual on-road tests. This process can help establish a series of steps to encourage an efficient and environmentally friendly driving style [1]. The highest potential for applying eco-driving policies seems to be in areas with moderate levels of traffic and little impact from geographic factors such as altitude or variation in road gradients. However, most studies focus on analyzing the impact of eco-driving in large cities with severe traffic problems, such as the Metropolitan District of Quito (DMQ). These metropolitan cities are characterized by fast-paced driving, with little regard for fuel consumption and environmental impact. Efficient driving makes it possible to take advantage of current vehicle technologies, achieving savings of up to 15 % of the fuel and CO_2 emissions to the atmosphere [5].

Understanding the characteristics and driving patterns in vehicles is an open problem to study, so authors such as Yuan et al. analyzed 160 vehicles with spark ignition engines on eight routes, using the Vehicle Specific Power (VSP) parameter to estimate emissions of CO_2, CO, HC and NO_x, finding potential reductions of between 6–40% [14]. In another study, Xu et al. evaluate driver behavior using GPS-collected data. Driving scores are determined based on the vehicle's CO_2 emissions and energy consumption [13]. Muslim et al. investigated the characteristics associated with driver behavior and their relationship to fuel consumption and found a strong correlation between driver personality and energy consumption demand [8]. The main goal of this study focuses on developing an eco-driving approach for a specific fleet of vehicles that encourages changes in driver behavior to reduce costs associated with fuel consumption and mitigate the environmental impact of mobile emissions. Recommendations include efficient driving patterns, smooth acceleration and deceleration, maintenance of optimal speeds, and strategic use of vehicle inertia [6].

The structure of this article is as follows: Sect. 2 describes the methodologies and materials employed from the experimental phase through to the selection of predictors. Section 3 details the results obtained by post-processing the data collected from the vehicles. Finally, Sect. 4 presents the key findings and conclusions of the study.

2   Methods and Materials

2.1   Test Vehicles

This study focuses on analyzing the driving patterns of a fleet of Chevrolet Sail sedan vehicles in the city of Quito, Ecuador. The fleet consists of 20 units used for transporting company employees. It is important to note that all vehicles

Table 1. Technical characteristics of Chevrolet Sail

Parameter	Value
Displacement	1485 cm^3
Power	109 HP
Torque	141 Nm

correspond to the same year and model and use fuel with an octane rating of RON 85. The vehicle characteristics are detailed in the Table 1.

Four units are randomly selected and monitored for 15 consecutive days. The drivers of each vehicle replicate their usual driving conditions and are not predisposed to any characteristic way of driving. After acquiring the data, it goes through a filtering stage where missing values are filled through linearization. Subsequently, an outlier removal stage is performed to smooth the signals. A correlation analysis then extracts the most important variables to train a model capable of predicting fuel consumption. Based on the fuel consumption, these values are optimized for each of the different gears of the vehicle. Finally, with these results, the drivers are categorized into timid, normal, and aggressive.

2.2 Data Acquisition

In order to gather the information, a device called Freematics ONE + was utilized. This device is connected to the OBD port without needing to add mass to the vehicle, making the connection non-intrusive. The device collects identifier parameters (PID) and records information such as speed (VSS) and accelerator pedal position (TPS), which are shown in the Table 2. Latitude and longitude data are stored by integrating a GPS antenna into the data logger. It should be noted that the routes in this study are not predefined; these trajectories correspond to the normal planning of the company's daily work.

Table 2. Acquired parameters

Parameter	Symbol	Unit
Throttle pedal position	TPS	%
Engine speed	RPM	1/min
Vehicle speed	VSS	km/h
Latitud	Lat	°
Longitud	Lon	°
Air temperature	IAT	°C
Intake manifold absolute pressure	MAP	KPa

Road Tests. The data was collected offline over a 15-day period, resulting in the acquisition of approximately 4.5 million data points. During this time, the vehicles traveled to their regular destinations within the DMQ. Figure 1 shows the routes and speeds of the four study vehicles on each map

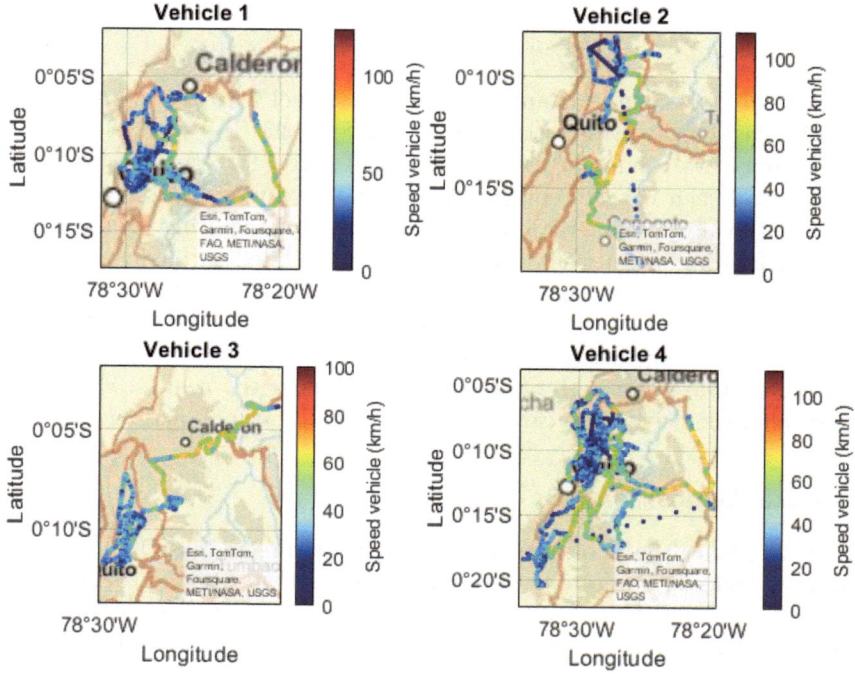

Fig. 1. Road tests

2.3   Fuel Consumption Estimation

The fuel consumption can be estimated through the ideal gas equation of the gases [2,7]. Where P represents the absolute pressure, V is the volume of the cylinder in cubic meters, R is the ideal gas constant, n represents the number of moles, and T is the ambient temperature, obtained from the IAT sensor.

$$PV = nRT \tag{1}$$

Once the number of moles has been calculated, an estimate of the mass of air can be applied using the molecular weight of the air. Then, using the volumetric efficiency of the engine, the airflow through the intake manifold and into the engine can be calculated using Eq. 2.

$$\dot{m} = \frac{PV}{RT} Mair \cdot \eta_{vol} \frac{rpm}{2 \cdot 60} \tag{2}$$

Finally, the fuel flow is calculated from the divided air mass flow rate for the stoichiometric mixture [3].

$$\dot{V} = \frac{\dot{m}}{AFR \cdot \rho_{fuel}} \tag{3}$$

2.4 Vehicle Specific Power Calculation

Vehicle Specific Power (VSP) is a parameter used to estimate driver behavior [10]. This variable describes all the resistive forces that the vehicle has to overcome to move forward, multiplied by the speed, and then divided by the mass of the vehicle. The calculation can be simplified as shown in Eq. 4 [4].

$$VSP = v \cdot (1.1 \cdot a + 0.132) + 0.000302 \cdot v^3 \tag{4}$$

According to the MOVES emission inventory, this parameter can be divided into 37 bins. Bin0 represents braking conditions with accelerations less than $0.89 \ m/s^2$, Bin1 represents idle conditions. Bin2 to Bin13 are considered low-stress conditions while Bin14 to Bin25 are considered medium-stress conditions and the remaining bins are considered high-stress conditions. Figure 2 shows the status of each of the different vehicles characterized by long stopping times in traffic. The second position indicates a lower number of speeds between 40 and 80 km/h, with very little time spent at speeds above 80 km/h.

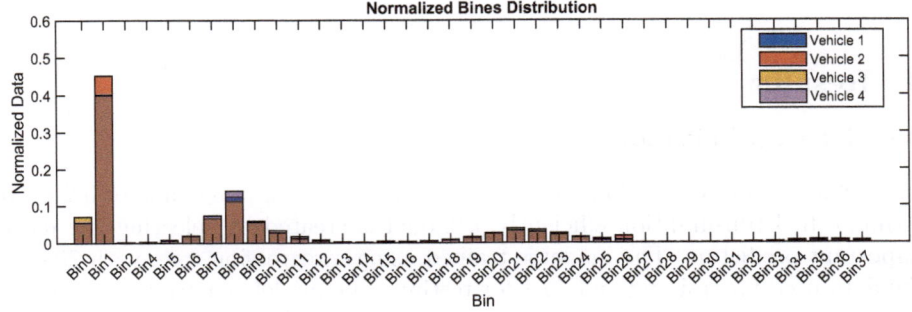

Fig. 2. Bin distribution according to the MOVES inventory

2.5 Exploratory Data Analysis

A correlation analysis is one way of verifying the relationship between two variables. For this reason, a Pearson correlation between the different variables obtained through the OBD port is depicted in Fig. 3. Additionally, variables such as acceleration (A), which represents the rate of change of speed over time, and the VSP parameter, are combined. From this analysis, it is evident that the key variables for estimating fuel flow are: VSS, RPM, and TPS, which is consistent with the conducted study in [11].

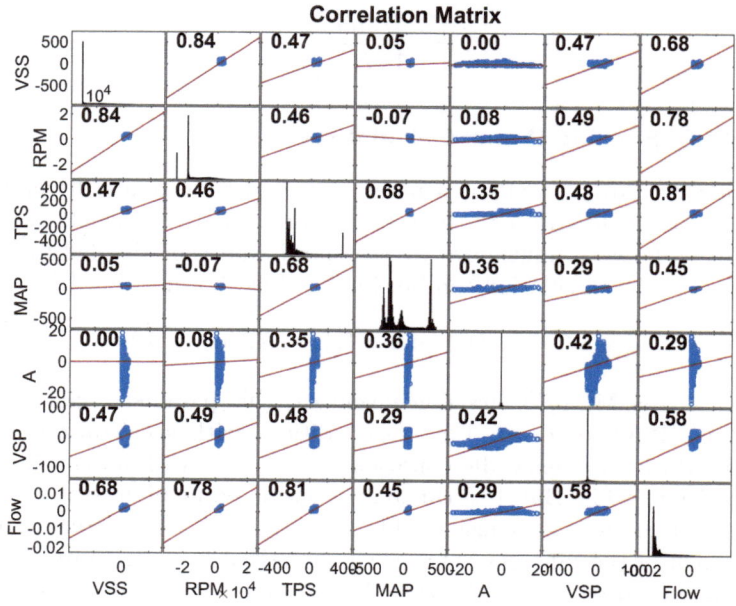

Fig. 3. Pearson correlation coefficient

3 Results

3.1 Fuel Optimization

The vehicles store approximately 25 h of data with a sampling time of 0.2 s. To estimate fuel consumption efficiently, a model is created by selecting the most important predictors. The model is trained using a regression tree, with 70% of the data used for training and 30% for testing. The model achieved an accuracy of 99.2% and an RMSE of $9.266 \cdot 10^{-5}$.

Decision Tree. The gearbox configuration of the study vehicle is manual, necessitating the separation of data according to the selected gear. To predict the gear, the linear relationship between vehicle speed and engine speed is used as follows $r = VSS/RPM$. In the Fig. 4 the vehicle displays different steps, starting from 0 for neutral to 5 for the fifth gear.

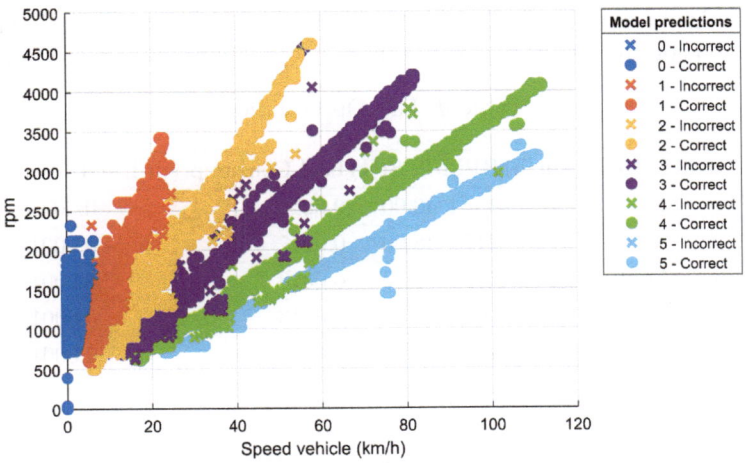

Fig. 4. Classification of vehicle gears

In Table 3 the number of observations between the true and predicted classes can be observed. An accuracy of 99.7% obtained among true positives compared to the sum of true positives and false positives, recall of 99.9% and 99.8% of F1-score.

Table 3. Confusion matrix (True Class - Predicted Class)

True/Predicted	N. Gear	1st Gear	2nd Gear	3rd Gear	4th Gear	5th Gear
Neutral Gear	32357	40	0	0	0	0
First Gear	21	10043	84	0	0	0
Second Gear	78	0	13587	125	0	0
Third Gear	0	116	0	29195	57	0
Fourth Gear	0	0	222	0	19372	11
Fifth Gear	0	0	118	0	0	13515

Minimization of Fuel Consumption. To carry out the fuel consumption optimization process, we start by defining a cost function given by: $flow_{comb} = f(VSS, TPS, RPM)$, then we look for the values that minimize this function, which can be expressed as follows:

$$\min_{VSS, TPS, RPM} = f(VSS, TPS, RPM) \tag{5}$$

These values should adhere to the physical constraints of the vehicle. For example, the accelerator pedal should be set between 0 and 100%. The maximum speed achieved during the tests was 120 km/h, with the engine speed maintained between 0 and 6000 rpm. Subsequently, the values obtained in Table 4 were optimized using the nonlinear optimization function of the Matlab fmincon software considering a tolerance value of $1 \cdot 10^{-6}$ y 1000 maximum stop iterations.

Table 4. Fuel Optimization

Gear	Velocity(km/h)	RPM	TPS(%)	Fuel Consumption(l/s)	RMSE
1	13.70	1954	48.00	0.0016	0.0001
2	34.88	2178	48.00	0.0017	0.0001
3	57.29	2100	47.86	0.0018	0.0001
4	69.30	2051	47.86	0.0013	0.0001
5	69.06	1719	48.50	0.0015	0.0001

3.2 Classification of Driving Styles

The fuel consumption of a vehicle depends on the driver's driving style. Therefore, it is important to categorize drivers into three groups for this study. Shy drivers could increase their driving speed to reduce travel times and maintain constant speeds. Normal drivers would serve as the model to replicate, and aggressive drivers, whose abrupt accelerations and decelerations would need to be decreased. The process starts with normalizing the data, followed by applying a k-means algorithm to divide the information into three distinct groups. Next, by using PCA, we aim to reduce the dimensions to decrease the number of variables in the set without losing information. The visualization of the classification can be seen in the Fig. 5.

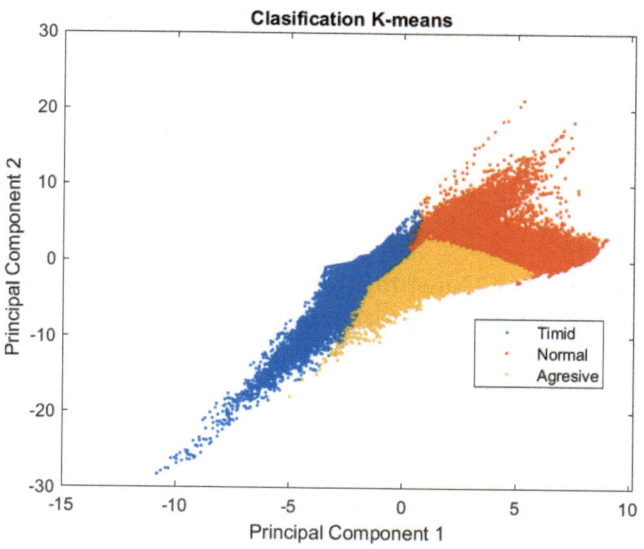

Fig. 5. Vehicle classification model

Finally, after identifying the centroids of each group that captures driving characteristics, a correlation between driving style and fuel consumption is established, as shown in the Table 5. From this section, it is evident that aggressive driving results in higher energy consumption compared to the other groups.

Table 5. Driving style

Driving Style	Average Flow (l/s)
Tímid	$9.271x10^{-4}$
Normal	$3.015x10^{-4}$
Aggressive	$1.832x10^{-3}$

4   Conclusions

Vehicle fuel consumption is impacted by various factors, such as route selection, and driving habits. As can be seen in this study, a normal driving style can represent a fuel flow for this vehicle fleet of $3.015 \cdot 10^{-4}$ l/s, while an aggressive one of $1.832 \cdot 10^{-3}$ l/s representing several orders of magnitude difference.

Decision trees were employed to categorize the various gears of the vehicle due to the absence of a sensor for predicting this condition. The model achieved a classification accuracy of 99.7%, a recall of 99.9%, and an F1-score of 99.8%, facilitating accurate estimations for a larger number of vehicles within the fleet.

The driving style of a fleet driver is important because it allows us to categorize the different drivers within an organization. This means that aggressive drivers must focus on more progressive starts and braking. In contrast, timid drivers can increase their driving speed to maintain a constant speed and avoid continuously braking, caused by the inertial resistance that comes with every start in a vehicle.

The operating intervals for each of the vehicle's gears were developed using mathematical optimization techniques. A tolerance value of $1x10^{-6}$ and a maximum of 1000 iterations were considered. In future work, these values can be used as driving parameters for this fleet of vehicles to measure the real-world variation in fuel consumption.

References

1. Alam, M.S., McNabola, A.: A critical review and assessment of eco-driving policy & technology: Benefits & limitations. Transp. Policy **35**, 42–49 (2014)
2. Andrade, P., Silva, I., Silva, M., Flores, T., Cassiano, J., Costa, D.G.: A tinyml soft-sensor approach for low-cost detection and monitoring of vehicular emissions. Sensors **22**(10), 3838 (2022)
3. Campoverde, P.M., Benavides, K., Montenegro, F., Molina, J.: Fuel consumption analysis of an mpi engine by varying fuel type, fuel filtering, and air filter employing a full-factor analysis. In: 2023 IEEE Seventh Ecuador Technical Chapters Meeting (ECTM), pp. 1–6. IEEE (2023)

4. He, W., et al.: Analysis of real-world fuel consumption characteristics of heavy-duty commercial diesel vehicle based on obd method. In: IOP Conference Series: Materials Science and Engineering. vol. 774, p. 012137. IOP Publishing (2020)
5. Ho, S.H., Wong, Y.D., Chang, V.: What can eco-driving do for sustainable road transport? perspectives from a city (singapore) eco-driving programme. Sustain. Urban Areas **14**, 82–88 (2015)
6. Mensing, F., Bideaux, E., Trigui, R., Ribet, J., Jeanneret, B.: Eco-driving: An economic or ecologic driving style? Transport. Res. Part C: Emerg. Technol. **38**, 110–121 (2014)
7. Meseguer, J.E., Toh, C.K., Calafate, C.T., Cano, J.C., Manzoni, P.: Drivingstyles: a mobile platform for driving styles and fuel consumption characterization. J. Commun. Netw. **19**(2), 162–168 (2017)
8. Muslim, N.H., Keyvanfar, A., Shafaghat, A., Abdullahi, M.M., Khorami, M.: Green driver: travel behaviors revisited on fuel saving and less emission. Sustainability **10**(2), 325 (2018)
9. Naula, A.B., Campoverde, P.M., Campoverde, N.R.: Methodological proposal for estimating polluting emissions: Case of cuenca, ecuador. In: IOP Conference Series: Earth and Environmental Science, vol. 1141, p. 012003. IOP Publishing (2023)
10. Ng, E.C., Huang, Y., Hong, G., Zhou, J.L., Surawski, N.C.: Reducing vehicle fuel consumption and exhaust emissions from the application of a green-safety device under real driving. Sci. Total Environ. **793**, 148602 (2021)
11. Rykała, M., Grzelak, M., Rykała, Ł, Voicu, D., Stoica, R.M.: Modeling vehicle fuel consumption using a low-cost obd-ii interface. Energies **16**(21), 7266 (2023)
12. Scott, A.J.: Industrialization and urbanization: a geographical agenda. Ann. Assoc. Am. Geogr. **76**(1), 25–37 (1986)
13. Xu, J., Tu, R., Ahmed, U., Amirjamshidi, G., Hatzopoulou, M., Roorda, M.J.: An eco-score system incorporating driving behavior, vehicle characteristics, and traffic conditions. Transp. Res. Part D: Transp. Environ. **95**, 102866 (2021)
14. Yuan, W., Frey, H.C., Wei, T.: Fuel use and emission rates reduction potential for light-duty gasoline vehicle eco-driving. Transp. Res. Part D: Transp. Environ. **109**, 103394 (2022)

A Practical and Cost-Effective Combination of GPS Data and Machine Learning Tools for Detecting Transportation Modes

Bryan Jachero$^{(\boxtimes)}$ (iD) and Karina Bermeo (iD)

Universidad Politécnica Salesiana, Grupo de Investigación en Ingeniería del Transporte, Vehicle Technology Society IEEE, Cuenca, Ecuador
bjachero@est.ups.edu.ec

Abstract. This work proposes a novel methodology to determine the modes of transportation used in the city of Cuenca, Ecuador, based on geolocation data and machine learning. For this purpose, 354,096 mobility samples from 40 people are collected via their mobile phones, with the respective identification of the transportation mode used: pedestrian, bicycle, bus, tram, taxi, and private vehicle. These samples are used to train and validate supervised learning architectures: classification trees, weighted k-nearest neighbor classifier, support vector machines, and two-layer neural networks. The classification tree achieved an accuracy of 99.5%, followed by 95.7% for the KNN model, 94.0% for SVM, and 93.2% for the BNN model. The trained, validated, and tested classification tree was applied to 110,242 samples obtained from the random mobility of 40 people, generating optimistic results. The findings indicate that the most used mode of transportation is the bus, followed by taxis and private vehicles. Pedestrians and bicycles, as well as trams, are predominantly used in the city center, while private transportation is more commonly used in rural areas. The obtained model can determine mobility patterns in the city, allowing for the effective establishment of origin-destination matrices. This facilitates public transportation planning, promotes alternatives to traditional mobility, and addresses the current problem of increasing traffic congestion, allowing for the reduction of pollutant emissions and mobilization costs, and aiding in the design of a new mobility plan.

Keywords: mobility patterns · transportation modes · machine learning · smart mobility

1 Introduction

Sustainable mobility is a comprehensive approach to transportation that meets mobility needs in work, educational, commercial, and recreational contexts [1, 2]. It manages urban development in environmental, economic, and health aspects [3, 4], allowing for transportation planning through various methodologies such as origin-destination (OD) surveys or information from mobile devices, enriched with diverse methods for understanding mobility patterns, flows, and transportation mode choices [5].

© The Author(s) 2025
E. M. Inga Ortega et al. (Eds.): CITIS 2024, LNNS 1331, pp. 195–204, 2025.
https://doi.org/10.1007/978-3-031-87065-1_18

In Rotterdam, OD matrices were obtained using phone data and compared with traditional roadside interview (RSI) matrices and travel surveys. Reasonable consistency in distance distribution and travel patterns was found between mobile data and RSI [5–7]. Correcting biases improved the accuracy and utility of OD matrices [8]. However, differences in patterns, underestimation, and biases in trips were noted between mobile OD matrices and traditional methods [9]. In Paris [10], OD flow analysis was combined with phone and sensor data to estimate transportation modes, understanding mobility patterns, and vehicle demand during peak hours. In the United Kingdom, algorithms [11] evaluated the alternation of transportation modes based on the emissions generated. Mobile technology monitored physical activities with 85% accuracy, proving crucial for improving sustainable mobility and reducing carbon emissions [12].

Cuenca, in southern Ecuador, has a growth rate of 1.9% [13]. Its urban area has 1,100 km of roads, divided into three main zones: the "Historic Center," "El Ejido," and a third zone to the south and west of the city [14]. The Mobility and Public Spaces Plan (2015) focuses on internal mobility [15], with 69% of trips made by motor vehicles and the remaining 31% by pedestrians and cyclists [15]. The growth and urban expansion of Cuenca have complicated mobility on the city's outskirts due to the limited public transportation coverage, which encompasses 77.5% in the urban area and 5% in expansion zones [16]. This has led to settlements in peripheral areas, where mobility is costly, slow, with unequal public transportation coverage, and underserved areas creating accessibility issues. Currently, the public transportation system via buses is not integrated with the tram system. The absence of studies and updated data is crucial for highlighting mobility patterns [17] and mobility indicators for transportation planning. Understanding these patterns is essential for efficient urban management and identifying contrasts in transportation use, costs, and accessibility [16]. Addressing these patterns highlights the urgency of an integral connection between urban design, public space, and transportation, providing valuable information for research, transportation planning, and services, as well as understanding Cuenca's current mobility system, offering smart services [18], and adopting a model more aligned with new technologies. Since 1970, local planners have relied on daily travel surveys [19] for data collection and transportation model creation. In Cuenca, OD surveys collected socio-demographic data and transportation preferences, highlighting buses as the primary mode with 42% of trips, private vehicles 37.16%, and bicycles 2.71% [20]. However, these surveys are costly, prone to biases, require more time, and are susceptible to recording out-of-sequence travel information [19, 21]. In Beijing, GPS data from mobile phones and machine learning [18, 22] have been used to acquire mobility data, daily travel demand, as well as urban planning, transportation design, and efficient city management free from traffic congestion.

Machine learning algorithms can identify and predict these patterns, promoting more accurate and anticipatory planning, enabling the prediction of variables and typologies [23]. In Chicago and Los Angeles [24, 25], transportation modes are inferred using GPS sensors from mobile devices and transportation network knowledge through decision trees, and a model [25] identifies transportation modes. Using a GPS receiver and accelerometers [26], a transportation flow behavior model is created using Python's MGO2 background elimination algorithm, whose prediction is unstable and not recommended in unfavorable lighting conditions. In Chicago [24], transportation modes are detected with spatial data of real-time bus locations, rail lines, and bus stops, highlighting the Random Forest method, with 93.42% prediction accuracy. This work expands on these existing approaches to classify the transportation modes of Cuenca's population, based on geolocation data from mobile devices and the application of machine learning architectures, achieving high accuracy rates in predicting the transportation mode used.

2   Methods and Materials

This work proposes a method to identify the six main modes of transportation in Cuenca: pedestrian, bicycle, tram, bus, taxi, and private vehicle, using GPS data from mobile devices. The proposed methodology, illustrated in Fig. 1, includes data collection for each mode of transportation, data processing, feature extraction, classification, and model evaluation. For data processing, a dataset of 354,096 samples is considered.

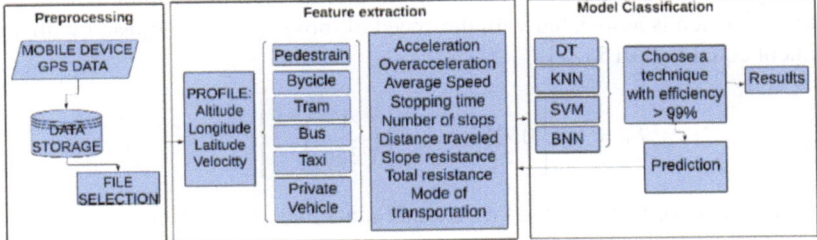

Fig. 1. Methodology

2.1   Data Acquisition and Predictor Extraction

For data acquisition of people's movements, mobile devices with the Android operating system were used. Through the background-running application DAMS, the movement of individual various modes of the transportation was collected. The information was recorded at a frequency of one second and stored in a text file format. A sample of 40 people, including students and random individuals traveling in the urban and rural areas of Cuenca, provided data on latitude (°), longitude (°), altitude (masl), speed (km/h), date, and time directly. Longitudinal acceleration (m/s^2) was determined by the variation of speeds (m/s) between consecutive points divided by the time variation (s) [27], using the formula (1).

$$a_x = \frac{V_{i+1} - V_i}{t_{i+1} - t_i} \tag{1}$$

Jerk, understood as the rate of change of acceleration [27], is the derivative of velocity or the third derivative of position with respect to time, calculated using Eq. (2).

$$Sa_x = \frac{a_{i+1} - a_i}{t_{i+1} - t_i} \tag{2}$$

To understand and analyze the dynamic characteristics of each transportation system, parameters and conditions are defined, considering the dynamics and geographical environment. All forces are calculated using unit values of: mass, density, aerodynamic coefficient, and frontal area.

Aerodynamic resistance (F_a) depends on velocity, frontal area (A_f), aerodynamic coefficient (C_x), and air density equal to 0.89 kg/m^3.

$$Fa = \frac{1}{2} * A_f * \rho * Cx * V^2 \tag{3}$$

The slope resistance (F_p) is determined based on the road grade angle (θ) [27] and is estimated using Eq. (4).

$$F_p = mg \left(\frac{s_{i+1} - s_i}{Alt_{i+1} - Alt_i} \right) \tag{4}$$

Rolling resistance (F_r) is given as a function of the unit mass and the rolling coefficient (f_r), which is associated with the speed of movement and adjusted with a static component equal to 0.015 [27].

$$F_r = mg \left[0,015 + 0,01 \left(\frac{V}{360} \right)^{2.5} \right] Cos(\theta) \tag{5}$$

Inertia resistance is determined by the unit mass and acceleration.

$$F_I = ma_x \tag{6}$$

Total resistance (Ft) is the sum of the forces opposing motion [27]: aerodynamic resistance, slope resistance, rolling resistance, and inertia resistance [29]. It is estimated using Eq. (7).

$$Ft = F_a + F_p + F_r + F_I \tag{7}$$

The traveled distance (dr) is determined from the rest until it returns to this state again using Eq. (8), with n being the number of samples while the individual is in motion [27].

$$dr = \frac{1}{3.6} \sum_{i=1}^{n} V_i \Delta t \tag{8}$$

The stop times (tp) are determined by summing the time intervals during which the body stopped. The number of stops is determined by summing the number of intervals where the speed is equal to 0 km/h.

2.2 Estimation of Predictor Importance

To perform the training of the machine learning architectures, it is necessary to verify the importance of each predictor on the mode of transport. In Fig. 2, the correlation matrix

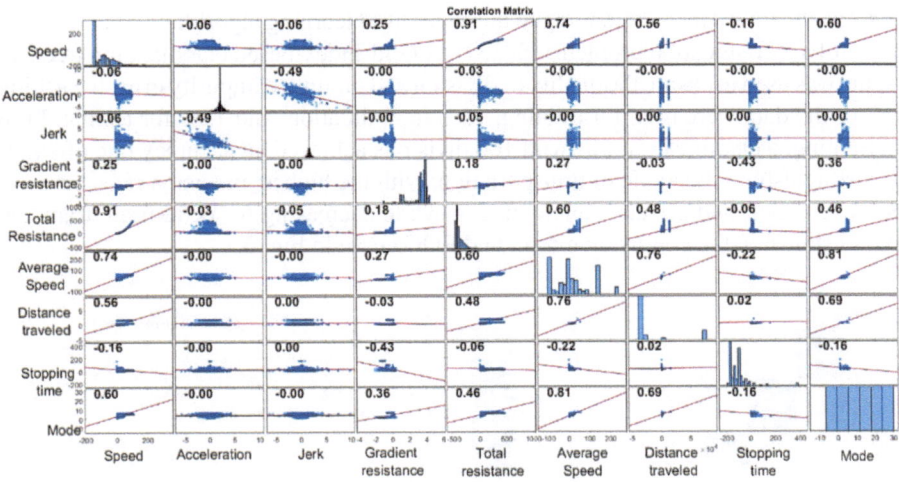

Fig. 2. Correlation Matrix of Variables

is observed, where values close to -1 and 1 of the Pearson coefficient denote a strong relationship between variables and the mode of transport (mode).

For model training, the predictors that have the greatest relationship with the mode of transportation are considered: average speed, stop time, instantaneous speed, traveled distance, and total resistance. In this way, based on the most influential variables, four classification models were applied to analyze their effectiveness and predict the modes of transportation.

2.3 Training of Machine Learning Algorithms

For the training, validation, and testing of the different machine learning algorithms, the predictor matrix $p = [Vm, tp, V, dr, F_T]$ is used, which consists of 354096 observations, the labels are represented by vector T, which takes values 1, 2, 3, 4, 5, and 6 for the modes of transport: pedestrian, bicycle, tram, bus, taxi, and private vehicle, respectively.

Weighted KNN: The Weighted K Nearest Neighbor model is a classifier that considers the weighting of the feature index [32]. Using a predictor matrix p and transportation modes T as the output variable, this algorithm was applied with a number of neighbors equal to 10. The weight of each nearest neighbor was weighted by the Euclidean distance considering the inverse square distance, with mean zero and standard deviation one standardization. This model achieved an accuracy of 95.7%. The highest precision rates are seen in predicting pedestrian, bicycle, and private vehicle modes.

SVM: The Support Vector Machine model makes distinctions between classes based on their characteristics in the hyperplane. This model uses a Gaussian kernel for better possible detection performance, an automatic kernel scale node in conjunction with a one-vs-all multiclass encoding due to its shorter training time. Additionally, a five-fold cross-validation was used for model development and evaluation, with 70% for training, 15% for validation, and 15% for testing. The overall accuracy rate was 94.0%, with the highest precision rates observed in pedestrian and bicycle modes.

Classification Trees: Belonging to the supervised learning algorithms, Classification Trees solve classification problems [30, 31]. A Gini-index-based classification tree with 5 splits was used to ensure the quality of division and minimize impurity in each partition. 70% of the data were used for training, 15% for validation, and 15% for testing. From the training, a classification tree with 100 divisions achieved an accuracy rate of 99.5% for distinguishing the mode of transportation, with the highest precision rates observed for bicycle, bus, and private vehicle modes, while decreasing for pedestrian, tram, and taxi, as shown in the confusion matrix and ROC curve in Fig. 3.

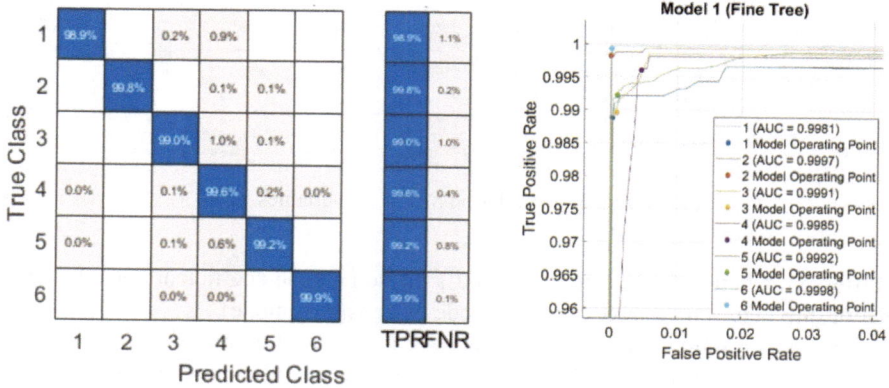

Fig. 3. Confusion Matrix.

Bilayered Neuronal Network: It is a multilayer neural network model that classifies multiclass data. This model consists of a two-layer network with 18 and 12 neurons in its hidden layers and uses the ReLU non-linear activation function to rectify negative values, enabling it to learn complex patterns. The training was conducted over 2750 epochs with an overall accuracy rate of 93.2%. For this model, 70% of the data was allocated for training, 15% for validation, and 15% for evaluation. The highest precision rates are achieved for pedestrian, bicycle, and private vehicle modes.

3   Result and Discussion

The classification of the four models presented very optimistic results. In the validation matrix shown in Table 1, the classification rate is displayed, demonstrating satisfactory percentages for each mode of transportation. The classification tree emerged as the best classification model, with an efficiency of 99.5% as reflected in Table 2, which, unlike the other models, only encountered minimal classification difficulty for the pedestrian and tram classes.

The classification tree obtained is fed with a random dataset of 110242 samples obtained from the same individuals in a completely random manner. Figure 4a shows the random mobility data in the urban and rural parts of the city, while Fig. 4b indicates the movements with their respective modes of transportation. It can be observed that

Table 1. Validation Matrix

Model	Transport Mode					
	Pedestrian	Bicycle	Tram	Bus	Taxi	Private Vehicle
DT	98,9%	99,8%	99,0%	99,6%	99,2%	99,9%
KNN	99,4%	99,9%	95,2%	96,0%	92,6%	97,1%
SVM	97,4%	99,8%	91,2%	95,7%	90,5%	93,7%
BNN	97,5%	99,5%	92,1%	92,3%	87,6%	98,9%

Table 2. Efficiency of Models

Model	Efficiency	Model	Efficiency
DT	99,5%	**SVM**	93,2%
KNN	95,7%	**BNN**	94,0%

movements in the rural area generally occur in private vehicles and taxis, while the rest of the modes of transportation are mostly used in the urban area of the ci.

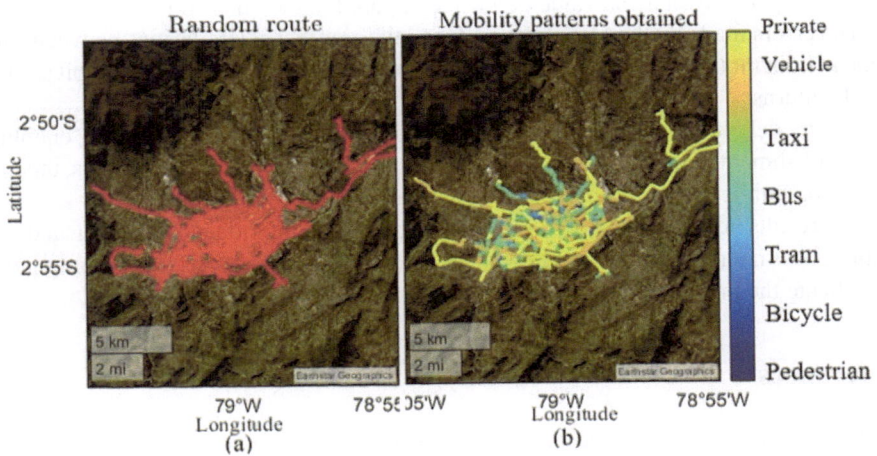

Fig. 4. Mobility patterns obtained in a random route.

In Fig. 5, the distribution of the use of different modes of transportation is observed. It is highlighted that the most used mode is the bus, accounting for 39.35% of the travel samples, followed by taxis and private vehicles with 28.82% and 19.52%, respectively. Walking, tram, and cycling are the least common modes with 5.54%, 4.56%, and 2.21%, respectively, mainly because they are predominantly used in the city center.

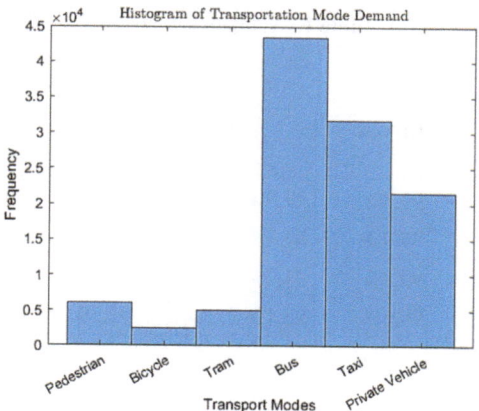

Fig. 5. Modes of transportation used.

4 Conclusion

Four classifiers based on geopositioning and machine learning were developed to identify modes of transportation: the most important variables for prediction are average speed, stop time, instant speed, distance traveled, and total resistance. Four models were trained and validated: KNN, SVM, BNN, and DT, with the DT model showing the best classification performance for modes of transportation, achieving an accuracy of 99.5%, which is superior by a mean value of 4.4% compared to the other models.

The errors made by the classifier are due to the significant similarities between trips made in light motorized vehicles such as taxis and private vehicles, which exhibit similar travel patterns.

The model obtained can be used to determine mobility patterns in the city, enabling the establishment of effective mobility matrices and origin-destination matrices, thereby enhancing public transport planning and alternative mobility solutions.

The results obtained are based on a sample that, while important for training and validation, does not represent the entire population of the city. Therefore, it is recommended to replicate the proposed method with a larger sample size.

References

1. GAD Municipal Cuenca. Categorización de la demanda de transporte de Cuenca, p. 1 (2014). http://www.cuenca.gob.ec/?q=system/files/planmovilidadcaracdemanda.pdf
2. Marquart, H., Schuppan, J.: Promoting sustainable mobility: to what extent is 'health' considered by mobility app studies? A review and a conceptual framework. Sustainability **14**(1) (2022). https://doi.org/10.3390/su14010047
3. Gillis, D., Semanjski, I., Lauwers, D.: How to monitor sustainable mobility in cities? Literature review in the frame of creating a set of sustainable mobility indicators. Sustainability **8**(1), 1–30 (2016). https://doi.org/10.3390/su8010029
4. Almeida, D., Rosas, M.: Estudio para implementar un Smart Mobility en la Universidad Central del Ecuador **1**(1,2) (2023)

5. Wismans, L.J.J., Friso, K., Rijsdijk, J., de Graaf, S.W., Keij, J.: Improving a priori demand estimates transport models using mobile phone data: a Rotterdam-region case. J. Urban Technol. **25**(2), 63–83 (2018). https://doi.org/10.1080/10630732.2018.1442075
6. Tolouei, R., Psarras, S., Prince, R.: Origin-destination trip matrix development: conventional methods versus mobile phone data. Transp. Res. Procedia **26**(2016), 39–52 (2017). https://doi.org/10.1016/j.trpro.2017.07.007
7. Larijani, A.N., Olteanu-Raimond, A.M., Perret, J., Brédif, M., Ziemlicki, C.: Investigating the mobile phone data to estimate the origin destination flow and analysis; case study: Paris region. Transp. Res. Procedia **6**, 64–78 (2015). https://doi.org/10.1016/j.trpro.2015.03.006
8. Huang, H., Cheng, Y., Weibel, R.: Transport mode detection based on mobile phone network data: a systematic review. Transp. Res. Part C: Emerg. Technol. **101**, 297–312 (2019). https://doi.org/10.1016/j.trc.2019.02.008
9. Bonnel, P., Hombourger, E., Olteanu-Raimond, A.M., Smoreda, Z.: Passive mobile phone dataset to construct origin-destination matrix: potentials and limitations. Transp. Res. Procedia **11**, 381–398 (2015). https://doi.org/10.1016/j.trpro.2015.12.032
10. Reul, J., Grube, T., Stolten, D.: Urban transportation at an inflection point: an analysis of potential influencing factors. Transp. Res. Part D: Transp. Environ. **92**, 102733 (2021). https://doi.org/10.1016/j.trd.2021.102733
11. Thomas, H., Serrenho, A.C.: Using different transport modes: an opportunity to reduce UK passenger transport emissions. Transp. Res. Part D: Transp. Environ. **126**, 103989 (2024). https://doi.org/10.1016/j.trd.2023.103989
12. Sohn, Timothy, et al.: Mobility detection using everyday GSM traces. In: Dourish, P., Friday, A. (eds.) UbiComp 2006: Ubiquitous Computing: 8th International Conference, UbiComp 2006 Orange County, CA, USA, September 17–21, 2006 Proceedings, pp. 212–224. Springer Berlin Heidelberg, Berlin, Heidelberg (2006). https://doi.org/10.1007/11853565_13
13. Flores-Juca, E., Carmona, J.C., Mora-Arias, E., Navarro, J.G.: Reinterpretando el papel de la movilidad en las zonas periurbanas: Un análisis multiescala en Cuenca – Ecuador. Revista de geografía Norte Grande **84**, 271–291 (2023). https://doi.org/10.4067/S0718-34022023000100271
14. Flores, J.: Modelo de caracterización de la movilidad vehicular en el Centro Histórico de Cuenca. Universidad del Azuay, Cuenca (2016)
15. Municipalidad de Cuenca. PLAN DE MOVILIDAD Y ESPACIOS PÚBLICOS, TOMO I. Cuenca (2015)
16. Flores, G.: MOVILIDAD SOSTENIBLE EN LOS SECTORES PERIFÉRICOS: UN APORTE A LAS METODOLOGÍAS DE PLANIFICACIÓN ESPACIAL Y DEL TERRITORIO.EL CASO CUENCA – ECUADOR (2021)
17. Sucuzhañay, H., Vázquez, C.: ESTUDIO DE LA PLANIFICACIÓN DEL TRANVÍA Y SU POSIBILIDAD DE INTEGRACIÓN A LA CONEXIÓN CON LOS VIAJES PERIURBANOS. UNIVERSIDAD DE CUENCA, Cuenca (2020)
18. Xiao, Z., Wang, Y., Fu, K., Wu, F.: Identifying different transportation modes from trajectory data using tree-based ensemble classifiers. ISPRS Int. J. Geo-Inf. **6**(2), 57 (2017). https://doi.org/10.3390/ijgi6020057
19. Gong, H., Chen, C., Bialostozky, E., Lawson, C.T.: A GPS/GIS method for travel mode detection in New York City. Comput. Environ. Urban Syst. **36**(2), 131–139 (2012). https://doi.org/10.1016/j.compenvurbsys.2011.05.003
20. Ilustre Municipalidad de Cuenca. Plan de movilidad y Espacios de Cuenca 2015–2025. Ilus. Munic. Cuenca, p. 118 (2015)
21. En, C., De, P., Física, L.A., Deporte, Y.: Universidad Central Del Ecuador Facultad De Cultura Física. Uvirtual. uce, pp. 1–5 (2018)
22. Xia, L., Huang, Q., Wu, D.: Decision tree-based contextual location prediction from mobile device logs. Mobile Inf. Syst. (2018). https://doi.org/10.1155/2018/1852861

23. Vázquez, J.: Aplicación de Técnicas de Aprendizaje Automático en el Sector Ferroviario. Universidad Pontifica Comillas, Madrid, Memoria (2016)
24. Stenneth, L., Wolfson, O., Philip, S., Xu, B.: Transportation mode detection using mobile phones and GIS information. Research (2011)
25. Reddy, S., Mun, M., Burke, J., Estrin, D., Hansen, M., Srivastava, M.: Using mobile phones to determine transportation modes. ACM Trans. Sensor Netw. **6**(2), 1–27 (2010). https://doi.org/10.1145/1689239.1689243
26. Capistran, O.: Sistema de Aprendizaje Automático para Detección y Análisis de Tráfico Vehicular. CENTRO DE INGENIERÍA Y DESARROLLO INDISTRIAL, Querétaro (2020)
27. Rivera-Campoverde, N., Sanz, J.M., Arenas-Ramirez, B.: Low-cost model for the estimation of pollutant emissions based on GPS and machine learning. In: Idoipe, A Vizán, Prada, J.C.G. (eds.) Proceedings of the XV Ibero-American Congress of Mechanical Engineering: CIBIM 22/CIBEM 22, pp. 182–188. Springer International Publishing, Cham (2023). https://doi.org/10.1007/978-3-031-38563-6_27
28. Alonso, M., García, C., Jiménez, F., Álvarez, D.: OPTIMIZACIÓN DE UNA MANIOBRA DE ADELANTAMIENTO APLICADA A VEHÍCULOS AUTÓNOMOS (2021)
29. Dávalos, D.: OBTENCIÓN DE UN CICLO TÍPICO DE CONDUCCIÓN PARA LOS VEHÍCULOS DE LA UNIÓN DE TAXISTAS DEL AZUAY. Cuenca (2017)
30. Bansal, M., Goyal, A., Choudhary, A.: A comparative analysis of K-Nearest Neighbor, Genetic, Support Vector Machine, Decision Tree, and Long Short Term Memory algorithms in machine learning. Decis. Analyt. J. **3**, 100071 (2022). https://doi.org/10.1016/j.dajour.2022.100071
31. Saber, M., El Rharras, A., Saadane, R., Kharraz, A.H., Chehri, A.: An optimized spectrum sensing implementation based on SVM, KNN and TREE algorithms. In: Proceedings – 15th International Conference on Signal Image Technology and Internet Based Systems, SISITS 2019, pp. 383–389. Institute of Electrical and Electronics Engineers Inc. (2019). https://doi.org/10.1109/SITIS.2019.00068
32. Tarakci, F., Ozkan, A.: Comparison of classification performance of kNN and WKNN algorithms. Selcuk University Journal of Engineering Sciences. **20**(2), 32–37 (2021). http://sujes.selcuk.edu.tr/sujes

Comprehensive Mapping and Simulation Approaches for Urban Scenarios for Autonomous Vehicles Operation

Juan D. Valladolid[1], Paul Ortiz[2]([⊠]), Steven Castillo[1], and Carlos Ochoa[1]

[1] Department of Automotive Engineering, Universidad Politécnica Salesiana, Cuenca 101007, Ecuador
jvalladolid@ups.edu.ec, {fcastillog2,cochoac2}@est.ups.edu.ec
[2] Mechatronics Engineering, Universidad Politécnica Salesiana, Cuenca 101007, Ecuador
jortizg@ups.edu.ec

Abstract. Autonomous driving has been one of the most popular and challenging topics in recent years. Researchers striving for full autonomy have employed a variety of sensors, such as Light Detection and Ranging (LiDAR), cameras, inertial measurement units (IMU), and GPS, along with developing intelligent algorithms for their effective application. In addition to these advancements, the creation of high-precision maps has gained significant attention. These maps play a crucial role in localization using point clouds and in detailing the operational environment by generating routes and identifying traffic elements. As a result, they have become critical components of autonomous driving systems. This paper explores a modern method for map generation specifically oriented towards autonomous driving and examines its evolution through simulations conducted on the Autoware software platform. Additionally, the limitations of current map generation technologies are discussed to inspire future research and development in this rapidly advancing field. Through these discussions, the paper aims to contribute to the ongoing progress in autonomous vehicle technology.

Keywords: Autonmous vehicle · LIO-SAM · LiDAR · HD maps · map generation

1 Introduction

The autonomous vehicle (AV), also known as a driverless vehicle, is capable of sensing its surroundings and navigating through defined lanes without human intervention. This innovative technology has transformed the automotive industry by automating vehicle control systems, primarily relying on advanced computer systems. AVs integrate a range of sensors, actuators, cameras, Light Detection and Ranging (LiDAR), and advanced computing systems to achieve efficient

E. M. Inga Ortega et al. (Eds.): CITIS 2024, LNNS 1331, pp. 205–214, 2025.
https://doi.org/10.1007/978-3-031-87065-1_19

and safe navigation. These sensors and cameras enable the vehicle to perceive its surroundings, detect potential hazards, and adjust its speed and direction accordingly. The integration of these advanced technologies has significantly enhanced the safety and efficiency of autonomous vehicles [1,4,5,7].

In the field of transportation and urban planning, developing a comprehensive roadmap has always been a crucial task. However, the advent of Intelligent Transportation System (ITS) applications has introduced an innovative concept known as Routable Digital Maps (RDM). These maps, designed as interconnected and directed graphs, offer a novel approach to navigation and transportation planning.

Recent years have seen a surge in interest in simultaneous localization and mapping (SLAM), both in indoor and outdoor environments. This technique relies on identifying key landmarks, such as building exteriors and traffic lights, for accurate mapping and positioning [11]. Visual SLAM, in particular, stands out for its ability to utilize various road features in the localization process, leveraging the richness of color information available. A simple explanation of SLAM, focusing on its geometric perspective and avoiding intricate algorithmic complexities, is provided in [1]. This approach offers a broad understanding of SLAM, elucidating the interaction between its various facets and stages, including the core elements of the front-end and back-end, and their connection to the overall SLAM paradigm. The speed of the robot and the traveled distance can be estimated from the IMU measurements, which are used for odometry. However, IMU-based navigation suffers from accumulated error over time due to the dead reckoning method [1].

Accurate localization is a critical task for the safe operation of autonomous vehicles. To achieve precise localization without Global Positioning System (GPS) signals, a Kernelized Rényi Distance (KRD)-based SLAM algorithm is presented in [9]. Digital maps have been significantly improved to meet the requirements of advanced driver assistance system (ADAS) functions such as lane-keeping assist [6] and adaptive cruise control (ACC) [3].

The current state of AV mapping is encouraging, with the field having matured to a point where detailed maps of complex environments are built in real-time and proven useful. Many existing techniques are robust to noise and can handle a wide range of environments. Nonetheless, there remain open problems for future research. AV mapping will continue to be a highly active research area, essential to achieving full autonomy [8].

Higher levels of autonomy require maps to be more refined in detail and adhere to quality standards. In this context, the solution for high-precision localization is to provide a unified representation that combines the agent dynamics, collected by perception and tracking systems, with the scene context, commonly provided as prior knowledge in the form of high-definition (HD) maps [2,10,14].

The structure of this article is organized as follows: Sect. 2 presents the methodology used for obtaining point clouds and generating maps specifically tailored for VE operations. Section 3 addresses the limitations of current HD map generation methods and identifies key challenges that remain unresolved. Finally, Sect. 4 provides the conclusions and future directions.

1.1 Contributions

The contributions of this article are the following:

- This article describes and compares some approaches to map representation and their applications, such as highly or moderately simplified map representations, which are mainly used in the autonomous driving domain.
- We provide a detailed bibliographic review on HD maps for autonomous vehicles, as well as the methodology to structure the structure of its various layers and the information contained in them, based on the definitions of a map.
- We evaluate the current limitations and challenges of the HD map, such as data storage and information updating in the point cloud, as well as related future research topics.

2 Methods

A three-step algorithmic methodology is proposed to convert the information provided by a point cloud into a map that can be used for the evaluation and operation of Autonomous Vehicles. The methodology is outlined in Fig. 1.

The first step involves generating and loading the point cloud of the sector using TIER IV software, resulting in a 3D representation of the sector [6]. This point cloud is then utilized to divide the road network into homogeneous, unidirectional road segments, each with a consistent width and number of lanes along its entire length. In the second step, road sections and additional elements are generated to build an environment suitable for the operation of autonomous vehicles. This includes creating road features such as lanes, intersections, and obstacles. The third and final step involves simulating the map layout using Autoware software [13], to validate the map and generate any necessary corrections. This comprehensive approach ensures the accuracy and effectiveness of the map generated for Autonomous Vehicle operation.

2.1 Step 1: Generation and Load Pintcloud Data

In the initial step, LiDAR sensors are employed to provide precise positioning, ensuring sufficient accuracy when the GPS signal is weak. This corrects GPS positioning errors, Fig. 2.

In order to enhance accuracy, the GPS data is refined using an RTK (Real-Time Kinematic) connection. The LIO-SAM algorithm is then utilized, which estimates the trajectory of the mobile robot and constructs high-precision maps in real time. The motion estimated from the preintegration of the Inertial Measurement Unit (IMU) corrects point clouds and provides an initial guess for optimizing LiDAR odometry. The resulting LiDAR odometry solution is used

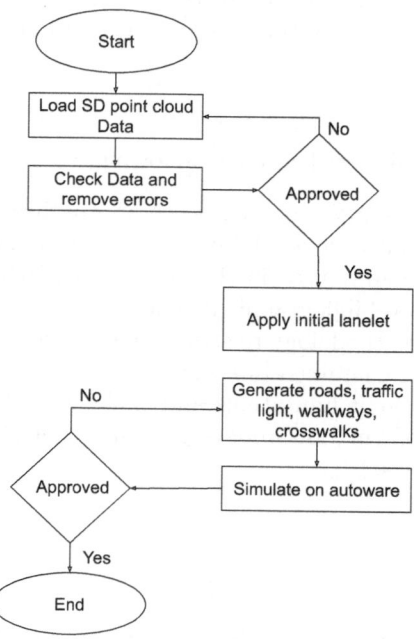

Fig. 1. Map planning flowchart.

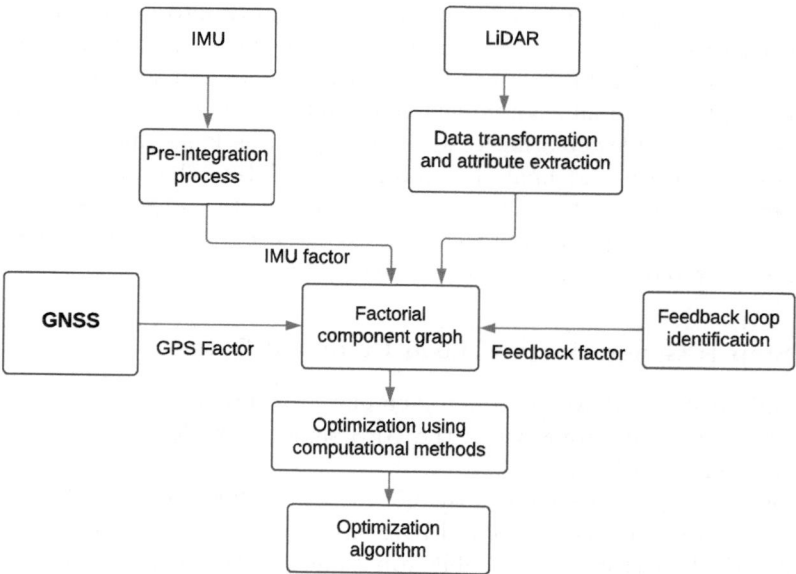

Fig. 2. Block diagram of the map generation methodology.

to estimate IMU bias. For maintaining high real-time performance, we prioritize optimizing the pose using sidelined older LiDAR scans, rather than matching the scans to a global map. The specific flowchart of the system is shown in Fig. 3.

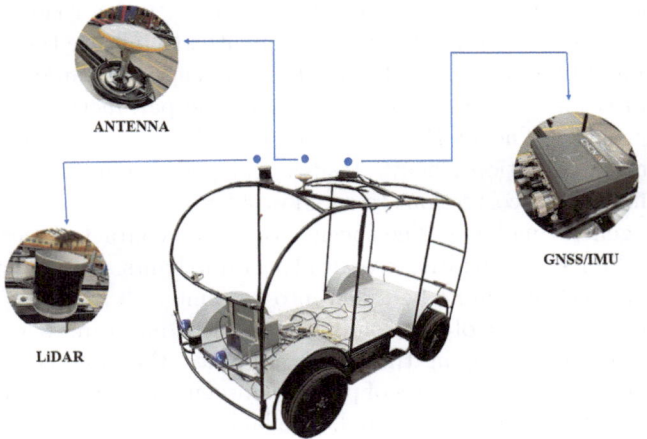

Fig. 3. Experimental vehicle platform.

The dataset was collected in the historic center of the city of Cuenca, Ecuador. Figure 4 displays the resulting point cloud after applying the LIO-SAM algorithm and incorporating sensor data.

Fig. 4. A point cloud image of an approaching vehicle captured by sensors and LIO-SAM algorithm.

In the pointcloud, the profiles of buildings and roads, which will be the focus of subsequent steps, can be clearly identified.

2.2 Step 2: Construction of Road Lines and Traffic Elements for AV Operation

After scanning and obtaining the PCD file containing the point cloud, the Tier IV program was used to create a map based on a series of points according to the PCD. Tier IV software is particularly useful for defining roads, lanes, traffic signals, crosswalks, and intersections. Subsequently, to initiate the mapping, the point cloud must be imported and a lanelet2Maps space created, which will be the storage area for the design and creation of the parameters required by an autonomous vehicle. The PCD, in some cases, presents issues due to roads in very unfavorable conditions, heavy traffic, and pedestrians. It is advisable to make adjustments and corrections to avoid problems such as misalignments that appear when generating lanes. The specifications and characteristics of the area, such as speed limits, lane widths, crosswalks, traffic lights, and angles present in curves and intersections, must be taken into account.

The detailed mapping of the roads was carried out using a series of tools stored in Tier IV. For mapping the lane boundaries, the "linestring" command was used, which consists of a series of points that generate a line used for creating the lane through the "add lanelet" tool. The lane orientation was modified using the "reverse direction" tool. Curves or intersections need to be smooth; for this, the "add joint lanelet" tool was used. To generate two-way lanes, the "copy lanelet" tool was utilized,Fig. 5.

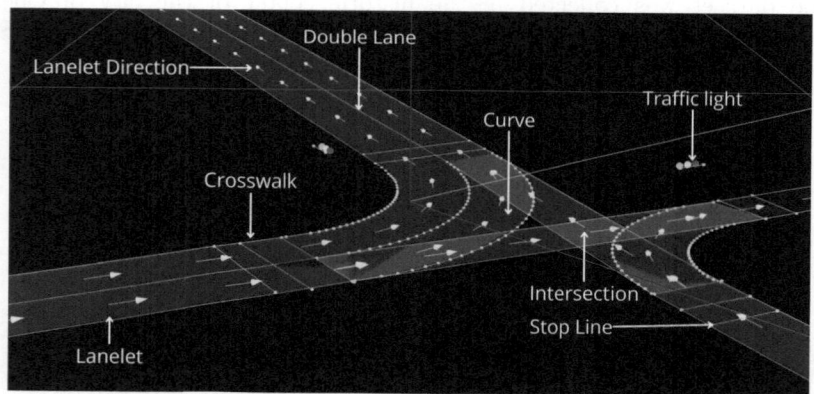

Fig. 5. Construction of the proposed map for AV navigation

The lane width depends on the working area and was established using the "straighten lanelet" tool. Lane changes were enabled using the "set to disable lane change" tool. For traffic signals and transit spaces, the commands "trafficlight", "crosswalk", "no parking area", "no stopping area", and "Stop line" were used.

An important parameter is the maximum vehicle speed, which is set according to the zone in which it is located. A standard maximum speed for urban areas in Ecuador, which is 30 km/h, was used. The speed was adjusted using the "speed limit" tool within the lane configuration. It is essential that, in curves, the maximum speed be set to 10 km/h to ensure safety.

Next, Fig. 6 shows the lanelet for Parque Calderón in Cantón Cuenca, Ecuador. The lanelet was constructed with all the aforementioned specifications.

Fig. 6. HD mapping representation of the road and point cloud.

2.3   Step 3: Map Evaluation Through Simulation on Autoware

Autoware is an open source software platform for developing autonomous mobility solutions. It operates under the ROS2 real-time operating system and was used to test routes created with Lanelet tools and other features available in TIER IV. The simulations verified that the vehicle navigates specific lanes, including turns and intersections, smoothly and correctly recognizes traffic signs. In future autonomous projects, vehicles will be simulated on congested streets, with adjustments to various vehicle parameters. Figure 6 shows that the simulation indicates that the autonomous vehicle can operate on the map designed according to the proposed methodology. During the simulation in the program, the start and end points for the vehicle are set, the trajectory is presented in green if there are no problems or obstacles.

3   Results and Discussion

Autonomous mobility promises to be highly efficient, but its performance depends on context and implementation. In Ecuador, autonomous mobility is practically non-existent, and the transition presents great challenges, due to the fact that autonomous vehicles face complications when related to ordinary vehicles. The proposed methodology for creating autonomous mobility maps is

(a) 2D representation of the simulation.   (b) 3D representation of the simulation.

Fig. 7. HD map evaluation using simulations in Autoware software.

based on parameters that consider the main traffic signals, allowing specific commands to be issued to the vehicle through advanced systems so that it operates
autonomously, improving its ability to adapt and respond. However, factors such
as vehicular traffic or the presence of pedestrians are not included, which can
compromise the efficiency and safety of the autonomous vehicle. To generate the
point cloud mentioned in step 1, the LIO-SAM algorithm was used, recognized
as one of the best 3D LiDAR SLAM algorithms according to [12]. This algorithm works correctly with favorable conditions; however, errors could occur if
certain geographical factors, climatic factors and large irregularities in the tracks
are not taken into account. One limitation is the amount of storage required to
store data in a point cloud.

4   Conclusions

In this paper, we introduce a comprehensive three-step methodology for constructing and evaluating lane-level digital maps. Our primary objective is to
establish a robust reference framework for extracting and developing lane-based
traffic databases from generated point clouds. This methodology is designed
to enhance the precision and reliability of digital maps, which are critical for
advanced driver-assistance systems and autonomous driving applications.

Our methodology consists of three key steps: data acquisition, map generation, and map validation. In the data acquisition phase, we collect high-resolution
point cloud data using LiDAR sensors mounted on vehicles. This data is then
processed in the map generation phase to create detailed lane-level maps that
accurately represent the road geometry and lane markings. Finally, in the map
validation phase, we verify the accuracy of the generated maps through extensive
simulations and real-world testing.

Although our simulations have yielded promising results, indicating the effectiveness of our approach, there is still potential for further refinement and enhancement. We tested this methodology primarily on urban roads, where the complexity of the road environment poses significant challenges. A promising direction for future research is to adapt and extend our methodology to highway networks, where the scale and dynamics of traffic are different.

In conclusion, our approach lays a solid foundation for lane-level digital map construction, providing a reliable framework for future developments. Its adaptability and potential for improvement open up new opportunities for applications in various types of road networks, from urban streets to highways. By continuing to refine and expand this methodology, we can contribute to the advancement of intelligent transportation systems and the realization of safer, more efficient roadways.

Acknowledgements. The authors are thankful for the project: *"Desarrollo de Estrategias de Movilidad Inteligente, Sostenible y Aceptación Social de Vehículos Autónomos en la Ciudad de Cuenca, Empleando Técnicas de Inteligencia Artificial y Realidad Virtual en Plataformas de Software y Hardware Especializados"* from the Group of Research in Transportation Engineering (GIIT) of the Salesian Polytechnic University for providing the data used in this document.

References

1. Alsadik, B., Karam, S.: The simultaneous localization and mapping (slam)-an overview. J. Appl. Sci. Technol. Trends **2**, 120–131 (11 2021). https://doi.org/10.38094/jastt204117
2. Bauer, S., Alkhorshid, Y., Wanielik, G.: Using high-definition maps for precise urban vehicle localization. In: 2016 IEEE 19th International Conference on Intelligent Transportation Systems (ITSC), pp. 492–497 (2016). https://doi.org/10.1109/ITSC.2016.7795600
3. Ebrahimi Soorchaei, B., Razzaghpour, M., Valiente, R., Raftari, A., Fallah, Y.P.: High-definition map representation techniques for automated vehicles. Electronics **11**(20) (2022). https://doi.org/10.3390/electronics11203374, https://www.mdpi.com/2079-9292/11/20/3374
4. Englot, s., Tixiao, B., Meyers, D., Wang, W., Ratti, C., Rus, D.: Lio-sam: Tightly-coupled lidar inertial odometry via smoothing and mapping (10 2020). https://doi.org/10.1109/IROS45743.2020.9341176
5. Giannaros, A., et al.: Autonomous vehicles: Sophisticated attacks, safety issues, challenges, open topics, blockchain, and future directions. J. Cybersecur. Privacy **3**(3), 493–543 (2023). https://doi.org/10.3390/jcp3030025, https://www.mdpi.com/2624-800X/3/3/25
6. Helfenstein, A., Mulder, V.L., Heuvelink, G.B., Okx, J.P.: Tier 4 maps of soil ph at 25m resolution for the netherlands. Geoderma **410**, 115659 (2022). https://doi.org/10.1016/j.geoderma.2021.115659, https://www.sciencedirect.com/science/article/pii/S0016706121007394
7. Idrovo-Berrezueta, P., Ortiz, J.P., Valladolid, J.D., Espinoza, F.E.: Development of an LSTM-based model with attention mechanism for detection of traffic's officer hand gestures for autonomous vehicles in ecuador. In: Salgado-Guerrero, J.P.,

Vega-Carrillo, H.R., García-Fernández, G., Robles-Bykbaev, V. (eds.) Systems, Smart Technologies and Innovation for Society, pp. 106–115. Springer Nature Switzerland, Cham (2024)

8. Ilci, V., Toth, C.: High definition 3d map creation using gnss/imu/lidar sensor integration to support autonomous vehicle navigation. Sensors **20**(3) (2020). https://doi.org/10.3390/s20030899, https://www.mdpi.com/1424-8220/20/3/899

9. Li, G., Bao, H., Wang, B., Wu, T.: Kernelised rényi distance for localization and mapping of autonomous vehicle. In: 2017 13th International Conference on Computational Intelligence and Security (CIS), pp. 69–72 (2017). https://doi.org/10.1109/CIS.2017.00023

10. Liu, L., Wu, T., Fang, Y., Hu, T., Song, J.: A smart map representation for autonomous vehicle navigation. In: 2015 12th International Conference on Fuzzy Systems and Knowledge Discovery (FSKD), pp. 2308–2313 (2015). https://doi.org/10.1109/FSKD.2015.7382313

11. Malakouti-Khah, H., Sadeghzadeh Nokhodberiz, N., Montazeri, A.: Simultaneous localization and mapping in a multi-robot system in a dynamic environment with unknown initial correspondence. Front. Robot. AI **10**, 1291672 (01 2024). https://doi.org/10.3389/frobt.2023.1291672

12. Wu, J., Huang, S., Yang, Y., Zhang, B.: Evaluation of 3d lidar slam algorithms based on the kitti dataset. J. Supercomput. **79**(14), 15760–15772 (2023)

13. Xu, J., Yao, Y.: Poster: A demo of autoware-based autonomous driving using depth sensing. In: 2023 IEEE/ACM Symposium on Edge Computing (SEC), pp. 261–263 (Dec 2023). https://doi.org/10.1145/3583740.3626627

14. Zheng, L., Li, B., Zhang, H., Shan, Y., Zhou, J.: A high-definition road-network model for self-driving vehicles. ISPRS Int. J. Geo-Inform. **7**(11) (2018). https://doi.org/10.3390/ijgi7110417, https://www.mdpi.com/2220-9964/7/11/417

Performance Analysis of the LiDAR-Based 3D LIO-SAM Algorithm in Urban Areas: Case Study Cuenca City in the Ecuadorian Andes

Juan P. Ortiz[1,2]([✉]) [iD], Juan D. Valladolid[1,2] [iD], and Soledad Gutiérrez[1,2] [iD]

[1] Universidad Politécnica Salesiana, Cuenca, Ecuador
[2] Grupo de Investigación En Ingeniería Del Transporte (GIIT), Cuenca, Ecuador
jortizg@ups.edu.ec

Abstract. Autonomous Vehicles (AVs) technology has increased dramatically over the last years. Accurate 3D point cloud maps are needed to perform this autonomous driving task. Mapping allows AVs to adapt to the surrounding environment and maneuver in complex situations. In this paper, a methodology for performance analysis of the LiDAR Inertia Odometry via Smoothing and Mapping (LIO-SAM) framework for the generation of 3D point cloud maps is proposed. The test scenario is intended for outdoor autonomous vehicles making use of an inertial measurement unit (IMU) sensor, a LiDAR sensor and a Global Navigation Satellite System (GNSS) module. The route selected for the collection of the data set is located in the city of Cuenca-Ecuador in the Ecuadorian Andes at an altitude of 2581 m a.s.l. The LIO-SAM framework was meticulously evaluated for accurate real-time vehicle trajectory estimation and mapping. The results obtained were highly precise with a Root Mean Squared Error (RMSE) of 0.04 m, 0.09 m and 0.08 m for the elevation, northing and easting, respectively. This results show a few centimeters of difference between the control points evaluated with cartographic precision equipment and Real-Time Kinematic (RTK) services versus LIO-SAM framework.

Keywords: LIO-SAM · IMU · LiDAR · Autonomous Vehicles (AVs) · Loop closure

1 Introduction

Over the past ten years, there has been great interest around the world in creating technology for autonomous vehicles. The goal is to make driving safer, more

Universidad Politécnica Salesiana Cuenca, Grupo de Investigación en Ingeniería del Transporte (GIIT), Carreras de Ingeniería Mecatrónica, Ingeniería Automotriz and Ingeniería Civil.

E. M. Inga Ortega et al. (Eds.): CITIS 2024, LNNS 1331, pp. 215–226, 2025.
https://doi.org/10.1007/978-3-031-87065-1_20

efficient, and lower energy consumption. Some of the challenges faced when operating in complex urban environments are dense and dynamic traffic, posing significant difficulty, rapid changes in road conditions and variable speeds between vehicles. Additionally, the unpredictability of pedestrians, who may cross at undesignated locations or ignore signs [8]. To achieve this, vehicles need sensors that can collect information from the environment with greater precision than humans [18]. For this reason, 3D Light Detection and Ranging (LiDAR), Inertial Measurement Unit (IMU), and Global Navigation Satellite System (GNSS) are frequently used. These sensors allow autonomous vehicles to perceive their surroundings and create accurate maps to aid in self-localization. In addition, together with SLAM algorithms, they provide three-dimensional data of the environment and their global position [13]. SLAM framework combines the high resolution of lidar with the frequent updating of inertial sensors, from this perspective problems such as precise localization and dynamic map construction, essential for the safety and efficiency of autonomous vehicles, could be solved. This article focuses on the analysis of the accuracy of the 3D point cloud map obtained through LIO-SAM framework, compared to traditional methods based on the use of GNSS (Global Navigation Satellite System) antennas. The structure of this article is organized as follows: Sect. 2 provides an overview of the related work, discussing previous studies in the field. Section 3 presents the proposed methods used for obtaining data sensors and the criteria for route selection. In addition, it addresses the data collection and preparation. Section 4 presents the generating point cloud maps using LIO-SAM framework. Section 5 provides the results obtained, and finally, Sect. 6 presents the conclusions of this paper.

1.1 Contributions

The contributions of this article are the following:

- This article describes and evaluates the performance of LIO-SAM framework in outdoor scenarios, which is mainly used for mapping in autonomous driving tasks.
- We provide a detailed state-of-the-art review on SLAM algorithms for 3D point cloud mapping.
- We propose an accurate methodology to evaluate SLAM algorithms for mapping in dynamic and variable conditions, allowing to identify and overcome specific challenges such as traffic flow, trees, and short buildings in urban and suburban areas.

2 Related Work

In recent years, several investigations have been carried out trying to obtain better results for the LIO-SAM framework presented at [14]. This framework uses IMU, GNSSS, and LiDAR sensors to estimate motion from IMU pre-integration, correct point clouds, and produce an initial guess for LiDAR odometry optimization. At the same time, it optimized LiDAR scans for better real-time performance. However, the loss of GPS data in areas with poor satellite coverage, as well as the poor coverage of RTK services for GPS correction, and the need for indoor mapping led to the development of new adaptations of this framework.

Simultaneous localization and mapping (SLAM) can be classified into Visual SLAM and LiDAR SLAM [7,16]. Visual SLAM allows capturing visual features of the environment and performing both localization and mapping simultaneously [10]. This makes it especially suitable for indoor environments where illumination is constant and visual features are prominent. In terms of hardware it may be less expensive, but it can be susceptible to illumination changes and the accuracy of the measured distance is limited due to the visual nature of the sensor. In contrast, LIO-SAM integration provides reliability in complex urban environments where accuracy and rapid map updates are important. However, the dependence on the GNSS signal and the complexity in integrating multiple sensors may limit its applicability in indoor environments or in areas with significant obstructions [17].

Currently, there are various applications of SLAM systems indoors and outdoors, achieving various results in the generation of maps. In [5] an outdoor 3D application for agriculture is shown with the implementation of SLAM in a mobile robot. In [12] GPS and LiDAR with Fast-SLAM are used to detect coherence in maps of petrochemical companies and build maps of fire robots. [6] shows a system intended to generate 3D terrain and tree maps to optimize harvest operations and forest inventory purposes at the individual tree level. 3D point cloud data of the surrounding forest environment is obtained using mobile laser scanning and combining the results with GNSS/INS (Inertial Navigation System) to optimize the trajectory with a graph optimization method. Other investigations to obtain a forest map have been developed in 3D environments [11] and in 2D [9], obtaining characteristics of the environment for mobile navigation based on graph-SLAM.

Despite the multiple works developed in the field, few or no investigations have been developed with a methodology that makes visible the precision of the error generated in SLAM frameworks. This research presents an efficient methodology for the evaluation of the LIO-SAM framework using high-precision mapping measurement equipment.

3   Proposed Method

Six-step techniques are the conduit for the development of this research (Fig. 1). The first fundamental step is the route selection criteria, this allows establishing the base parameters of the route for reliability in data collection, guaranteeing applicability of the method in urban and suburban areas. In the second and third steps, Lidar, GNSS, and RTK equipment is configured for data collection and logging. For the control points, post-processing is necessary to compare the results with the LIO-SAM framework. In the fourth step, a 3D point cloud map is generated with the LIO-SAM framework. Finally, in the fifth and sixth steps, the comparative analysis of the resulting data is carried out where the error generated between both methods is quantified through the use of the RMSE index.

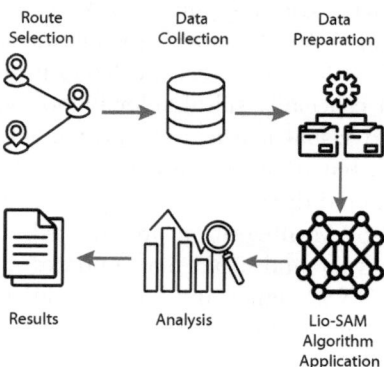

Fig. 1. Proposed analysis method.

3.1   Route Selection Criteria

The following criteria were considered for route selection: the presence of surrounding vegetation, medium height buildings and a mixed road with low vehicular flow and intermediate slope. The surrounding vegetation naturalizes the spaces of urban and suburban areas, and at the same time provides visual reference points for obstacle detection, without interfering with the satellite reception of GNSS devices for positioning correction. Medium height buildings are characteristic of the study area, so it has been considered that the selected route has this type of surrounding buildings. A low traffic density on a mixed road reduces the complexity of the driving environment, providing more predictable and safer conditions for the operation of autonomous vehicles, especially in a later testing phase.

Based on the above, the analysis was carried out in several sections of the "Av. de los Migrantes", the route is presented in Fig. 2. This route has a total length of 3372.491 m. The maximum and minimum altitudes of the route are 2559.61 m and 2523.37 m, respectively.

3.2   Data Collection and Preparation

Figure 3a shows the LiDARS, IMU, and GNSS equipment used in the vehicle considered as the mobile system. In addition, the placement of static differential GPS stations that fulfill the function of total station for the validation of the data obtained by the vehicle. A Robosense LiDAR with sixteen laser beams and

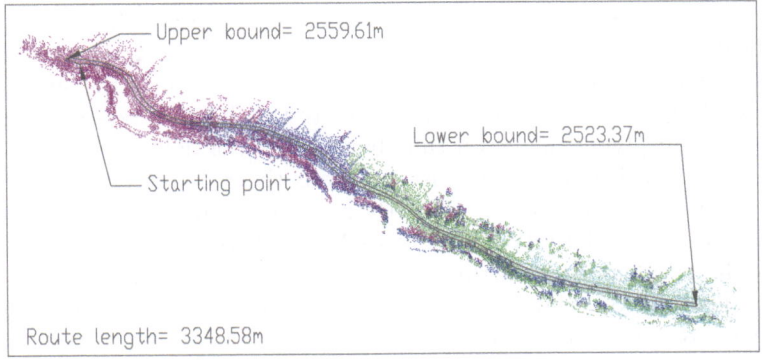

Fig. 2. Selected route "Av. de los Migrantes" in Cuenca, Ecuador.

CHCNAV GNSS are used. The network architecture of the devices used for data collection is shown in Fig. 3b. The data update rates for the different devices are IMU at 100 Hz, LiDAR at 10 Hz, and GPS-RTK at 10 Hz. The Robosense Lidar device is connected to a GNSS antenna which receives signals via internet in real time from the nearest base station to perform the positioning corrections when generating the point cloud.

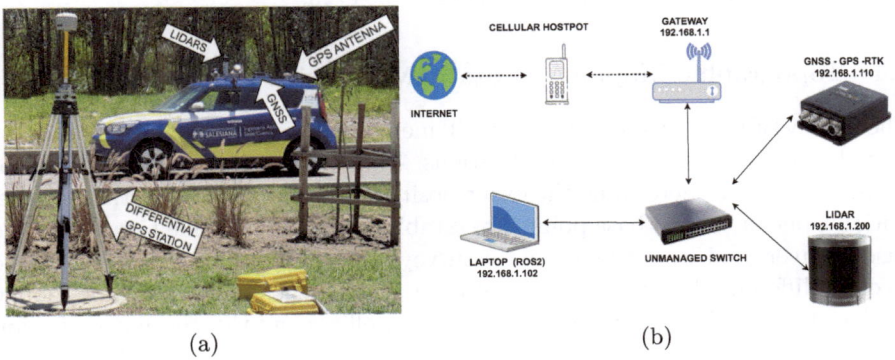

Fig. 3. (a) Mobile laboratory with LiDARS, GNSS, GPS and static differential GPS station (RTK). (b) Network architecture of the mobile laboratory.

3.3 Topographic Mapping in Static Mode

GNSS receivers store spatial coordinate data with high accuracy, these devices receive signals from multiple satellites in space. Each satellite constantly emits signals containing information about its position and time. This information is transmitted via radio frequency signals that are picked up by the GNSS receiver on the Earth's surface. To determine its position, the GNSS receiver uses a process called trilateration. This process involves measuring the time it takes for a signal emitted by a satellite to reach the receiver. Since the signal travels at the speed of light, the receiver can calculate the distance between it and the satellite by multiplying the travel time by the speed of light.

However, to calculate the precise position, the receiver needs information from at least three satellites or more. By receiving signals from multiple satellites and measuring the travel time of each signal, the receiver can determine its three-dimensional position through a process of intersecting spheres (trilateration). The GNSS receiver was linked to the GPS, Galileo (European), and GLONASS (Russian). The equipment was located taking into account an adequate triangulation and coverage of the area, it was positioned with a level tripod and the survey was carried out in static mode.

For the static survey time several important factors were considered, the survey was performed in the morning to have good reception of signals from the systems, the weather was sunny and clear, in the environment there are no high buildings which allows no interference with the satellites, the distance to the base station is less than 5 km in a straight line, so the antennas were left parked for 1:30 h, to obtain good horizontal and vertical accuracy. A suitable observation interval was used to ensure optimal temporal resolution of each antenna with respect to the base station. The horizontal accuracy obtained from the data processing of the deployed GNSS antennas is between 0.007 m to 0.012 m and the vertical accuracy is between 0.013 m to 0.049 m.

3.4 Topographic Mapping in Static Mode with RTK Station

The survey of points using RTK (Real Time Kinematic) is a very useful process in high-precision topographic applications. This method is based on the use of GNSS receivers to determine the exact position of points on the earth's surface. Initially, one or several base points are established in static mode to position the base receiver, and from this the data survey is performed, for which the device receives differential corrections in real time. During the data survey, the mobile receiver tracked the signals from the GNSS satellites and the permanent station and the permanent station of the Instituto Geografico Militar (IGM) in Cuenca to correct the atmospheric and clock errors, thus improving the accuracy of the position obtained from the measured points. The points surveyed with the RTK equipment are shown in Fig. 4 for comparison with the point cloud.

4 LIO-SAM Algorithm

Tightly-Coupled LIO-SAM algorithm uses measurements from sensors including a 3D LiDAR, an IMU, and optionally a GPS to estimate the state and the trajectory of the vehicle. LIO-SAM transforms the raw data from the IMU frame to the LiDAR frame [14]. Apply the proposed IMU integration method in [2] to obtain the relative motion between two-time steps, the approach of the LiDAR

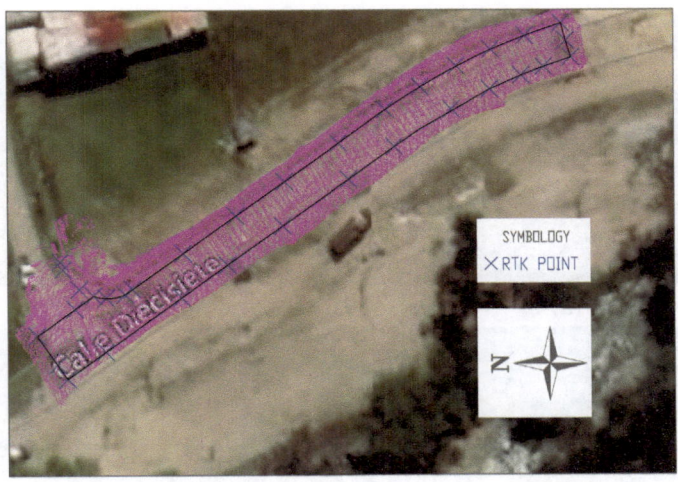

Fig. 4. RTK Points.

odometry proposed in [13], as well as the loop closure factor, for loop closure detection that has a result compatible with [3,4], the aforementioned loop closure corrects deviations, when the vehicle returns to a previously visited location, the algorithm recognizes the environment and adjusts the trajectory.

Table 1 shows the configuration used for the generation of the maps. GPS data plays an important role in correcting the altitude measurements of the point cloud, taking a maximum distance of 0.1 m to update the altitude of the points on the route taken. The LiDAR is configured for a measurement range of 1.5 m to 100 m away. The default LOAM values are used and loop closure is activated with a key frame that is within a 15 m radius for altimetry correction. Figure 5 presents the entire route (Fig. 5a), initial position point (Fig. 5b), mapping result with trees environment (5c) and mapping result with buildings environment (5d).

5   Results and Discussion

The location of the post-processed control points are shown in Fig. 6 as P1, P2, and P3, displaying the distance and triangulation of the points to the IGM base. The control points allow triangulation in the field that allows to have a controlled spatial reference concerning northing, easting, and elevation. The point cloud obtained was processed in the Cloud Compare software, georeferenced in UTM WGS84-17 South coordinates, and exported to the Civil 3D software to compare the coordinates with the control points and with the points surveyed with the RTK station.

The coordinates of the points obtained through the static (GNSS antennas) and dynamic (RTK) survey were compared with the coordinates of the points of the point cloud. Thirty-three points were analyzed in the corresponding section.

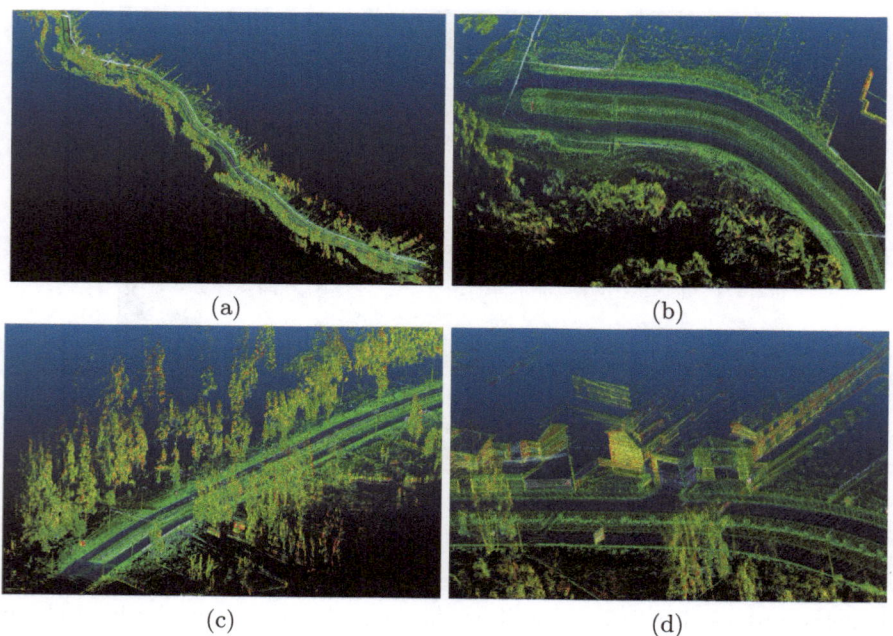

(a)   (b)

(c)   (d)

Fig. 5. 3D point cloud map obtained from LIO-SAM algorithm. (a) Pointcloud map of
the entire route. (b) Initial position for LIO-SAM Loop Closure. (c) Part of the route
with trees environment. (d) A sector of the route with buildings.

Table 1. LIO-SAM Algorithm Configuration.

GPS Settings		LiDAR Settings	
Parameter	Value	Parameter	Value
useImuHeadingInitialization:	false	sensor:	robosense
useGpsElevation:	true	N_SCAN:	16
gpsCovThreshold:	1.0	Horizon_SCAN:	1800
poseCovThreshold:	0.01	downsampleRate:	1
gpsDistance:	0.1	lidarMinRange:	1.5
		lidarMaxRange:	100.0
LOAM feature threshold		Loop closure	
Parameter	Value	Parameter	Value
edgeThreshold:	1.0	loopClosureEnableFlag:	true
surfThreshold:	0.1	loopClosureFrequency:	1.0
edgeFeatureMinValidNum:	10	surroundingKeyframeSize:	50
surfFeatureMinValidNum:	100	historyKeyframeSearchRadius:	15.0
		historyKeyframeSearchTimeDiff:	30.0
		historyKeyframeSearchNum:	25
		historyKeyframeFitnessScore:	0.3

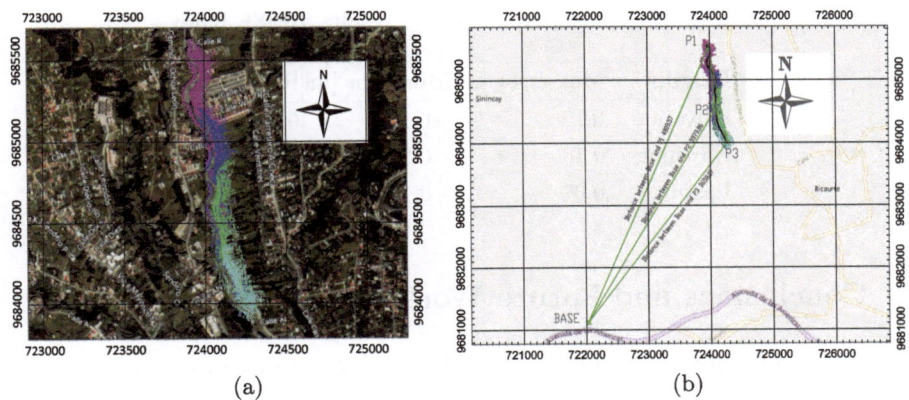

Fig. 6. 3D point cloud map in UTM coordinates (a) Point cloud map obtained from LIO-SAM algorithm (b) Point cloud map obtained from GNSS Antennas

For this study, a comparison was made for northing, easting and elevation. These comparative points are presented in the Fig. 7.

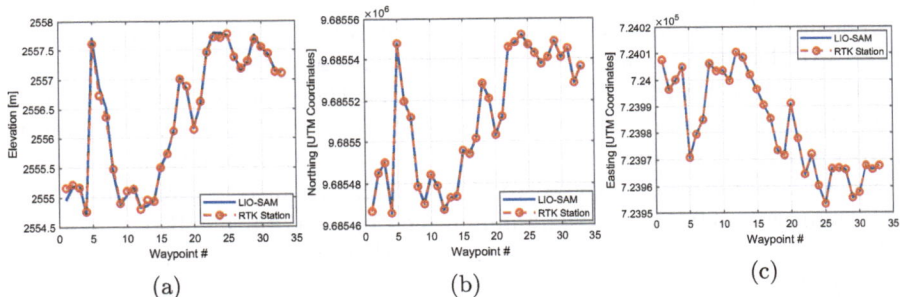

Fig. 7. Pointcloud data comparison in UTM Coordinates. (a) Elevation comparison. (b) Northing comparison. (c) Easting comparison.

To evaluate the accuracy of the data and determine the error, the root-mean-square error (RMSE) calculation was used as a metric based on the differences obtained between the northing, easting and elevation coordinates from the point cloud and the points measured with RTK receiver. This indicator is given by:

$$RMSE = \sqrt{\frac{\sum_{i=1}^{n} (P_{RTK,i} - P_{PCD,i})}{n}} \qquad (1)$$

where, $P_{RTK,i}$ is the ith measured point data, $P_{PCD,i}$ is the ith point from point cloud data and n is the number of measurements available for analysis. These errors allow us to validate the methods used or implement corrections in engineering applications such as road mapping for autonomous vehicle navigation [15]. Table 2 presents the aforementioned comparison parameters.

Table 2. Quantitative odometry results, RMSE index.

Description	Max error [m]	Min error [m]	RMSE [m]
Elevation	0.15	−0.15	0.04
Northing	0.16	−0.2	0.09
Easting	0.07	−0.17	0.08

6   Conclusions and Future Works

This paper presents an accurate methodology for performance analysis of the LiDAR-based 3D LIO-SAM algorithm in urban areas for future applications in autonomous vehicles. Integration of LiDAR, IMU, and GNSS sensors is very important to improve the accuracy of generated 3D point cloud map. By the other hand, the control points generated with RTK base station are necessary to make the comparisons of the error and to obtain the RMSE index. Based on the experimental results obtained from the RMSE index (Table 2), great proximity can be observed between the point cloud data created by the LIO-SAM framework and the geopositioned points of the RTK station in terms of elevation, northing, and easting.

The elevation error shows a symmetric range of ± 0.15 m, indicating that the LIO-SAM algorithm maintains consistency in height detection with a relatively small error. The RMSE of 0.04 m suggests that the elevation accuracy is quite high. Regarding the northing error, it is slightly larger, with a maximum error of 0.16 m and a minimum error of −0.20 m. The RMSE of 0.09 m indicates a larger spread of errors compared to elevation. In the easting direction, it has a maximum error of 0.07 m and a minimum error of −0.17 m. The RMSE of 0.08 m, although slightly lower than northing, shows that there is a similar deviation in the algorithm's accuracy in this direction. These results show a better behavior in comparison with the reference [1, 7] that obtained a RMSE index of 0.3339 m and 0.55m respectively.

In conclusion, the high effectiveness of the LIO-SAM framework for the creation of 3D point clouds is established as a powerful tool for mapping the environment, being used to date for the generation of Lanelet HD autonomous driving maps. However, some aspects are not yet covered in autonomous driving and map generation, these are how to correct the loss of precision in the measurements when there is interference or loss in the satellite signal. The loss or fall of the RTK service to correct the vehicle's geopositioning is another aspect to attack. Therefore, it is recommended to evaluate the incorporation of additional or more advanced sensors that can reduce interference without affecting the precision of GNSS measurements. It is recommended in future work to focus efforts on evaluating other frameworks such as FAST-LIO, visual-LIO, ORB-SLAM, among others, with the method proposed in this paper to determine if they can offer improvements in odometry accuracy.

References

1. Chen, K., Zhan, K., Pang, F., Yang, X., Zhang, D.: R-lio: rotating lidar inertial odometry and mapping. Sustainability **14**(17), 10833 (2022)
2. Forster, C., Carlone, L., Dellaert, F., Scaramuzza, D.: On-manifold preintegration for real-time visual-inertial odometry. IEEE Trans. Rob. **33**(1), 1–21 (2016)
3. Guo, J., Borges, P.V., Park, C., Gawel, A.: Local descriptor for robust place recognition using lidar intensity. IEEE Robot. Autom. Lett. **4**(2), 1470–1477 (2019)
4. Kim, G., Kim, A.: Scan context: egocentric spatial descriptor for place recognition within 3d point cloud map. In: 2018 IEEE/RSJ International Conference on Intelligent Robots and Systems (IROS), pp. 4802–4809. IEEE (2018)
5. Kim, H.G., Lee, H.M., Lee, S.H.: A new covariance intersection based integrated slam framework for 3d outdoor agricultural applications. Electron. Lett. **60**(9), e13206 (2024)
6. Kukko, A., Kaijaluoto, R., Kaartinen, H., Lehtola, V.V., Jaakkola, A., Hyyppä, J.: Graph slam correction for single scanner mls forest data under boreal forest canopy. ISPRS J. Photogramm. Remote. Sens. **132**, 199–209 (2017)
7. Li, S., He, R., Guan, H., Shen, Y., Ma, X., Liu, H.: A 3d lidar-inertial tightly-coupled slam for mobile robots on indoor environment. IEEE Access (2024)
8. Malik, S., Khan, M.A., El-Sayed, H., Khan, M.J.: Should autonomous vehicles collaborate in a complex urban environment or not? Smart Cities **6**(5), 2447–2483 (2023). https://doi.org/10.3390/smartcities6050111, https://www.mdpi.com/2624-6511/6/5/111
9. Miettinen, M., Ohman, M., Visala, A., Forsman, P.: Simultaneous localization and mapping for forest harvesters. In: Proceedings 2007 IEEE International Conference on Robotics and Automation, pp. 517–522. IEEE (2007)
10. Mur-Artal, R., Montiel, J., Tardos, J.D.: ORB-SLAM: a versatile and accurate monocular SLAM system. IEEE Trans. Rob. **31**(5), 1147–1163 (2015)
11. Pierzchała, M., Giguère, P., Astrup, R.: Mapping forests using an unmanned ground vehicle with 3d lidar and graph-slam. Comput. Electron. Agric. **145**, 217–225 (2018)
12. Shamsudin, A.U., et al.: Consistent map building in petrochemical complexes for firefighter robots using slam based on gps and lidar. Robomech J. **5**, 1–13 (2018)
13. Shan, T., Englot, B.: Lego-loam: Lightweight and ground-optimized lidar odometry and mapping on variable terrain. In: 2018 IEEE/RSJ International Conference on Intelligent Robots and Systems (IROS). pp. 4758–4765. IEEE (2018)
14. Shan, T., Englot, B., Meyers, D., Wang, W., Ratti, C., Daniela, R.: Lio-sam: tightly-coupled lidar inertial odometry via smoothing and mapping. In: IEEE/RSJ International Conference on Intelligent Robots and Systems (IROS), pp. 5135–5142. IEEE (2020)
15. Valladolid, J.D., Patiño, D., Ortiz, J.P., Minchala, I., Gruosso, G.: Proposal for modeling electric vehicle battery using experimental data and considering temperature effects. In: 2019 IEEE Milan PowerTech, pp. 1–6 (2019). https://doi.org/10.1109/PTC.2019.8810611
16. Wang, J., Rünz, M., Agapito, L.: Dsp-slam: object oriented slam with deep shape priors. In: 2021 International Conference on 3D Vision (3DV), pp. 1362–1371. IEEE (2021)

17. Zhang, J., Singh, S., Kaess, M.: LIO-SAM: tightly-coupled Lidar-inertial odometry and mapping. In: 2014 IEEE International Conference on Robotics and Automation (ICRA), pp. 2327–2333. IEEE (2014)
18. Zhao, C., Li, L., Pei, X., Li, Z., Wang, F.Y., Wu, X.: A comparative study of state-of-the-art driving strategies for autonomous vehicles. Accident Anal. Prevent. **150**, 105937 (2021)

Effect of Oxyhydrogen Gas (HHO) Addition on Fuel Consumption of M2 Category Vehicle by Road Tests

Marcelo Estrella-Guayasamín[1]([⊠]) [ID], Victor Vivar Quiroz[1] [ID],
Aaron Delgado Quinto[1] [ID], and Fernando Gomez Berrezueta[2] [ID]

[1] GMovInt, Universidad Politécnica Salesiana, Guayaquil, Ecuador
`mestrellag@ups.edu.ec`, {`vvivarq,adelgadoq1`}`@est.ups.edu.ec`
[2] Universidad Internacional del Ecuador, Guayaquil, Ecuador
`magomezbe@uide.edu.ec`

Abstract. A change in fossil fuel consumption within the automotive sector is crucial due to its significant impact on climate change. This study assessed the effect of electrolyte concentration and HHO flow rate on fuel consumption in an M2 vehicle. A 2023 KARRY Q22L AC 1.2 5P 4X2 TM van, classified as M2, was equipped with a wet cell HHO gas generator powered by a variable DC source. The cell consisted of two 316 stainless steel plates, each 10 cm by 10 cm by 1.5 mm, containing 250 cm of electrolyte. The electrolyte was made from 1 liter of distilled water and KOH concentrations ranging from 0.5% to 1.5%. HHO flows of 1.03, 1.31, and 1.69 slpm were achieved with currents of 2, 7, and 17 A, respectively. Fuel consumption was measured gravimetrically during 40 km road tests on the Virgen de Ftima - Puerto Inca route, following the extra-urban driving cycle (EUDC) for low-power vehicles. Fuel consumption reductions of 5–8% were observed with increased HHO flow and higher KOH concentrations.

Keywords: HHO gas · Wet cell · Fuel consumption

1 Introduction

Oil, natural gas and coal are the most used and exploited fossil fuels around the world, leaving as a result of their use a deadly footprint on the planet of a bionatural and social nature, evident as: changes and destruction of ecosystems, displacement and transformation of ancestral cultures, over-exploitation and poverty, among others [13].

The Intergovernmental Panel on Climate Change - IPCC, of the UN, estimates the effects of the increase in global temperature, such as changes in biodiversity, shortages of drinking water, lack of access to healthy food and social conflicts over obtain natural resources and, therefore, increase in migration; and insists that every tenth of a degree that rises or falls is crucial for our planet. Since the pre-industrial era, the world has warmed +1.2° and it is presumed,

E. M. Inga Ortega et al. (Eds.): CITIS 2024, LNNS 1331, pp. 227–237, 2025.
https://doi.org/10.1007/978-3-031-87065-1_21

thanks to its studies, that there will be an increase between 2030–2035 by +1.5°, therefore, one of the most prevailing needs is to modify fossil energy production and close its existing facilities as soon as possible, as recommended by the Sixth Assessment Cycle on climate change, which was proposed in Egypt in March 2023 [14].

The use of alternative and renewable energy sources or fuels is imperative, because in addition to the damage that fossil fuels do to the environment, the deposits of these resources are being diminished by their continuous and indiscriminate extraction [7,13].

The automotive industry is the main responsible for producing the largest proportion of polluting gases, especially carbon emissions, on a global scale, which makes the search for clean energies urgent, [10] thus leading the automotive industry to develop new lines of research to increase engine efficiency and reduce emissions of polluting gases [8].

Alternative fuels, such as biofuel and green hydrogen, are an option for the automotive industry [5],as it helps reduce greenhouse gas emissions and toxic gases such as carbon monoxide (CO), sulfur dioxide, lead, hydrocarbons (HC) and soot particles [4,20].

For several researchers, hydrogen as an energy vector is an option to reduce the polluting gases of fossil energy, because, in addition to being the most abundant element on Earth, it cannot be destroyed and during its consumption it changes. State is from water to hydrogen and back to water [2,17]. In addition, they highlight its attractive long-term storage capacity, but the most important thing is that it is capable of transferring energy for all energy sectors, both transportation and the chemical and industrial industry, although the manufacture of engines that integrate hydrogen as fuel represents a large commercial cost in its manufacture [8]

Among the main properties of hydrogen that demonstrate its superiority over fossil fuels is its octane rating of 130 compared to 92 and 97 for gasoline distributed in the country [12]. Another advantage is that its combustion has a short duration, this is due to the speed of propagation of its flame which is approximately 60 times higher than that of gasoline [6], burning the hydrogen much faster, which can reduce the ignition times in spark ignition engines. In addition, having a greater calorific value and being one of the most abundant compounds on the planet [11] These perspectives support the position of hydrogen as a promising and efficient alternative in the search for cleaner and more sustainable energy. Currently, hydrogen technology is quite expensive both to obtain and to implement in vehicles with internal combustion engines or with fuel cells [23]. However, a short-term application option and Relatively economical is the use of hydrogen as a secondary fuel [20] obtained through electrolyzers with fossil fuels [15,19], of this manera takes advantage of the properties of hydrogen to reduce fuel consumption and polluting emissions.

Several studies carried out on internal combustion engines with the combination of fossil fuels and Hydroxy gas (HHO) have evaluated some of the advantages of their use, among them are:

1. Improvement in engine performance, obtaining increases in engine torque of up to 19.1% [25] and improvements in power of up to 14% [6].
2. Reduction of specific fuel consumption, between 5% and 14% for gasoline engines ?,[1,6,25] and up to 20% for diesel engines
3. Reduction of polluting gases, approximate reductions of 10% to 14% were evident for CO2 [16,25] and of 20% in HC and CO hydrocarbon emissions [6,16].
4. Reduction of opacity in Diesel engines, between 8% and 25% depending on the engine speed [1].

These advantages are possible due to the properties of the HHO gas that is obtained from the dissociation of the hydrogen and oxygen atoms of water, producing better combustion, generating a rapid increase in temperature and pressure, which minimizes detonation and thus the reduction in engine noise due to the effect of ignition delay; thus showing that the use of HHO as a secondary fuel is a promising solution to mitigate greenhouse gas emissions.

Hydroxy gas, also known as Brown gas, is obtained from electrolyzers. Electrolysis is the process by which electricity is used to split water into hydrogen and oxygen, this reaction takes place in a unit called an electrolyzer. Electrolyzers consist of an anode and a cathode separated by an electrolyte that, when applying direct current, Hydrogen ions in water undergo a reduction reaction at the cathode to precipitate hydrogen, and hydroxide ions undergo an oxidation reaction at the anode to precipitate oxygen. The most common electrolyzers in vehicle applications are dry cell and wet cell electrolyzers [1,9,26]. The difference between them is that dry cell electrolyzers have a tank external, while in wet cell ones its reservoir is the same as the electrolyzer, very similar to a lead acid battery.

Category M2 vehicles are used to transport people with more than nine seats, including the driver, and their maximum mass does not exceed 5 tons of weight, such as minibuses and cargo vans, they are essential in urban transport and rural. Due to their long distances, the emissions rate of this type of vehicle is highly representative, so any improvement in the combustion process in the engine would generate reductions in fuel consumption and therefore emissions. Thus, considering this reality and the previous studies, this study planned a set of experiments to evaluate how the Hydroxy - Gasoline mixtures and the electrolyte concentration influence the fuel consumption of an M2 type vehicle. The experiment was carried out in the city of Guayaquil - Ecuador.

2   Methods

2.1   Experimental Setup

The equipment and instruments used in the experimental setup of this study are shown in Fig. 1. The M2-type vehicle used for the experiment was a Karry van model Q22L, 1200 cc, 4X2 MT, 2023 (1) with a capacity of 11 passengers. Prior to the study, complete corrective maintenance of the fuel supply and ignition system was carried out to ensure the operability of the engine for the study.

A HHO gas generation system of the wet cell type (3) was installed on the engine of the test vehicle, which is characterized by its greater generation of HHO gas compared to the dry cell [22] . The generator was powered by a PSW 80–27 variable DC power supply from the Instek brand (2), capable of regulating voltages from 0 to 80 V and currents from 0 to 27 A. The cell structure consists of two plates 316L stainless steel, with dimensions of 10 cm wide by 10 cm long and 1.5 mm thick. The electrolyte (4) is composed of 1 liter of distilled water with a concentration of KOH, this was used because according to studies carried out is the most effective in the production of HHO [1,3,24].

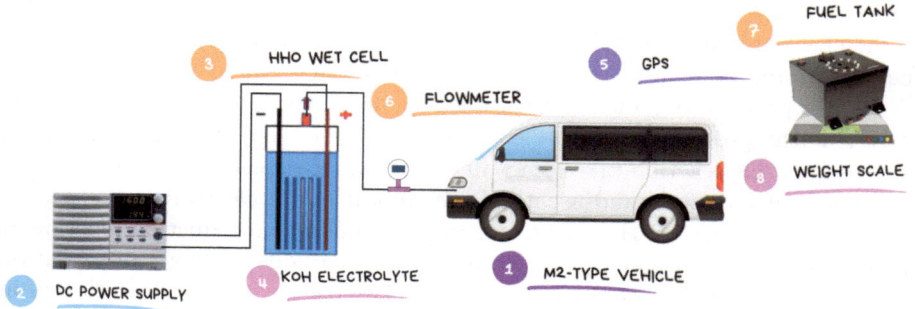

Fig. 1. Components and experimental setup of the study.

The HHO gas flow was dosed directly to the engine intake manifold constantly and was controlled by a SIARGO brand flowmeter (6), model MF5708, which has an operating range of 0–100 slpm with a response speed of less than 2 s [22]. To measure fuel consumption, the gravimetric method was used, which consists of quantifying the difference in weight between the fuel with which the test begins and ends [27,28]. For this purpose, an external fuel tank (7) with capacity for 4 gallons of fuel was built, which has the vehicle's original fuel pump for its correct operation. To quantify the weight of the fuel, a digital scale with a precision of 0.5 g was used (8); its calibration was performed using standard weights. The recording of the route traveled and the speed profile was carried out with a GPS speedometer application installed on a mobile phone (5). This application allows data on latitude, longitude, speed, and time to be obtained with a frequency of 2 Hz at .cvs format for further analysis

2.2   Experimental Methodology

The experimental methodology adopted herein aimed to assess the impact of electrolyte concentration and HHO flow rate, utilized as a supplementary fuel, on the overall fuel consumption of an internal combustion engine fueled with regular 87-octane gasoline. The HHO flow rates were obtained by adjusting the electrical current input to the electrolytic cell. Currents of 2 A, 7 A, and 17 A were employed to attain HHO flow rates denoted as, $HHO_1 = 1{,}03$ standard liters per minute (slpm), $HHO_2 = 1.31$ slpm, and $HHO_3 = 1.69$ slpm, correspondingly. These currents were chosen based on the minimum and maximum current values allowed for the HHO wet cell under study. Currents higher than 17A can produce high temperatures that would lead to evaporation of the electrolyte and generation of water vapor. Concurrently, the influence of varying concentrations of potassium hydroxide (KOH) on the electrolyte was examined. Concentrations of 0.5%, 1%, and 1.5% by mass of KOH were scrutinized, with a consistent electrical current of 17 A maintained throughout the experiments. These concentrations were chosen based on a previous study carried out by Aydin [3]. The experimental road tests were conducted along the Virgen de Ftima - Puerto Inca - Virgen de Ftima route, encompassing a total round trip distance of 40 km. This particular route was selected due to its alignment with the Extra-Urban Driving Cycles for low-power vehicles (EUDC) standard [30], ensuring optimal testing conditions.

The protocol employed to conduct the test begins with a comprehensive assessment of the vehicles conditions before the test. This includes verifying the tire pressure (30 psi), ensuring the engine temperature is within the normal operating range (85–92 °C), and confirming that the air conditioning is turned off. Subsequently, the test conditions are established according to the experimental design. The HHO cell is then filled with the appropriate electrolyte solution, ensuring there are no leaks in the system. The flow rate of the HHO gas is determined by setting the electric current applied to the HHO cell, which is verified using a flowmeter. Next, the initial weight of the fuel tank filled with fuel is measured to establish a baseline. The GPS system is then activated, and the test drive is conducted along the predefined route. During the test drive, it is crucial to maintain the vehicle speed within the specified range. Upon completion of the test drive, the weight of the fuel tank is measured again to determine the fuel consumption. This is calculated based on the difference in the fuel tanks weight before and after the test drive, considering the fuels density.

3   Results

Figure 2 shows the installation of the HHO generator in the test vehicle. The results of the six replications of the completely randomized set of experiments are shown in Table 1. The normality analysis was conducted on the data obtained for each treatment using the Ryan-Joiner (RJ) statistic. The RJ value was found to be 0.978, which is very close to 1, suggesting that the population likely follows a normal distribution. This inference is supported by a p-value greater than 0.1, exceeding the significance level of 0.05. Additionally, Levene's test was utilized to assess the homogeneity of variance in our data, resulting in a p-value of

0.131, which surpasses the 0.05 threshold, thereby confirming the assumption of homogeneity of variance.

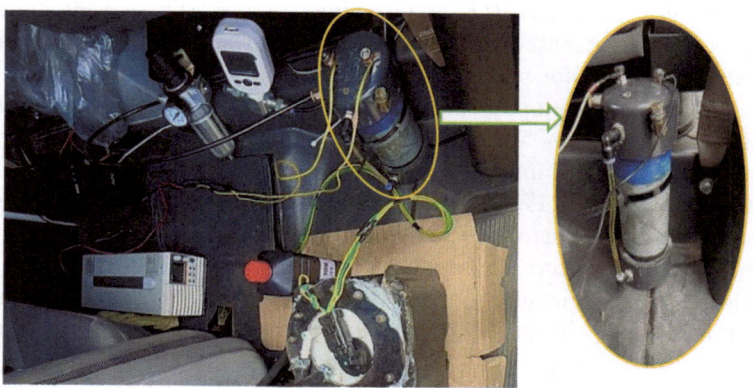

Fig. 2. Installing the HHO generator on the test vehicle engine.

Table 1. Fuel consumption for different treatments

Treatment	A	B	Fuel consumption [lt]						Mean
			R1	R2	R3	R4	R5	R6	
0	0	-	3,300	3,140	3,270	3,100	3,300	3,240	3,225
1	1	3	3,050	3,060	3,050	3,060	3,050	3,050	3,053
2	2	3	3,023	3,037	3,023	3,023	3,023	3,037	3,028
3	3	3	2,950	2,920	2,980	2,940	2,950	2,980	2,953
4	-	1	3,090	3,110	3,120	3,090	3,110	3,110	3,105
5	-	2	3,060	3,080	3,060	3,080	3,090	3,060	3,072
6	-	3	2,950	2,920	2,980	2,940	2,950	2,980	2,953

Factor A (Fuel, Level: 0=Gasoline only, 1=Gasoline+HHO$_1$, 2=Gasoline+HHO$_2$ and 3=Gasoline+HHO$_3$), Factor B (KOH concentration in electrolyte, Level: 1=0.5%, 2=1% and 3=1.5%).

An ANOVA analysis was performed to determine if there are significant differences in fuel consumption means when varying the flow of HHO supplied to the engine as a secondary fuel along with gasoline. The analysis revealed that variations in HHO flow significantly affect fuel consumption, as indicated by a p-value less than the significance threshold. Figure 3 displays a decreasing trend in fuel consumption as the flow of HHO increases. Specifically, the mean consumption decreases from 3.225 liters (T0 - Gasoline only) to 2.953 liters (T3 - Gasoline + HHO_3). A comparison of the different treatments using the Tukey

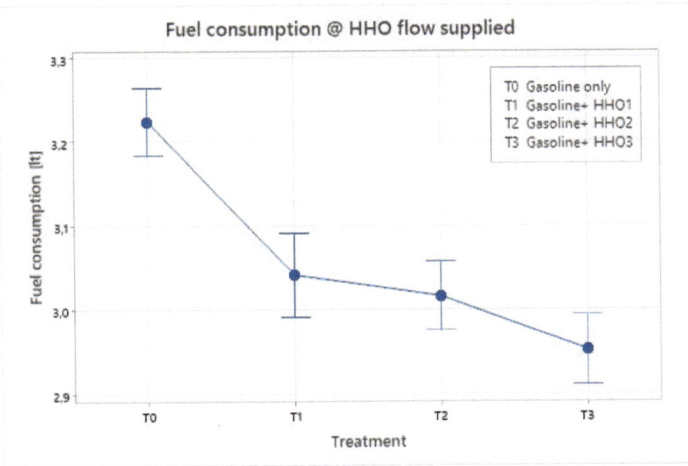

Fig. 3. Fuel consumption vs HHO flow supplied.

method shows that the mean fuel consumption for the control treatment T0 differs significantly from treatments T1, T2, and T3.

A similar analysis was performed to evaluate the effects on fuel consumption by varying the concentration of KOH in the electrolyte, using a constant current of 17 A. Changes in KOH concentration were found to influence the fuel consumption mean, as indicated by a p-value below the significance level.

Figure 4 demonstrates that increasing the KOH concentration in the electrolyte causes a decrease in fuel consumption. The fuel consumption mean for the treatments was 3.225 l for treatment T0 (gasoline only) and 3.105 l, 3.072 l, and 2.953 l for treatments T4, T5, and T6, respectively. In addition, the comparison of the different treatments using Tukey's method showed that significant differences were observed when comparing the control treatment T0 with treatments T4, T5, and T6.

4   Discussion

Based on the values obtained from the experimental tests, it can be observed that the increase in the flow of HHO supplied as a secondary fuel to the category M2 vehicle's engine reduces fuel consumption. Specifically, reductions of 5.32%, 6.11%, and 8.42% were experienced with HHO gas contributions to the engine of 1.03, 1.31, and 1.69 slpm, respectively (Fig. 3). Various studies [1,6,25] demonstrate that this behavior is due to the addition of HHO gas improving the combustion process, thereby increasing engine efficiency, which translates into

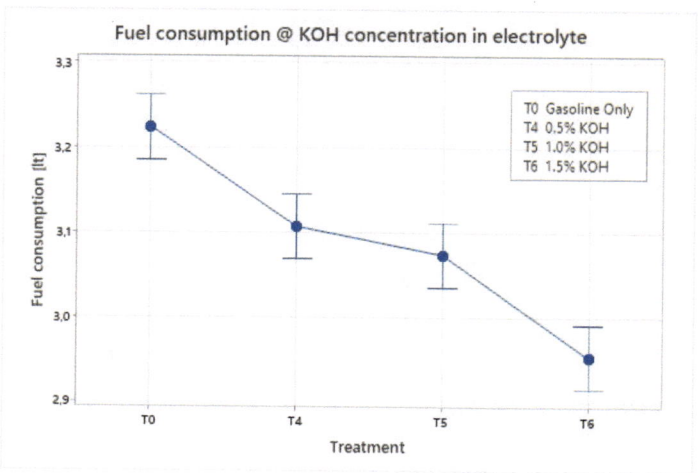

Fig. 4. Fuel consumption vs KOH concentration in electrolyte.

reduced fuel consumption during the test route performed by the category M2 vehicle.

Moreover, when analyzing the effect of KOH electrolyte concentration on fuel consumption, it is evident that the use of electrolyte concentrations of 0.5%, 1%, and 1.5% KOH by mass and a current of 17 A applied to the electrolytic cell resulted in reductions of 3.72%, 4.75%, and 8.42%, respectively (Fig. 4). According to Rusdianasari [18] and Sivakumar [21], this is because the electrolyte concentration affects HHO gas production; higher concentrations generate a greater flow, leading to more significant reductions in fuel consumption. The maximum reduction values achieved in this study (8%) are lower compared to those obtained in other studies (15–30%). This discrepancy may be attributed to several factors: first, the type of cell used; this experiment employed a wet cell, whereas the values reported in [1] were obtained using a dry cell. Second, the size and number of plates or electrodes; in this study, two plates of 10 x 10 cm (one positive and one negative) were used, whereas in [1], seven plates of 16 x 20 cm (one positive, two negative, and four neutral) were used.

5   Conclusions

This study was conducted to assess the impact on fuel consumption within a type M2 vehicle when utilizing varied Hydroxy-Gasoline mixtures as supplementary fuel, alongside diverse concentrations of KOH in the electrolyte, through road trials.

The conducted road trials adhered to the EUDC standard and were executed along the Virgen de Ftima - Puerto Inca - Virgen de Ftima Route, spanning a distance of 40 km.

Notably, three distinct levels of HHO flow were administered, resulting in appreciable reductions in fuel consumption ranging from 5% to 8% as the HHO flow rate was adjusted from 1.03 slpm to 1.69 slpm. Concomitantly, analogous trends were observed upon escalating the concentration of KOH within the electrolyte from 0.5% to 1.5% by mass. These results emphasize the potential efficacy of incorporating Hydroxy-Gasoline mixtures and optimizing electrolyte composition as a means to reduce fuel consumption dynamics in vehicular applications.

References

1. Al-Dawody, M.F., et al.: Using oxy-hydrogen gas to enhance efficacy and reduce emissions of diesel engine. Ain Shams Eng. J. (2023). https://doi.org/10.1016/j.asej.2023.102217
2. Ares, J., Leardini, F., Sánchez, C., Fernández, J., Ferrer, I.: El hidrógeno como vector energético: Mucho hecho pero casi todo por hacer, pp. 1–10 (2019). http://www.encuentros-multidisciplinares.org/revista-62/jose-ares_y_otros.pdf
3. Aydin, K., Kenanoglu, R.: Effects of hydrogenation of fossil fuels with hydrogen and hydroxy gas on performance and emissions of internal combustion engines **43**, 14047–14058 (2018). https://doi.org/10.1016/j.ijhydene.2018.04.026
4. Cepal: Lanzamiento de la plataforma h2lac: el poder del hidrógeno verde de latinoamérica para la transición energética mundial (2023). https://www.cepal.org/es/notas/lanzamiento-la-plataforma-h2lac-poder-hidrogeno-verde-latinoamerica-la-transicion-energetica
5. Dawud, F., Andá, M., Shafiullah: Producción de hidrógeno para energía: una visión general **45**, 3847–3869 (2020). https://doi.org/10.1016/j.ijhydene.2019.12.059
6. Dewangan, A., et al.: Production of oxy-hydrogen gas and the impact of its usability on ci engine combustion, performance, and emission behaviors. Energy **278** (2023). https://doi.org/10.1016/j.energy.2023.127937
7. do Sacramento, E., de Lima, L., Oliveira, C., Veziroglu, T.N.: A hydrogen energy system and prospects for reducing emissions of fossil fuels pollutants in the ceará state-brazil. Int. J. Hydrogen Energy **33**(9), 2132–2137 (2008). https://doi.org/10.1016/j.ijhydene.2008.02.018
8. Díaz-Rey, A., González-Gil, J.: Análisis de un generador de hho de celda seca para su aplicación en motores de combustión interna analysis of a dry cell hho generator for application in internal combustion engines. Revista UIS Ingenierías **17**, 143–154 (2018). https://www.redalyc.org/journal/5537/553756967014/
9. EERE: Hydrogen production: Electrolysis — department of energy (2). https://www.energy.gov/eere/fuelcells/hydrogen-production-electrolysis
10. EPA, U.: Emisiones de dióxido de carbono (2017). https://espanol.epa.gov/la-energia-y-el-medioambiente/emisiones-de-dioxido-de-carbono
11. Gutiérrez, L.: El hidrógeno, combustible del futuro. Real Academia de Ciencias Exactas, Físicas y Naturales **99**, 49–67 (2005). https://rac.es/ficheros/doc/00447.pdf
12. Hurtado, J.I.L., Soria, B.Y.M.: El hidrógeno y la energía. Asociación Nacional de Ingenieros del ICAI (2007)
13. IPCC: 5 gráficos claves del último informe del ipcc (2023). https://porelclima.org/actua/ambicioncop/actualidad/5615-5-graficos-claves-del-ultimo-informe-del-ipcc

14. IPCC: Nuestras decisiones de hoy repercutirán en todo el mundo durante cientos de años (2023). https://porelclima.org/actua/ambicioncop/actualidad/5610-nuestras-decisiones-de-hoy-repercutiran-en-todo-el-mundo-durante-cientos-de-anos

15. Jakliński, P., Czarnigowski, J.: An experimental investigation of the impact of added hho gas on automotive emissions under idle conditions. Int. J. Hydrogen Energy **45**, 13119–13128 (2020). https://doi.org/10.1016/j.ijhydene.2020.02.225

16. Mendoza-Casseres, D., Valencia-Ochoa, G., Duarte-Forero, J.: Evaluación experimental del rendimiento de combustión en motores estacionarios de baja cilindrada que funcionan con mezclas de biodiesel e hidroxi (2021). https://doi.org/10.1016/j.tsep.2021.100883

17. Momirlan, M., Veziroglu, T.: Las propiedades del hidrógeno como combustible del mañana en un sistema energético sostenible para un planeta más limpio. Revista internacional de energía del hidrógeno **30**, 795–802 (2005). https://www.sciencedirect.com/science/article/abs/pii/S0360319904003398

18. Rusdianasari, Bow, Y., Dewi, T.: Hho gas generation in hydrogen generator using electrolysis. IOP Conf. Series: Earth Environ. Sci. **258** (2019). https://doi.org/10.1088/1755-1315/258/1/012007

19. Salek, F., Zamen, M., Hosseini, S.V.: Experimental study, energy assessment and improvement of hydroxy generator coupled with a gasoline engine. Energy Reports **6**, 146–156 (11 2020). https://doi.org/10.1016/J.EGYR.2019.12.009

20. Shadidi, B., Najafi, G., Yusaf, T.: A review of hydrogen as a fuel in internal combustion engines. Energies **14** (2021). https://doi.org/10.3390/en14196209

21. Sivakumar, B., Navakrishnan, S., Cibi, M., Senthil, R.: Generation of brown gas from a dry cell hho generator using chemical decomposition reaction. IOP Conf. Series: Materials Sci. Eng. **1130**, 012002 (4 2021). https://doi.org/10.1088/1757-899X/1130/1/012002, https://iopscience.iop.org/article/10.1088/1757-899X/1130/1/012002

22. Soly, A.E., Kady, M.E., Farrag, A.E.F., Gad, M.: Comparative experimental investigation of oxyhydrogen (hho) production rate using dry and wet cells. Int. J. Hydrogen Energy **46**, 12639–12653 (4 2021). https://doi.org/10.1016/j.ijhydene.2021.01.110

23. Subramanian, B., Ismail, S.: Production and use of HHO gas in IC engines. Int. J. Hydrogen Energy **43**, 7140–7154 (4 2018). https://doi.org/10.1016/J.IJHYDENE.2018.02.120

24. Tamayo, E., Rosales, C., Guzman, A., Pasmiño, P.: Efecto del uso de hidrógeno en la potencia y rendimiento de un motor de combustión interna **7**, 43–54 (2016). https://www.redalyc.org/journal/5722/572261626004/html/

25. Yilmaz, A.C., Uludamar, E., Aydin, K.: Effect of hydroxy (hho) gas addition on performance and exhaust emissions in compression ignition engines. Int. J. Hydrogen Energy **35**, 11366–11372 (10 2010). https://doi.org/10.1016/j.ijhydene.2010.07.040

26. Zárate, H., Silva, R.D., Martins, M., Rodrigues, C., Lima, M.: Investigación experimental de la adición de hidrógeno en el aire de admisión de motores de encendido comprimido que funcionan con una mezcla de biodiesel **42**, 4530–4539 (2017). https://doi.org/10.1016/j.ijhydene.2016.11.032

27. Kang, J.Y., Choi, J.G., Lee, J.H., Lee, C.G., Ko, S.C., Lee, D.: A platform study of fuel consumption measurements for an excavator in motion. J. Drive Control **14**, 35–40 (2017). https://doi.org/10.7839/ksfc.2017.14.1.035
28. Tibaquirá, J., Huertas, J., Ospina, S., Quirama, L., Niño, J.: The effect of using ethanol-gasoline blends on the mechanical, energy and environmental performance of in-use vehicles. Energies **11**, 221 (2018). https://doi.org/10.3390/en11010221

Sustainability and Environment

Enhanced Reactive Power Compensation for Flicker Mitigation in Wind Farm-Integrated Distribution Networks Using Advanced D-STATCOM Control

Kevin Joel Vela Palaquibay⑩, Manuel Darío Jaramillo$^{(\boxtimes)}$⑩, and Diego Carrión⑩

Electrical Engineering Department, Universidad Politcnica Salesiana, Quito, Ecuador
mjaramillo@ups.edu.ec

Abstract. This research aims to demonstrate the importance of reactive power compensation in electrical systems for mitigating fast Flicker voltage disturbances that commonly occur in systems with wind generation; thereby, this paper's main goal is to improve the quality of distribution systems. The 34-node IEEE test system was chosen as the base system for this study. Simulations were conducted using MATLAB to verify system stability and ensure the system's nominal voltage matches the design specifications set by the IEEE. The analysis identified bus 27 as having the greatest voltage fluctuation throughout the system. Case studies were conducted on this bus, including the connection of dynamic load and a wind farm (WF) from the MATLAB library, which consists of six 1.5 MW turbines. The Flickermeter in MATLAB, modeled according to IEC-61000 for measuring the short-term Flicker severity index (Pst), was used to evaluate the increase in Pst caused by WF and dynamic loading. To mitigate the flicker severity index, a D-STATCOM of $+/-3MVAR$ based on PID control was implemented. This device was connected to the 34-node IEEE system to reduce the voltage variations produced by both loads.

Keywords: Electrical Power System · FACTS · Flicker · Static Synchronous Compensator · Wind farm

1 Introduction

Distribution systems of three-phase, four-wire configurations widely supply energy to low-voltage loads, including industrial, commercial, and residential sectors. These loads are often unbalanced and non-linear, causing various problems that negatively affect power quality (PQ) [1]. Smart Grids (SG) growth introduces additional PQ issues such as poor voltage regulation, harmonic currents, considerable neutral currents, high reactive power demand, and load imbalance [1].

© The Author(s) 2025
E. M. Inga Ortega et al. (Eds.): CITIS 2024, LNNS 1331, pp. 241–251, 2025.
https://doi.org/10.1007/978-3-031-87065-1_22

SG growth has led to increased use of distributed generation (DG), including renewable energies like wind and photovoltaic (PV) generation, which require relatively lower investment and space compared to hydroelectric generation [2]. The demand for renewable energy is driven by their regenerative capacity, constant availability, pollution-free nature, and low maintenance costs [3].

However, renewable energies also introduce challenges. Wind energy, in particular, has inconsistent and stochastic wind speeds, leading to dynamic instability in wind farms and network instability due to load variations and wind turbine intermittencies [4–6]. Voltage instability from abrupt load changes necessitates solutions to stabilize voltage [7].

Reactive power control stabilizes voltage and power supply, with FACT devices being the most efficient for reactive power management [8,9]. Power electronics-based devices, known as Flexible Alternating Current Transmission Systems (FACTS), can rapidly control voltage and reactive power [10]. Among these, the unified power flow controller (UPFC) functions as a STATCOM for voltage control [11].

FACTS devices can also control voltages in series to influence active power. Early devices used thyristors, while later ones use voltage source converters (VSC) [12]. The Distribution Static Compensator (DSTATCOM) is popular for addressing PQ issues in distribution systems, [3], [12]. Although DSTAT-COM systems are more complex and expensive than Static Var Compensators (SVC), [3], they offer enhanced control capabilities, operating through inverters or transformer-based topologies.

2 Methodology

Electrical distribution systems integrated with PV energy sources or wind generation often face power quality issues, notably Flicker, which are rapid voltage disturbances affecting equipment performance and customer well-being [13]. To address these issues, this study evaluates Flicker severity in the IEEE 34-bus system, which is a test system from the Institute of Electrical and Electronics Engineers under three different scenarios using a Flickermeter designed per IEC 61000-4-15 standards [14,15].

The IEEE 34-bus system represents a feeder in Arizona with a nominal voltage of 24.9 kV, one generating unit, no transformers, and 29 loads distributed across different nodes.

2.1 Flicker Measurement Process

Flicker measurement involves five procedural modules, grouped into three simulation stages to simulate the lamp-eye-brain response [15,16]:

Stage 1: Signal Normalization - Module 1: Normalizes and converts the input signal to a reference signal based on the system's Vrms.

Stage 2: Eye-Brain Perception Simulation - Module 2: Applies a quadratic multiplier to recover voltage fluctuation by squaring the input voltage, simulating a lamp's behavior. - **Module 3:** Consists of two filters: - **First**

filter: A low-pass filter eliminates double-frequency ripple components at 42 Hz for 120 V lamps at 60 Hz. - **Second filter:** A second-order high-pass filter with an fc of 0.05 Hz removes any DC voltage component. - **Module 4:** Comprises a quadratic multiplier and a first-order low-pass filter to simulate human eye perception, producing the Pst (instantaneous Flicker sensation); this process is fully detailed in Fig. 1.

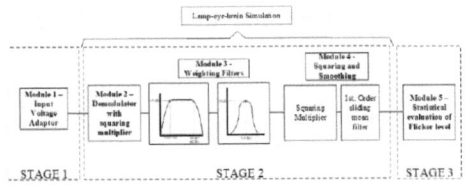

Fig. 1. Lamp-eye-brain simulation

Stage 3: Statistical Analysis - Module 5: Conducts statistical analysis of the Pst, sampling the Pst level into typically 256 class samples and obtaining percentile values (P0.1 s, P1s, P3s, P10s, and P50s), representing flicker levels exceeded by 0.1%, 1%, 3%, 10%, and 50% of the observation time.

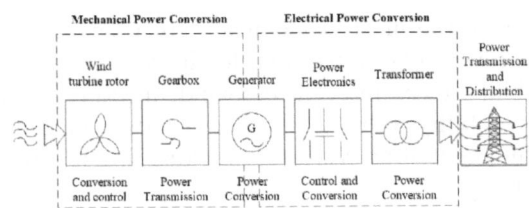

Fig. 2. General schematic of power conversion in wind turbines

2.2   Wind Energy Systems

In wind energy systems, kinetic energy from the wind is converted to mechanical energy by wind turbine blades, typically made of fiberglass or carbon fiber with epoxy resin. This mechanical energy is then converted to electrical energy through generators coupled to variable-speed wind turbines. Wind turbines operate between 5 and 16 rpm, often requiring gearboxes to increase efficiency. The double-fed induction generator (DFIG), combined with gearboxes and high-speed power converters, is widely used in wind generation technologies. A general schematic of power conversion for wind turbines can be seen in Fig. 2.

2.3   Role of D-STATCOM

The STATCOM, or D-STATCOM when connected to electrical distribution systems, mitigates rapid voltage variations caused by wind farm (WF) integration as shown in 3. This study uses the IEEE 34-bus system with a WF consisting of six 1.5 MW turbines, totaling 9 MW [8]. The D-STATCOM reduces excess reactive power, enhancing system robustness and flexibility [17]. Previous studies, such as [18], have shown that D-STATCOMs can maintain voltage control at the PCC, minimizing the impact of sudden load changes.

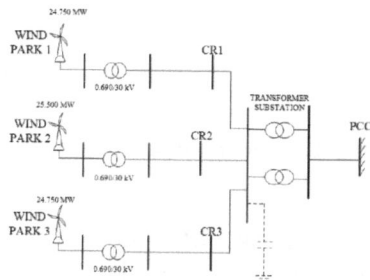

Fig. 3. Diagram of WF connected to the network

Wind energy, being the most economical "green energy," has become competitive with other renewable sources and fossil fuels. However, wind farms can cause system instability due to intermittent wind, affecting energy supply quality. Ensuring transient stability, voltage, and frequency regulation is crucial for stable operation, as highlighted by researchers addressing instability issues caused by fluctuating loads [8].

2.4   Case Studies

Case Study: Dynamic Load Connected to the System. For the first case study, the system's flow was carried out to find the bar with the greatest voltage drop and, therefore, the greatest probability of being affected by flicker. Once the observation above was made, it was identified that the bus with the highest voltage drop corresponds to bus 27 of the system. Subsequently, a breaker was added to automatically "switch" the bar load to simulate a dynamic load, as would normally be the behavior of a wind farm (WF). The load L27 is connected to its respective bus with a power of 4.5 MW. For the connection and disconnection of the load, a scenario was defined in which 100 switching instances were generated in the breaker, which will disconnect the linked load ten times to the circuit breaker, thus simulating the behavior that a WF would have due to the intermittency of the nature of the wind.

Case Study: Distribution System Connected to Wind Farm. Once the voltage profiles in the bars were analyzed, the Flicker severity indices (Pst) of the system were analyzed, and how the dynamic load influenced bus 27 of the system was concluded. The 9 MW WF wind farm (6×1.5 MW) will then be connected to the system at bar 27 to check how wind and PV energy can generate Flicker in the system and whose problem will be solved later with the connection of a Static Synchronous Distribution Compensator (D-STATCOM).

Case Study: Inclusion of D-STATCOM to the System Connected to the WF. In the previous case with the wind farm connection, a greater variation was generated in the bus voltages of the branch where the WF is connected (Bus 27); this, therefore, increased the measurement of short-term Flicker severity indices-Pst term on bars. The schematic for this case is shown in Fig. 4.

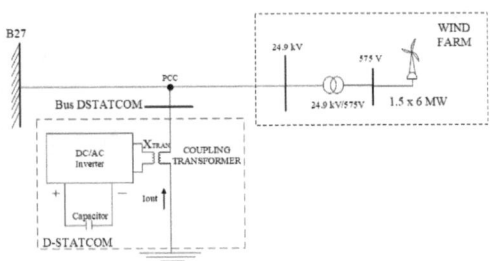

Fig. 4. Case 3: D-STATCOM entry diagram to the Distribution system.

To do this, a DSTATCOM will be connected to said bus that will be responsible for controlling the voltage variations generated by the wind farm and, consequently, reducing the Flicker Pst severity indices in the bars of the branch affected by the Flicker incidence due to the WF.

3   Analysis of Results

For each case study, the section of RMS voltages will be presented in each of the bars to corroborate that the voltage levels indicate a higher flicker severity index Pst depending on the system bus; the flicker severity indices will be presented for each of the bars to complete the analysis above and to finish, boxplots will be displayed for a better visualization and interpretation of the results.

Dynamic Load and its Effect on the Distribution System In Fig. 5, it can be seen in the boxplot that the base system has a low Flicker severity index because no large loads are changing over time, having an average value of Pst = 0.4271 and an upper extreme with a value of Pst = 0.4276, without taking into account the atypical values of the measurements.

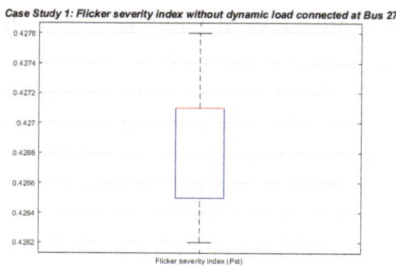
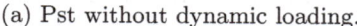
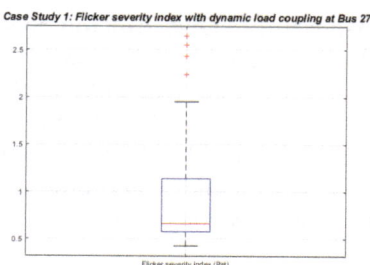

(a) Pst without dynamic loading.   (b) Pst with dynamic loading.

Fig. 5. Boxplot comparisons of system Pst values without and with dynamic loading, showing the impact of dynamic loading on flicker severity.

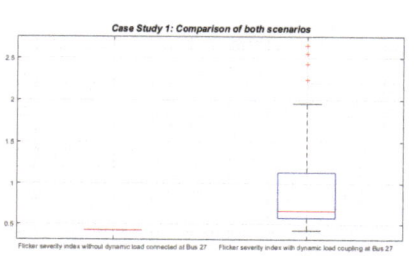
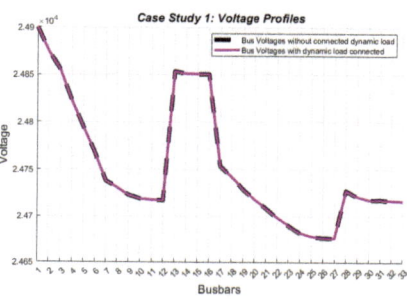

(a) Pst for both scenarios.   (b) VP with/without dynamic load.

Fig. 6. Comparison of Pst values and voltage profiles under different loading conditions.

A substantial increase in the Pst indices is observed when the dynamic load is connected to bus 27 of the 34-bus IEEE system. The boxplot shows an average value of $Pst = 0.6626$ and an upper-end value of $Pst = 1.9532$, excluding the outliers in the boxplot.

The graphs of both boxplots in Fig. 6a illustrate the growth scale at the Flicker level for both cases in which the system was evaluated without dynamic loading and with dynamic loading.

In Fig. 6b, it can be seen that the variation of the voltage profiles for both scenarios (with dynamic load and without dynamic load) is not a considerable variation because only 100 Connection and disconnection instances were generated. Additionally, the load has a power of 4.5 MW, corresponding to half of the power connected to the system through the WF for study case 2.

Discussion of Results: Pst Values for Case Study 1. The Pst values for the case study show significant differences with and without load. Without load, the Pst values are 0.4276 (upper extreme), 0.4271 (median), and 0.4262 (lower extreme), indicating minimal flicker severity and stable voltage. With load, the upper extreme rises to 2.6428 and the median to 0.6626, while the lower extreme remains at 0.4262. This increase indicates a considerable rise in flicker severity and a negative impact on voltage stability.

Thus, the variations in voltages (dV) of each node are not so far from the reference point, in this case, a nominal voltage of 24.9 kV, giving the impression that the voltages in both scenarios are superimposed. This will not happen in the following scenarios, where the load will increase and fluctuate more over time, as is typical of wind energy.

Distribution System with Connection to Wind Farm. The third case study presents a scenario aimed at mitigating the level of Flicker generated in the previous case due to the wind farm's (WF) entry into the test system. The D-STATCOM, with a capacity of $+/-3$ MVAR, utilizes PID control logic to reduce the Flicker level throughout the system to a certain degree.

In Fig. 7a, the decrease in the Pst values in the system is evident compared to the scenario in case 2, which experienced an increase in Flicker due to the wind farm. The upper limit values for the scenario with the D-STATCOM connected to the network were $Pst = 1.0459$ and $Pst = 3.2491$.

Figure 7b illustrates a considerable difference in the voltages of the bars, corresponding to the increase in Flicker shown in Fig. 14 for this specific case study.

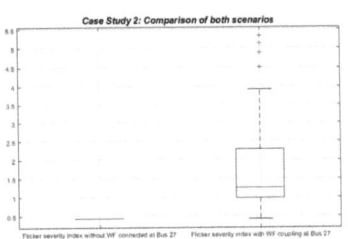
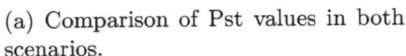
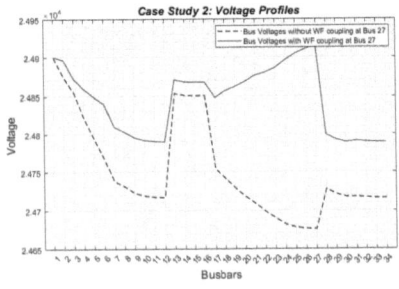

(a) Comparison of Pst values in both scenarios.

(b) Voltage profiles with wind farm connection (Case 2).

Fig. 7. Comparison of Pst values and voltage profiles for the system under different loading scenarios, highlighting the impact of wind farm integration.

Discussion of Results: Pst Values for Case Study 2. The boxplot data for Case Study 2 shows that without the D-STATCOM, the Pst values are 0.4276 (upper extreme), 0.4271 (median), and 0.4262 (lower extreme), indicating minimal flicker severity. With the D-STATCOM, the upper extreme rises to 5.3168 and the median to 1.2506, while the lower extreme remains at 0.4262. This indicates that the D-STATCOM increases flicker severity and variability in the system's voltage stability.

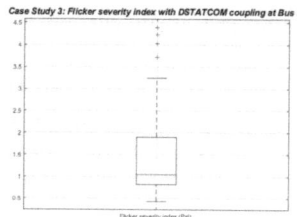
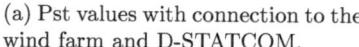
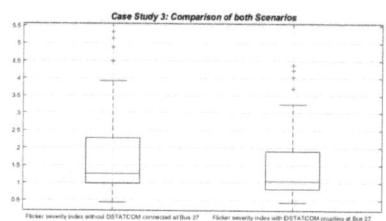

(a) Pst values with connection to the wind farm and D-STATCOM.

(b) Comparison of Pst values for both scenarios in case 3.

Fig. 8. Boxplot comparisons of Pst values in the system: (a) with connection to the wind farm and D-STATCOM, and (b) comparison of both scenarios in case 3, highlighting the impact of D-STATCOM on flicker severity.

Distribution System with Connection to WF and D-STATCOM for Flicker Mitigation. The third case study presents a scenario to mitigate the Flicker generated in the previous case due to the wind farm's (WF) entry into the test system. The D-STATCOM, with a capacity of $+/-3$ MVAR, utilizes PID control logic to mitigate the Flicker level throughout the system to a certain degree.

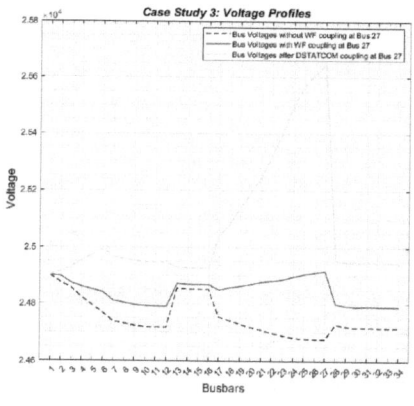

Fig. 9. Voltage Profiles in Scenario with DSTATCOM Coupling, Case 3.

Figures 8 and 9 illustrate the decrease in the Pst values in the system compared to the scenario in case 2, which experienced an increase in Flicker due to the wind farm. The upper limit values obtained for the D-STATCOM scenario connected to the network were $Pst = 1.0459$ and $Pst = 3.2491$.

Discussion of Results: Pst Values for Case Study 3. The boxplot data for Case Study 3 compares the flicker severity index (Pst) values for the wind farm (WF) with and without using a D-STATCOM. Without the D-STATCOM, the Pst values show an upper extreme (LS) of 5.3168, a median of 1.2506, and a lower extreme (LI) of 0.4262. These values indicate a higher flicker severity and greater variability in voltage stability. However, with the introduction of D-STATCOM, the upper extreme (LS) decreases to 4.386, and the median reduces to 1.0459, while the lower extreme (LI) remains constant at 0.4262. This reduction in the upper extreme and median Pst values demonstrates that the D-STATCOM effectively mitigates flicker severity, improving the system's voltage stability.

4   Conclusions

Through the simulation of scenario 1, it was verified how load fluctuations can affect the quality of the network, specifically by increasing the Flicker severity index. The average value of Pst was scaled from 0.4271 to 0.6626, with Pst growth equal to +55.14%, and the maximum value of Pst was scaled from 0.4276 to 2.6428, representing an increase of 518.1 For case study 2, the median Pst value increased from 0.4271 to 1.2506, or 192.8%; meanwhile, the maximum value increased from 1134.4%. In case study 3, it was confirmed that the D-STATCOM, through PID control logic, can reduce disturbances in the network due to intermittent loads. It managed to mitigate an average Pst value (with WF) of 1.2506 and reduce it to 1.0459, a decrease or mitigation of 16.37%; for maximum Pst values of 5.3168 to 4.368, a 17.85% reduction in Pst was calculated. Bus 27 has the greatest presence of Flicker, sharing the maximum value of Pst throughout the system equal to 5.3168. In case study 3, it was mitigated through the DSTATCOM reactive power control, reducing it to 4,386.

References

1. Jaramillo, M., Carrión, D., Muñoz, J.: A deep neural network as a strategy for optimal sizing and location of reactive compensation considering power consumption uncertainties. Energies **15**(24) (2022). https://doi.org/10.3390/en15249367.
2. Jaramillo, M., Pavón, W., Jaramillo, L.: Adaptive forecasting in energy consumption: a bibliometric analysis and review. Data **9**(1) (2024). https://doi.org/10.3390/data9010013.
3. Kumar, A., Kumar, P.: Power quality improvement for grid-connected PV system based on distribution static compensator with fuzzy logic controller and uvt/adaline-based least mean square controller. J. Mod. Power Syst. Clean Energy **9**, 1289–1299 (2021). https://doi.org/10.35833/MPCE.2021.000285.
4. Jaramillo, M., Carrión, D.: An adaptive strategy for medium-term electricity consumption forecasting for highly unpredictable scenarios: case study quito, ecuador during the two first years of COVID-19. Energies **15**(22) (2022). https://doi.org/10.3390/en15228380.
5. Jaramillo, M., Tipán, L., Muñoz, J.: A novel methodology for optimal location of reactive compensation through deep neural networks. Heliyon **8**(10), e11097 (2022). https://doi.org/10.1016/j.heliyon.2022.e11097

6. Jaramillo, M.D., Carrión, D.F.: Optimizing critical overloaded power transmission lines with a novel unified SVC deployment approach based on FVSI analysis. Energies **17**(9) (2024). https://doi.org/10.3390/en17092063.
7. Gavilanez, C., Jaramillo, M., Pavón, W., Barrera-Singaña, C., Ruíz, M.: Power factor improvement in a distribution system by implementing adaptive control for reactive power filters. In: 2022 IEEE Global Conference on Computing, Power and Communication Technologies (GlobConPT), pp. 1–6 (2022)
8. Naz, M.N., Imtiaz, S., Bhatti, M.K.L., Awan, W.Q., Siddique, M., Riaz, A.: Dynamic stability improvement of decentralized wind farms by effective distribution static compensator. J. Mod. Power Syst. Clean Energy **9**, 516–525 (2021). https://doi.org/10.35833/MPCE.2018.00422.
9. Zúñiga, M., Jaramillo, M., Pavón, W., Muñoz, J.: Artificial neural networks as a methodology for optimal location of static synchronous series compensator in transmission systems. In: 2022 IEEE Global Conference on Computing, Power and Communication Technologies (GlobConPT), pp. 1–6 (2022)
10. Jaramillo, M.D., Carrión, D.F., Muñoz, J.P.: A novel methodology for strengthening stability in electrical power systems by considering fast voltage stability index under N minus; 1 scenarios. Energies **16**(8) (2023). https://doi.org/10.3390/en16083396.
11. Vaca, F., Jaramillo, M., Muñoz, J., Barrera-Singaña, C., Pavón, W.: THD Minimizationin electrical distribution networks through vector space control implementation in power inverters. In: 2023 IEEE IAS Global Conference on Renewable Energy and Hydrogen Technologies, GlobConHT 2023, pp. 1–6 (2023). https://doi.org/10.1109/GlobConHT56829.2023.10087359.
12. Expósito, A., Conejo, A.J., Cañizares, C.A.: Electric Energy Systems Analysis and Operation Second Edition (2018)
13. Benítez, K., Jaramillo, M., Muñoz, J., Barrera-Singaña, C., Pavón, W.: Multiobjective analysis for optimal location and location of distributed generation focused on improving power quality. In: 2023 IEEE IAS Global Conference on Renewable Energy and Hydrogen Technologies, GlobConHT 2023, pp. 7–12 (2023). https://doi.org/10.1109/GlobConHT56829.2023.10087886.
14. Majchrzak, J., Wiczyński, G.: Basic characteristics of IEC Flickermeter processing. Model. Simul. Eng. **2012** (2012). https://doi.org/10.1155/2012/362849.
15. Ketabipour, S., Samet, H., Mohammadi, M., Li, Q., Terzija, V.: Modelling of extremely short-time power variations of wound rotor induction machines wind farms for flicker studies. IEEE Access (2023). https://doi.org/10.1109/ACCESS.2023.3337388
16. IEC: Flickermeter-functional and design specifications, standard 61000-4-15, IEC. Basic EMC Publ. **2003**, 1–24 (2003)
17. Bai, H., Mi, C.: Power electronic devices. Topology, and control, circuits (2011)
18. Hock, R.T., de Novaes, Y.R., Batschauer, A.L.: A voltage regulator for power quality improvement in low-voltage distribution grids. IEEE Trans. Power Electron. **33**, 2050–2060 (2018). https://doi.org/10.1109/TPEL.2017.2693239.

In vitro Evaluation of Antimicrobial Activity from the Methanolic Extracts of *Taraxacum officinale* and *Melissa officinalis* on *Staphylococcus aureus*

Noelia Gómez[✉] ⓘ, Sandy Gavilanes ⓘ, Inés Malo ⓘ, and Mónica Espadero ⓘ

Universidad Politécnica Salesiana, Cuenca, Ecuador
ggomeza@est.ups.edu.ec, {sgavilanes,imalo,mespadero}@ups.edu.ec

Abstract. The use of plants has been increasing by the presence of bioactives, which is an opportunity in the creation of products such as medicines, food and cosmetics. *Taraxacum officinale* (Dandelion) and *Melissa officinalis* (Toronjil), are cosmopolitan species with great therapeutic potential, traditionally used for their anti-inflammatory, antioxidant and antimicrobial activity. Therefore, this research evaluates the antimicrobial activity of the methanol extracts of *Taraxacum officinale* and *Melissa officinalis* on *Staphylococcus aureus* using *in vitro* tests. Analyzes of secondary metabolites were performed by phytochemical screening tests, as well as the identification of possible compounds responsible for antimicrobial activity using the Bioautography technique. The phenolic compounds were quantified by the Folin-Ciocalteu assay, and the antimicrobial capacity of the extracts was evaluated by the Kirby-Bauer method. Phytochemical screening revealed the presence of alkaloids, flavonoids, saponins, phenols and tannins in both extracts. Using Bioautography, potentially inhibitory metabolites were identified, including chlorogenic acid (Rf 0.45), taraxasterol (Rf 0.65), caffeic acid (Rf 0.87), geraniol (Rf 0.2), citral (0.42) and citronellal (0.8). In addition, a greater number of phenolic compounds (112.20 ± 10.13 mg EAG/g MS) was observed in the extract of *Melissa officinalis*, while the extract of *Taraxacum officinale* showed a greater bacterial inhibition capacity, reaching 65.49%.

Keywords: *Taraxacum officinale* · *Melissa officinalis* · *Staphylococcus aureus* · antimicrobial activity · phenolic compounds

1 Introduction

Plants with healing properties and their derived products have been commonly used by traditional medicine, thus their relevance as raw material in the production of medicines in the pharmaceutical industry; hence, their use has been increasing in recent years [1]. There are several medicinal plants in Ecuador with great therapeutic potential, mainly with anti-inflammatory, antibacterial, antioxidant, analgesic activity, among others. However, some of the species with antimicrobial properties have not been thoroughly investigated to determine all their benefits [2].

© The Author(s) 2025
E. M. Inga Ortega et al. (Eds.): CITIS 2024, LNNS 1331, pp. 252–260, 2025.
https://doi.org/10.1007/978-3-031-87065-1_23

Staphylococcus aureus, is considered a pathogenic microorganism associated with localized, invasive and toxin-mediated infections at the level of the dermis, respiratory tract, bone and less common in the urinary tract, and can cause serious infections if it enters the bloodstream, raising concern because strains of *Staphylococcus aureus* are resistant to several commercial antibiotics [3]. Facing these challenges, contemporary medicine has directed its efforts to exploring and creating innovative treatments, allowing the investigation of several bioactive compounds present in plant species, which have been extracted by different methods to be applied in various cosmetic, food and pharmaceutical formulations.

Due to the growing market demand in the use of medicinal plants, there is the need to conduct research of introduced plants such as *Taraxacum officinale* commonly called "dandelion" and *Melissa officinalis* known as "toronjil" and compare the antimicrobial activity of its methanolic extracts on the bacterium *Staphylococcus aureus*, enabling the use of existing resources in the country. In addition, it is intended to generate and provide essential data that may be useful for future research. In this way, the idea is to take advantage of the richness of the country's biodiversity, maximizing its possible applications in various industries, as well as promoting its conservation and use as alternatives for treating various conditions.

2 Materials and Methods

2.1 Plant Material Conditioning

Twelve kilograms of aerial parts of *Taraxacum officinale* and *Melissa officinalis* were collected and were washed with distilled water. Then, were dried at room temperature and put in a stove at 40 °C ± 2; both processes were carried out for 24 h. Once dried, the leaves were pulverized, weighed and placed in paper cases to preserve their biological characteristics until their use.

2.2 Preparation of Extracts

Cold maceration was done with constant stirring of dry and ground plant material with absolute methanol in a 1:6 ratio for 72 h. Subsequently, a double filtration was performed; the macerates concentrated in rotavapor at 337 mbar at 50 °C until the complete elimination of the solvent and the raw extracts were obtained [2]. The concentrated pure extracts were stored in amber jars and placed in refrigeration for 7 days to ensure the stability of the secondary metabolites [2, 4].

2.3 Qualitative Characterization

The phytochemical screening of *Taraxacum officinale* and *Melissa officinalis* extracts was carried out by preliminary tests that qualitatively determined the presence of bioactive compounds that were identified by their coloration and precipitation [5]; the following process was carried out from the tests described in Table 1.

Table 1. Phytochemical testing of extracts of *Taraxacum officinale* and *Melissa officinalis* leaves.

Compounds	Testing
Alkaloids	Dragendroff's essay
Flavonoids	Shinoda's essay
Saponins	Fehling reagent test + Na_2CO_3
Phenols	Folin reagent + Na_2CO_3 *al* 5%
Tannins	Ferric Chloride Test

2.4 Quantification of Phenolic Compounds

The phenolic content of the extracts was determined by the Folin-Ciocalteu assay [6, 7]; a calibration curve was carried out with different concentrations of gallic acid from 10 to 100 ppm. The *Taraxacum officinale* and *Melissa officinalis* extracts were diluted with distilled water, the Folin-Ciocalteu 1 N reagent was placed, and the sodium carbonate solution Na_2CO_3 (20% w/v) was added. It was kept in the dark for two hours; absorbance at 760 nm was measured on a UV-Vis spectrophotometer. Results are expressed as (mg EAG/100 g DM).

2.5 Antimicrobial Activity Assessment Using Bioautography Method

A strain of *Staphylococcus aureus* ATCC 25923 was used in nutrient broth + agar agar 0.3% [8]. It was proceeded to read the optical density of the culture with wavelength of 600 nm; silica gel plates were prepared, using the thin layer chromatography (TLC) procedure, they were arranged in saturated chromatographic cameras with mobile phase of ethyl acetate + methanol (9:1 v/v) [9]. The bands were observed by fluorescence at 254 nm and 366 nm. The plates were prepared in humidity chambers, incubating them at 25 °C for 90 min. After this time, they were immersed in the bacterial suspension of *Staphylococcus aureus* ATCC 25923 for 10 s. They were placed in steam chambers at 37 ± 2 °C of incubation for 24 h. Finally, the plates were immersed in an aqueous solution of 0.1% MTT + Triton X-100 dye for 60 s and were incubated in the water vapor chambers at 37 °C for 1 h. An immersion in 70% ethanol was carried out for 10 s.

2.6 Determination of Antimicrobial Capacity by Kirby-Bauer Method

The antimicrobial sensitivity test was carried out by means of the disk diffusion technique [10, 11]. A pure strain of *Staphylococcus aureus* ATCC 25923 was used in blood agar, colonies were taken and suspended in sterile physiological serum up to a concentration of 0.5 on the McFarland scale. Sowing was performed with the standardized sample in Mueller Hinton agar. The disks were placed in the pre-seeded Petri boxes, including positive control (Vancomycin), and negative control with sterile distilled water. Finally, it was incubated at 37 °C ± 2 for 24 h. The diameter of the halos and the percentage of the inhibition effect were expressed.

2.7 Statistical Analysis

The normal distribution of the data was evaluated by the Shapiro-Wilk normality test and Q-Q graph; the ANOVA model was applied with $p < 0.05$. The assumptions of the model, assumption of normality and assumptions of homocedasticity of variances were evaluated by the Shapiro-Wilk and Levene test. When the assumptions were met, the best treatment was finally determined by Tukey Test.

3 Results

3.1 Extraction of Methanol Extracts

Out of the 100 g of the dried and ground leaves of *Taraxacum officinale* and *Melissa officinalis* were obtained by cold maceration with methanol 30 mL of the extract with greenish brown coloring and 20 mL of the extract with dark green appearance, respectively.

3.2 Qualitative Characterization

The presence of alkaloids in the extracts was identified in *Taraxacum officinale*, as well as a moderate presence of saponins, and low presence of flavonoids, and high presence of flavonoids was observed in *Melissa officinalis*. Additionally, moderate presence of phenolic compounds and tannins was identified in both extracts.

The results obtained from the different phytochemical screening assays are presented in Table 2.

Table 2. Results of Phytochemical Screening

Compounds	Methanolic Extracts	
	Taraxacum officinale	*Melissa officinalis*
Alkaloids	+++	+++
Flavonoids	+	+++
Saponins	+ ++	++
Phenols	++	++
Tannins	+++	+++

Note: (+++): high evidence; (++): medium evidence; (+): low evidence; (−): negative.

3.3 Antimicrobial Activity Assessment on *Staphylococcus aureus* by the Bioautography Method

Using the Bioautography method it was possible to identify the compounds potentially responsible for the biological activity. Tables 3 and 4 detail the distances traveled by the substances, the solvent front, the reference front (Rf), the suggested antimicrobial compound and the color detected at a wavelength of 366 nm (Figs. 1 and 2).

Compounds showing antimicrobial activity were identified using the reference front value (Rf) [9, 12, 13].

Table 3. Results obtained in the Bioautography with extract of *Taraxacum officinale*.

	Distance traveled (cm)	Solvent Front	Reference Front (RF)	Compound	Color Detected
1	2.7	6	0.45	Chlorogenic acid	Blue
2	3.9	6	0.65	Taraxasterol	Orange
3	5.22	6	0.87	Caffeic acid	Reddish brown

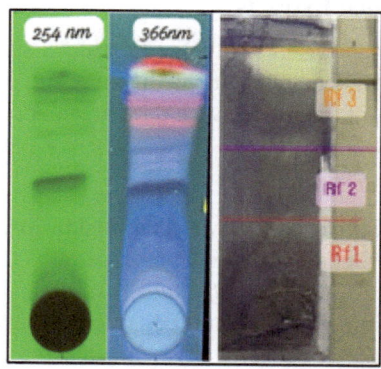

Fig. 1. TLC and Bioautography of *Taraxacum officinale* extract.

Table 4. Results obtained in the Bioautography with extract of *Melissa officinalis*.

	Distance traveled (cm)	Solvent front	Reference Front (RF)	Compound	Color Detected
1	1.2	6	0.2	Geraniol	Blue
2	2.52	6	0.42	Citral	Blue violet
3	4.8	6	0.8	Citronella	Reddish coffee

3.4 Quantification of Polyphenolic Compounds

The following results were obtained (Table 5) from the quantification of phenolic compounds.

3.5 Determination of Antimicrobial Capacity (Kirby-Bauer)

The results of the antimicrobial susceptibility assays revealed that both the disks containing the extract of *Taraxacum officinale* and *Melissa officinalis* generated inhibition halos in the presence of *Staphylococcus aureus* ATCC 25923 (Table 6), indicating an inhibitory activity on the bacterium.

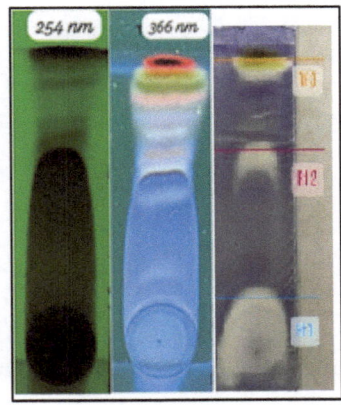

Fig. 2. TLC and Bioautography of *Melissa officinalis* extract

Table 5. Quantification of polyphenolic compounds in methanol extracts.

Extracts	Content of phenolic compounds (mg EAG/g DM)
Melissa officinalis	112,20 ± 10,13
Taraxacum officinale	23,07 ± 6,87

Note: the values used are the average of 15 samples (±DS).

Table 6. Antimicrobial activity of extracts on *Staphylococcus aureus*.

Treatments (Methanol extracts)	Inhibition Halo Diameter (mm)	% Inhibition	Groups
Taraxacum officinale	11,22 ± 0,50	65,49 ± 5,61	a
Melissa officinalis	10,44 ± 1,26	53,39 ± 1,63	b

Note: the values used are the average of 15 samples (±DS).

The analysis of variance (ANOVA) showed significant difference between treatments (P value = 0.0231). There were no significant deviations from the model assumptions (normality and homogeneity of variances). The treatment with *Taraxacum officinale* extract (Fig. 3) obtained the highest inhibition percentage, because *Staphylococcus aureus* is sensitive to the group of phenolic compounds present in the extract.

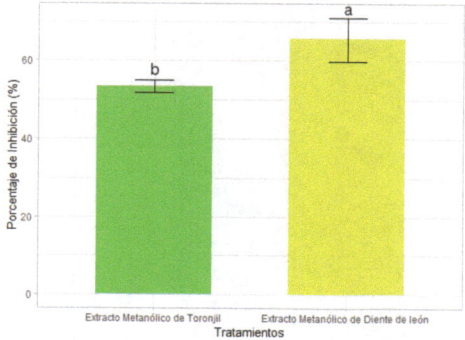

Fig. 3. Effect of methanol extracts of *Melissa officinalis* and *Taraxacum officinale* on *Staphylococcus aureus*.

4   Discussion

Methanol is known to be a highly effective solvent for extracting a wide range of secondary metabolites from plant material, and it presents a high solubility for a wide variety of compounds, concluding that a higher extraction efficiency is achieved on the third day of maceration [14, 15].

According to studies, the use of solvents such as methanol, ethanol and chloroform favor the entrainment of the alkaloid taraxine, which are found in both species [16]. In a research that evaluates plant species against pathogenic microorganisms [17], flavonoids were observed in *Taraxacum officinale* extracts, results that are consistent with the ones obtained in this paper, but in low concentrations. This variation could be due to various factors such as the percentage of the solvent, the chemical composition of the extracts, the age of the plant or even climatic conditions. On the other hand, in the trial conducted with *Melissa officinalis*, a high number of flavonoids was detected, also supported by studies where they were identified as modifiers of antibiotic activity [14].

From the detection of compounds with antimicrobial properties, in the chemical analysis of *Taraxacum officinale* [12] chlorogenic acid is classified as a flavonoid and is credited with numerous properties, including anti-inflammatory, enzyme inhibitor and antimicrobial properties. Taraxasterol, a compound mostly present in *Taraxacum officinale*, belongs to the group of triterpenoids. This compound has anti-inflammatory, antioxidant and antimicrobial activity, especially against *Staphylococcus aureus* with significant value [18]. In addition, another investigation determined that the main components of *Melissa officinalis* extract were Geraniol, Citronella and Citral [19]. These components are responsible for antimicrobial activity, indicating the potential of *Melissa officinalis* as a promising natural preservative for the food industry.

According to the quantification of polyphenolic compounds, values of 90.1 mg of GAE/g of dry matter are exposed in extracts of *Melissa officinalis* obtained by maceration, being similar to the results presented in this research [20]. When *Melissa officinalis* is compared with other species studied such as *Agastache foeniculum* (27.19 mg AG/g), *Lavandula angustifolia* (12.44 mg AG/g) and *Nepeta cataria* (14.66 mg AG/g), this extract presents a higher concentration [21]. This suggests that the antioxidant properties of *Melissa officinalis* leaves may be because of the presence of various phenolic compounds (rosmarinic acid, caffeic acid, among others), and vary according to regions. Therefore, it is essential to consider factors such as the variety used, the region of origin, extraction conditions, and temperature, as these influence the presence or absence of compounds with antioxidant capacity.

5   Conclusions

The extracts of *Taraxacum officinale* and *Melissa officinalis* contain a variety of secondary metabolites, including alkaloids, flavonoids, saponins, phenols and tannins, in varying proportions. From the tests carried out by the Bioautography method, it was determined that the possible compounds with antimicrobial activity are: chlorogenic acid, taraxasterol and caffeic acid, in the extract of *Taraxacum officinale*. On the other hand, in the extract of *Melissa officinalis* (grapefruit), geraniol, citral and citronellal were identified.

Melissa officinalis extract had the highest content of phenolic compounds. In addition, it was determined that methanol extracts of *Taraxacum officinale* and *Melissa officinalis* showed antimicrobial activity on *Staphylococcus aureus*. However, through statistical analysis with a 95% confidence level, it was determined that the methanolic extract of *Taraxacum officinale* achieved the highest inhibition percentage.

References

1. Lozada, M.: ESTUDIO FIQOQUÍMICO Y EVALUACIÓN DE ACTIVIDAD ANTIBACTERIANA SOBRE Staphyloccosus aureus ATCC:25923, Streptoccocus mutans ATCC: 25175. Streptpccocus pneumoniae ATCC: 49619, Streptoccocus pygenes ATCC: 19615DE EXTRACTOS APOLARES (CLOROFORMO-HEXANO) DE Croton elegans KUNTH (mosquera), Quito (2016)
2. Azuero, A., Jaramillo, C.J., San Martín, D., D'Armas, H.: Análisis del efecto antimicrobiano de doce plantas medicinales de uso ancestral en Ecuador/Analysis of antimicrobial effect of twelve medicinal plants of ancient use in Ecuador. CIENCIA UNEMI 9(20), 11–18 (2016). https://doi.org/10.29076/issn.2528-7737vol9iss20.2016pp11-18p.
3. Cervantes-García, E., García-González, R., María Salazar-Schettino, P.: Características generales del Staphylococcus aureus. Rev. Latinoam. Patol. Clin. Med. Lab. 61(1), 28–40 (2014). www.medigraphic.com/patologiaclinicawww.medigraphic.org.mx
4. Vélez, R., D'Armas, H., Jaramillo-Jaramillo, C., Vélez, E.: Metabolitos secundarios, actividad antimicrobiana y letalidad de las hojas de Cymbopogon citratus (hierba luisa) y *Melissa officinalis* (toronjil). FACSALUD-UNEMI 2(2), 31–39 (2018). https://doi.org/10.29076/issn. 2602-8360vol2iss2.2018pp31-39p
5. Rodríguez, J., Hernández, M., Méndez, L.: Manual de prácticas de farmacognosia. Xalapa (2020)
6. Noreen, H., Semmar, N., Farman, M., McCullagh, J.S.O.: Measurement of total phenolic content and antioxidant activity of aerial parts of medicinal plant Coronopus didymus. Asian Pac. J. Trop. Med. 10(8), 792–801 (2017). https://doi.org/10.1016/j.apjtm.2017.07.024
7. Vargas, C.: Determinación y cuantificación de compuestos fenólicos en flores de *Taraxacum officinale*, mediante HPLC-DAD-MS y ensayos colorimetricos UV-vis. Bogotá (2020)
8. Dewanjee, S., Gangopadhyay, M., Bhattacharya, N., Khanra, R., Dua, T.K.: Bioautography and its scope in the field of natural product chemistry. J. Pharm. Anal. 5(2), 75–84 (2015). https://doi.org/10.1016/J.JPHA.2014.06.002

9. Wagner, H., Bladt, S.: Plant Drug Analysis, 2nd edn. Springer, New York (1996)
10. Bernal, M., Guzman, M.: EL ANTIBIOGRAMA DE DISCOS. NORMALIZACION DE LA TÉCNICA DE KIRBY-BAUER. BIOMEDICA **4**(3–4), 114–121 (1984)
11. Hudzicki, J.: Kirby-Bauer Disk Diffusion Susceptibility Test Protocol (2009). www.atcc.org
12. Cortés, N., et al.: Microscopical descriptions and chemical analysis by HPTLC of *Taraxacum officinale* in comparison to Hypochaeris radicata: a solution for mis-identification. Rev. Bras. **24**(4), 381–388 (2014). https://doi.org/10.1016/J.BJP.2014.07.018
13. Senguttuvan, J., Subramaniam, P.: HPTLC fingerprints of various secondary metabolites in the traditional medicinal herb Hypochaeris radicata L. J. Bot. 1–11 (2016). https://doi.org/10.1155/2016/5429625
14. De Sousa, A.T.L., et al.: Evaluation of the antibacterial activity of methanolic and hexanic extracts of puff pastry stem, *Melissa Officinalis* L. Revista Ciencias de la Salud **14**(2), 201–210 (2016). https://doi.org/10.12804/revsalud14.02.2016.05
15. Najda, A., Sugier, D.: Selected secondary metabolites content in the roots of *Taraxacum officinale* depending on the method of plantation establishment. Herbal Polonica **53**(3), 152–156 (2007). https://www.researchgate.net/publication/258519306
16. Muñoz, M., Montes, M., Wilkomirsky, T.: Plantas medicinales de uso en Chile: química y farmacología, 2nd ed. Universidad de Chile, Vicerrectoría de Asuntos Académicos, Universitaria, Chile (2004). https://bibliotecadigital.infor.cl/handle/20.500.12220/19111. Accessed 04 February 2024
17. Rodríguez, C., Zarate, A., Sánchez, L.: Actividad antimicrobiana de cuatro variedades de plantas frente a patógenos de importancia clínica en Colombia. NOVA **15**(27), 119–129 (2017)
18. Jiao, F., Tan, Z., Yu, Z., Zhou, B., Meng, L., Shi, X.: The phytochemical and pharmacological profile of taraxasterol. Front. Pharmacol. **13**(2022). https://doi.org/10.3389/FPHAR.2022.927365
19. Yu, H., Pei, J., Qiu, W., Mei, J., Xie, J.: The antimicrobial effect of *Melissa officinalis* L. essential oil on Vibrio parahaemolyticus: insights based on the cell membrane and external structure. Front. Microbiol. **13**(2022). https://doi.org/10.3389/FMICB.2022.812792
20. Moacă, E.A., et al.: A comparative study of *melissa officinalis* leaves and stems ethanolic extracts in terms of antioxidant, cytotoxic, and antiproliferative potential. Hindawi **2018**, 1–12 (2018). https://doi.org/10.1155/2018/7860456
21. Duda, S.C., Mărghitaş, L.A., Dezmirean, D., Duda, M., Mărgăoan, R., Bobiş, O.: Changes in major bioactive compounds with antioxidant activity of *Agastache foeniculum*, *Lavandula angustifolia*, *Melissa officinalis* and *Nepeta cataria*: effect of harvest time and plant species. Ind. Crops Prod. **77**, 499–507 (2015). https://doi.org/10.1016/J.INDCROP.2015.09.045

Smart Waters: Harnessing Machine Learning to Predict Water Quality in a Tropical Andean Watershed

Paola Duque-Sarango[1]([✉]) [iD], Cristina Cárdenas[1], Bryam Crespo[1],
Christian Mera-Parra[1] [iD], and Sebastián Cedillo[2] [iD]

[1] Universidad Politécnica Salesiana. Grupo de Investigación en Recursos Hídricos
(GIRH-UPS), Campus El Vecino. Calle Vieja 12-30 y Elia Liut., Cuenca, Ecuador
pduque@ups.edu.ec
[2] Facultad de Ingeniería, Universidad de Cuenca, Av. 12 de Abril S/N, 010203 Cuenca, Ecuador

Abstract. This study is based on the application of machine learning to compare regression tree models based on data with a reduced set of parameters using the National Sanitation Foundation Index (WQI NSF) to estimate water quality in the Yanuncay River watershed. Physical, chemical, and biological parameter data were collected from the watershed and equations derived from curve fits were used to calculate the WQI NSF. The results showed that the regression random forest model trained with three parameters: fecal coliforms, pH and nitrates, was the most suitable option. This model demonstrated consistent performance, with an R^2 of 0.930 and a standard deviation of 0.026. The importance of fecal coliforms and nitrates as key indicators of contamination were highlighted, and pH was considered crucial due to its ease of sampling in the field and low requirement of specialized equipment. Thus, this study highlights the importance of continuous and long-term water quality monitoring in the Yanuncay River watershed and suggests that regression tree-based models can optimize monitoring requirements without compromising accuracy in estimating the WQI NSF.

Keywords: Machine Learning · Yanuncay watershed · water quality · Water Quality Index · Regression Random Forest

1 Introduction

El agua es esencial para los ecosistemas y la salud humana. La industrialización, la agricultura intensiva y la urbanización han aumentado la contaminación y el deterioro del agua, afectando los ecosistemas y la salud pública. Es fundamental supervisar y evaluar la calidad del agua para prevenir y abordar la contaminación. Tradicionalmente, esto se ha hecho mediante el análisis de parámetros físicos, químicos y biológicos, utilizando Índices de Calidad del Agua (WQIs), aunque estos métodos pueden ser costosos y difíciles de implementar [1–3].

In response to these limitations, advances in technology and science have driven the development of data-driven models employing machine learning techniques and

© The Author(s) 2025
E. M. Inga Ortega et al. (Eds.): CITIS 2024, LNNS 1331, pp. 261–270, 2025.
https://doi.org/10.1007/978-3-031-87065-1_24

statistics [4]. These models have the potential to overcome the limitations of WQIs by considering nonlinear and complex relationships between water quality parameters [5].

Recent research, such as the study by Koranga, et al. [6] on Lake Nainital, Uttarakhand, has demonstrated the effectiveness of algorithms such as Random Forest in predicting water quality. In addition, research like that of Cañar Uyaguari [7] has explored innovative approaches using technologies such as computer vision and neural networks to assess water conditions rapidly.

This study focused on the Yanuncay River watershed, using data collected by monitoring stations established by ETAPA E. P. The main objective was to develop water quality models based on Regression Trees with a reduced number of parameters compared to those required by the WQI NSF, using data analysis and machine learning techniques for the optimization of monitoring requirements in the watershed, thus contributing to a more efficient and effective management of water resources.

The paper describes the study area and methods for collecting hydrological data and delimiting the Yanuncay river watershed. It then presents the calculation of WQI NSF and the methodology for creating models based on regression trees. The results are discussed, including water quality interpolation and model performance, and it is concluded that it is possible to optimise monitoring requirements without compromising accuracy in estimating the NSF Water Quality Index.

2 Materials and Methods

2.1 Study Area

The Yanuncay River watershed is located in the mountainous region of Ecuador and forms an integral part of the Paute River watershed. To the north, it borders the Cañar River watershed and the Tomebamba River sub-watershed; to the south, it borders the Jubones River watershed and the Tarqui River sub-watershed; to the east, it shares boundaries with the Tomebamba and Tarqui River sub-watersheds; and to the west, it is bounded by the Balao River watershed (Fig. 1).

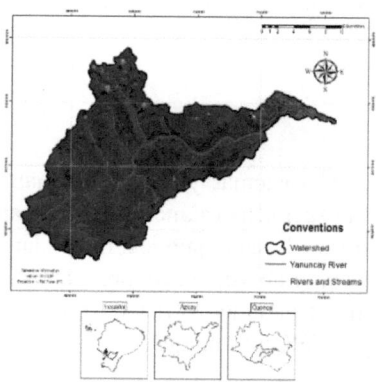

Fig. 1. Location of the study area: Yanuncay River watershed.

2.2 Hydrological Data Collection

The Yanuncay River watershed was delimited using a Digital Elevation Model (DEM) provided by the United States Geological Survey. The DEM provided detailed topography data, facilitating the watershed's delimitation. ArcMap 10.5 software was used to clean up the study area, employing tools such as Hydrology and Map Algebra [8] (Fig. 2). These tools were essential to clean and prepare the geospatial data for the hydrological analysis.

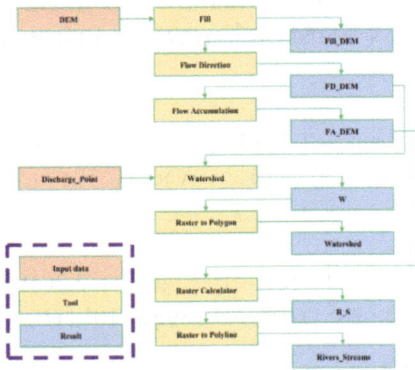

Fig. 2. Diagram for the delineation of a watershed.

2.3 National Sanitation Foundation Index Calculation

The calculation of the WQI NSF in the Yanuncay River watershed was carried out through a process that involved the treatment of the data provided by ETAPA E. P. Cleaning, and filtering techniques were employed using Excel macros to eliminate null and outliers to ensure the accuracy of the results.

Determination of Scaling Factors Q_i. The methodology of Jiménez and Vélez [9] proposes the use of equations instead of standardization curves to simplify the process. These equations were implemented in Python to calculate the scale factors, automating and improving the accuracy of the calculations.

Dissolved Oxygen (Saturation %)

$$Q_{DO} = 3.1615 \cdot 10^{-8} \cdot (DO)^5 - 1.0304 \cdot 10^{-5} \cdot (DO)^4 + 1.0076 \cdot 10^{-3} \cdot (DO)^3$$
$$- 2.7883 \cdot 10^{-2} \cdot (DO)^2 + 8.4068 \cdot 10^{-1} \cdot (DO) - 1.612 \cdot 10^{-1} \quad (1)$$

Fecal Coliforms (MPN/100mL)

$$Q_{FC} = e^{-0.0152 \cdot (\ln(FC))^2 - 0.1063 \cdot (\ln(FC)) + 4.5922} \quad (2)$$

Hydrogen Potential (pH units)

$$Q_{pH>7.5} = -1.11429 \cdot (pH)^4 + 44.50952 \cdot (pH)^3 - 656.6 \cdot (pH)^2 + 4215.34762 \cdot (pH) - 9840.14286 \tag{3}$$

$$Q_{pH:7.5} = -0.1789 \cdot (pH)^5 + 3.7932 \cdot (pH)^4 - 30.517 \cdot (pH)^3 + 119.75 \cdot (pH)^2 - 224.58 \cdot (pH) + 159.46 \tag{4}$$

Biochemical Oxygen Demand over 5 days (mg/L)

$$Q_{BOD5} = 1.8677 \cdot 10^{-4} \cdot (BOD)^4 - 1.6615 \cdot 10^{-2} \cdot (BOD)^3 + 5.9636 \cdot 10^{-1} \cdot (BOD)^2 - 1.1152 \cdot 10^1 \cdot (BOD) + 1.0019 \cdot 10^2 \tag{5}$$

Nitrates (mg/L)

$$Q_N = 3.5603 \cdot 10^{-9} \cdot (N)^6 - 1.2183 \cdot 10^{-6} \cdot (N)^5 + 1.6238 \cdot 10^{-4} \cdot (N)^4 - 1.0693 \cdot 10^{-2} \cdot (N)^3 + 3.7304 \cdot 10^{-1} \cdot (N)^2 - 7.521 \cdot (N) + 1.0095 \cdot 10^2 \tag{6}$$

Phosphates (mg/L)

$$Q_P = 4.6732 \cdot 10^{-3} \cdot (P)^6 - 1.6167 \cdot 10^{-1} \cdot (P)^5 + 2.20595 \cdot (P)^4 - 1.50504 \cdot 10^1 \cdot (P)^3 + 5.38893 \cdot 10^1 \cdot (P)^2 - 9.98933 \cdot 10^1 \cdot (P) + 9.98311 \cdot 10^1 \tag{7}$$

Temperature Change (°C)

$$Q_{\Delta T} = 1.9619 \cdot 10^{-6} \cdot (\Delta T)^6 - 1.3964 \cdot 10^{-4} \cdot (\Delta T)^5 + 2.5908 \cdot 10^{-3} \cdot (\Delta T)^4 + 1.5398 \cdot 10^{-2} \cdot (\Delta T)^3 - 6.7952 \cdot 10^{-1} \cdot (\Delta T)^2 - 6.7204 \cdot 10^{-1} \cdot (\Delta T) + 9.0392 \cdot 10^1 \tag{8}$$

Turbidity (NTU)

$$Q_T = 1.8939 \cdot 10^{-6} \cdot (T)^4 - 4.9942 \cdot 10^{-4} \cdot (T)^3 + 4.9181 \cdot 10^{-2} \cdot (T)^2 - 2.6284 \cdot (T) + 9.8098 \cdot 10^1 \tag{9}$$

Total Solids (mg/L)

$$Q_{TS} = -4.4289 \cdot 10^{-9} \cdot (TS)^4 + 4.65 \cdot 10^{-6} \cdot (TS)^3 - 1.9591 \cdot 10^{-3} \cdot (TS)^2 + 1.8973 \cdot 10^{-1} \cdot (TS) + 8.0608 \cdot 10^1 \tag{10}$$

Ranking Ranges for the WQI NSF. Once the scaling factors Q_i calculations were completed, using Eq. 11 and the relative weights W_i (Table 1), the WQI NSF was calculated. Then, a function was created that assigns categories to the numerical values obtained, simplifying the interpretation and communication of the results. This allows the identification of different water quality levels according to the assigned categories (Table 2).

$$WQI = \sum_{i=1}^{n} Q_i \cdot W_i \tag{11}$$

Table 1. Relative weights for each parameter of the WQI.

n	Parameter i	Units	W_i
1	Dissolved oxygen	Saturation %	0.17
2	Fecal coliforms	MPN/100mL	0.15
3	pH	pH	0.12
4	BOD_5	mg/L	0.10
5	Nitrates	mg/L	0.10
6	Phosphates	mg/L	0.10
7	Temperature	°C	0.10
8	Turbidity	NTU	0.08
9	Total Solids	mg/L	0.08

Table 2. Relative weights for each parameter of the WQI.

Water quality	Color	Value
Excellent		91-100
Good		71-90
Fair		51-70
Poor		26-50
Very poor		0-25

2.4 Design of Data-Driven Models

With the WQI NSF calculated, we created models based on regression trees. For this purpose, there were 4 stages (Fig. 3), as follows:

- Creation of the dataset
- Definition of the base model
- Creation and modeling of parameter subsets
- Determination of similarity to the base model

Creation of the Dataset. The dataset for the Regression Trees models was constructed from information from 23 monitoring stations. This information was organized in a matrix, where each row corresponds to a specific monitoring and the columns contain the characteristics, including parameters and WQI NSF.

Definition of the Base Model. Five models were developed to establish a base model: one Regression Tree model and four Random Forest models for regression, three of which have modifications in their hyperparameters. The objective is to identify the best fitting model for estimating water quality using the WQI NSF as the primary variable.

Cross-validation with 10 splits was used to evaluate each model's performance objectively. Summary statistics, such as mean and standard deviation, were also calculated. In addition, hypothesis testing was performed among all these regression models to determine the base model.

Creation and Modeling of Parameter Subsets. To generate models with reduced parameter subsets, the same data set was used as for the base models. From the 9 available parameters, all possible combinations of subsets were created. The models were built following the Random Forest 1 procedure, known for its simplicity in applying parameter values. At each iteration of model building, performance was evaluated by cross-validation with 10 splits. Finally, summary statistics, such as average and standard deviation, were calculated for the scores obtained in each iteration.

Determination of Similarity to the Base Model. Finally, comparisons and hypothesis tests were performed to evaluate if there is evidence of significant differences between the proposed models and the base model. This allows for a better understanding of the effectiveness and potential of regression tree-based models to assess and estimate water quality in the Yanuncay River watershed.

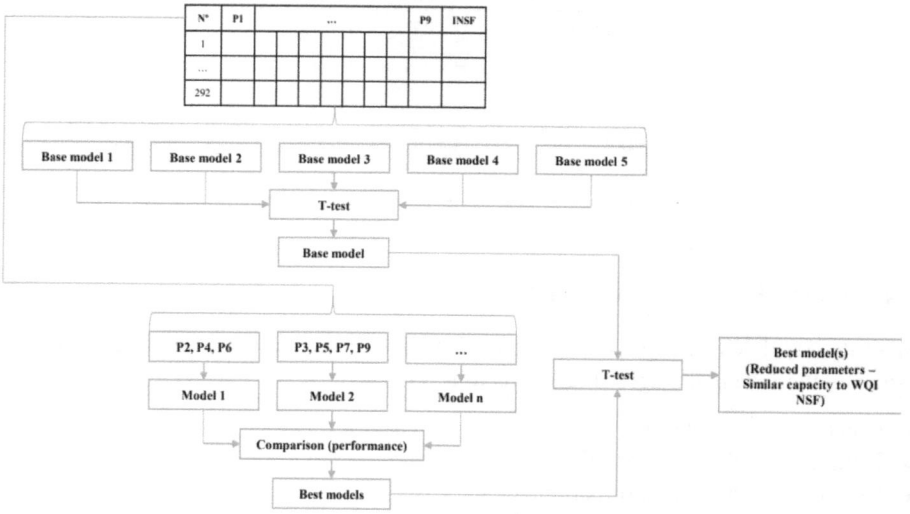

Fig. 3. Stages for creating models based on Regression Trees.

3   Results

3.1   Interpolation of Water Quality Deterioration.

Figure 4 shows a detailed analysis of how water quality varies with altitude at 23 monitoring stations. The results indicate that various natural and human factors affect water quality as rivers and streams travel through the watershed to their discharge point, interacting with soil, vegetation, and other environmental elements.

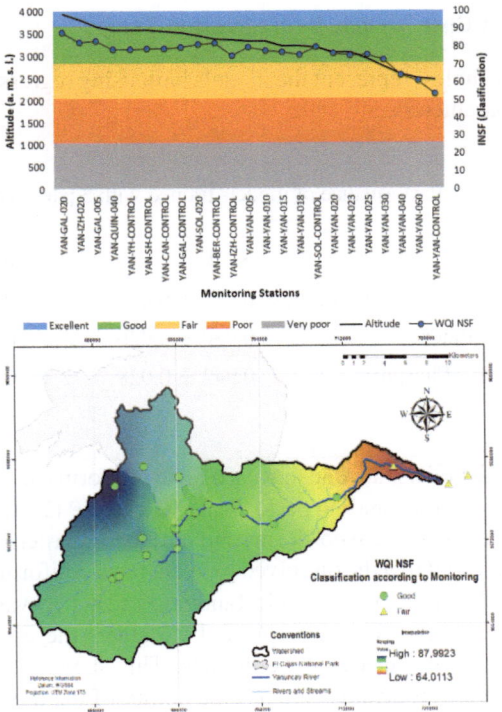

Fig. 4. Behavior of the WQI NSF depending on the altitude variable.

3.2 Performance of Base Models by Regression Trees

The results of the hypothesis tests showed that there was no evidence of significant differences in performance between the regression Random Forest models. However, there were significant differences between Random Forest-1 and Regression Tree (Table 3). Since there were no significant differences between the Random Forest models, Random Forest-1 was chosen as the base model based on its performance advantages.

Table 3. Performance of base models.

Model	$R^2 \pm$	p-value (0.05)
Regression Tree	0.888 ± 0.045	0.002
Random Forest-1	0.951 ± 0.024	*
Random Forest-3	0.954 ± 0.018	0.706

* Base model used for T-tests (comparison)

3.3 Performance of Base Models by Regression Trees

The results shown in Table 4 represent the models built using regression Random Forests with three specific parameters.

Table 4. Performance of models trained with 3 parameters.

Parameter subsets	$R^2 \pm$	p-value (0.05)
Dissolved Oxygen, Fecal Coliforms, Nitrates	0.922 ± 0.032	0.047
Fecal Coliforms, pH, Nitrates	0.930 ± 0.026	0.101
Fecal Coliforms, Nitrates, Phosphates	0.942 ± 0.028	0.511

Among the trained models, the one that used fecal coliforms, nitrates and phosphates as input variables showed outstanding performance ($R^2 = 0.942$). In contrast, the model employing dissolved oxygen, fecal coliforms and nitrates underperformed ($R^2 = 0.922$). Hypothesis testing revealed that the dissolved oxygen, fecal coliforms and nitrates model differed significantly from the base model, Random Forest-1 (p-value = 0.047). These results highlight the importance of the selected variables; the exclusion of dissolved oxygen improved the performance of the models. This suggests that the presence or absence of this variable significantly impacts the ability of the models. In addition, the other variables used are crucial to accurately assess water quality, achieving a more complete representation and better overall model performance.

4 Discussion

In addition to the statistical indicators used to select the models, the logistics required for the laboratory analysis of each parameter (physical, chemical and biological) were considered. Since each requires a specific methodology, a qualitative rating was performed to determine the most appropriate selection (Table 5) [10].

In the analysis, several predictive water quality models were evaluated, focusing on the parameters of fecal coliforms and nitrates, which were considered key indicators of contamination. The most adequate model was the Random Forest by regression, trained with fecal coliforms, pH and nitrates [11]. The pH is practical since it does not require laboratory analysis or specialized equipment, facilitating its implementation and reducing costs.

Another model evaluated included fecal coliforms, nitrates and phosphates, showing slightly better performance but with greater complexity and resources needed due to the requirements for Phosphate analysis. Therefore, the model with pH was prioritized.

Models with four parameters were also considered, highlighting the one trained with fecal coliforms, pH, nitrates and temperature due to its ease of sampling and absence of laboratory analysis. Although the model with fecal coliforms, nitrates, phosphates and total solids had better performance, its complexity increased due to the laboratory analyses required for phosphates and total solids.

Table 5. Qualitative qualification for the 9 parameters required by WQI NSF.

Parameter	Field sampling ease	Laboratory analysis ease	Requirement for specialized equipment
Dissolved Oxygen	Moderate	Moderate	Medium
Fecal Coliforms	Moderate	Moderate	Medium
pH	Simple	Not applicable	Low
BOD_5	Moderate	Complex	High
Nitrates	Simple	Simple	Medium
Phosphates	Simple	Simple	Medium
Temperature	Simple	Not applicable	Low
Turbidity	Simple	Not applicable	Medium
Total Solids	Simple	Moderate	High

5   Conclusion

The results of this study have demonstrated the effectiveness of Regression Tree-based models for predicting water quality in the Yanuncay River watershed with a limited number of parameters compared to the National Sanitation Foundation Index. These findings suggest the possibility of optimizing monitoring requirements without compromising accuracy in estimating the WQI NSF.

Although 4-parameter models can offer greater accuracy, the fecal coliforms, pH and Nitrate model is preferable due to its balance between accuracy and logistical ease. The simplicity of sampling and analysis of pH and nitrates, together with the importance of fecal coliforms as an indicator of contamination, makes this model the most practical and efficient option for assessing water quality in the field.

References

1. Boyle, D.P., Gupta, H.V., Sorooshian, S.: Toward improved calibration of hydrologic models: combining the strengths of manual and automatic methods. Water Resour. Res. **36**(12), 3663–3674 (2000). https://doi.org/10.1029/2000WR900207
2. Duque-Sarango, P., Hernández, B.: Estudio integral del recurso hídrico de la microcuenca del rio guarango, cuenca – ecuador. Rev. Ibérica Sist. e Tecnol. Informação **30**, 240–252 (2020). https://search.proquest.com/docview/2404399354?accountid=32861
3. Duque-Sarango, P., Patiño, D.M., López, X.E.: Evaluación del Sistema de Modelamiento Hidrológico HEC-HMS para la Simulación Hidrológica de una Microcuenca Andina Tropical. Inf. tecnológica **30**(6), 351–362 (2019). https://doi.org/10.4067/s0718-07642019000600351
4. Bui, D.T., Khosravi, K., Tiefenbacher, J., Nguyen, H., Kazakis, N.: Improving prediction of water quality indices using novel hybrid machine-learning algorithms. Sci. Total. Environ. **721**, 137612 (2020). https://doi.org/10.1016/j.scitotenv.2020.137612
5. Rahat, S.H., et al.: Remote sensing-enabled machine learning for river water quality modeling under multidimensional uncertainty. Sci. Total. Environ. **898**, 165504 (2023). https://doi.org/10.1016/j.scitotenv.2023.165504

6. Koranga, M., Pant, P., Kumar, T., Pant, D., Bhatt, A.K., Pant, R.P.: Efficient water quality prediction models based on machine learning algorithms for Nainital Lake, Uttarakhand. Mater. Today Proc. **57**, 1706–1712 (2022). https://doi.org/10.1016/j.matpr.2021.12.334

7. Cañar Uyaguari, M.I.: Desarrollo de un sistema para determinar la calidad de agua en la cuenca del río Paute, basado en la identificación de macroinvertebrados acuáticos como indicadores de calidad de agua, utilizando técnicas de visión por computador. (2023). http://dspace.ups.edu.ec/handle/123456789/26620

8. Duque-Sarango, P., Hernández, B., Cando, G.: Flood risk analysis in an Andean Watershed by integrating satellite data and multicriteria analysis. In: CITIS. LNNS, vol. 871, pp. 25–35 (2024). https://doi.org/10.1007/978-3-031-52090-7_3

9. Jiménez, M., Vélez, M.: Análisis comparativo de indicadores de la calidad de agua superficial. *Av. en Recur. hidráulicos*, 53–69 (2006). http://www.redalyc.org/articulo.oa?id=145020399004

10. Montalvo-Ochoa, F., Robles-Bykbaev, V., Duque-Sarango, P., Gonzalez-Arias, K.: An educational rule-based expert system to determine water quality for environmental engineering and biotechnology students. In: EDUNINE 2020 - 4th IEEE World Engineering Education Conference: The Challenges of Education in Engineering, Computing and Technology without Exclusions: Innovation in the Era of the Industrial Revolution 4.0, Proceedings, p. 9149502 (2020). https://doi.org/10.1109/EDUNINE48860.2020.9149502

11. Cárdenas Patiño, C. M., Crespo Cuzco, B.A.: Estimación de la calidad del agua mediante técnicas de modelización con árboles de regresión y el Índice de la Fundación Nacional de Saneamiento para la cuenca hidrográfica del Río Yanuncay (2024). http://dspace.ups.edu.ec/handle/123456789/27246

From Rice Husk to Graphene: Exploring the Effect of Calcination and Leaching

Angela Pacheco Flores de Valgaz[1]([⊠]) [iD], Gabriela Salcedo Cajas[2],
and Ana Rivas Fermín[3] [iD]

[1] Ingeniería en Biotecnología, Grupo de Investigación en Aplicaciones de Biotecnología
(GIAB), Universidad Politécnica Salesiana, Campus María Auxiliadora, Km 19.5 Vía a la Costa,
Guayaquil, Ecuador
apachecof@ups.edu.ec
[2] Facultad de Ciencias Naturales y Matemáticas, Escuela Superior Politécnica del Litoral,
Campus Gustavo Galindo, Km 30.5 Vía Perimetral, Guayaquil, Ecuador
[3] Facultad de Ingeniería Mecánica y Ciencias de la Producción, Escuela Superior Politécnica
del Litoral, Campus Gustavo Galindo, Km 30.5 Vía Perimetral, Guayaquil, Ecuador

Abstract. This study aims to assess the impact of leaching with HCl and calcination at varying temperatures (300–600 °C) on the production of graphene from rice husks. The samples underwent calcination at different temperatures, were mixed with KOH, and synthesized at 850 °C for two hours. SEM-EDS, FTIR, TGA, XRD, and Raman were used for characterization. The resulting synthesized products had carbon content higher than 96% and morphology similar to corrugated paper with a multilayered structure. The temperature range of 450–600 °C was found to be the most favorable for successful synthesis. The use of XRD and Raman spectroscopies confirmed the synthesis of graphene. These findings provide valuable insights for optimizing graphene production, paving the way for a sustainable and potentially profitable approach.

Keywords: graphene · rice husk · leaching · calcination

1 Introduction

Graphene, a two-dimensional material composed of carbon atoms arranged in a hexagonal structure, has captured the attention of the scientific and technological community due to its exceptional physical, mechanical, electrical and optical properties [1, 2]. Its thermal stability, high mechanical strength, and high specific surface area make it a material with enormous potential for diverse applications in electronics, energy, biomedicine, environmental remediation, and many other fields [3, 4].

Despite its great potential, the large-scale production of high-quality graphene is still a challenge. Traditional methods, such as graphite peeling, are costly and inefficient1. In this context, the search for sustainable and cost-effective production methods that make use of agricultural by-products and industrial waste materials has become a field of great interest [5].

E. M. Inga Ortega et al. (Eds.): CITIS 2024, LNNS 1331, pp. 271–281, 2025.
https://doi.org/10.1007/978-3-031-87065-1_25

Rice production and consumption are on the rise globally, and new measures are necessary to mitigate the significant impact of rice by-products [6]. Harvesting companies aggregate large volumes of rice husks, with over 75% being disposed of through burning, resulting in harmful carbon dioxide emissions. In Ecuador alone, more than 500 tons of rice husks are generated annually, making it one of the largest waste products of the rice industry. Its lignocellulosic composition and hydrocarbons make it an ideal precursor for synthesizing carbonaceous materials, such as graphene [7].

Recently, several methods have been developed to extract graphene from rice husks, including chemical, physical, and biological methods [8–10]. These methods rely on breaking down the lignocellulosic structure of the rice husk and transforming the carbon it contains into graphene [11].

Extracting graphene from rice husks presents an opportunity to add value to this waste and contribute to the circular economy [12]. However, challenges remain in improving the efficiency and quality of graphene obtained from rice husk by-products. This study aims to evaluate the effect of leaching and calcination on obtaining high-quality graphene from rice husks.

2 Methods

The methodology used in this investigation was adapted from Muramatsu et al. (2014), Rhee et al. (2016), and Singh et al. (2017).

2.1 Raw Materials

The rice husk used in the present investigation was obtained from a rice mill located in the province of Guayas, Ecuador.

2.2 Pretreatment

The rice husks were divided into two groups. Group 1 received a chemical wash with 1N HCl to remove impurities, while Group 2 underwent a simpler wash with just distilled water. The use of chemical treatments before calcination is a common practice to eliminate impurities from rice husk. This is due to the presence of certain elements in the husk.

2.3 Calcination

After washing, the rice husk was calcined in air at temperatures ranging from 300 °C to 600 °C [13].

2.4 Graphene Synthesis

Rice husk ash was mixed with analytical grade KOH in a 1:4 ratio and triturated for 5 min. The resulting mixture was deposited in a mullite crucible, covered with ceramic wool, and fixed in a larger crucible made of fused silicon carbide [10].

The top of the large crucible was coated with activated carbon to prevent oxidation of the sample inside the mullite crucible. The sample was annealed at 850 °C for 2 h in a muffle furnace [9].

2.5 Post-Treatment

Finally, the sample was washed multiple times with distilled water to eliminate excess KOH and dried.

2.6 Characterization Techniques

Thermogravimetric Analysis (TGA) was performed using a Simultaneous Differential Technique thermal analyzer model Q600 and the TA Universal Analysis 2000 program (version 4.5A). This allowed for the determination of the mass loss of the sample as a function of temperature, enabling quantification of the percentage of moisture, volatile compounds, and ash (residues) present.

Fourier Transform Infrared Spectroscopy (FTIR) was used to identify the structural changes and vibrational frequencies of the functional groups responsible for metal adsorption on the material's surface.

Scanning Electron Microscopy (SEM) and Energy Dispersive X-Ray Spectroscopy (EDX) were used to analyze the microstructure morphology and chemical composition of the material using an FEI brand scanning electron microscope, model INSPECT S. The xT Microscope Server and xT Microscope Control soft-ware controlled the microscope, while the EDAX Genesis software analyzed the sample's elemental composition. This technique allowed for the determination of both the overall elemental composition of the sample and the composition of its components.

X-Ray Diffraction (XRD) was used to perform a qualitative analysis of the crystalline phases of the material on a Panalytical X'Pert diffractometer (40 kV, 30 mA). Scans with a 2θ step size of 0.017 and a counting time per step of 0.1 s were used, utilizing a Cu-K α radiation source ($\lambda = 1.542$ Å).

Raman Spectroscopy is a technique used to analyze the structure of carbon materials, including graphene nanolayers. The Raman spectra were obtained using a Renishaw InVia Reflex confocal microscope with a glass substrate, at a wavelength of 514 nm, 100x magnification, numerical aperture of 0.9, and incident power of 10 mW.

3 Results

3.1 Calcination and Chemical Synthesis

A decreasing trend in the residual content of the calcined rice husk was obtained as the temperature increased. At 400 °C and above, the residual content stabilizes at around 22% by weight, indicating significant combustion of the organic matter present. This reduction is attributed to the decomposition and elimination of volatiles, such as cellulose, hemicellulose, and lignin, present in the husk. Between 400 °C and 600 °C, there is no significant variation in residual weight, indicating the complete elimination of volatile organic matter. However, the color of the calcined scale changes noticeably with increasing temperature. At a temperature of 300 °C, the ash appears intensely black, indicating incomplete combustion. At 400 °C and higher, the ash takes on varying degrees of gray tones, often with black spots. At 600 °C, the ash is primarily gray with a slight pinkish hue (Fig. 1).

The colors observed during calcination reflect the various stages of organic matter decomposition and transformation, as well as ash formation. Black colors indicate incomplete combustion with residual carbon present. Gray and pink tones suggest more complete combustion and the formation of different inorganic compounds, such as silica, alumina, and iron oxides. 84 syntheses of graphene from rice husk ash were performed, 7 per each calcination temperature.

(a)   **(b)**   **(c)**   **(d)**

Fig. 1. Ashes with different shades obtained at various temperatures: (a) 300 °C, (b) 500 °C, (c) 550 °C, and (d) 600 °C.

3.2 Thermal Analysis

The rice husk showed a weight loss of 5.94% in the range of 50 °C to 140 °C. Thermal decomposition, responsible for the loss of 50.13% of the total weight, occurred between 140 °C and 900 °C (Fig. 2a). The residual ash content, mainly composed of silica, represented 31.70% of the total weight.

The sample synthesized from rice husk calcined at 450 °C without treatment had a water content of 7.4%, which evaporated rapidly around 47 °C. The volatile components were eliminated up to 900 °C and represented 11% of the total weight, while the residues constituted 81.5% (Fig. 2b).

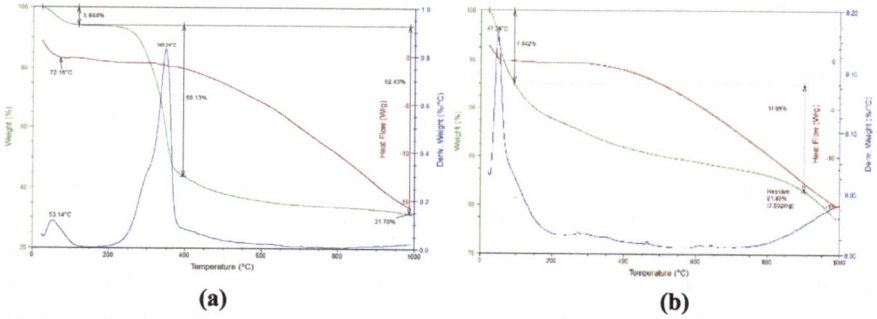

(a)   **(b)**

Fig. 2. TGA curves of synthesized sample from (a) rice husk without chemical treatment and (b) rice husk calcined at 450 °C without treatment.

3.3 FTIR Spectroscopic Analysis

Figure 3 shows a significant reduction of functional groups in the calcined rice husk, with the disappearance of peaks related to oxygen (carbonyl and others) in the synthesized sample.

Studies report limitations in the interpretation of FTIR spectra for the characterization of graphene due to the low sensitivity of the bands, especially in the case of monolayers.

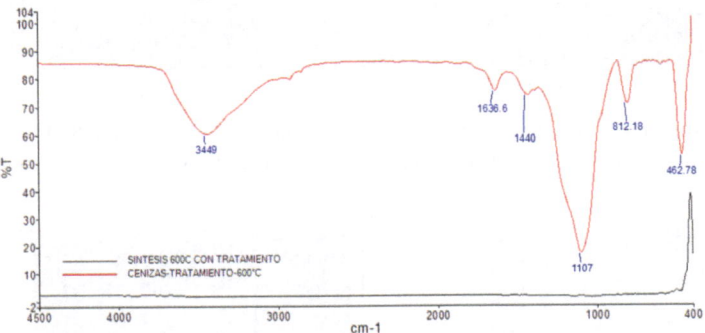

Fig. 3. FTIR comparison of rice husk calcined at 600 °C with treatment (red line) vs. its respective synthesis at 850 °C (black line).

3.4 Morphological Analysis by Scanning Electron Microscopy and Elemental Composition by Energy Dispersive X-Ray Spectroscopy

Figure 4 shows the synthesized product and reveals the presence of sheets with morphological features compatible with graphene. A multilayer structure similar to corrugated paper is observed, including silica nanoparticles dispersed on its surface. The microstructure has pores on the front surface that act as channels through the material, similar to the structure of a honeycomb.

EDX analysis revealed a high carbon content, close to 90%, in all samples, especially in those synthesized from rice husks calcined at 600 °C without treatment, at 450 °C with treatment, and at 450 °C without treatment (Fig. 5a).

In addition, the presence of other elements such as Al, Si, K, Fe and Mg was detected (Fig. 5b). These elements were more abundant in the samples synthesized at lower temperatures, such as 400 °C and 450 °C without treatment. It should be noted that impurities such as Mg and Fe were not observed in the treated samples.

3.5 X-Ray Diffraction Analysis

In the diffractograms of the synthesized samples, peaks or broad bands centered around 27.0° and 44° are observed (Fig. 6b). It should be noted that the intensity of these peaks is weak.

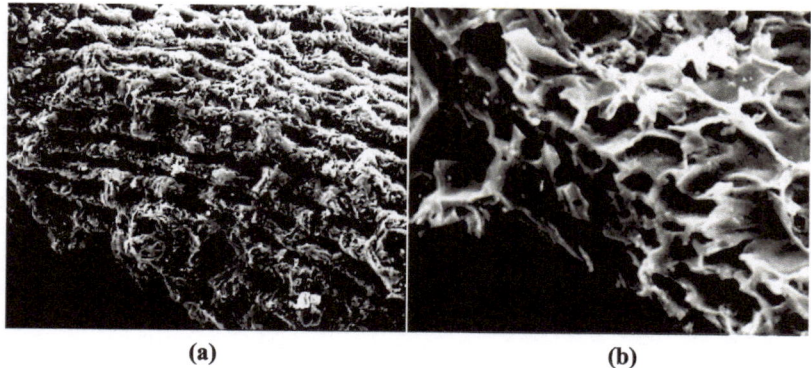

Fig. 4. Micrograph at 100 µm using Scanning Electron Microscopy of the 850 °C synthesis of rice husks calcined at 600 °C (a) without chemical treatment (b) with chemical treatment.

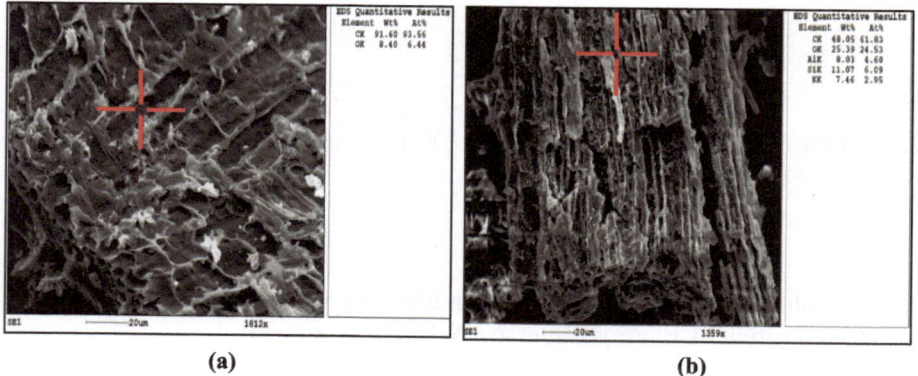

Fig. 5. Spot EDX characterization of the sample synthesized from rice husk calcined at (a) 600 °C without treatment and (b) 450 °C without treatment.

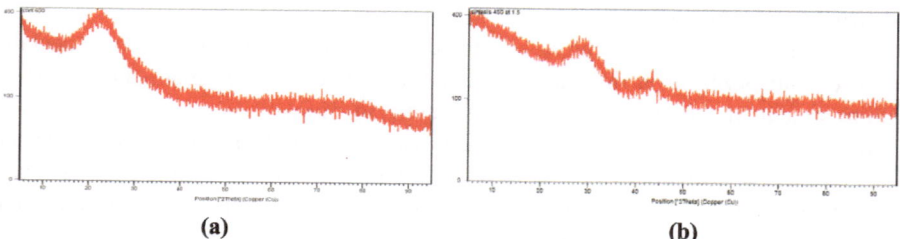

Fig. 6. X-ray diffractograms of: (a) rice husk calcined at different temperatures; (b) synthesis product from rice husk calcined at 450 °C without chemical treatment.

3.6 Raman Spectroscopic Analysis

The spectra of samples m1, m2 and m8 (calcined at 450 °C without chemical treatment) and m3 and m6 (calcined at 600 °C without chemical treatment) show a distinct 2D peak, while this peak is not observed in the other samples (Fig. 7). The intensity of the D band (ID) relative to the G band (IG) provides information about the degree of defects and disorder in the graphene structure. A low ID/IG value indicates a lower degree of defects in the obtained graphene.

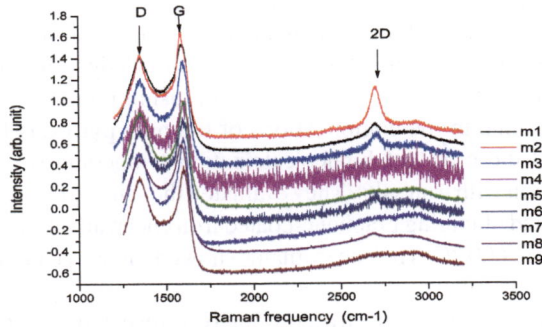

Fig. 7. Raman spectra of graphene from calcined rice husk at 450 °C without chemical treatment (m1, m2 and m8). CCA at 600 °C without chemical treatment (m3 and m6), at 600 °C with chemical treatment (m4 and m9) and at 300 °C with chemical treatment (m5) and without chemical treatment (m7).

4 Discussion

This study focuses on explore the variables for producing graphene from rice husks. At 400 °C, a significant combustion of the organic matter results in a residual content of about 22%. Between 400 °C and 600 °C, the color changes from black to greyish, indicating the evolution of the chemical structure of the material. 15 successful syntheses were obtained between 450 °C and 600 °C, suggesting that there is a greater likelihood of successful syntheses at this temperatures range. Figure 4 seem to indicate that higher temperatures produce hexagonal surfaces with clean edges, which improves the quality of the graphene.

Figure 2a shows an initial loss (5.94%, 50 °C–140 °C) due to the evaporation of moisture retained in the porous structure of the rice husk and possible light volatiles. The second loss (about 50%, 140 °C–370 °C) corresponds to the thermal decomposition of the organic matter present in the husk. This organic matter consists mainly of hemicellulose (decomposes between 150 °C and 250 °C), cellulose (de-composes between 250 °C and 350 °C) and lignin (decomposes between 250 °C and 500 °C) [14, 15].

During decomposition, the carbon present in the organic matter is released in the form of gases such as carbon dioxide and carbon monoxide. Based on this, the main source of carbon above 450 °C is residual lignin and biochar formed from pyrolysis, also known as enhanced charcoal or high-temperature charcoal, which may react with KOH, promoting the decomposition of lignin and the formation of a porous nanoarchitecture characteristic of graphene. Qi et al. (2020) claim that the high activation temperature could induce the formation of a graphitized carbon skeleton of biochar with high surface functionalities and porosity, in contrast to that of biochar prepared at low temperature. Thus, it is possible that the high synthesis temperature (850 °C) would provide the necessary energy for carbon rearrangement and graphene formation. Current research shows that it is possible to obtain graphene from biochar using different activators and nitrogen gas atmospheres [16, 17].

In their research, Wang et al. (2022) applied deionized water-only washing and calcination at 400 °C in an N_2 atmosphere to bamboo samples, from which they obtained a graphene-biochar mixture. Other authors have acknowledged that the production of graphene from biochar by carbo-thermal reduction using biomass as a carbon source remains a challenge due to the complications of biomass pyrolysis [18]. Karunaratne et al. (2022) tested the effect of higher calcination temperatures (600–1000 °C) on bamboo and activation with iron salts. They evaluated the salt application in pyrolysis and synthesis and reported obtaining graphene-coated iron encapsulates after salt application following pyrolysis at 600 °C. The encapsulates showed improved contaminant removal capabilities.

The X-ray energy dispersive spectroscopy analysis revealed the main components of the calcined rice husk, which consisted of a high carbon content of approximately 90% in all samples. Other compounds, including Al, Si, K, Fe, and Mg, were also present. These compounds were more prevalent at lower temperatures, such as 400 °C and 450 °C, without HCl (1N) treatment (Fig. 5b). No impurities in the final product, such as Mg and Fe, were detected from the treated samples. In terms of chemical treatment, some reports suggest that the presence of certain metals, such as iron, which is a natural mineral found in rice husk, can provide sites for the precipitation of graphitic carbon [14, 18].

According to Ghogia et al. (2022), iron is the most effective transition metal for catalytic graphitization of biochar. The ineffectiveness of the syntheses may be due to the acid washing of the raw materials to re-move impurities, which affects the presence of metal ions such as iron. However, it is not recommended to completely discard this treatment. Depending on the application, the contact time with the raw material could be optimized because the graphene with the highest carbon content was obtained from the leached samples. Furthermore, the size and structure of the pores are modified by HCl leaching, while the amorphousness of the silica remains unaffected. Additionally, although the success of the synthesis does not solely depend on the thermal and chemical treatments, they do play a crucial role in the process.

The X-ray diffraction spectra confirm the presence of silica in both the calcined and synthesized samples, as shown by the peak at 27° (Fig. 6). The X-ray diffractograms also indicate that the silica in the calcined samples is in an amorphous state and remains unchanged, indicating that the amorphous state of the silica is not affected by temperature or washing method (Fig. 6). The diffractograms of the synthesized samples show a peak at 44°, which is attributed to the graphitic structure of the carbon. The low intensity of this peak suggests that the material has low crystallinity. The diffraction pattern of the synthesis product presents characteristic peaks in specific positions similar to those reported for graphene [10].

Figure 3 illustrates the distinctions between the FTIR spectra of the calcined scale and the synthesis product, revealing a considerable reduction of functional groups in the calcined rice husk. The FTIR spectrum's lack of additional peaks may suggest the absence of impurities, such as oxides, carboxyls, or amines. This absence could be due to the graphene synthesis process, which produces a material with few defects or impurities. The high peak at 400 cm^{-1} corresponds to the G-band of graphene, which is associated with the vibration of carbon atoms in the sp2 plane.

The analysis of the synthesized samples using Raman confirms the presence of graphene. The Raman spectrum confirms the presence of G, D and 2D bands characteristic of graphene. However, the characteristics of the graphene vary depending on the leaching and calcination conditions, as shown in Fig. 7. It is important to note that the intensity of the G-band is higher than that of the 2D-band in all spectra, indicating the presence of thicker graphene layers rather than monolayers. The characteristic D- and G-band peaks correspond to imperfect, well-defined carbon structures. These results are similar to those obtained by other authors through the formation of biochar from biomass by thermolysis [19].

The production of graphene from rice husks represents a promising alternative due to its simplicity and sustainability. This method exploits an abundant and renewable agro-industrial waste as a raw material. This strategy presents significant advantages compared to other graphene production techniques, as it may reduce costs, minimize environmental impact, and promote a circular eco-economy.

5 Conclusion

The investigation confirms the presence of thick layered graphene compiled from rice husk, with variable characteristics depending on leaching and calcination conditions. Significant combustion of the organic matter in the rice husk requires calcination at 400 °C or higher. The most favorable conditions for graphene synthesis were identified as calcination temperatures of 450 °C and 600 °C without HCl (1N) pre-treatment.

While HCl (1N) pretreatment effectively removes metallic impurities, it is believed that this step also removes metals that promote graphene production. Therefore, it is recommended to investigate the impact of minerals such as Al, Si, K, and Mg on the formation of a graphitic structure from rice husk at calcination temperatures ranging from 450 °C to 600 °C. The purpose of this study might be to determine which minerals should be retained and which should be removed from the raw material. K_2CO_3 is another reagent commonly used in the chemical activation of natural carbonaceous sources. Future research should include a comparison of its effectiveness in activating rice husk ash with the results obtained using KOH.

This research confirms the value of using Ecuadorian rice husks for producing new materials. It also provides valuable information for optimizing parameters to obtain graphene at a low cost and for large-scale production. The results represent a significant advancement in the development of sustainable technologies for producing advanced materials. However, additional research is required to gain a better understanding of the impact of leaching and calcination variables on the process of synthesizing graphene.

References

1. Safian, M.T., Haron, U.S., Nasir, M., Ibrahim, M.: A review on bio-based graphene derived from biomass wastes. BioResources **15**, 9756–9785 (2020)

2. Xu, Z., et al.: Green synthesis of nitrogen-doped porous carbon derived from rice straw for high-performance supercapacitor application. Energy Fuels **34**(7), 8966–8976 (2020)
3. Zhang, M., Gao, B., Yao, Y., Xue, Y., Inyang, M.: Synthesis, characterization, and environmental implications of graphene-coated biochar. Sci. Total. Environ. **435–436**, 567–572 (2012)
4. Asiri, M., et al.: Rice straw derived graphene-silica based nanocomposite and its application in improved co-fermentative microbial enzyme production and functional stability. Sci. Total Environ. **876**, 162765 (2023)
5. Abdel-Aal, S.K., Hassan, M.L., Abou Elseoud, W.S., Ward, A.: High-lignin-content rice straw cellulose nanofibers/graphene oxide nanocomposites films: electrical and mechanical properties. J. Appl. Polym. Sci. **139**(45) (2022)
6. Colombo, R., Moretto, G., Barberis, M., Frosi, I., Papetti, A.: Rice byproduct compounds: from green extraction to antioxidant properties. Antioxidants **13**(1) (2024)
7. Naranjo, J., et al.: Preparation of adsorbent materials from rice husk via hydrothermal carbonization: optimization of operating conditions and alkali activation. Resources **12**(12), 145 (2023)
8. Uda, M.N.A., et al.: Simple and green approach strategy to synthesis graphene using rice straw ash. IOP Conf. Ser. Mater. Sci. Eng. **864**(1) (2020)
9. Muramatsu, H., et al.: Rice husk-derived graphene with nano-sized domains and clean edges. Small **10**(14), 2766–2770 (2014)
10. Singh, P., Bahadur, J., Pal, K.: One-step one chemical synthesis process of graphene from rice husk for energy storage applications. Graphene **06**(03), 61–71 (2017)
11. Seitzhanova, M., et al.: Production of graphene membranes from rice husk biomass waste for improved desalination. Nanomaterials **14**(2) (2024)
12. Uda, M.N.A., et al.: Production and characterization of graphene from carbonaceous rice straw by cost-effect extraction. 3 Biotech **11**(5), 1–11 (2021)
13. Rhee, I., Lee, J.S., Kim, Y.A., Kim, J.H., Kim, J.H.: Electrically conductive cement mortar: incorporating rice husk-derived high-surface-area graphene. Constr. Build. Mater. **125**, 632–642 (2016)
14. Ghogia, A.C., Romero L.M., White, C.E., Nzihou, A.: Synthesis and growth of green graphene from biochar revealed by magnetic properties of iron catalyst. ChemSusChem **16**(3) (2023)
15. Lotfy, V.F., Fathy, N.A., Basta, A.H.: Novel approach for synthesizing different shapes of carbon nanotubes from rice straw residue. J. Environ. Chem. Eng. **6**(5), 6263–6274 (2018)
16. Thomas, B., et al.: Electrochemical properties of biobased carbon aerogels decorated with graphene dots synthesized from biochar. ACS Appl. Electron. Mater. **3**(11), 4699–4710 (2021)
17. Wang, Y., et al.: Preparation of novel biochar containing graphene from waste bamboo with high methylene blue adsorption capacity. Diam. Relat. Mater. **125**, 109034 (2022)
18. Karunaratne, T.N., et al.: Pyrolytic synthesis of graphene-encapsulated zero-valent iron nanoparticles supported on biochar for heavy metal removal. Biochar **4**(1) (2022)
19. Qi, Y., et al.: Three-dimensional porous graphene-like biochar derived from enteromorpha as a persulfate activator for sulfamethoxazole degradation: role of graphitic N and radicals transformation. J. Hazard. Mater. **399**, 123039 (2020)

Hyperaccumulation of Cadmium and Lead in Ruderal Species of the Maria Auxiliadora Campus of the Universidad Politécnica Salesiana Guayaquil, Ecuador

Jaime Naranjo-Moran$^{(\boxtimes)}$ ⓘ, Gianna Mawyin-Mora, Tannia Reyes-Vilches, and José Luis Ballesteros-Lara ⓘ

Grupo de Investigación en Aplicaciones Biotecnológicas, GIAB, Carrera de Biotecnología, Universidad Politécnica Salesiana, UPS, Campus María Auxiliadora, kilómetro 19.5 Vía a La Costa, 090901 Guayaquil, Ecuador
jnaranjo@ups.edu.ec

Abstract. Heavy metal contamination in urban soils is a growing concern. Hyper-accumulating ruderal species offer an alternative for bioremediation of these soils. To characterize Cd and Pb hyper-accumulating ruderal species in the soil adjacent to the María Auxiliadora campus of the Universidad Politécnica Salesiana Guayaquil, Ecuador. Soil samples and more than 50 ruderal species were collected. The three most representative species (*Distimake aegyptius*, *Bidens bipinnata* and *Chenchrus echinatus*) were selected for translocation factor, bioconcentration, germination and mortality analyses. Cd and Pb concentrations were determined by inductively coupled plasma spectroscopy. The soil is contaminated by Cd and Pb, due to a nearby company. Cd concentrations ranged from 1 to 2 ppm. The three selected species showed hyperaccumulation capacity, with *Chenchrus echinatus* standing out as an accumulator and *Distimake aegyptius* as a transporter. The results confirm the presence of ruderal hyper-accumulating species in the study area with potential for bioremediation. The implementation of phytoextraction strategies with *Distimake aegyptius* and phytostabilization with the other species is suggested. The presence of hyperaccumulating ruderal species offers a viable alternative for the bioremediation of soils contaminated by heavy metals.

Keywords: Ruderal species · phytoextraction · heavy metals

1 Introduction

The growing industrialization and urbanization of cities in general have intensified environmental pollution by heavy metals, negatively impacting ecosystems and human health. Among the various pollutants, heavy metals such as cadmium (Cd), lead (Pb), chromium (Cr) and arsenic (As) stand out for their persistent toxicity and bioaccumulation [1].

On the other hand, ruderal plant species are plants that colonize soils disturbed by human activities and have emerged as a promising tool for the bioremediation of soils

© The Author(s) 2025
E. M. Inga Ortega et al. (Eds.): CITIS 2024, LNNS 1331, pp. 282–291, 2025.
https://doi.org/10.1007/978-3-031-87065-1_26

contaminated with heavy metals. Unlike cultivated species, these plants can absorb and accumulate heavy metals in their tissues without suffering adverse effects [2]. This capacity for hyperaccumulation is attributed to various physiological and genetic mechanisms that allow ruderal plants to tolerate and store high concentrations of heavy metals in their various parts, such as leaves, roots and stems [3, 4]. Hyperaccumulation not only facilitates the removal of heavy metals from the soil but can also contribute to the stabilization of contaminants and the reduction of their bioavailability to other organisms [5].

The implementation of bioremediation strategies based on ruderal plant species offers multiple advantages, due to their low cost, effectiveness in degraded environments, and their potential for the ecological restoration of contaminated areas such as roads, landfills or mining tailings [6, 7]. Ecuador is a country with many environmental problems; this nation has exuberant ecosystems such as the Tumbesian dry forests that cover 35% of the coastal region, which have suffered a significant reduction and degradation. Today, only a fraction of this natural treasure remains intact, while most of it has fallen victim to deforestation, urban development, and agriculture [8]. Within this context, herbaceous ruderal species that colonize disturbed land have gained new importance [9]. These species, far from being undesirable intruders, can be valuable tools for environmental assessment and bioremediation [10].

2   Materials and Methods

Study Site. This research was carried out in the province of Guayas, in an area adjacent to the María Auxiliadora campus located on the Vía a "La Costa", kilometer 19.5. The study area was delimited on the ground adjacent to the campus, with a total extension of 500 m. The geographic coordinates of the study area are 2°11′33.4 "S and 80°02′43.7" W. For a better understanding of the location, Fig. 1, with the area of the investigation, is attached.

Census of Ruderal Floristic Composition. The circular plant census is a traditional technique to evaluate the composition and structure of plant communities [11], in addition the use of digital tools facilitated the identification for the collection of species remotely, by optimizing the research process and making it more accessible to a wide range of users on the live internet. The cellular applications used for automated plant recognition through images were iNaturalit Ecuador (https://ecuador.inaturalist.org/) and Pl@ntNet (https://identify.plantnet.org/es), these applications have been very useful for researchers in recent years, even where there is not much plant diversity [12]. The circular plant census incorporates a global positioning system (GPS) to determine the exact location of the sampling plots. This will allow the creation of an interactive map that provides the digital repositories with a visualization of the spatial distribution of plant species, and this information will be updated in future surveys. In addition, many apps have tools to record detailed information about each species, such as scientific name, height, stem diameter and conservation status. The counting of plant species, an essential component of the circular census, was simplified by the use of apps. The circular plot consisted of the numbering and geographic location of the PSUs (Permanent Sampling Units) in a flat field. The PSU was 500 × 15 m, where the ruderal species were evaluated. The PSU was divided into eight sampling subunits with a diameter of 10 m, where the different ruderal species were collected in a 5 m radius. From each central subunit were placed squares with the dimension of 1 m2 × 1 m2, and then three subsamples for each sample were analyzed randomly (Fig. 1). Instead of taking manual notes, users captured images of the plants to identify remotely with the image recognition algorithm. This not only saves time and effort but also reduces the possibility of errors in species identification. A validation with three recognized databases "Herbario Rapid Reference" (https://plantidtools.fie ldmuseum.org/es/rrc/5581), "Trópicos" (https://www.tropicos.org/home), and "Bioweb Ecuador" (https://bioweb.bio/) was carried out to contrast the information collected.

Fig. 1. A) Sampling sites in ruderal zone at km 19 via La Costa; B) Circular sampling installation; C-D) Photographic shots for remote reconnaissance; E) Sample processing in acid digestion.

Elemental Determination Of Metals (Cd and Pb). To determine the concentration of Cd and Pb in soil samples, roots, and leaves, the technique of inductively coupled plasma optical emission spectroscopy (ICP-OES) was used. The samples were prepared by drying, grinding, and sieving according to the EPA-3051A method, developed by the United States Environmental Protection Agency [13]. The leaves were dried, ground, and sieved before acid digestion. A microwave acid digestion was used, 0.5 g of each type of sample (soil, roots, and leaves) were weighed and placed in digestion vessels. Ultrapure nitric acid was added (10 ml for soil and 10 ml for roots and leaves) and different heating-digestion programs were applied according to the type of sample, for sue-lo: 175 °C for 5:30 min. And for roots and leaves: 200 °C for 15:00 min. Once the digestion was finished, the solutions were filtered and diluted to a final volume of 50 ml with type 1 water. The ICP-OES analysis was performed by calibrating the equipment, where a multi-element solution was prepared with concentrations of 0.2; 0.7; 1.1; 1.1; 1.1; 1.2; and 4.8 ug/g for each element (Cd and Pb) from an individual 1000 mg/L standard solution. For the reading, the diluted samples were introduced into an autosampler and analyzed in a Thermo Scientific ICP-OES equipment model iCAP 7400 Duo. The following wavelengths were used for the reading of each element Cd: 226,502 nm and Pb: 220,353 nm, the validation of the calibration curves for each element showed a coefficient of determination (R2) of 0.9970 for both elements analyzed (Cd and Pb).

Accumulation and Translocation Factors for Cd and Pb. To evaluate the bioaccumulation capacity of plants, the concentration of metal present in their tissues was divided by the amount of metal present in the ruderal soil [14]. This index indicates how much metal the plant has absorbed about the amount available in the soil. As for translocation, the concentration of metal present in the plant biomass (roots and leaves) was divided by the concentration of metal present in the roots. This indicator reflects the efficiency with which the plant transports the metal absorbed from the roots to the aerial parts, i.e. bioaccumulation measures the total amount of metal that the plant has absorbed and retained, while translocation evaluates the capacity of the plant to distribute the metal absorbed by its different parts. Consideration should be given to measuring the quantities of plants and agricultural products in Ecuador [15].

Data Analysis. Once the significant differences were identified by ANOVA, Tukey's test was used to establish which of these differences were statistically relevant. A significance level of $p < 0.05$ was established for all tests.

3   Results

As a result of this research, a total of 56 species belonging to 21 plant families were identified in the eight transects, as shown in Table 1. These species are considered ruderal, due to the high anthropogenic intervention in the area.

As a result of the circular plant census, three representative species were found in several individuals in the eight transects: *Distimake aegyptius*, *Chenchrus echinatus*, and *Bidens bipinnata*, belonging to the families Convolvulaceae, Poaceae, and Asteraceae. From these species, root, leaf, and soil samples were collected for each of the three selected plant species. It is observed that point eight had a higher level of Cd, while point two had a higher level of Pb, as shown in Fig. 2 concerning the other sampling points, which means that the accumulation is well determined by anthropogenic activity.

Cd values in leaves ranged from 0.10 ppm (*D. aegyptius*) to 0.19 ppm (*B. bipinnata*). For Pb, values ranged from 0.10 ppm (*D. aegyptius*) to 0.20 ppm (*B. bipinnata*). No significant differences were found in the accumulation of heavy metals among the leaves of the three species. As for the accumulation in roots, the species *B. bipinnata* and *D. aegyptius* presented values of Cd and Pb uptake in roots lower than 1 ppm. *C. echinatus* showed values greater than 1 ppm, with 1.03 ppm for Cd and 1.40 ppm for Pb. In general, roots of the three species accumulated more heavy metals than leaves. Accumulation in rhizospheric soil showed a higher accumulation of heavy metals for all three plants, with

Table 1. Ruderal species identified at each sampling point at km 19 via La Costa.

Sampling points	Scientific name	Family
First point	*Rhynchosia minima*	Fabaceae
	Argentina adenophora	Asteraceae
	Eclipta prostrata	Asteraceae
	Ludwigia peruviana	Onagraceae
	Rhynchosia minima	Fabaceae
	Ptelea trifoliata	Rutaceae
	Cenchrus echinatus	Poaceae
	Sida rhombifolia	Malvaceae
Second point	*Echinochloa colona*	Poaceae
	Dactyloctenium aegyptium	Poaceae
	Chloris barbata	Poaceae
	Sorghum halepense	Poaceae
	Merremia umbellata	Convolvulaceae
Third point	*Sporobolus pyramidatus*	Poaceae
	Melochia lupulina	Malvaceae
	Achyranthes aspera	Amaranthaceae
	Commelina diffusa	Commelinaceae
	Bidens bipinnata	Asteraceae
	Echinochloa colona	Poaceae
	Cucumis dipsaceus	Cucurbitaceae
	Ageratum conyzoides	Asteraceae
	Distimake aegyptius	Convolvulaceae
	Cyperus rotundus	Cyperaceae
	Mitracarpus hirtus	Rubiaceae
	Sida acuta	Malvaceae
Fourth point	*Tephrosia purpurea*	Fabaceae
	Melochia lupulina	Malvaceae
	Albizia multiflora	Fabaceae
	Ipomoea coccinea	Convolvulaceae
	Rhynchosia minima	Fabaceae
Fifth point	*Portulaca oleracea*	Portulacaceae
	Convolvulus althaeoides	Convolvulaceae

(*continued*)

Table 1. (*continued*)

Sampling points	Scientific name	Family
	Tetramerium nervosum	Acanthaceae
	Cenchrus echinatus	Poaceae
	Mimosa quadrivalvis	Fabaceae
	Ruellia blechum	Acanthaceae
	Sorghum halepense	Poaceae
	Amaranthus retroflexus	Amaranthaceae
	Distimake aegyptius	Convolvulaceae
	Sida rhombifolia	Malvaceae
	Sida acuta	Malvaceae
Sixth point	*Distimake aegyptius*	Convolvulaceae
	Prestonia mollis	Apocynaceae
	Sida rhombifolia	Malvaceae
	Tecoma castanifolia	Bignoniaceae
	Cenchrus echinatus	Poaceae
	Sorghum vulgare	Poaceae
	Senecio viscosus	Solanaceae
Seventh point	*Cynodon dactylon*	Poaceae
	Solanum nigrum	Solanaceae
	Euphorbia prostrata	Euphorbiaceae
	Alysicarpus vaginalis	Fabaceae
Eighth point	*Murraya paniculata*	Rutaceae
	Synedrella nodiflora	Asteraceae
	Coccoloba uvifera	Polygonaceae
	Cyperus rotundus	Cyperaceae

values above 1 ppm for all metals. Cd values in soil were similar for the three species, ranging from 1.53 ppm (*B. bipinnata*) to 1.60 ppm (*C. echinatus*). For Pb, *C. echinatus* presented the lowest value (1.10 ppm), while *B. bipinnata* (3.70 ppm) and *D. aegyptius* (3.90 ppm) showed the highest values (Fig. 3).

The results for the bioconcentration factor obtained for Cd were as follows: 0.96 for *D. aegyptius*, 0.43 for *B. bipinnata*, and 1.46 for *C. echinatus* (Fig. 4). For Pb they are detailed as follows: 0.62 for *D. aegyptius*, 0.75 for *B. bipinnata*, and 1.20 for *C. echinatus*. The species *C. echinatus* for both metals have the highest bioconcentration factor value, so that, being greater than 1, this species is considered an accumulator of heavy metals. On the other hand, *B. bipinnata* and *D. aegyptius* have values lower than 1, which indicates that they are classified as heavy metal exclusion species. The results

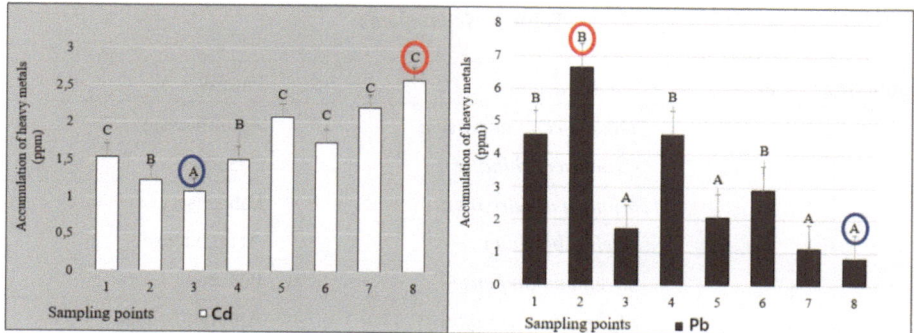

Fig. 2. Quantification of Cd and Pb in the ruderal sampling points of Km 19 Via "La Costa". Different letters in the same row indicate significant statistical differences according to Tukey's test (p < 0.05); ± standard error.

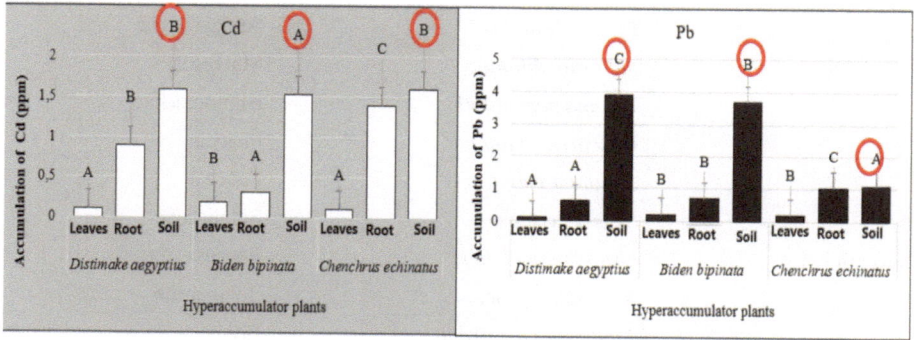

Fig. 3. Cd and Pb accumulation in soil and vegetative structures in the three hyper-accumulating plant species. Different letters in the same row indicate significant statistical differences according to Tukey's test (p < 0.05); ± standard error.

obtained for the translocation factor for Cd were as follows: 1.40 in *D. aegyptius,* 0.65 in *B. bipinnata,* and 0.08 in *C. echinatus.* On the other hand, for Pb, 0.18 was obtained in *D. aegyptius,* 0.29 in *B. bipinnata,* and 0.19 in *C. echinatus.* The species *D. aegyptius* was the only one that reported a value higher than 1 for the translocation factor, specifically for Cd. This means that the plant has a greater capacity to accumulate and transport metals from the soil. It could be used for phytoextraction purposes in soils contaminated by pe-rated metals. However, the analyzed species present Pb values lower than 1.

4 Discussion

Research on heavy metal contamination in soil adjacent to an industrial zone has shown that Cd and Pb levels range from 0.13–.88 ppm for Cd and 4.64–18.21 ppm for Pb [16]. Although these values do not exceed the national average for uncontaminated soils [17], like the results of this investigation, they represent a warning sign. The results

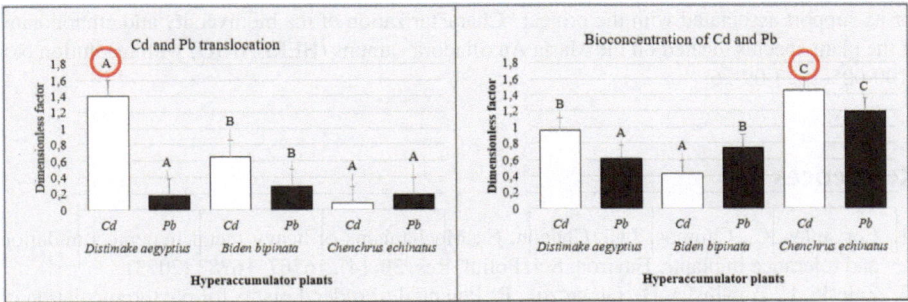

Fig. 4. Cd and Pb accumulation and translocation factors for three hyperaccumulating plants. Different letters in the same row indicate significant statistical differences according to Tukey's test (p < 0.05); ± standard error.

suggest the need to implement remediation and prevention strategies for soil contamination at km 19 via La Costa. The accumulation of heavy metals in the leaves of the three species studied (*Bidens bipinnata*, *Distimake aegyptius* and *Chenchrus echinatus*) showed a low accumulation of heavy metals. Other research carried out on the species *Ambrosia ambrosioides* (Asteraceae), found accumulation values of 249.6 ppm for Pb and 2227.8 ppm for Cd in roots, and 420.1 ppm for Cd and 710.7 ppm for Pb in leaves [18]. The root presented the highest accumulation of Cd. Researchers indicate that in the species of the genus *Brachiaria spp.* (Poaceae), values of 0.01 to 0.024 ppm for Cd and 1.01 to 1.32 ppm for Pb in leaves, and 0.02 to 0.034 ppm for Cd and 1.72 to 1.92 ppm for Pb in roots were recorded. Values in soil ranged from 0.1 to 1.11 ppm for Cd and 12.90 to 23.43 ppm for Pb [19]. It is commented that in *Ipomoea batatas* (Convolvulaceae), values from 4.08 to 77.1 ppm for Pb in roots and 0.99 to 43.43 ppm for Pb in leaves were found. These values exceed the maximum limits established for human consumption [20]. For these reasons, it is of vital importance to monitor the ruderal plant community and, finally, to promote bioprospecting of ruderal plant species in bioremediation processes.

5 Conclusions

The study shows that the soil adjacent to the Salesian Polytechnic University is contaminated with heavy metals, which negatively affects the germination as well as survival of the ruderal plant species present in the area. *Chenchrus echinatus* and *Distimake aegyptius* species, due to their ability to accumulate and tolerate heavy metals, could be used for phytoextraction and phytostabilization strategies, respectively. Further research is needed to determine the optimal conditions for germination and growth of these species in contaminated soils and to evaluate their potential for the remediation of contaminated sites.

Acknowledgments. We, the authors of this work, express our gratitude to Anggie Auz, Kevin Cedeño and the research project director, Carina Hidalgo, whose trust, logistical support and commitment to the research inspired our work. We thank the Universidad Politecnica Salesiana for its support associated with the project "Characterization of the biodiversity and ethnobotany of the plant species located on the María Auxiliadora campus (HERBARIO) with resolution No. 0100-005-2023-09-26.

References

1. Cervantes, C., Campos, J.C., Chácon, E.: Mechanisms of heavy metal hyperaccumulation and tolerance in plants. Environ. Sci. Pollut. Res. **30**(14), 16767–16783 (2023)
2. Zaleski, I., Trawińska, D., Oleszczuk, P.: Potential of ruderal plants for phytoremediation of heavy metals-contaminated soils. Int. J. Environ. Res. Public Health **18**(17), 8662 (2021)
3. Verkleij, H.J., Schat, H., Verbeek, F.J.: A review of mechanisms involved in plant metal hyperaccumulation. ACS Publ. **120**(51), 12871–12906 (2020)
4. Mal, J., Rangabhashiyam, S.: Bioremediation and phytoremediation as the environmentally sustainable approach for the elimination of toxic heavy metals. In: Sustainable Technologies for Water and Wastewater Treatment, pp. 227–258. CRC Press (2021)
5. Ali, H., Ismail, Z.R., Khan, M.A.: Phytoremediation of heavy metals-contaminated soils: efficiency mechanisms and challenges. Environ. Sci. Pollut. Res. **26**(28), 23275–23310 (2019)
6. Arias-Trinidad, A., Rivera-Cruz, M.C., Roldán-Garrigós, A., Aceves-Navarro, L.A., Quintero-Lizaola, R., Hernández-Guzmán, J.: Uso de Leersia hecandra (Poaceae) en la fitorremediación de suelos contaminados con petróleo fresco e intemperizado [Use of Leersia hexandra (Poaceae) for soil phytoremediation in soils contaminated with fresh and weathered oil]. Rev. Biol. Trop. **65**(1), 21–30 (2017)
7. Pérez-López, U., Espínola, J., Campos, J.C.: Ruderal plants for phytoremediation of heavy metal-contaminated soils: a review. Sci. Total. Environ. **849**, 158029 (2022)
8. Mena, P.A., Navarrete, L., Pacheco, L.: Deforestation and fragmentation of the Tumbesian dry forests: a review of threats and conservation strategies. Biol. Conserv. **257**, 105085 (2021)
9. Espinosa-García, F.J., Villaseñor, J.L.: Biodiversity, distribution, ecology and management of non-native weeds in Mexico: a review. Revista mexicana de biodiversidad **88**, 76–96 (2017)
10. Chacón, E., Pérez-López, U., Campos, J.C.: Ruderal plants as bioindicators of heavy metal pollution in urban soils. Environ. Sci. Pollut. Res. **30**(6), 9046–9060 (2023)
11. Muñoz, J., Erazo, S., Armijos, D.: Composición florística y estructura del bosque seco de la quinta experimental "El Chilco" en el suroccidente del Ecuador. DOAJ (DOAJ: Directory Of Open Access Journals) (2017)
12. Jones, H.G.: What plant is that? Tests of automated image recognition apps for plant identification on plants from the British flora. AoB Plants **12**(6), plaa052 (2020)
13. EPA Method 3051A: Microwave Assisted Acid Digestion of Sediments, Sludges, and Oils I US EPA, 31 julio 2023. US EPA. https://www.epa.gov/esam/us-epa-method-3051a-microwave-assisted-acid-digestion-sediments-sludges-and-oils
14. Maldonado-Magaña, A., Favela-Torres, E., Rivera-Cabrera, F., Volke-Sepúlveda, T.: Lead bioaccumulation in Acacia farnesiana and its effect on lipid peroxidation and glutathione production. Plant Soil **339**(1–2), 377–389 (2010)
15. Romero-Estévez, D., Yánez-Jácome, G.S., Navarrete, H.: Non-essential metal contamination in Ecuadorian agricultural production: a critical review. J. Food Compos. Anal. **115**, 104932 (2023)
16. Quero-Jiménez, Z.V., Fernández, M., Pequeño, R.: Determinación de la contaminación por metales pesados en suelos aledaños a la Empresa Electroquímica de Sagua. Centro Azúcar **44**(3), 53–62 (2017)
17. Alfaro, M.R., et al.: Background concentrations and reference values for heavy metals in soils of Cuba. Environ. Monit. Assess. **187**(1) (2014). https://doi.org/10.1007/s10661-014-4198-3

18. Ramírez Gottfried, R.I., García Carrillo, M., Álvarez Reyna, V.D.P., González Cervantes, G., Hernández Hernández, V.: Potencial fitorremediador de la chicura (Ambrosia ambrosioides) en suelos contaminados por metales pesados. Revista mexicana de ciencias agrícolas **10**(7), 1529–1540 (2019)
19. Peláez-Peláez, M.J., Bustamante Cano, J.J., Gómez López, E.D.: Presencia de cadmio y plomo en suelos y su bioacumulación en tejidos vegetales en especies de brachiaria en el Magdalena Medio colombiano. Luna Azul **43**, 82–101 (2016)
20. Cáceres Atencia, M.J., Ramos Caballero, E.M.: Determinación de Cadmio, Plomo, Mercurio y Arsénico en Beta vulgaris, Ipomoea batatas y Beta vulgaris var. cicla, Santa (2023)

In vitro Antibacterial Activity of Ethanolic and Acetonic Extracts of *Curcuma longa* Against *Propionibacterium acnes* (*Cutibacterium acnes*)

Patricia Pacheco[✉] [iD], Dévora Carrión [iD], Sandy Gavilanes [iD], and Mónica Espadero [iD]

Universidad Politécnica Salesiana, Cuenca, Ecuador
{ppachecos1,dcarrionn}@est.ups.edu.ec, {sgavilanes, mespadero}@ups.edu.ec

Abstract. Throughout history, plants have been indispensable in the development of society due to their high content of bioactive substances, which are used in different industries such as food, pharmaceutical, and cosmetics. *Curcuma longa* is a perennial species that belongs to the *Zingiberaceae* family, which is used for therapeutic purposes due to its anti-oxidant, anti-inflammatory and antibacterial properties, among others. This research evaluates the antibacterial activity of ethanolic and acetonic extracts of Curcuma longa against *Propionibacterium acnes*, recently renamed *Cutibacterium acnes*, that has long been implicated in the pathogenesis of acne. Secondary metabolites were identified by phytochemical screening tests. Curcuminoids determination was carried out by using TLC (thin-layer chromatography) and the phenolic compounds were quantified by the Folin-Ciocalteu assay. The antibacterial activity was evaluated using the Kirby-Bauer method with different treatments (50,75 and 100%) and broth microdilution method was used to determine the minimum inhibitory concentration (MIC). The extracts showed the presence of phenolic compounds, flavonoids, terpenoids, and tannins, confirming the presence of curcuminoids. In the quantification of phenolic compounds, no significant differences in their composition were observed. The treatment 100% of acetonic extract of *Curcuma longa* exhibited an inhibition percentage of 48.79% compared to the ethanolic ex tract 75% treatment, which reached 31.67%. Additionally, the acetonic extract showed a minimum inhibitory concentration of 15.62 μg/mL, while for the ethanolic extract, it was 31.25 μg/mL. These findings highlight the potential antibacterial activity of the acetonic extract against *Cutibacterium acnes*, which could make it a promising natural option for acne control.

Keywords: *Curcuma longa* · *Cutibacterium acnes* · antibacterial activity · curcuminoids

© The Author(s) 2025
E. M. Inga Ortega et al. (Eds.): CITIS 2024, LNNS 1331, pp. 292–302, 2025.
https://doi.org/10.1007/978-3-031-87065-1_27

1 Introduction

In traditional medicine, the use of plants has been an extended practice throughout history, where secondary metabolites stand out for their chemical diversity and relevance in various areas [1]. Studies have shown that many of these compounds, when combined with others, enhance biological activity and are used in the formulation of drugs, cosmetics and foods [2].

There is a wide variety of plants from the *Zingiberaceae* family whose bioactive content characterizes them with different medicinal properties [3]. One of them, *Curcuma longa*, known as "turmeric", a perennial herbaceous plant, distributed in tropical areas and is characterized by its orange rhizome [4, 5]. It is rich in bioactive compounds such as tannins, alkaloids, flavonoids, terpenoids, triterpenes, saponins, and polyphenols, including curcuminoids, which grant its medicinal properties, such as anti-inflammatory, antiviral, antibacterial, antiprotozoal, antineoplastic, antioxidant, and anthelmintic activities [6, 7].

On the other hand, skin diseases such as atopic dermatitis, psoriasis, seborrheic dermatitis, rosacea, and acne affect millions of people worldwide. Acne is commonly associated with bacteria such as *Staphylococcus aureus*, *Staphylococcus epidermidis*, *Streptococcus pyogenes*, *Corynebacterium*, and especially *Cutibacterium acnes*, it is a Gram-positive anaerobic bacillus [8, 9]. Therefore, many treatments involve using products that inhibit bacterial growth [10].

Consequently, *Curcuma longa*, with its antimicrobial and anti-inflammatory properties, is a viable alternative for inhibiting bacterias. In this way, this study focuses on evaluating the antibacterial susceptibility of ethanolic and acetonic extracts of *Curcuma longa* using *in vitro* techniques on *Cutibacterium acnes*.

2 Methods

2.1 Preparation of Extracts

The plant extracts were prepared according to the protocol described by Manasa et al. [11]. This methodology consisted of the ethanolic and acetonic extraction of the *Curcuma longa* rhizome. The conditions used were 160 mL of each solvent with 20 g of turmeric rhizome for 5 h at 70 °C. Subsequently, the extracts were filtered and concentrated in rotavapor until the complete elimination of the solvent.

2.2 Qualitative Characterization

Preliminary tests were conducted to qualitatively determine the presence of bioactive compounds identified by coloration, following the procedures described by Zulmardi et al. and Pawar [12, 13]. The following process was carried out from the tests described in Table 1.

2.3 Determination of Curcuminoid Compounds

Curcuminoids compounds were determinate by thin layer chromatography using a mobile phase prepared with chloroform, methanol, and glacial acetic acid in a 95:5:1 ratio [14]. The fluorescent substances were observed by UV light at 254 nm and 366 nm.

Table 1. Phytochemical testing of extracts of *Curcuma longa*

Compounds	Testing
Triterpenes/steroids	Liebermann-Burchard
Flavonoids	Shinoda
Tannins	Ferric Chloride Test
Phenols	Folin reagent + $Na_2CO_3$5%

2.4 Quantification of Phenolic Compounds

The phenolic content of the extracts was determined by the Folin-Ciocalteu assay [15]. A calibration curve was carried out with different concentrations with ten dilutions of gallic (2–20 µg/mL), extracts were diluted with distilled water (1:10), starting from 3.5 mg of the extract. The absorbance was measured on a UV-Vis spectrophotometer at 750 nm.

2.5 Determination of Antimicrobial Capacity

Kirby-bauer method [16] or disk diffusion technique was used. A 0.5 McFarland standard bacterial suspension corresponding to 1.5×10^8 CFU/mL with an absorbance of 0.08 at 625 nm was used. Petri dishes with Mueller-Hinton medium were inoculated, and disks impregnated with the extracts at different concentrations (100%, 75%, 50%) were subsequently applied. Erythromycin was used as a positive control, and distilled water as a negative control. After 24 h of incubation at 37 °C, halos were measured, and the inhibition percentage was calculated.

2.6 Determination of Minimum Inhibitory Concentrations (MIC)

The minimum inhibitory concentration was determined using the broth microdilution method [17]. In a 96-well microtiter plate with TSB broth (Trypticase Soy Broth), the antimicrobial agent and the extracts were placed. A concentration of 250 µg/mL of the extract was used [18, 19] and 6 serial dilutions of the extract were made. The bacterial suspension was inoculated and incubated for 24 h at 35 ± 2 °C. Finally, TTC (triphenyl-tetrazolium chloride) was applied for 30 min at 35 ± 2 °C to observe bacterial growth based on the color change.

2.7 Statistical Analysis

An analysis of variance was conducted to compare the different extracts in terms of inhi-bition percentage. Two extracts were selected: pure acetone extract and 75% ethanolic extract of *Curcuma longa* that previously showed the best inhibitory effect. The Shapiro-Wilk test was applied to evaluate data normality with a significance level of 0.05. Then, the ANOVA model assumptions were evaluated, ensuring no significant deviations in independence and homogeneity of variances. Finally, to determine which extract pre-sented a higher inhibition percentage, a multiple comparison analysis was performed using the Tukey test.

3 Results

3.1 Preparation of Acetonic and Ethanolic Extracts

According to the methodology, the results obtained shown in Table 2.

Table 2. Results of extraction with acetone and ethanol.

Solvent	Dry sample weight (g)	Initial solvent Volume (mL)	Extract final volume (mL)	concentrated extract Volume (mL)
Acetone	20	160	110	13
Ethanol	20	160	130	8

The acetonic extract showed a slight yellow color and an aqueous consistency, in contrast to the ethanolic extract, which exhibited a dark brown tone and a semi-fluid consistency. This disparity is attributed to the different affinities and quantities of the compounds present in turmeric.

Fig. 1. Acetonic and ethanolic extracts of *Curcuma longa*.

3.2 Qualitative Characterization

The results of the phytochemical analyses of both extracts are shown in Table 3. A high presence of flavonoids, phenols, and terpenoids was observed in the acetonic extract, and tannins and phenols in the ethanolic extract.

3.3 Determination of Curcuminoid Compounds

TLC results shown in Figs. 2 and 3, yellow bands characteristic of curcuminoid compounds were observed. Additionally, the separation of three compounds, which can be attributed to the three main curcuminoids: curcumin, demethoxycurcumin, and bisdemethoxycurcumin, was observed. The *Rf* values obtained for the acetonic and ethanolic extracts are shown in Table 4 [20].

Table 3. Results of Phytochemical Screening.

Compounds	Acetone Extract	Ethanol Extract
Tannins	+	+++
Flavonoids	+++	++
Phenols	+++	+++
Terpenoids	+++	++

Note: (+++): high evidence; (++): medium evidence; (+): low evidence; (-): negative.

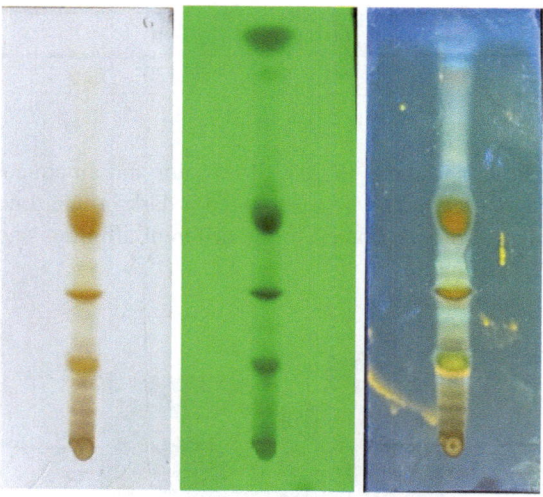

Fig. 2. Determination of curcuminoid at UV light, ethanolic extract of *Curcuma longa. Note.* Visible light (a), UV light λ 254 nm (b) UV light λ 366 nm (c).

Table 4. *Rf* values of curcuminoids in Acetonic and ethanolic extracts

Extract	T	(*Rf values*)	theoretical *Rf* values	Compound
Acetonic	1	0,33	0,33	Bisdemethoxycurcumin
	2	0,57	0,55	Desmethoxycurcumin
	3	0,61	0,62	Curcumin
Ethanolic	1	0,16	0,18	Bisdemethoxycurcumin
	2	0,30	0,26	Desmethoxycurcumin
	3	0,45	0,45	Curcumin

*Note: * Rf: Retardation factor.*

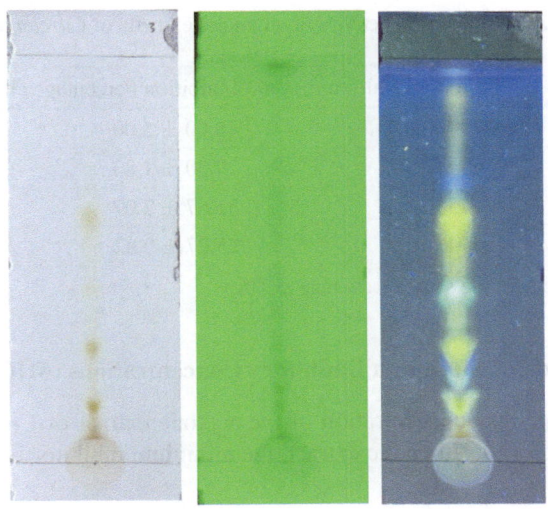

Fig. 3. Determination of curcuminoid at UV light, acetonic extract of *Curcuma longa. Note.* Visible light (a), UV light λ 254 nm (b) UV light λ 366 nm (c).

3.4   Quantification of Phenolic Compounds

Phenolic compounds in the acetonic extract was 0.1323 ± 0.0312 mg GAE/g DW and in the ethanolic extract, 0.1078 ± 0.0268 mg GAE/g DW. The ANOVA analysis showed a p-value of 0.361, indicating no statistically significant difference, suggesting that both extracts of *Curcuma longa* (Turmeric) contain phenolic compounds.

3.5   Determination of Antimicrobial Capacity

Acetone extract at 100% concentration obtained the highest inhibition percentage against *Propionibacterium acnes* (*Cutibacterium acnes*) reaching 48.79% and the 75% ethanolic extract inhibited the growth of the bacteria by 31.67%. The results are de-tailed in Tables 5 and 6.

Table 5. Antimicrobial activity of acetonic extracts of *Curcuma longa.*

Treatments	n	Inhibition Halo (mm)	Inhibition Percentage (%)	Groups
Antibiotic	3	$27{,}48 \pm 1{,}24$	$55{,}00 \pm 5{,}00$	a
100%	3	$16{,}10 \pm 3{,}15$	$48{,}79 \pm 3{,}92$	ab
75%	3	$13{,}46 \pm 2{,}98$	$40{,}79 \pm 4{,}44$	b
50%	3	$9{,}22 \pm 1{,}23$	$27{,}93 \pm 0{,}72$	c

Note. The values used are the samples average $\pm$ DS.

Table 6. Antimicrobial activity of ethanolic extracts of *Curcuma longa*.

Treatments	n	Inhibition Halo (mm)	Inhibition Percentage (%)	Groups
Antibiotic	3	14,94 ± 0,61	45,00 ± 5,00	a
100%	3	9,67 ± 1,02	29,30 ± 1,80	b
75%	3	10,45 ± 1,23	31,67 ± 2,07	b
50%	3	8,34 ± 1,08	25,27 ± 0,82	b

Note. The values used are the samples average ± DS.

3.6 Determination of Minimum Inhibitory Concentrations (MIC)

The minimum inhibitory concentration of the acetone extract of *Curcuma longa* was 15.62 μg/ mL and for the ethanolic extract, the minimum inhibitory concentration was 31.25 μg/mL.

3.7 Comparison of the Antibacterial Activity of the Acetonic and Ethanolic Extracts of *Curcuma Longa*

The data obtained follow a normal distribution, through analysis of variance (ANOVA) it was determined that there is a significant difference between the treatments (ρ = 0.00262). Therefore, Tukey's multiple comparison was carried out where it was observed that the 100% acetone extract presented better antibacterial capacity than the rest of the treatments. Result that can be seen in Fig. 4.

4 Discussion

The difference in color and consistency of the turmeric extracts obtained with different solvents, such as acetone and ethanol, aligns with previous research [21] which determined that acetone is the most effective solvent for extracting all phenolic compounds from *Curcuma longa* compared to ethanol. The characteristic yellow hue of the acetone extract is primarily due to phenolic compounds and flavonoids, while the dark brown color of the ethanol extract is indicative of other compounds present in the species.

The low presence of tannins and high content of flavonoids, phenols and terpenoids identified qualitatively in the acetone extract agree with previous studies [14, 22, 23], where flavonoids, phenols and terpenoids were identified in acetone extract of turmeric. On the other hand, in the ethanolic extract the presence of tannins and phenols in a greater proportion than flavonoids and terpenoids was observed. These results coincide with the findings of studies in which they worked with ethanol as an extraction solvent [23, 24], where a high concentration of tannins and phenols was found. However, the moderate presence of flavonoids and terpenoids in the ethanol extract, as described in another study [22], suggests a different chemical composition than that of the acetone extract.

Contrasting the results of Sepahpour [21] Regarding the quantification of polyphenolic compounds, it was established that the highest concentration of total phenolics in

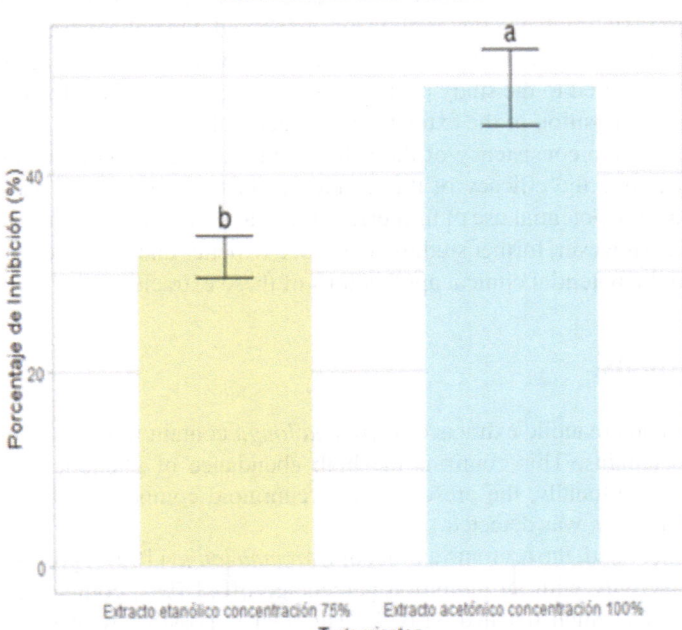

Fig. 4. Comparison of the antibacterial activity of the acetonic and ethanolic extracts of *Curcuma longa. a: Acetonic extract 100%, b: ethanolic extract 75%.*

turmeric is obtained with acetone, this is due to the fact that the main phenolic compounds of turmeric (curcuminoids) are formed by a long non-polar chain of bonds. Carbon-carbon covalent with a phenolic group attached to both ends, as a consequence of this structure, dissolve better in acetone, due to its low polarity, than in ethanol and methanol, and slightly in water, since the latter has a high polarity.

The antibacterial activity of the 100% acetone extract of *Curcuma longa* reached the highest percentage of inhibition. Similar results were obtained in a study of the antimicrobial activity of the rhizome of *Curcuma longa* [24], a strong antibacterial activity with the 100% acetone extract, presenting a broader zone of inhibition for Gram-positive bacteria with an average value of 17 mm.

Regarding treatments based on the ethanolic extract, the results show that the 75% and 100% concentrations present the highest inhibition percentage. These results contrast with other study [4], who found that the ethanolic extract showed the highest inhibition capacity starting from a 75% concentration with inhibition zones of 9 mm against Gram-positive bacteria.

Additionally, the phenolic components of turmeric can enhance cellular destruction and inhibition by creating hydrophobic and hydrogen bonds with membrane proteins, resulting in the disruption of the lipid bilayer. [25] indicate that these phenolic compounds influence bacterial metabolism after entering the bacterial cell following cell wall destruction.

The results obtained in this study for the determination of the minimum inhibitory concentration (MIC) indicate that both the acetonic and ethanolic extracts of *Curcuma longa* exhibit considerable antibacterial activity. The lower MIC of the acetonic extract (15 μg/mL) compared to the study [26] (32 μg/mL) may be attributed to differences in the chemical composition of the extracts or the bacterial strains used.

Furthermore, the consistency of the MIC results for the ethanolic extract with the study [27] supports the efficacy of this extract against Gram-positive bacteria. These results indicate the potential use of turmeric extracts as therapeutic agents against bacterial infections. However, further studies are needed to better understand the mechanisms of action and the potential clinical applications of these extracts.

5   Conclusions

The acetonic and ethanolic extracts of *Curcuma longa* contain flavonoids, phenols, terpenoids, and tannins. This confirms the high abundance of phenolic compounds in the extracts. Additionally, the presence of curcuminoid compounds, known for their antibacterial activity, was detected.

On the other hand, the acetonic extract of *Curcuma longa* (100%) presented a higher antibacterial activity with an inhibition percentage of 48.79%, compared to the 75% ethanolic extract, which reached 31.67%. This fact is possibly related to the greater interaction of curcuminoids with the thicker structural compounds of Gram-positive bacteria.

The acetone extract of *Curcuma longa* at a concentration of 15.62 μg/mL, and the ethanolic extract at 31.25 μg/mL inhibited the growth of *Cutibacterium acnes*, with the acetone extract having the best antibacterial activity. In conclusion, the investigation confirms the remarkable efficacy of *Curcuma longa* acetonic extract in inhibiting the growth of *Cutibacterium acnes,* this being a preliminary study to evaluate this species as an adjuvant for the treatment of acne.

References

1. Sayantani, C.: Fitoquímica y potencial terapéutico de la cúrcuma (Curcuma longa) (2021). https://www.eolss.net/outlinecomponents/mundo-de-plantas-aromaticas-medicinales.aspx
2. Bachheti, A., Deepti, Bachheti, R.K., Husen, A.: Medicinal plants and their pharmaceutical properties under adverse environmental conditions. In: Husen, A. (ed.) Harsh Environment and Plant Resilience, pp. 457–502. Springer, Cham (2021). https://doi.org/10.1007/978-3-030-65912-7_19
3. Norajit, K., Laohakunjit, N., Kerdchoechuen, O.: Antibacterial effect of five Zingi-beraceae essential oils. Molecules 12(8), 2047–2060 (2007)
4. Persaud, N., Ragobeer, P., Daniel, R.: The antibacterial activity of locally grown turmeric (Guyana) using ethanol extracts at different concentrations against Escherichia coli, Proteus vulgaris, Pseudomonas aeruginosa and Staphylococcus aureus. Asia J. Appl. Microbiol. 7(1), 8–18 (2020)
5. Rodriguez, L.A.O.: Actividad antioxidante, antibacteriana y citostática de extractos de cúrcuma (Curcuma longa). Gaceta Médica Boliviana 45(1), 12–16 (2022)

6. Jyotirmayee, B., Mahalik, G.: A review on selected pharmacological activities of Curcuma longa L. Int. J. Food Prop. **25**(1), 1377–1398 (2022)
7. Cahyani, A., Anggraini, D., Soleha, T.: Uji Efektivitas Antiba-kteri Ekstrak Rimpang Kunyit (Curcuma domestica Val.) terhadap Pertumbuhan Propioni-bacterium acnes In Vitro. Jurnal Kesehatan **11**(3), 414–421 (2020)
8. Mustari, A.P., Agarwal, I., Das, A., Vinay, K.: Role of cutaneous microbiome in dermatology. Indian J. Dermatol. **68**(3), 303–312 (2023)
9. Legiawati, L., Halim, P.A., Fitriani, M., Hikmahrachim, H.G., Lim, H.W.: Microbiomes in acne vulgaris and their susceptibility to antibiotics in Indonesia: a systematic review and meta-analysis. Antibiotics **12**(1), 145 (2023)
10. Organización Mundial de la Salud: Resistencia a los antimicrobianos (2021). https://www.who.int/es/news-room/fact-sheets/detail/antimicrobial-resistance
11. Manasa, P.S., Kamble, A.D., Chilakamarthi, U.: Various extraction techniques of curcumina comprehensive review. ACS Omega **8**(38), 34868–34878 (2023)
12. Zulmardi, Mitayani, Febriyanti, Soo, K.-M.: Phytochemical screening of turmeric (Curcuma longa Linn.) extract with 97% ethanol solution. Jundishapur J. Microbiol. **15**(2), 926–931 (2022)
13. Pawar, H.: Phytochemical evaluation and curcumin content determination of turmeric rhizomes collected from Bhandara district of Maharashtra (India). Med. Chem. **4**(8) (2014)
14. Espinoza Gómez, A., La Fuente Rios, K.: Efecto antimicrobiano, in vitro del Extracto de Curcuma longa L. (Palillo) sobre cepas de staphylococcus aureus, escherichia coli y candida albicans (2017). https://repositorio.ucsm.edu.pe/items/b1fbd065-0bde-45ae-8d12-d917bf1387d9
15. García Martínez, E., Fernández Segovia, I., Fuentes López, A.: Determinación de polifenoles totales por el método de Folin-Ciocalteu (2015). http://hdl.handle.net/10251/52056
16. Cavalieri, S.J.: Manual de pruebas de susceptibilidad antimicrobiano. ASM Press (2009)
17. Alcaciega, A., Pazmiño, M.: Evaluación de la actividad antibacteriana de extractos de uvilla (physalis peruviana) y diente de león (taraxacum officinale) en una formulación para desinfección de alimentos (2022). http://dspace.ups.edu.ec/handle/123456789/22660
18. Teow, S.Y., Liew, K., Ali, S., Khoo, A., Peh, S.: Antibacterial action of curcumin against staphylococcus aureus: a brief review. J. Trop. Med., 1–10 (2016)
19. Hamdy, A., et al.: In-vitro evaluation of certain Egyptian traditional medicinal plants against Propionibacterium acnes. S. Afr. J. Bot. **109**, 90–95 (2017)
20. Aguilera Martínez, L.E., Arellano Martínez, L.A., Penieres Carrillo, J.G., García Estrada, J.G., Ortega Jiménez, F.: Purificación de curcumina por cromatografía en columna. Propuesta para la enseñanza experimental en química orgánica, Ciudad de México (2017)
21. Sepahpour, S., Selamat, J., Manap, M., Khatib, A., Razis, A.F.: Comparative analysis of chemical composition, antioxidant activity and quantitative characterization of some phenolic compounds in selected herbs and spices in different solvent extraction systems. Molecules **23**(2), 402 (2018)
22. Charan, T.R., Bhutto, M.A.: Análisis comparativo por rendimiento total, evaluación antimicrobiana y fitoquímica de curcuminoide del distrito de Kasur: con su potencial usao y caracterización en nanofibras de electrohilado. Revistas de textiles industriales **52** (2022)
23. Agidew, M.G.: Phytochemical analysis of some selected traditional medicinal plants in Ethiopia. Bull. Natl. Res. Cent. **46**(1) (2022)
24. Chauhan, P., Sharma, A., Kumar, R., Sengar, R., Purushottam, Gupta, S., Kapoor, N.: Phytochemical studies and antimicrobial activity of curcuma longa Linn rhizomes. Biotech Today Int. J. Biol. Sci. **10**(1), 39 (2020)
25. Nath, S., Chatterjee, P., Chowdhury, S., Ray, N., Mukherjee, S.: Antimicrobial activity of turmeric (Curcuma longa) extract and its potential use in fish preservation. Indian J. Anim. Health **60**(2) (2021)

26. Nguyen, H.T., et al.: Effects of roasting conditions on antibacterial properties of Vietnamese turmeric (Curcuma longa) rhizomes. Molecules **28**(21), 7242 (2023)
27. Boeder, A.M., et al.: Antimycoplasma activity of Curcuma longa extracts and your isolated compound, the curcumin. Revista Fitos **12**(2) (2018)

Carbon Footprint Estimation at the Salesian Continuous Training Center - San Bartolo Using System Dynamics

Thalia Elvira Timbiano Feraud[1]([envelope]) [iD], Juan Gabriel Mollocana Lara[2] [iD],
Pedro José Calderón Coba[1] [iD], and César Iván Álvarez Mendoza[2] [iD]

[1] Carrera de Ingeniería Ambiental, Universidad Politécnica Salesiana, Quito 170702, Ecuador
ttimbiano@est.ups.edu.ec
[2] Grupo de Investigación Ambiental en el Desarrollo Sustentable GIADES, Carrera de
Ingenieria Ambiental , Universidad Politécnica Salesiana, Quito 170702, Ecuador

Abstract. The Carbon Footprint measures greenhouse gas emissions, which are harmful to the environment and contribute to climate change, and is a solution to evaluate the production of GHGs, thus implementing mitigation actions to counteract their environmental impacts. This study focused on evaluating GHG emissions mitigation strategies at the Salesian Center for Continuing Education in San Bartolo. Data, obtained through interviews and utility bills, covered electricity, water and fuel consumption, as well as paper, plastic and organic waste. A system dynamics model was applied to identify the major contributors to greenhouse gas emissions, using Vensim PLE software, and the carbon footprint was simulated for one year. The results revealed annual emissions of 88.28 tons of CO_2 equivalent, mainly from solid waste production (73.92 tons). A mitigation scenario was simulated representing the use of LED lighting, flow control pumps, waste sorting containers, and the adoption of clean energy for transportation. The results suggest a potential reduction of 27.78 tons of CO_2 equivalent, representing a significant decrease in greenhouse gases.

Keywords: System Dynamics · carbon footprint · modeling

1 Introduction

The Carbon Footprint (CF) is defined as the amount of greenhouse gases (GHG) produced, which are considered pollutants to the environment and contribute to climate change [1]. The CF arises as a solution to evaluate the production of GHG, thus implementing mitigation actions to counteract their environmental impacts [2].

There are several methodologies that allow estimating the CF in an institution among them is the Bilan Carbone methodology [3] designed by the French Agency for Environment and Energy (ADEME) in 2012. This tool quickly converts data related to production activities into emissions using emission factors such as different types of energy and water consumption, etc. [4]; once the sources of greenhouse gas emissions are known,

E. M. Inga Ortega et al. (Eds.): CITIS 2024, LNNS 1331, pp. 303–313, 2025.
https://doi.org/10.1007/978-3-031-87065-1_28

the area that emits the largest number of pollutants during the year of study can be identified [5]. In addition, strategies can be proposed for the reduction of emissions produced in an institution [6].

System Dynamics (SD) is based on analyzing evolution of the state variables of a system and allows the generation of models based on observations and experiences; therefore, it is an adequate methodology to analyze the behavior of main factors that contribute to the CF. The research conducted by [7], presented a System Dynamics model where the Box Counting method was implemented to determine the CF of several types of asphalt mixtures. System dynamics is an approach developed for the evaluation of variables that allows obtaining reliable results. The study of CF using SD relay on an optimal selection of state variables affecting GHG emissions where quantitative regression methods could be used [8]. Arroyo Lopez developed a SD model following two stages: qualitative and quantitative. In the qualitative phase, relevant variables associated with the system under analysis were identified. In the quantitative phase, these variables were expressed explicitly through equations or tabular functions supported by empirical data [9].

This article aims to develop a System Dynamics model to assess greenhouse gas emissions resulting from electricity, water, paper, plastic, and transportation consumption [10]. The Salesian Continuous Training Center - San Bartolo serves as the case study, with data collected from interviews and utility bills being utilized for model generation. Once the Carbon Footprint (CF) is estimated, proposed strategies for reducing greenhouse gas (GHG) emissions are formulated based on scenario comparisons [11].

2   Methodology

2.1   Case Study

The Salesian Continuous Training Center - San Bartolo is located at UTM 17S, Longitude 775380 and Latitude 9970938 and covers 8534.70 m². There three educational programs are managed, these are: Talleres Escuela San Patricio (TESPA) for technical training, the Escuela de Educación Básica Fiscomisional San Patricio (UESPA) for educational support to vulnerable children, and Start Labs with laboratories specialized in design and 3D printing for research projects (Fig. 1).

2.2   Carbon Footprint Estimation

To estimate the CF in the Salesian Continuous Training Center - San Bartolo, the same methodology applied in the Salesian Project Foundation Zona Norte in 2023 will be used. This study focused directly on the operation of the institution, collecting accurate data provided by the establishment. In addition, interviews were conducted with administrative staff and teachers, focusing on the consumption of electrical energy (EEQ electricity bills), drinking water (EPMAPS water bills), paper (reams of A4 bond paper, fomix, etc.) and fuel (diesel). The latter was recorded through previous use in the van used for visits or transfers planned by the institution. In this way, a complete collection of data provided by all personnel was obtained. Subsequently, key greenhouse gases were identified for

Fig. 1. Location Salesian Work - San Bartolo

the calculation of the carbon footprint, such as CO_2, CH_4 and N_2O [12]. In addition, the research managed three scopes: Scope 1, direct water and transport emissions from the institution's own sources such as fuel consumption; Scope 2, indirect emissions from energy consumption; and Scope 3, indirect emissions from waste generation [4] (Fig. 2).

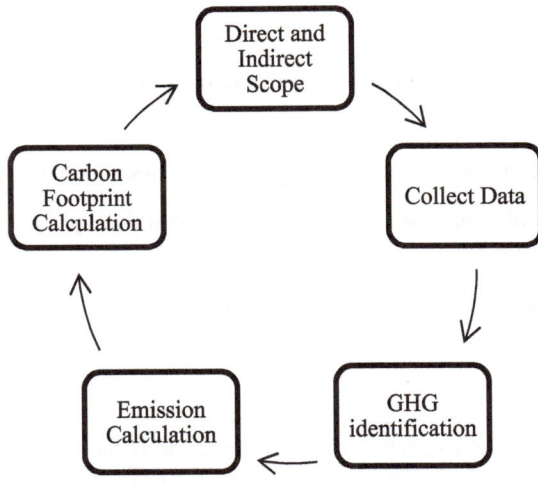

Fig. 2. Steps for estimating CF.

2.3 Modeling

To simulate the behavior of the carbon footprint, System Dynamics was applied, and a Forrester diagram was developed in the Vensim PLE program. Forrester diagrams represent the state variables of the system through Levels or Stocks elements, while its variations are represented through Flow elements. In this study, energy consumption, water consumption, fuel consumption for transportation and GHG generated by the waste produced were considered as state variables (Stocks). In addition, GHG emission factors for energy, water and fuel consumption, as well as the Global Warming Potential (GWP) of the different GHG generated (CO_2, CH_4 and N_2O) were considered as model parameters (Table 1).

Table 1. Emission factors

Parameters	Emission factor			Units
	CO_2	CH_4	N_2O	
Energy	0,000197			ton/kWh
Water	0,000272			ton/m^3
Transportation	0,00268	0,0000051	0,00000022	ton/l
Waste		0,055		ton

On the other hand, the following equation was used to calculate GHG emissions measured in tons of CO_2 equivalents (Ton CO_2eq):

$$GHG\ emissions = Activity\ data * FE * PCG \tag{1}$$

where:

- Activity data represents fuel consumption, energy consumption, water consumption or waste generated.
- EF is the emission factor for each activity
- GWP is the global warming potential of each GHG generated (CO_2, CH_4 and N_2O).

And for the calculation of the total carbon footprint the equation presented in ISO 14064:2019 was applied:

$$Carbon\ Footprint = Indirect\ Emissions + Direct\ Emissions \tag{2}$$

2.4 Mitigation Scenarios

Simulations of a mitigation scenario where strategies to reduce the carbon footprint are implemented were carried out. These were the use of LED lighting, flow control pumps, and waste sorting garbage cans, as well as the adoption of clean energy for transportation.

3 Results

The Forrester diagram obtained can be presented in several parts that represent the dynamics of each state variable, i.e. electricity consumption, water consumption, fuel consumption and GHG emissions from waste generated.

From the monthly electricity consumption data, the annual consumption was calculated and promised as 5212.37 kWh. The Forrester diagram modeling this process is shown in Fig. 3.

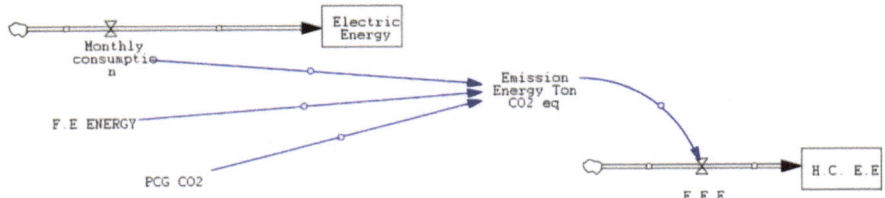

Fig. 3. Forrester diagram for electricity consumption.

Monthly potable water consumption was estimated at 148,6153 m^3 and corresponds to the average consumption from January to December 2023. The Forrester diagram modeling this process is shown in Fig. 4.

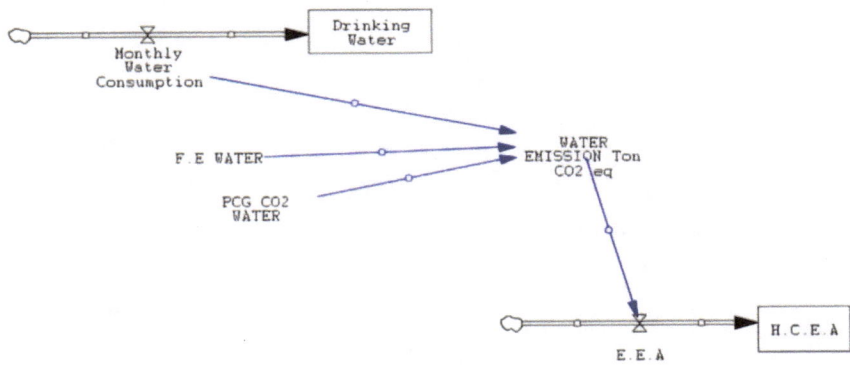

Fig. 4. Forrester diagram for water consumption.

The data on fuel consumption per transport covers the period from January to July, and a monthly consumption of 47.04 L has been estimated for the calculation of the carbon footprint. The Forrester diagram modeling this process is shown in Fig. 5.

The dynamic model representing GHG emissions due to waste generation considers the generation of organic and inorganic waste (plastic and cardboard) with their different emission factors and GWP. The Forrester diagram modeling this process is shown in Fig. 6.

Finally, the four contribution sources modeled above are summed to calculate the total carbon footprint, as shown in Fig. 7.

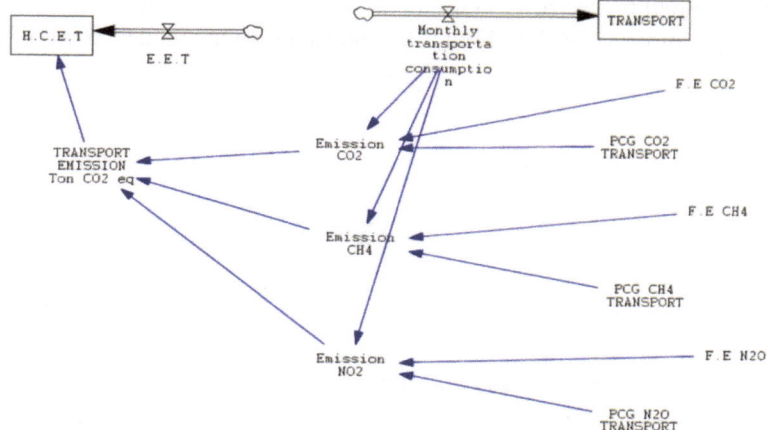

Fig. 5. Forrester diagram for fuel consumption.

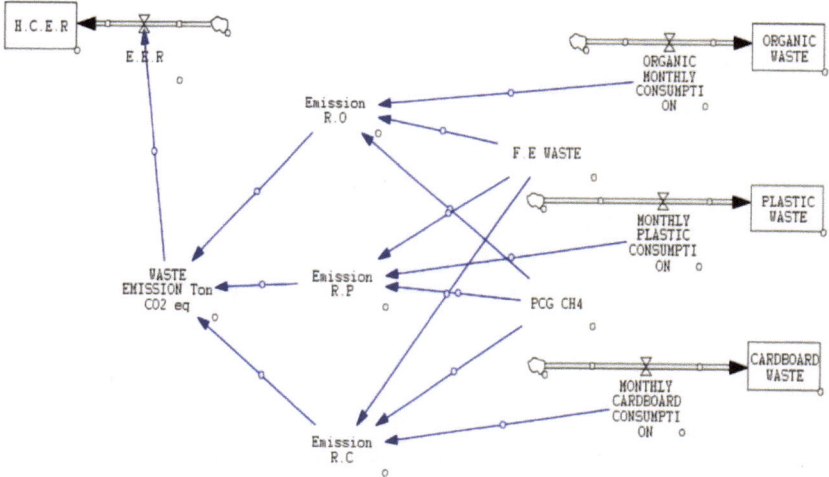

Fig. 6. Forrester diagram for GHG emissions due solid waste production.

Figure 8 shows the GHG emissions measured in tons of CO_2 equivalent during the year 2022 for the following variables: C.E.E.E. (Carbon Footprint - Energy Emission), C.E.E.P. (Carbon Footprint - Potable Water Emission), C.E.T. (Carbon Footprint - Transport Emission) and C.E.R. (Carbon Footprint - Waste Emission) [11].

On the other hand, Fig. 9 shows the carbon footprint for the year 2023 of the sum of the different emissions, which is energy, water, transportation and waste, as shown in Fig. 8.

To evaluate the effect of the implementation of mitigation measures in the reduction of the carbon footprint, especially considering that waste generation is one of the main contributors, a simulation was performed in Vensim where it was assumed that the

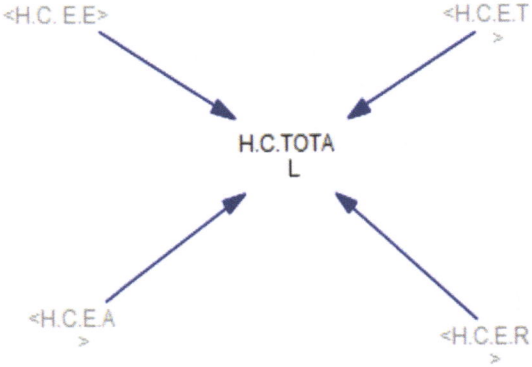

Fig. 7. Forrester diagram for total carbon footprint.

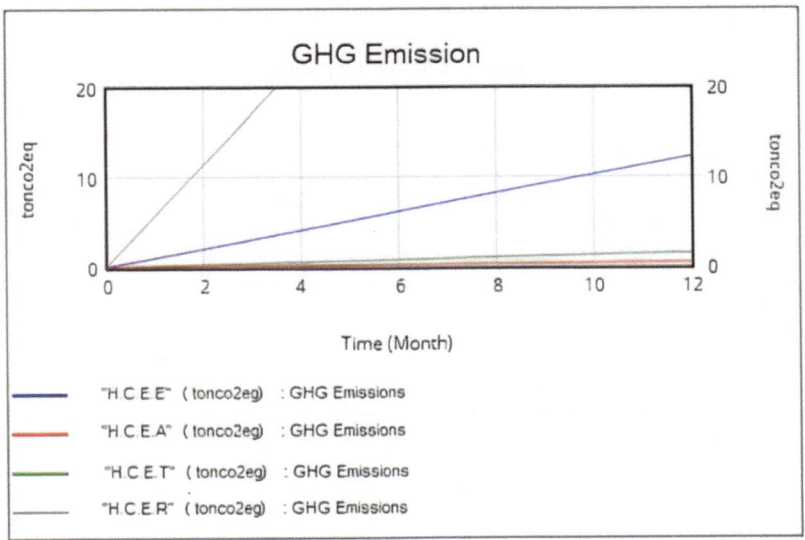

Fig. 8. GHG emissions

implementation of LED lighting for electrical energy stands out as a highly efficient measure, achieving a reduction of 80%. In the case of water, the installation of a flow control pump results in a 20% reduction. Transportation for alternative fuel generates a 30% reduction, waste management through sorting achieves a 70% reduction in organic waste, 50% in plastics and 30% in cardboard, each of the proposed strategies reduces the GHG of each footprint [14].

It can be seen that effective waste sorting can lead to a significant reduction of 16% of GHG emissions, as graphically represented in Fig. 10, which illustrates the carbon footprint with the application of these mitigation measures (Fig. 11).

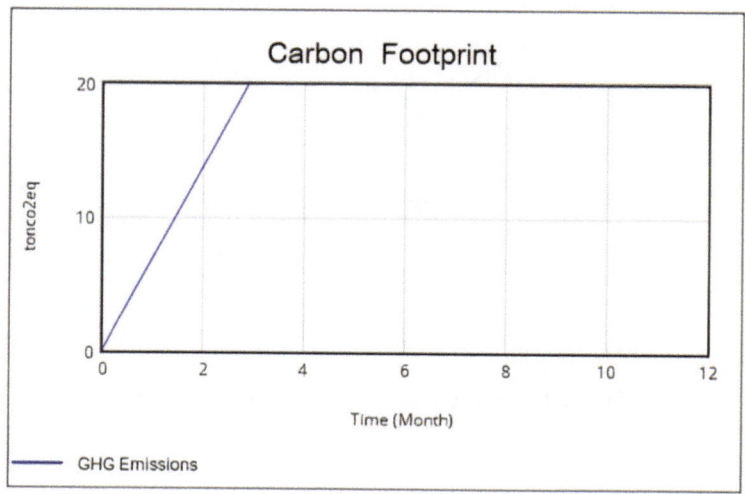

Fig. 9. Carbon footprint

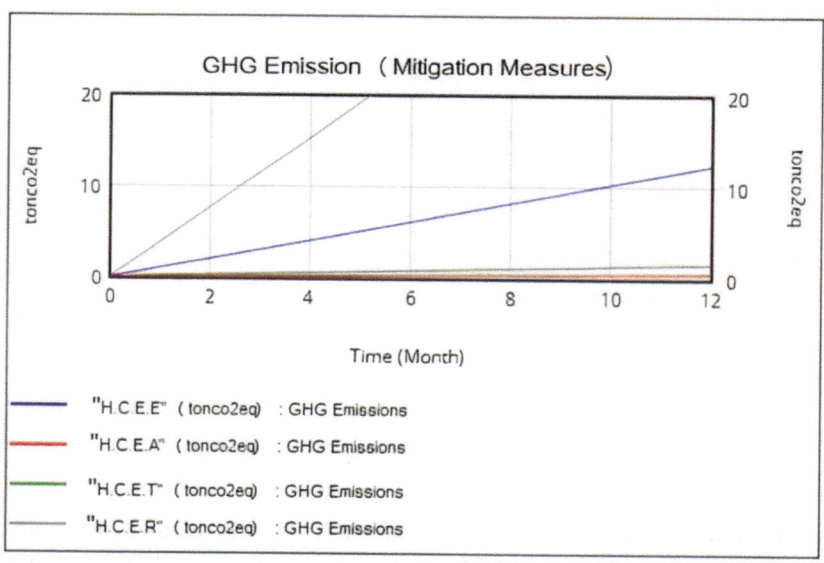

Fig. 10. Mitigation measures for carbon footprint reduction

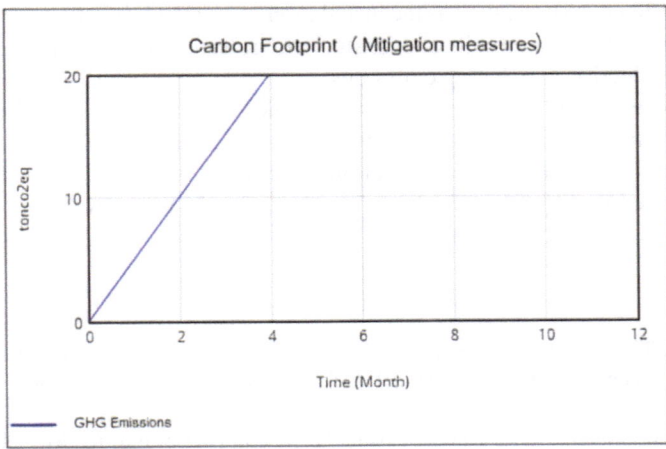

Fig. 11. Carbon footprint reduction.

4 Discusion

The results of the simulation reveal that the carbon footprint of the Salesian Continuous Training Center - San Bartolo amounts to 88.28 tons of CO_2 equivalents, exceeding the acceptable range of greenhouse gas emissions. The largest contribution comes from waste, totaling 73.92 tons of CO_2 equivalents. In second place is electric power, generating 12.32 tons of CO_2 equivalents, while transportation and water have the lowest emissions, with 1.55 and 0.49 tons of CO_2 equivalents, respectively. These findings highlight the importance of implementing reduction strategies, especially focused on waste management, to mitigate the environmental impact of the Salesian facility [15].

For carbon footprint reduction, mitigation measures have been proposed that prove to be effective in reducing emissions.[16]. The implementation of LED lighting for electrical energy stands out as a highly efficient measure, achieving an 80% reduction. In the case of water, the installation of a flow control pump results in a 20% reduction. For transportation, the adoption of alternative fuels can generate a 30% reduction. Finally, addressing the most significant emission, waste management through sorting achieves a 70% reduction in organic, 50% in plastics and 30% in cardboard, each of the proposed strategies reduces the GHG of each footprint. These strategies prove to be fundamental to achieve a more sustainable and responsible environmental management [17].

With the implementation of mitigation measures, the carbon footprint was reduced to 27.7325 tons of CO_2 equivalent. These results significantly indicate that the mitigation measures adopted are effective in decreasing the greenhouse gas emissions that the institution emits into the environment [19].

5 Conclusions

In conclusion, the carbon footprint of the institution amounts to a total of 88,2776 tons of CO_2eq, exceeding acceptable emissions limits. Waste accounts for most of these emissions, with a total of 73.92 tons of CO_2eq, followed by energy, transportation, and water. Mitigation measures, such as the installation of LED lighting, flow control pumps, the adoption of alternative fuels, and waste sorting, have been implemented and have proven to be effective in reducing emissions.

For the development of these reduction strategies, system dynamics were used in the Vensim program, which facilitated the integration of mitigation measures to reduce the carbon footprint generated by the institution. As a result of these strategies, the carbon footprint was reduced by 68%, reaching a total of 27,7325 tons of CO_2eq.

References

1. Anquetin, T., Coqueret, G., Tavin, B., Welgryn, L.: Scopes of carbon emissions and their impact on green portfolios. Econ. Model. **115**, 105951 (2022). https://doi.org/10.1016/J.ECO NMOD.2022.105951
2. Shahbaz, M., Nuta, A.C., Mishra, P., Ayad, H.: The impact of informality and institutional quality on environmental footprint: the case of emerging economies in a comparative approach. J. Environ. Manag. **348**, 119325 (2023). https://doi.org/10.1016/J.JENVMAN.2023.119325
3. Alcázar Navarrete, B., Márquez Martín, E., Alcázar Navarrete, B.: Pneumosure paper on carbon footprint and climate change. Rev. Esp. Patol. Torac. **34**(2), 121–127 (2022)
4. Valderrama, J.O., Espíndola, C., Quezada, R.: Carbon footprint, a concept that cannot be absent in engineering and science courses. Univ. Train. **4**(3), 3–12 (2011). https://doi.org/10.4067/S0718-50062011000300002
5. Aristizábal Alzate, C.E., González Manosalva, J.L.: Application of the NTC-ISO 14064 standard to calculate Greenhouse Gas (GHG) emissions and Carbon Footprint (CF) at the ITM Robledo Campus. Dyna J. Fac. Mines. Natl. Univ. Colomb. Sede Medellín **88**(218), 88–94(2021). ISSN 0012-7353. https://doi.org/10.15446/dyna.v88n218.88989
6. Udemba, E.N., Tosun, M.: Moderating effect of institutional policies on energy and technology towards a better environment quality: a two dimensional approach to China's sustainable development. Technol. Forecast. Soc. Change **183**, 121964 (2022). https://doi.org/10.1016/J.TECHFORE.2022.121964
7. Colombiana De Cardiología, R., et al.: Dynamical systems and probability theory applied to the diagnosis of cardiac dynamics in sixteen hours. Rev. Colomb. Cardiol. **27**(1), 29–35 (2020). https://doi.org/10.1016/j.rccar.2019.04.008
8. Wahab, S., Imran, M., Ahmed, B., Rahim, S., Hassan, T.: Navigating environmental concerns: unveiling the role of economic growth, trade, resources and institutional quality on greenhouse gas emissions in OECD countries. J. Clean. Prod. **434**, 139851 (2024). https://doi.org/10.1016/J.JCLEPRO.2023.139851
9. López, P.A., Bringas, M.V., Iniestra, J.G., Vargas, M.G.: Simulation of the recycling rate of electronic products. A system dynamics model for the reverse logistics network. Account. Manag. **59**(1), 9–41 (2014). https://doi.org/10.1016/S0186-1042(14)71242-2
10. Tarcaya, H.R., Arenas, A.N.: Greenhouse gas emission in urban passenger transport. Transp. Res. Procedia **58**, 158–164 (2021). https://doi.org/10.1016/J.TRPRO.2021.11.022
11. Yang, J.H., Huang, H., Sanyal, S., Khan, S., Alam, M.M., Murshed, M.: Heterogeneous effects of energy productivity improvement on consumption-based carbon footprints in developed and developing countries: the relevance of improving institutional quality. Gondwana Res. **124**, 61–76 (2023). https://doi.org/10.1016/J.GR.2023.06.013
12. de Godoy, S.G.M.: Greenhouse gas emission reduction projects: performance and transaction costs. Revista de Administração **48**, 310–326 (2013). https://doi.org/10.5700/RAUSP1090

13. Saavedra, E., Saavedra-Farfán, E.: Carbon footprint: GHG emissions from the use of the lighting system of the environmental engineering faculty of the Universidad Nacional de Ingeniería, Lima-Peru. TECNIA **30**(1), 121–136 (2020). https://doi.org/10.21754/TECNIA. V30I1.827
14. Avellaneda, I.L., Asesora, P., Libre, U.: Proposal for the reduction of the carbon footprint in the facilities of the regional direction of Magdalena Centro-Car Anteproyecto Liliana Andrea Antury Torres Laura Marcela Lara Castellanos research seminar
15. Lysiak, E.: The carbon footprint of agricultural production of certified tea buds in Argentina. Revista de Investigaciones Agropecuarias **44**(3), 6 (2018). ISSN 0325-8718, ISSN-e1669-2314. https://dialnet.unirioja.es/servlet/articulo?codigo=8308245&info= resumen&idioma=SPA. Accessed 28 June 2024
16. Bongiovanni, R., Tuninetti, L., Garrido, G.: Carbon footprint of the Argentine peanut chain. RIA Revista de in-vestigaciones agropecuarias **42**(3), 324–336 (2016). http://www.scielo.org.ar/scielo.php?script=sci_arttext&pid=S1669-231420160 00300013&lng=es&nrm=iso&tlng=es. Accessed 28 June 2024
17. García Ochoa, J.A., Quito Rodríguez, J.C., Perdomo Moreno, J.A.: Analysis of the carbon footprint in construction and its impact on the environment. Civ. Eng. (2020). Universidad Cooperativa de Colombia, Villavicencio. https://hdl.handle.net/20.500.12494/16031. Accessed 28 June 2024

Sustainable Energy in Ecuador and Latin America: A Review of the Energy Landscape of the Type of Renewable Sources

Lizbeth Flores-Bastidas[1]([✉]) [ID], Juan Lata-García[1] [ID], Sandro C. S. Jucá[2] [ID], and Gary Ampuño[1] [ID]

[1] GIPI Group, Salesian Polytechnic University, Guayaquil, Ecuador
lfloresb@ups.edu.ec

[2] LAESE Laboratory, Telematics, Federal Institute of Ceará, Industrial District I, Maracanaú 61939-140, Brazil

Abstract. The development of a region is closely related to the consumption of electrical energy as long as the generation is able to meet the needs of industry, commerce and residence. This paper reviews the production, consumption of traditional energy and especially renewable generation in Latin America, detailing the energy trend in recent years in Ecuador. On the other hand, it shows the general overview of the development of renewable energies, such as hydroelectric, biomass, wind, solar, and biogas in the South American country. The figures show that the renewable energy installed in the country went from 20,382.76 GWh in 2010 to 37,036.70 in 2024, while generation from renewable sources went from 43.54% to 73.19% in the same period of time. Non-conventional renewable energies have not had an expected growth as is the case of solar despite the abundant solar resource that on average global insolation is 4,575 Wh/m^2/day, in 2010 the installed solar and wind power went from 0.02 MW and 2.4 MW to 29.06 MW and 71.13 MW respectively. Another of the main findings is the decrease in post-pandemic energy consumption in Latin America, in 2017 consumption was 2477,207 GWh of which 55.7% comes from renewable sources while in 2021 consumption was 1637,868 GWh corresponding to 58.97 from renewable sources. The document opens other lines of research as proposals for the legislation that regulates renewable energies in Ecuador, from distributed generation for the self-supply of regulated consumers to the methodology for the term and prices of generation and self-generation projects.

Keywords: Energy · Generators · Latin America · Outlook · Renewable

1 Introduction

By 2030, the United Nations (UN) proposed 17 Sustainable Development Goals (SDGs), the seventh goal is about access to clean and affordable energy for communities [1]. Five challenges were identified to achieve the seventh SDG, especially in developing countries, such as cutting the use of fossil fuels, decentralizing energy generation and distribution, reducing the costs of energy generation processes, increasing energy efficiency and

E. M. Inga Ortega et al. (Eds.): CITIS 2024, LNNS 1331, pp. 314–323, 2025.
https://doi.org/10.1007/978-3-031-87065-1_29

storage, and migrating to renewable and diversified energy sources. These challenges can be addressed through the increase of non-fossil fuel energy sources, reformulation of global energy policies, investment support from technologically developed countries and implementation of optimization and automation technologies, however, knowing in detail the energy development of Latin American countries remains a complex task. Countries should take advantage of opportunities to increase renewable energies such as data logging to recognize patterns of energy consumption and seek to empower the population on saving energy consumption [2].

Various studies in Latin American and Caribbean (LAC) countries have identified factors that must be taken into account to increase the use of renewable energies, detailing the factors that have a positive impact on the installed capacity of non-hydroelectric renewable energies (wind, solar, marine, geothermal and residual biofuel, etc.). Solid and liquid) such as the increase in financing flows, reduction of costs of renewable energy components, the existence of financial policies that encourage private and public financing in projects of this nature, the liberalization of foreign, commercial and financial investment and the reduction of import barriers in LAC countries due to restructuring projects that increase energy production and seek to reduce the use of Non-renewable energy sources [3–5].

Mulali et al. in their research indicate that in the LAC region more than 50% of total energy consumption comes from renewable energies [6]. In 2021, 59.07% of electricity generation in Latin America and the Caribbean comes from renewable sources and 40.93% from non-renewable sources, made up of 42.83% from hydroelectric, 38.70% from non-renewable thermal, 7.70% from wind, 4.47% from renewable thermal and 3.42% from other energy sources. In addition, countries such as Paraguay depends almost entirely on hydroelectric energy, having 99.99% of the installed capacity, and Brazil and Costa Rica have more than 60% of the installed capacity. By the end of the year 2023, global renewable energy capacity grew to 3,870 GW, comprised of most of solar with a capacity of 1,419 GW, followed by renewable hydropower and wind with a capacity of 1,268 GW and 1,017 GW respectively [7].

Despite efforts to increase renewable energy capacity in the LAC region, it does not reach other markets such as Asia, which accounted for the largest global share of installed capacity in renewable energy with 50.7% and renewable energy growth of 327.8 GW, followed by Europe with 20.3% and an increase of 71.2 GW. North America with 7.5% and an increase of 34.9 GW, in fourth place is South America with 7.5% and a growth of 22.4 GW, followed by Eurasia with 3.2% and an increase of 3.0 GW, Oceania with 1.7% and a progression of 5.5 GW, Africa with 1.6% and a development of 2.7 GW, The Middle East with 0.9% and an increase of 5.1 GW and in last place Central America and the Caribbean with 0.5% of the world share and 0.8 GW with respect to renewable energy growth (GW) [8].

Despite the little industrial development that occurs in the regional and local community, several studies propose to increase renewable energy generation to meet the growing energy demand and reduce the barriers to energy access that arise in disadvantaged communities. Therefore, it is important to know how renewable energies have been developing, especially in underdeveloped countries such as those in Latin America and the Caribbean [9] where the implementation of renewable generation creates jobs, in search of the transition to net zero greenhouse gas emissions. The objective of this research is to analyze the energy landscape of Latin America in terms of the growth of clean energy, consumption, and new projects developed with an emphasis on Ecuador.

2   Energy Consumption in Latin America

Latin America in the last decade has not had a constant growth in energy consumption, in 2010 of energy harnessed and from various sources reached 1319,113 GWh of which 59.37% is of renewable origin, in 2017 it reached the record figure of 2477,207 GWh of which 55.7% is from renewable origin, in the post-pandemic period energy consumption was reduced and with a slow recovery, in 2021 consumption was 1637,868 GWh, corresponding to 58.97% from renewable sources as shown in Fig. 1. Despite this growth, they have had to face challenges such as access to capital, the impact on nature, the absence of policies that guarantee their effectiveness, and access to clean energy technologies.

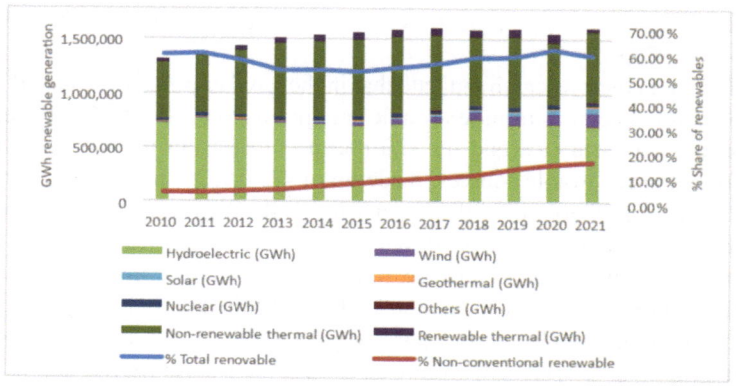

Fig. 1. Renewable generation trend in Latin America and the Caribbean

The Latin American Energy Organization (OLADE) [10] releases in the Energy Panorama of Latin America and the Caribbean for the year 2023 the state of the energy sources of various countries in the LAC region, in which it indicates:

For the year 2022, **Argentina** reached 31.33% of demand supply from renewable energies, composed in the first place of wind energy with 71.1%, solar energy with 17.3%, in a lesser proportion hydroelectric with 5.78% and bioenergy with 5.7% corresponding to biomass and biogas and inaugurated the first Arauco III Wind Farm that will generate 150 MW for the grid. Supplying 81% of homes in La Rioja.

Bolivia has 1,161 MW of installed clean energy capacity, reducing the use of gas for electricity generation by 50% by using environmentally friendly energies such as biomass, solar, hydroelectric and wind. In addition, it has the Electricity to Live with Dignity Program (PEVD) in place, which allows access to electricity to communities in poverty through (SFV).

In 2022, **Brazil** reduced thermal generation with coal and natural gas by more than 50% due to its increase in various renewable energies compared to 2021 such as solar photovoltaic with a growth of more than 78%, making it the eighth country worldwide with respect to the installed capacity to generate energy from solar sources. Hydropower with 16% and wind with more than 12%.

Chile reached 33% of electricity generation from clean and environmentally friendly energy, reduced greenhouse gas emissions by 22% and reached a record growth of 56.8% in energy self-generation through SF (Acesol).

Colombia inaugurated the country's largest wind farm, Guajira 1, which will contribute 20 MW to the energy matrix and issue emission reduction certificates.

Costa Rica reached more than 98% of electricity generation from its national resources, among the main energy resource is water with 75.16%, followed by geothermal with 12.97%, wind with 10.65%, biomass and solar together added 0.47%.

Cuba presented the Bioenergy Atlas that focuses on three sources of bioenergy (biogas, biomass and biodiesel), making known the energy potential of bioenergy sources and serving as an aid for decision-making in the creation of projects and programs that benefit the production chain.

Ecuador with respect to renewable energy, hydroelectric power predominates because of the production of electricity, which was 33,292 GWh in 2022, of the 78% from renewable energy, 76.8% belonged to this type of energy. In addition, it has the largest wind farm in the country with 50 MW of power, helping to reduce carbon dioxide emissions by about 76,625 tons per year.

El Salvador reported that 59.4% of its installed capacity comes from renewable energies, increasing biomass by 1.8 MW and biogas by 0.7%. Also, energy generation through sun increased by 5.71%, geothermal by 7.92% and wind by 40.97%.

Guatemala achieved a record by achieving that 78% of energy generation is from renewable sources and thanks to favorable weather conditions, 52% of generation comes from hydroelectric plants.

Mexico has ranked sixth in the world with respect to installed geothermal capacity since 2016. In 2022, the share of clean energy increased by 31.16%, inaugurated the solar park, La Pimienta, which will generate 300 MW, and the largest solar park in Latin America, Puerto Peñasco, with a capacity of 1,000 MW, is under construction.

Based on the 2022 Generation Matrix report, **Nicaragua** pointed out that 70.87% of electricity generation was from renewable energy such as hydroelectric, geothermal, biomass, wind and solar.

Panama in its national energy transition reached 97% in renewable energy generation, of which 91.09% was represented by hydroelectric sources, 5.82% by wind and photovoltaic sources.

Paraguay is the only country to generate 100% clean and renewable energy, naming it the world leader in sustainable energy production. Its hydroelectric plants cover the entire electricity demand, which are the Itaipu, Yacyretá and Acaray power plants.

Peru launched the pilot project to assemble 13 photovoltaic panels for Distributed Generation on Grid in order to improve the quality of service and reduce the purchase of energy.

The Dominican Republic has developed various clean energy projects such as the micro-hydroelectric and photovoltaic project in the Los Limoncitos community that has an installed power of 192.5 kVA, Santanasol and El Socol of photovoltaic energy with 50MWh each, Los Guzamancitos of wind energy with 50 MWh and the solar roof program that favored 18 homes isolated from the grid.

Trinidad and Tobago seek to reduce the use of non-renewable energy by 3.5% through the implementation of a solar park at Piarco International Airport.

Uruguay implemented the Renewable Energy Innovation Fund to promote energy transition projects through UN funds and private capital.

3 Renewable Energy Generation in Ecuador

One of the countries that make up South America is Ecuador, a sovereign, democratic, legal and plurinational country, it is made up of twenty-four provinces. It is bordered to the north by Colombia, to the east and south by Peru, and to the west by the Pacific Ocean, in addition to the Galapagos Islands, the country has a high concentration of rivers per square kilometer in the world, the area is 256,370 km^2, the population is 16,938,986 inhabitants according to the 2022 census, the gross domestic product in 2024 is USD 268,199 million. In the last 15 years, energy consumption has increased considerably, in 2023 final consumption in the transport sector is 6.81 Mtoe, final consumption in the industrial sector is 3.73 Metp.

The final consumption in the residential sector is 1.85 Metp, in Fig. 2 you can see the behavior of electricity consumption, in 2014 Ecuadorians required 25,143.95 GWh while in 2023 it is 36,682.97 GWh which corresponds to an increase of 45.89% [11].

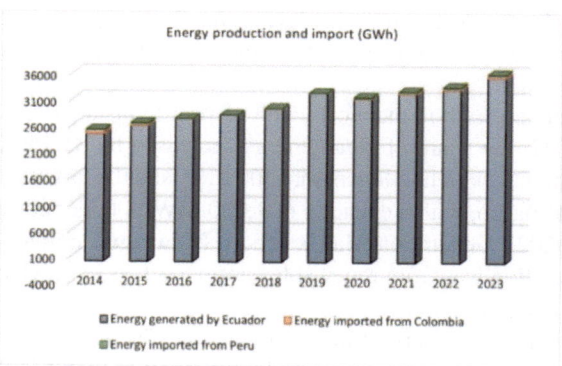

Fig. 2. Trend in energy production and importance for Ecuador

In 2020, energy consumption was 790.85 GWh less than in 2019, which reached 32,293.91 GWh, this decrease in consumption is due to the COVID 2020 pandemic. The country's energy matrix is divided into two segments: a renewable one, with an effective power of 5,401.71MW, and non-renewable, with 2,907.27 MW, which corresponds to 65.01% and 34,99%, respectively. The subdivision of each of these segments is shown in Table 1, in the case of renewables they are divided into the different types of generation sources, hydropower represents 58.01% of the total with an installed power of 5,198.80 MW, wind energy represents 0.79% with 71.13 MW installed, energy from the sun with just 0.32% which represents 29.06 MW despite all the photovoltaic potential that is available in the country due to its geographical location near the equator, energy from biomass contributes 1.61% with an installed power of 144.3 MW, especially from sugar mills, while energy from garbage landfills contributes 0.09% with 8.32 MW.

On the other hand, non-renewable generation is made up of internal combustion engines with 23.47% representing 2103 MW of installed power, turbogas generation reaches 10.54% with 944.85 MW, to complete it turbosteam engines are used with a contribution of 5.15% representing 461.63 MW of installed power [12].

Table 1. Detail of the Installed Generation Power in Ecuador

Installed power in generation				
Electrical energy	Power Rating		Effective Power	
	MW	%	MW	%
(Ren + No Ren)	8.961,73	100,00%	8.308,98	100,00%
Renewable	5.451,60	60,83%	5.401,71	65,01%
Hydraulics	5.198,80	58,01%	5.158,81	62,09%
Wind	71,13	0,79%	71,13	0,86%
Photovoltaic	29,06	0,32%	28,17	0,34%
Biomass	144,30	1,61%	136,40	1,64%
Biogas	8,32	0,09%	7,20	0,09%
Non-Renewable	3.510,13	39,17%	2.907,27	34,99%
MCI	2.103,66	23,47%	1.684,42	20,27%
Turbogas	944,85	10,54%	791,35	9,52%
Turbovapor	461,63	5,15%	431,50	5,19%

The generating plants that came into operation in 2023 are the Huascachaca plant (wind) with a contribution of 49.98 MW, while the hydroelectric plant is Sarapullo and Ulba with 48.45 and 1 MW respectively.

Other thermal self-generating plants that started production are Andes Petro and Sipec with 2.4 MW and 1.95 MW, this type of plant produces the energy to be used in the production process of the company itself [13]. The infrastructure of generation plants, transmission lines, voltage levels, lifting, reducing, and sectioning transformers are shown in Fig. 3. The use of renewable sources entails a series of benefits, including economic development due to the creation of jobs and business opportunities, the reduction of greenhouse gas emissions and therefore the improvement of air quality.

3.1 Hydropower

Hydroelectric power plants are the ones that supply the greatest power to the national interconnected system with 5,198 MW of installed power and 5,158 MW of effective

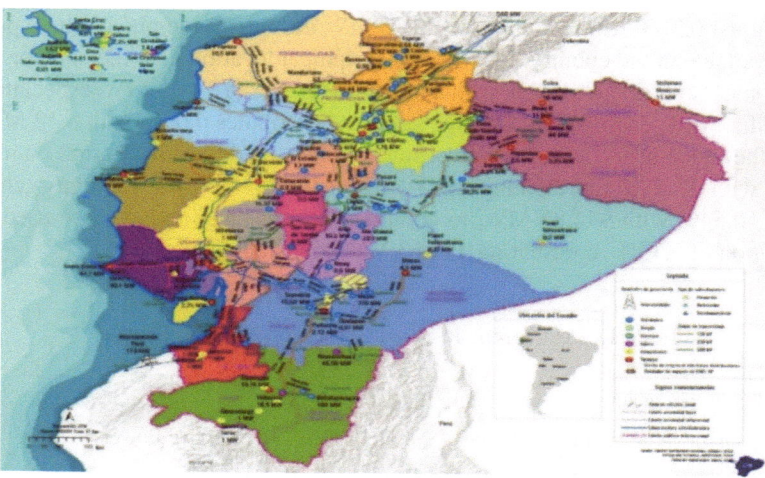

Fig. 3. Infrastructure of the Ecuadorian electricity sector.

power, due to the high hydraulic potential of the rivers, the average calculated flow is 15,123 m3/s, the theoretical power that can be reached is 91,000 MW, equivalent to 615,175 GWh/year, while the technically feasible power is 31,000 MW in 11 river basins, while economically feasible up to 22,000 MW can be installed in the same number of river basins. The country has 12 hydroelectric plants with a capacity of more than 50 MW, among the main ones is Coca codo Sinclair with 1500 MW, Paute mill 1075 MW and blow molding 487 MW. Figure 4 shows the dams and powerhouses of Paute integral with an installed capacity of 2,352.6 MW producing 13,000 GWh per year [14, 15].

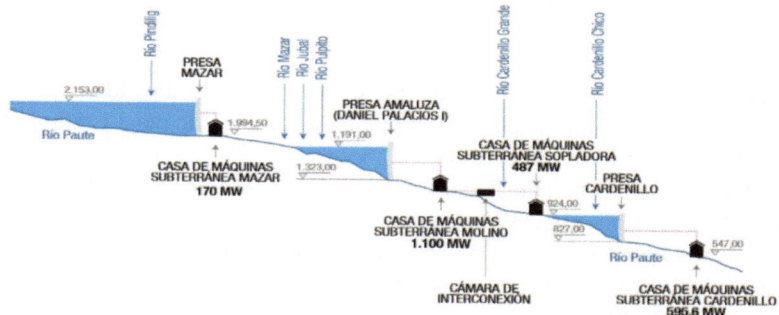

Fig. 4. Hydroelectric use of the Paute river basin.

Until 2018, the country had 71 plants, of which 5 have a reservoir with a capacity of 1,598 MW, while 66 plants do not have a reservoir, the installed capacity is 3,443 MW [14].

3.2 Biomass Energy

One way of generating electricity from organic matter is known as biomass. The origin as its nature is diverse, sugarcane bagasse is used to produce electricity. The country has three Biomass plants installed in sugar mills with an installed capacity of 144.30 MW, the effective power is 136.4 MW, which corresponds to 1.64% of the supply in the energy matrix in the country. In 2005 the first plant came into operation with a power of 28 MW, in 2006 it reached 55.60 MW, in 2008 another generator came into operation reaching 94.5 MW from 2010 to 2013 the installed power of 101.3 MW, in 2014 the last 43 MW was entered reaching 144.3 MW to date [16].

3.3 Wind Energy

The energy from wind speed is known as wind, the MEER in 2013 presents the "Wind Atlas of Ecuador for electricity generation purposes", the main results indicate that the gross wind potential is 1,691 MW, especially in the Andean area where there is an average annual wind speed greater than 7 m/s, by implementing wind turbines, an average of 2,869 GWh would be generated. Feasibility studies carried out by the ministerial entity show that in the short term the installable power is 884 MW and the average annual energy of 1,518 GWh, which can be increased if equipment is installed that operates at low wind speeds between 5 and 6 m/s. In 2007, the first wind turbines were installed in the Galapagos archipelago with a power of 2.4 MW, remaining so until 2013, when the installed power increased to 19.56 MW with the commissioning of 11 wind turbines of the Villonaco wind farm in the province of Loja. In 2022, the first wind turbines of the Huascachaca mines project will be generated, reaching a power of 53.15 MW, and by 2023 the wind farm will start up the 14 turbines providing 50 MW, having a total power at the country level of 71.13 MW, which translates to 208.58 GWh [17].

3.4 Solar Energy

According to the "Solar Atlas of Ecuador for electricity generation purposes", the average direct insolation value of continental Ecuador is 2,543 Wh/m^2/day; the average diffuse insolation is 2,032 Wh/m^2/day; and the average global insolation is 4,575 Wh/m^2/day. In 2005, the first solar plants were installed with an installed capacity of 0.02 MW until 2011 when the power increased to 0.04 MW, currently there are 34 plants with an effective power of 26.74 MW.

To complete the energy matrix, Ecuador has two Biogas plants with an installed capacity of 6.50 MW, which barely represents 0.09%, however, it is a great step in the diversification of the energy matrix and giving energy use to urban solid landfills.

4 Conclusions

The development of a region is linked to the consumption of reliable, safe energy at a competitive price, taking into account the generation of renewable origin avoiding the emission of greenhouse gases, in the document presented the generation trend of the different generation sources in Latin America for each of the countries is reviewed, the energy indicators are reviewed, Ecuador is taken as a case study where the production of energy is reviewed in detail, especially its hydroelectric, biomass, solar and wind renewable energies, the trend in the last 10 years, the number of plants and their evolution over time.

Hydroelectric energy contributes 58.01% of the country's total consumption, being the main source of generation, non-conventional renewable energies lack a leading role, despite the fact that the first plants were installed in 2005, solar photovoltaic barely reaches 29.06 MW which corresponds to 0.32%, wind power took a boost in the last year with the commissioning of the 50 MW Huascachaca plant contributing a total 0.79%, the 3 biomass plants have a significant contribution of generation reaching 1.61%, on the other hand a relevant fact in the study is the use of urban solid waste in landfills to produce electricity through Biogas, the installed power of 8.32 MW.

References

1. United Nations (UN): La Asamblea General adopta la Agenda 2030 para el Desarrollo Sostenible - Desarrollo Sostenible (2015). https://www.un.org/sustainabledevelopment/es/2015/09/la-asamblea-general-adopta-la-agenda-2030-para-el-desarrollo-sostenible/. Accessed 20 May 2024
2. Kay Lup, A.N., et al.: Sustainable energy technologies for the Global South: challenges and solutions toward achieving SDG 7. Environ. Sci. Adv. **2**, 570–585 (2023). https://doi.org/10.1039/D2VA00247G
3. Washburn, C., Pablo-Romero, M.: Measures to promote renewable energies for electricity generation in Latin American countries. Energy Policy **128**, 212–222 (2019). https://doi.org/10.1016/J.ENPOL.2018.12.059
4. Bersalli, G., Menanteau, P., El-Methni, J.: Renewable energy policy effectiveness: a panel data analysis across Europe and Latin America. Renew. Sustain. Energy Rev. **133**, 110351 (2020). https://doi.org/10.1016/J.RSER.2020.110351
5. Silva, N., Fuinhas, J.A., Koengkan, M.: Assessing the advancement of new renewable energy sources in Latin American and Caribbean countries. Energy **237**, 121611 (2021). https://doi.org/10.1016/J.ENERGY.2021.121611
6. Al-Mulali, U., Fereidouni, H.G., Lee, J.Y.M.: Electricity consumption from renewable and non-renewable sources and economic growth: evidence from Latin American countries. Renew. Sustain. Energy Rev. **30**, 290–298 (2014). https://doi.org/10.1016/J.RSER.2013.10.006
7. Hub de Energía para América Latina y El Caribe: Capacidad, generación y consumo de electricidad | IADB (2021). https://hubenergia.org/es/indicators/capacidad-generacion-y-consumo-de-electricidad. Accessed 18 May 2024
8. International Renewable Energy Agency (IRENA): Renewable capacity statistics 2024 (2024). https://www.irena.org/Publications/2024/Mar/Renewable-capacity-statistics-2024. Accessed 20 May 2024
9. Zhang, J.: Energy access challenge and the role of fossil fuels in meeting electricity demand: promoting renewable energy capacity for sustainable development. Geosci. Front. **101873** (2024). https://doi.org/10.1016/J.GSF.2024.101873
10. Panorama energético de América Latina y el Caribe 2023 – OLADE. https://www.olade.org/publicaciones/panorama-energetico-de-america-latina-y-el-caribe-2023/. Accessed 26 May 2024

11. Ecuador, E., Eras, A.A., Barragán, E.A.: Mecanismos de Promoción y Financiación de las Energías Renovables en El Ecuador. Revista Técnica "energía" **9**, 128–135 (2013). https://doi.org/10.37116/REVISTAENERGIA.V9.N1.2013.142
12. Balance Nacional de Energía Eléctrica – Agencia de Regulación y Control de Energía y Recursos Naturales no Renovables. https://www.controlrecursosyenergia.gob.ec/balance-nacional-de-energia-electrica/. Accessed 26 May 2024
13. Agencia de Regulación y Control de Energía y Recursos Naturales no Renovables: Estadísticas del sector eléctrico ecuatoriano buscar – Agencia de Regulación y Control de Energía y Recursos Naturales no Renovables (2023). https://www.controlrecursosyenergia.gob.ec/estadisticas-del-sector-electrico-ecuatoriano-buscar/. Accessed 26 May 2024
14. Ministerio de Energía y Minas - Ecuador Expansión de la Generación. https://www.recursosyenergia.gob.ec/wp-content/uploads/2020/01/4.-EXPANSION-DE-LA-GENERACION.pdf. Accessed 23 May 2024
15. Icaza, D., Borge-Diez, D., Galindo, S.P.: Proposal of 100% renewable energy production for the City of Cuenca- Ecuador by 2050. Renew. Energy **170**, 1324–1341 (2021). https://doi.org/10.1016/J.RENENE.2021.02.067
16. Boletín estadístico sector eléctrico ecuatoriano (2012)
17. Inventario de recursos energéticos del Ecuador | ARIAE. https://www.ariae.org/servicio-documental/inventario-de-recursos-energeticos-del-ecuador. Accessed 26 May 2024

Analysis of UV Radiation Monitoring: Distribution of Stations in South American Countries During the Summer Season

Byron Xavier Bone Moncayo[1](✉) ⓘ, Carlos Andrés Ulloa Vaca[1,2](✉) ⓘ,
and Alexandra Karina Pazmiño Pacheco[1,2](✉) ⓘ

[1] Carrera de Ingeniería Ambiental, Universidad Politécnica Salesiana, Quito 170702, Ecuador
{culloa,apazminop}@ups.edu.ec
[2] Grupo de investigación en ciencias ambientales GRICAM, Carrera de Ingeniería Ambiental,
Universidad Politécnica Salesiana, Quito 170702, Ecuador

Abstract. The present study examines the distribution of ultraviolet (UV) radiation monitoring stations in South American countries during the months of December to February. Data were collected from Argentina, Brazil, Chile, Colombia, Ecuador, Peru, Uruguay, Venezuela, Bolivia, and Paraguay, covering a total of 89 stations. The results reveal significant variability in the number of stations among countries, with Brazil, Chile, Colombia, Ecuador, and Peru standing out with a higher number of stations (range: 9–11), while Uruguay, Venezuela, and Bolivia show a lower number (range: 6–9). This variability may be related to geographical, demographic, and funding factors. Possible implications for human health and the environment are discussed, highlighting the importance of UV radiation monitoring to understand its effects on health and the environment. The need for awareness and preventive measures, including the use of sunscreen and avoiding direct sun exposure during peak UV radiation hours, is emphasized. This study underscores the importance of an equitable distribution of UV radiation monitoring stations to ensure adequate surveillance and promote health and well-being in the South American region.

Keywords: UV Index · Health Implications · UV monitoring stations

1 Introduction

Environmental radiation is an omnipresent component in our surroundings, originating from both natural and artificial sources. Although exposure to this radiation is inevitable, its accumulation can pose a risk to humans. The electromagnetic spectrum of ultraviolet (UV) radiation encompasses wavelengths shorter than visible light but longer than X-rays. UV radiation is divided into three main subcategories, each with its own characteristics and effects on health: UV-A (320–400 nm), UV-B (280–320 nm), and UV-C (100–280 nm).

According to Dedios (2016), UV radiation has enough energy to produce damage and harm terrestrial systems, affecting plants, microorganisms, and humans. UV-A (320–400 nm) has the lowest energy, can penetrate deeply into the skin, and is responsible

© The Author(s) 2025
E. M. Inga Ortega et al. (Eds.): CITIS 2024, LNNS 1331, pp. 324–335, 2025.
https://doi.org/10.1007/978-3-031-87065-1_30

for premature aging and some eye damage. UV-B (280–320 nm) has higher energy and causes sunburn, damages the DNA of skin cells, and increases the risk of skin cancer. UV-C (100–280 nm) has the highest energy and is very harmful to living organisms, but it is almost completely absorbed by the Earth's atmosphere and does not reach the surface (Cendros & Durante 2013)

The use of the UV Index was proposed in 1992 by the World Health Organization (WHO) and other entities such as the World Meteorological Organization (WMO), the United Nations Environment Programme (UNEP), and the International Commission on Non-Ionizing Radiation Protection (ICNIRP). The scale proposed for the UV Index is: Low in green [0 to 2]; Moderate in yellow [3 to 5]; High in orange [6 to 7]; Very high in red [8 to 10]; Extremely high in violet [11 to 20].

According to Madrid (2021), long-term exposure to ultraviolet radiation can cause health problems such as skin damage, sunburn, premature aging, and more serious conditions like skin cancer or eye damage. One of the most common effects is sunburn, especially in children and young people, which increases the risk of skin growths such as moles and melanomas.

Cañarte (2010) indicates that intermittent and cumulative exposure to UV radiation causes degenerative alterations in the skin, accelerating its aging and manifesting as fine wrinkles. According to Fernandez (2006), chronic exposure to UVB rays, and to a lesser extent to UVA, induces the appearance of in situ and invasive squamous cell carcinomas, basal cell carcinomas, melanoma, and probably other skin carcinomas as a result of DNA damage and interference with molecular or immunological mechanisms.

Consequently, research has been conducted on the types of sensors capable of measuring the UV index, these being Sensor A, Sensor B, and Sensor C. Sensor A covers a spectral range of 280–400 nm, with a sensitivity of 0.1 mW/m^2/nm, spatial resolution of 1×1 m^2, temporal resolution of 60 s, and is small in size. Sensor B operates in a spectral range of 290–320 nm, with a sensitivity of 0.05 mW/m^2/nm, spatial resolution of 0.5×0.5 m^2, temporal resolution of 30 s, and is medium in size. Sensor C covers a spectral range of 280–380 nm, with a sensitivity of 0.08 mW/m^2/nm, spatial resolution of 2×2 m^2, temporal resolution of 120 s, and is large in size.

In South American countries, prevention of the effects of solar radiation has been prioritized. Consequently, different solmeters have been installed. A "solmeter" is a UV radiation traffic light system designed to inform the public about current UV index levels and provide guidance on protective measures. The installation of solmeters in South America varies by country, with specific policies and locations depending on national and regional initiatives to promote public awareness of UV radiation risks.

For example, in Chile, the Chilean Meteorological Office has implemented solmeters in several cities, especially in areas with high UV exposure, such as Santiago and regions in the north like Antofagasta. These installations aim to increase awareness and promote protective behaviors among residents and visitors. In Argentina, solmeters have been installed in major urban centers such as Buenos Aires and coastal areas like Mar del Plata. The National Meteorological Service oversees these installations as part of broader public health campaigns. Peru has also adopted solmeters in key locations, including Lima and other high-altitude regions where UV radiation levels are particularly intense. These efforts are part of the government's strategy to reduce rates of skin cancer and other health issues associated with excessive UV radiation exposure. In Ecuador, solmeters have also been installed in some public parks to promote the culture of sun protection.

These systems typically use a color-coded scheme aligned with the international UV index scale: green for low (0–2), yellow for moderate (3–5), orange for high (6–7), red for very high (8–10), and violet for extreme (11+), providing a direct and effective means of communicating UV risk levels to the public (World Health Organization (WHO), United States Environmental Protection Agency (EPA), European Space Agency) (Orozco & Ordoñez et al. 2019).

2 Matherials and Methods

The research was based on a review of 25 bibliographic sources. Scientific databases were accessed, and a search strategy was employed using well-known academic databases such as Google Scholar and Scopus to gather relevant articles on the topic. Keywords and initial filters were used, including a temporal range of the last ten years. A review of abstracts was conducted to assess the relevance of the articles, and boolean operators were used to enhance the searches. To ensure the relevance and quality of the selected studies, additional inclusion and exclusion criteria were established, including geographical focus and methodologies used. Finally, 25 articles demonstrating greater relevance and methodological rigor were selected. Microsoft Excel was used to visualize and process quantitative data.

The proposed methodology focused on the review of scientific documents to analyze advances in the development and application of environmental radiation sensors, with particular emphasis on sensors designed for the detection of ultraviolet (UV) radiation. The scientific databases used were SCOPUS due to its broad scope and coverage in various branches of science. The search strategy included relevant terms such as "ultraviolet radiation sensors," "effects of ultraviolet radiation on humans," and "environmental monitoring."

A comprehensive review of the selected documents was conducted, extracting relevant data on the classification principle of sensors, environmental radiation, and UV radiation. For the results, research was conducted on platforms and meteorological institutions from various countries, manually extracting data over a period of four years (2020, 2021, 2022, and 2023) on the radiation index in the months of December, January, and February, when ultraviolet radiation is most prominent. The data were averaged by reviewing the daily and monthly meteorological reports from each meteorological entity.

3   Results

The table presents a specific breakdown of South American countries, accompanied by data on the quantity of sensors deployed in each one (Table 1).

Table 1. Stations and UV Radiation Monitoring Institutions

Country	UV monitoring station	Number	ORGANIZATION
Argentina		8	Servicio Meteorológico Nacional (SMN) y Comisión Nacional de Energía Atómica (CNEA)
Brasil		9	Instituto Nacional de Meteorología (INMET) y Centro de Previsión del Tiempo y Estudios Climáticos (CPTEC)
Chile		11	Dirección Meteorológica de Chile

(*continued*)

Table 1. (*continued*)

Colombia   10   Instituto de Hidrología,
                                                                    Meteorología y Estu-
                                                                    dios   Ambientales
                                                                    (IDEAM)

Ecuador   10   Instituto Nacional de
                                                                    Meteorología e Hi-
                                                                    drolo- gía (INAMHI)

Perú   10   Servicio Nacional de
                                                                    Meteorología e Hi-
                                                                    drolo- gía (SENAMHI)

(*continued*)

Table 1. (*continued*)

Uruguay   6   Instituto Uruguayo de
                                                                            Meteoro-
                                                                          logía
                                                                          (INUMET)

Venezuela   6   Instituto Nacional de
                                                                            Meteorología e Hi-
                                                                          drolo- gía (INAMEH)

Bolivia   9   Servicio Nacional de
                                                                            Meteorología e Hi-
                                                                          drolo- gía (SENAMHI)

(*continued*)

Table 1. (*continued*)

Paraguay   10   Dirección de Meteorolo-

gía e Hidrología

In South American countries, ultraviolet radiation was analyzed over an average of 4 years, especially during the months of December, January, and February, and with which type of sensor it was measured (Table 2).

Table 2. Average Radiation Index in South America and Possible Health Effects.

Countries	Range of UV Index (Annual Average)	Possible health effects
Argentina	5–6	Moderate risk of keratitis, eye damage
Paraguay	5–6	Mild sunburns, premature aging
Uruguay	5–6	Moderate incidence of cancer, possible eye damage, and mild burns
Chile	6–7	Skin cancer, sunburns
Colombia	9–10	Sunburns, carcinomas, eye damage
Brazil	10–11	Sunburn, eye damage, melanomas
Bolivia	10–12	Keratitis, eye damage, malignant melanomas
Ecuador	10–12	Sunburns, basal and squamous cell carcinomas, melanomas
Peru	11–12	Keratitis, eye damage, malignant melanomas
Venezuela	11–12	Sunburns, melanomas, carcinomas

For four years, radiation was analyzed in South American countries during the months of December, January, and February. During this period, a monthly comparative study was conducted to calculate the average UV radiation index (Table 3).

Table 3. Radiation Index in December for 4 years

PAIS	DECEMBER					JANUARY					FEBRUARY					FINAL
	2020	2021	2022	2023	AVG	2020	2021	2022	2023	AVG	2020	2021	2022	2023	AVG	AVG
Argentina	3	5	6	6	5	4	5	5	6	5	3	4	5	7	4,75	5–6
Brasil	10	9	11	11	10,3	9	8	10	11	9,5	9	7	11	11	9,5	10–11
Chile	4	6	7	6	5,75	4	6	5	6	5,25	6	6	7	6	6,25	6–7
Colombia	7	8	8	10	8,25	7	7	10	11	8,75	7	8	10	10	8,75	9–10
Ecuador	9	10	10	12	10,3	10	9	11	12	10,5	9	10	9	11	9,75	10–12
Perú	10	11	12	12	11,3	11	11	12	12	11,5	11	10	11	12	11	11–12
Uruguay	5	5	6	7	5,75	6	6	5	7	6	6	5	4	6	5,25	6–7
Venezuela	10	11	12	12	11,3	11	11	12	12	11,5	10	9	11	11	10,3	11–12
Bolivia	10	10	9	12	10,3	12	11	10	11	11	10	9	10	12	10,3	11–12
Paraguay	5	4	5	5	4,75	7	5	6	7	6,25	6	6	7	7	6,5	6–7

Below are graphically presented the averages of ultraviolet radiation index for each country. As can be observed, Peru, Venezuela, Bolivia, and Ecuador have the highest ultraviolet radiation indexes compared to the rest of South American countries. See (Fig. 1).

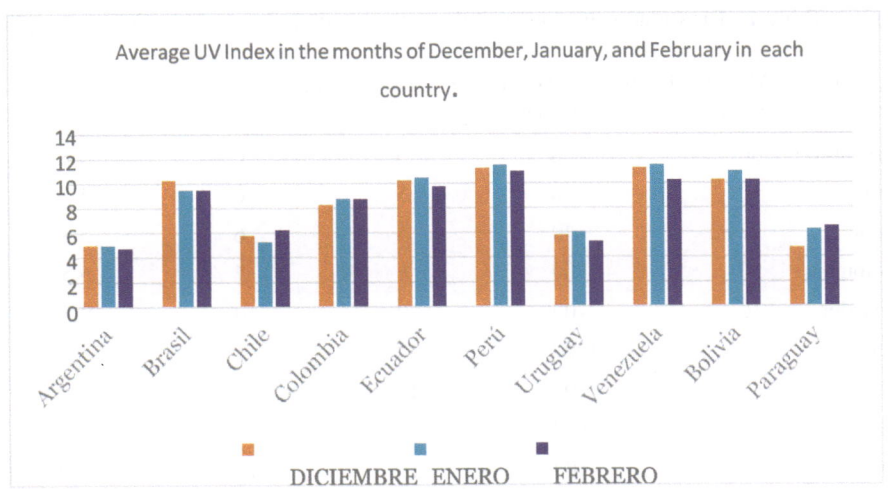

Fig. 1. Average UV Index in the months of December, January, and February in each country.

With the following type of line graph, it can be observed that the highest peaks correspond to Brazil, Peru, and Venezuela (Fig. 2).

This table presents comparative data on altitude and UV indices across South American countries, detailing ranges of altitude and minimum, maximum, and average UV indices. It provides insights into how these factors vary across the region (Table 4).

Countries such as Colombia, Ecuador, Venezuela, Peru, and Brazil, located near the equator, generally exhibit higher UV indices. The proximity to the equator results in

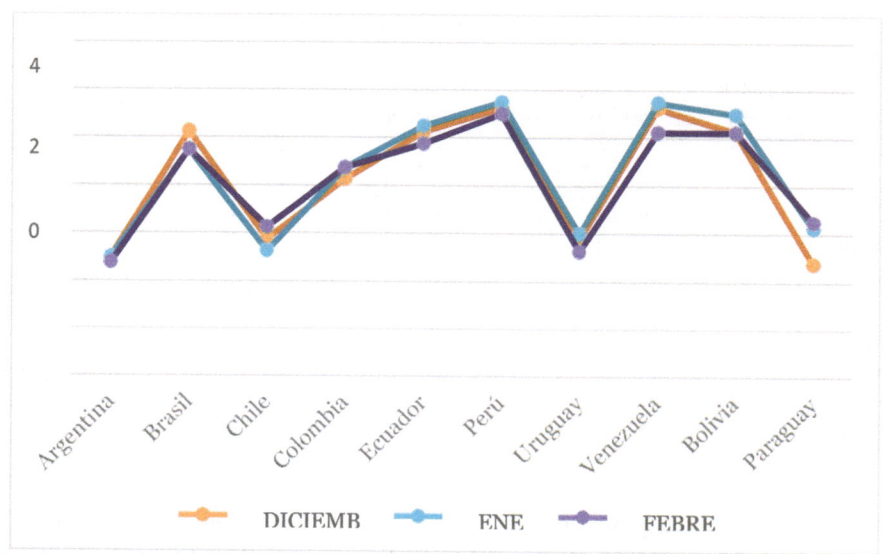

Fig. 2. Graphical trend of UV Index in the months of December, January, and February in each country

Table 4. The variation in altitude of the sensors with respect to their UV index

Countries	Altitude Min in (m.s.n.m)	Altitude Max in (m.s.n.m)	Index Uv Min	Index Uv Max	Index Uv AVG
Argentina	50	3200	5	6	5,5
Brasil	10	170	5	6	5,5
Chile	23	40	5	6	5,5
Colombia	50	700	6	7	6,5
Ecuador	10	2700	9	10	9,5
Perú	10	1200	10	1	5,5
Uruguay	157	5300	10	12	11
Venezuela	24	2900	10	12	11
Bolivia	20	4300	11	12	11,5
Paraguay	900	1600	11	12	11,5

more direct solar incidence throughout the year, increasing the intensity of UV radiation. In contrast, countries such as Argentina, Paraguay, and Uruguay, situated further south, have lower UV indices, regardless of altitude. These countries experience less direct solar exposure due to their greater distance from the equator (Fig. 3).

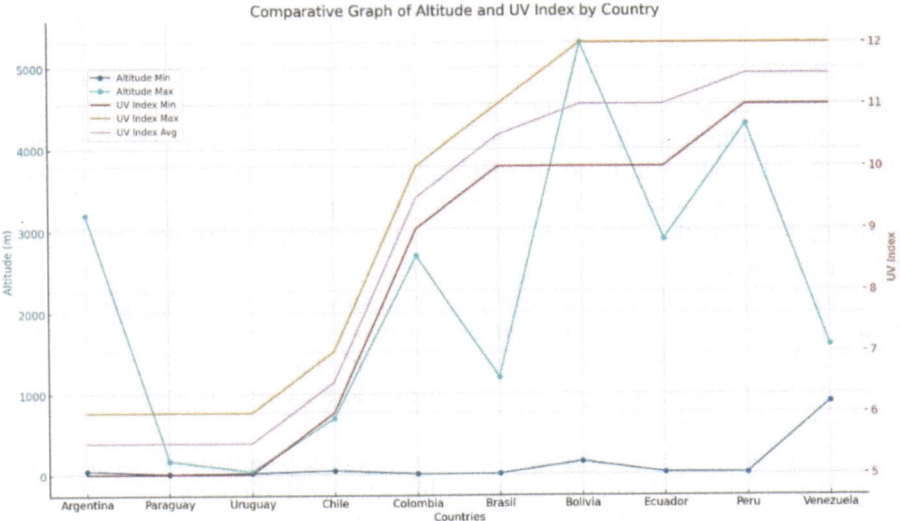

Fig. 3. Comparative graph of the altitude range with respect to the min, max and average UV index

4 Discussion

Solar radiation during the months of December to February in South American countries is influenced by various factors such as the sun's inclination, cloud coverage, and regional climatic patterns. These factors can contribute to variations in observed levels of solar radiation between countries and years. The inclination of the sun varies during these months due to the Earth's position relative to the sun, affecting the amount of solar radiation reaching the Earth's surface. Additionally, the presence of clouds can modify the amount of radiation reaching the surface, as clouds can reflect, absorb, or block solar radiation. Regional climatic patterns, such as high and low-pressure systems, can also influence the amount and distribution of solar radiation in the region. These climatic patterns may vary from year to year, which could contribute to the fluctuations observed in solar radiation levels over time.

Regarding the number of UV radiation monitoring stations in each country, we can observe variations that may be related to factors such as geographic extent, population density, and funding priorities in health and the environment. For example, larger countries or those with more diverse geographic conditions may require a broader network of stations to capture the variability of solar radiation. Additionally, countries with greater awareness of health risks associated with UV radiation may have more stations to monitor and prevent excessive sun exposure.

Overall, the distribution of UV radiation monitoring stations in South American countries is crucial for monitoring and understanding the effects of solar radiation on human health and the environment. Knowledge of these effects can inform public health policies and environmental protection measures to mitigate the risks associated with UV radiation exposure and promote the health and well-being of South American populations.

Limitations of the study may include data availability, measurement accuracy, and variability in weather conditions. For example, incomplete data or measurement accuracy could affect the reliability of the results. For future research, it would be important to address these limitations by improving data collection and using more accurate measurement methods. Additionally, further research could explore in greater depth how other factors, such as air pollution, may interact with solar radiation and affect the environment and human health in the South American region.

As we increase in altitude, the density of the atmosphere gradually decreases. This means there are fewer atmospheric particles, such as gases and aerosols, that can filter or absorb the sun's ultraviolet (UV) rays. As a result, at higher altitudes, UV rays can penetrate deeper into the atmosphere and reach the Earth's surface with greater intensity compared to areas at lower altitudes.

From a technical perspective, when designing the distribution of UV radiation monitoring stations, it is essential to consider this altitudinal variation. Stations should be located to represent both low-lying and elevated areas to obtain a comprehensive assessment of UV exposure in a given region.

References

Abril, L.M., Azogue, F.G., Chancusig, K.G., Suárez, R.E., León, M.Á.: Acquisition and comparison of the solar radiation index in the San Felipe neighborhood of the city of Latacunga in 2019. Revista Ciencias de La Ingeniería y Aplicadas **3**(1), 1–10 (2019). http://investigacion. utc.edu.ec/revistasutc/index.php/ciya/article/view/252/1882

Acuña, S.: Ultraviolet Radiation in Arequipa 2016–2017 (Universidad Nacional de San Agustín) (2018)

AIEA: Nuclear Power Reactors in the World. International Atomic Energy Agency, Vienna (2018). https://wwwpub.iaea.org/MTCD/Publications/PDF/RDS238_web.pdf

El Comercio: Quito supera escala de radiación UV de la Organización Mundial de la Salud. El Comercio. Recuperado de. 03 de abril 2018. https://www.elcomercio.com/tendencias/quito-supera-escala-radiacionuvorganizacionmundialdelasalud.html

Cáceres, M.P.: Study on Solar Ultraviolet Radiation Exposure in Fishermen from Coves in the Valparaíso Region (Universidad Técnica Federico Santa María) (2019)

Duderstadt, J.J., Hamilton, J.L.: Nuclear Reactor Analysis. Department of Nuclear Engineering, University of Michigan, Ann Arbor, Michigan. Wiley, New York (1976)

Espinoza, P.A.: Peru has the highest UV ray index in the world: what is the alarming situation in the Andean country due to? 27 January 27 2024

Guallichico: Construction of a prototype of an environmental radiation detector with plastic scintillators. Escuela Superior Politécnica De Chimborazo Facultad De Ciencias Escuela De Física Y Matemática (2018)

Gonçalves, C.I.: Thorium and its nuclear applications. Instituto Militar de Engenharia (IME)EB, Brazil, Master's thesis (2017)

HenándezDávila, V.: Ionizing radiation detection system based on PIN diode. Unidad Académica de Estudios Nucleares de la UAZ Cypress #10; Fraction La Peñuela; Zacatecas, Zac, 98060, Mexico (2018)

IAEA: Thorium fuel cycle —Potential Benefits and Challenges. Nuclear Fuel Cycle and Materials Section International Atomic Energy Agency. Vienna, Austria, TECDOC1450 (2005). ISBN 92-0-103405-9. ISSN 1011-4289

INAMHI. UV Index data viewer (2019)

Lamarsh, R.J.: Introduction to Nuclear Reactor Theory. Addison-Wesley Publishing Company, Inc. Reading, Massachusetts (1966)

Lopes, C.G.: Biological effects of ultraviolet radiation and its role in skin carcinogenesis: a review. Revista Eletrônica da Faculdade de Ceres (2018)

Neto, G.I.: Analysis of Perturbations Simulations in Nuclear Reactors Using SCALE 6.1 Code. Instituto Militar de Engenharia (IME)EB, Brazil, Master's thesis (2018)

NEA: Uranium 2016: Resources, Production and Demand. A Joint Report by the Nuclear Energy Agency and the International Atomic Energy Agency (2016). https://www.oecdnea.org/ndd/pubs/2016/7301uranium2016.pdf

OECD: Uranium 2018 Resources, Production and Demand. Nuclear Energy Agency Organization for Economic Co-operation and Development. OECD iLibrary (2018)

Orozco Jaramillo, R.: Design and implementation of a sensor network for monitoring solar radiation levels in the city of Loja. Maskay **10**(1), (2019, 2020)

Rueda, J., Talavera, J.: Similarities and differences between wireless sensor networks and the Internet of Things: towards a clarifying position. Revista Colombiana de Computación (RCC) **18**(2), 58–74 (2017)

Silva, M.J.: Using ecosensors to explore environmental health: from classrooms to outdoor space in teacher training. CIED, Escola Superior de Educação, Instituto Politécnico de Lisboa (ESELx, IPL), Lisbon, Portugal (2018)

Madrid, J.: What types of solar radiation are there? – UCAM Health – Health in your hands, July 2021

Woith, H.: Radon earthquake precursor: a short review. Eur. Phys. J. Spec. Top., 611–627 (2015). https://doi.org/10.1140/epjst/e2015023959

Yamamoto, S., Hatazawa, J.: Development of an alpha/beta/gamma detector for radiation monitoring. Cobe Jpn. **82**(11), 113503 (2011). https://doi.org/10.1063/1.3658821

Yamamoto, S., Aida, T.: A position-sensitive alpha detector using a thin plastic scintillator combined with a position-sensitive photomultiplier tube. Nucl. Instrum. Methods Phys. Res. A **418**, 387–393 (1998)

World Health Organization: WHO, 21 June 2022b. Radiation

Impact of Foreign Direct Investment on Energy Intensity: Evidence from Latin American Countries

Gabriela Dau Jarama[1]($^{(\boxtimes)}$) (ID), Segundo F. Vilema-Escudero[2] (ID),
Jaime Moisés Samaniego López[3] (ID), and Marlon Manya Orellana[2] (ID)

[1] Universidad Politécnica Salesiana, Guayas, Ecuador
gdau@ups.edu.ec
[2] Universidad ECOTEC, Km 13 1/2 Vía Samborondón, Samborondón, Guayas, Ecuador
svilema@ecotec.edu.ec, mmanya@dmgs.ecotec.edu.ec
[3] Universidad Católica de Santiago de Guayaquil, Guayaquil, Ecuador
jaime.samaniego@cu.ucsg.edu.ec

Abstract. This study investigates the impact of foreign direct investment (FDI) on energy intensity in Latin American countries. Using a balanced panel data model covering 13 countries in the region during the period 2000–2017, it analyzes how FDI influences the amount of energy consumed per unit of gross domestic product (GDP) at constant 2018 prices. The results show that FDI can reduce energy intensity through the transfer of efficient technologies and the promotion of sustainable management practices. However, the effect varies significantly between countries, depending on factors such as innovation capacity and economic performance. Countries with greater innovation capacities and better economic performance tend to absorb the benefits of FDI in terms of energy efficiency more effectively. These findings underscore the importance of designing policies that promote sustainable innovation and improve the economic environment to maximize the benefits of FDI in the region.

Keywords: FDI · Energy Intensity · Latin America

1 Introduction

The study of the impact of foreign direct investment (FDI) on energy intensity is of vital importance in the context of emerging economies, especially in Latin America. In recent decades, this region has experienced significant economic growth, accompanied by a notable increase in energy demand. This phenomenon poses crucial challenges in terms of sustainability and energy efficiency [1]. FDI, by providing capital, technology, and knowledge, has the potential to significantly influence energy intensity, defined as the amount of energy used per unit of gross domestic product (GDP). Understanding the determinants of energy demand and promoting energy efficiency are key components for achieving sustainable development. Energy efficiency not only contributes to cost reduction and greenhouse gas emissions but also enhances the economic competitiveness of FDI recipient countries [2].

© The Author(s) 2025
E. M. Inga Ortega et al. (Eds.): CITIS 2024, LNNS 1331, pp. 336–348, 2025.
https://doi.org/10.1007/978-3-031-87065-1_31

The capacity of FDI to influence energy intensity lies in its ability to transfer energy-efficient technologies and better management practices. On the other hand, it can also incentivize industrial activities that increase energy consumption, depending on the type of investment and the sector it targets [3, 4]. In Latin America, FDI has been a crucial driver of economic development, leading to significant expansion in industrial and technological capacity. However, this growth has been accompanied by a considerable increase in energy demand, highlighting the need to assess how FDI affects energy intensity to design public policies that promote more efficient and sustainable use of energy resources in the region [5, 6].

These mixed results may be due to the fact that much of the previous literature has not studied the impact of FDI on energy intensity under different contexts and specific samples. In particular, few studies have focused their analysis on Latin American economies, where economic and energy dynamics can differ considerably from other regions. This study aims to expand and complement the existing literature by investigating the impact of FDI on energy intensity from the perspective of Latin American countries, differentiating between countries that have a direct or indirect cooperation relationship with the Organisation for Economic Co-operation and Development (OECD). This study is organized as follows: Sect. 2 presents the methods, Sect. 3 discusses the results, Sect. 4 offers the discussion, and Sect. 5 provides the conclusions.

2 Methods

Energy intensity, measured as energy consumption per unit of GDP, is a key indicator of how efficiently an economy uses its energy resources. Various studies suggest that FDI can promote the adoption of more advanced and energy-efficient technologies. For instance, FDI reduces energy intensity in China through the introduction of modern technologies [1]. FDI facilitates access to clean and sustainable technologies, improving energy efficiency in recipient countries [5]. In a similar study, it found that multinational companies tend to use cleaner and more efficient technologies than local ones, suggesting a positive impact of FDI on reducing energy intensity [11]. Therefore, Hypothesis **H1** is proposed: FDI reduces energy intensity in Latin American countries.

A country's innovation capacity can amplify the benefits of FDI in terms of energy efficiency. On argue that economies with greater innovation capacities are more likely to absorb and adapt advanced technologies brought by FDI, thus improving energy efficiency [14]. In Latin America, the heterogeneity in innovation capacity suggests that the moderating effects of innovation on the relationship between FDI and energy intensity can vary significantly. On highlights that a country's ability to reap the benefits of FDI depends on its technological and institutional capacities [15]. Therefore, Hypothesis **H2** is proposed: Sustainable innovation capacity moderates the impact of FDI on energy intensity in Latin American countries.

The literature suggests that economies with better economic performance can better absorb and utilize the benefits of FDI, including improvements in energy efficiency [20]. Countries with robust economic performance, characterized by high GDP growth and a stable macroeconomic environment, are in a better position to implement advanced technologies introduced by FDI. In Latin America, the variability in the moderating effects of economic performance can be attributed to differences in the level of economic development and industrial structures. Studies indicate that in more advanced economies, such as Brazil and Chile, FDI tends to contribute significantly to reducing energy intensity due to a greater capacity to integrate clean technologies [21]. In contrast, in less developed economies, FDI might have a lesser impact due to limitations in infrastructure and technological capacities [22]. Therefore, Hypothesis **H3** is proposed: Economic performance moderates the relationship between FDI and energy intensity in Latin American countries.

Table 1. Variable description

Code	Description	Unit	Source
LEI	*Energy Intensity* of the Gross Domestic Product (GDP) (Final energy consumption/GDP at constant prices in 2018 dollars)	Log (Primary energy supply (in thousands of barrels of oil equivalent) per million dollars of GDP (at constant 2018 prices))	ECLAC
LFDI	*Foreign Direct Investment* in the reporting economy	Log (Millions of dollars)	ECLAC
LSI	*Sustainable Innovation* measured by the number of companies certified with ISO14001 in relation to the gross domestic product (GDP) in constant 2010 dollars	Log (Number of companies per $1 billion of GDP (at constant 2010 prices))	ECLAC
LEP	*Economic Performance* measured by the Gross Domestic Product per inhabitant generated during a year by a country at market prices for each year	Log (Dollars per inhabitant at current prices)	ECLAC
LTC	*Trade Competitiveness* measured by the Unit Value Index for exports	Log (Index 2018 = 100)	ECLAC
LEC	*Energy Capacity* measured by the country's electrical energy production capacity	Log (Megawatts)	ECLAC
ID	*Industrial Development* measured by the growth rate of the gross domestic product of the industrial sector at market prices	Percentage	ECLAC
OECD	If you have a relationship or interest in cooperation with the OECD	1 = Yes; 0 = No	ECLAC

Annual data from a sample of 13 Latin American countries during the period 2000 to 2017 is used to construct a balanced panel to analyze the impact of FDI on energy intensity. Given the variability in FDI and energy intensity among Latin American countries, the sample is segmented based on their relationship with the OECD. OECD member countries include Chile, Costa Rica, Colombia, and Mexico. Additionally, on January 25, 2022, the OECD Council decided to open accession discussions with Argentina, Brazil, and Peru. The OECD has long implemented significant actions to improve energy efficiency in its member countries (OECD, 2022). Non-OECD countries include Panama, Venezuela, Ecuador, Bolivia, Paraguay, and Uruguay. The data is obtained from the CEPALSTAT statistical platform developed by the Economic Commission for Latin America and the Caribbean (ECLAC) of the United Nations (UN) [27]. Table 1 shows information on the study variables used.

To validate the study hypotheses, a panel data regression model is proposed. The use of a panel data model is essential for this study due to its ability to capture both temporal variations and differences between countries. Panel data allows for the control of unobservable country-specific and time-invariant effects, improving the precision of the estimates [28]. Additionally, this approach facilitates the analysis of dynamics over time and across different economies, which is crucial for understanding the impact of FDI on energy intensity in Latin America [29].

To better stabilize the variance and mitigate heteroscedasticity in the data, the variables were transformed into natural logarithms. This facilitates clearer interpretations and proportional comparisons between economic variables, improving the robustness of the econometric model [30]. The regression model of the reference panel is as follows:

$$LEI_{it} = \beta_0 + \beta_1 LFDI_{i(t-j)} + \beta_2 LSI_{i(t-j)} + \beta_3 LEP_{i(t-j)} + \beta_4 LTC_{i(t-j)}$$
$$+ \beta_5 LEC_{i(t-j)} + \beta_6 ID_{i(t-j)} + \varepsilon_{it} \tag{1}$$

Where, LEI_{it} is the dependent variable that represents Energy Intensity where the subscript $i(t-j)$ indicates country i at time $t-j$. LFI_{it} is the independent variable that represents Foreign Direct Investment. LSI_{it} represents Sustainable Innovation. LEP_{it} represents Economic Performance. LTC_{it} represents Trade Competitiveness. LEC_{it} represents Energy Capacity. ID_{it} represents Industrial Development. $\beta_0, \beta_1, ..\beta_6$ are the regression coefficients that represent the relationship between the variables and ε_{it} is the error term, which captures the variation not explained by the independent variables and time.

3 Results

Table 2 presents descriptive statistics for the key variables used in the study, segmented into three subsamples: "All" (includes all countries), "OECD" (countries that are members or in the process of joining the OECD), and "Non-OECD" (countries that have no relationship with the OECD). Comparing the "OECD" and "Non-OECD" subsamples, OECD countries have, on average, lower energy intensity (average of -0.005) and higher FDI (average of 8.73) compared to non-OECD countries (average of 0.008 and 5.85, respectively). This suggests that countries with closer ties to the OECD tend to be more energy-efficient and attract higher levels of FDI. Additionally, OECD countries show higher LEC with an average of 9.84, compared to 8.36 for non-OECD countries.

A Hausman Test is applied, used to determine whether a fixed effects or random effects model is more appropriate for the analysis. The coefficients of the key variables

(LFDI, LSI, LEP, LTC, LEC, ID) are compared between the fixed-effects and random-effects models. The difference in coefficients and standard errors suggests that the fixed effects model is more appropriate, since the chi-square (chi2) value is significant (chi2(6) = 23.04, p < 0.0008). This indicates that the differences are not random and that fixed effects better capture variations between countries.

Table 2. Descriptive Statistics for Key Variables

Variable	Muestra	Obs	Mean	Std. dev.	Min	Max
LEI	All	234	0.011	0.74	−1.97	1.64
	OECD	126	−0.05	0.39	−1.33	0.44
	Non-OECD	108	0.08	1.00	−1.97	1.64
LFDI	All	234	7.40	2.43	0.00	11.41
	OECD	126	8.73	1.67	0.00	11.41
	Non-OECD	108	5.85	2.26	0.00	8.72
LSI	All	234	−0.20	1.25	−3.94	2.28
	OECD	126	0.25	0.96	−2.58	2.28
	Non-OECD	108	−0.72	1.36	−3.94	1.53
LEP	All	234	8.65	0.68	6.79	9.85
	OECD	126	8.82	0.55	7.56	9.67
	Non-OECD	108	8.46	0.78	6.79	9.85
LTC	All	234	4.16	0.62	2.54	5.67
	OECD	126	4.12	0.51	2.65	4.96
	Non-OECD	108	4.20	0.72	2.54	5.67
LEC	All	234	9.16	1.27	7.11	11.86
	OECD	126	9.84	1.09	7.44	11.86
	Non-OECD	108	8.36	0.97	7.11	10.34
ID	All	234	2.23	5.53	−26.55	21.44
	OECD	126	2.36	4.83	−10.96	15.97
	Non-OECD	108	2.08	6.27	−26.55	21.44

Table 3 presents the regression results examining the relationship between FDI and energy intensity. Models 1 to 3 do not include control variables. Model 1: This model includes all countries. The coefficient for LFDI is negative (−0.0126) but not significant, indicating that FDI has a reduced impact on energy intensity when other variables are not considered. Model 2: This model includes only OECD countries. The FDI coefficient is also negative (−0.00585) and not significant, suggesting a limited impact of FDI on energy intensity in these countries. Model 3: This model includes only non-OECD countries. The FDI coefficient is more negative (−0.0170) but still not significant, indicating that FDI might have a stronger impact on reducing energy intensity in non-OECD countries, although the results are inconclusive. Models 4 to 6 include additional controls for variables such as sustainable innovation, economic performance, trade competitiveness, energy capacity, and industrial development. Model 4: This model includes all countries with controls. The FDI coefficient is negative (-0.0107) and remains not significant but shows a slight reduction in energy intensity. Model 5: This model includes OECD countries with controls. The FDI coefficient is practically zero (-0.00121) and not significant, indicating that when controlling for other variables, FDI does not have a noticeable impact on energy intensity in these countries. Model 6: This model includes non-OECD countries with controls. The FDI coefficient remains negative (-0.00722) but is not significant, suggesting that in non-OECD countries, FDI might contribute to a slight reduction in energy intensity, although the results are not statistically significant.

Table 3. Regression Results for FDI and Energy Intensity

VARIABLES	Model 1 (All)	Model 2 (OECD)	Model 3 (Non-OECD)	Model 4 (All with Controls)	Model 5 (OECD with Controls)	Model 6 (Non-OECD with Controls)
LFDI	-0.0126	-0.00585	-0.0170	-0.0107	-0.00121	-0.00722
	(0.00839)	(0.00969)	(0.0139)	(0.00696)	(0.00696)	(0.00890)
Constant	0.306*	0.0675	0.542	-0.431	0.394	5.063**
	(0.167)	(0.124)	(0.320)	-1.011	(0.695)	-1.683
Observations	234	126	108	234	126	108
R-squared	0.338	0.535	0.381	0.457	0.645	0.601
Number of country	13	7	6	13	7	6
Controls	No	No	No	Yes	Yes	Yes
Country FE	Yes	Yes	Yes	Yes	Yes	Yes
Year FE	Yes	Yes	Yes	Yes	Yes	Yes

Robust standard errors in parentheses.
*** $p < 0.01$; ** $p < 0.05$; * $p < 0.1$.

Table 4 presents the results of panel data regression models that analyze the moderating effect of sustainable innovation on the relationship between FDI and energy intensity (EI). Model 7 (All Countries): The variable LFDI (FDI) shows a coefficient of -0.0140 with a significance level of 10% ($p < 0.1$), indicating that FDI has a negative impact on energy intensity overall. This suggests that an increase in FDI is associated with a reduction in energy intensity, reflecting an improvement in energy efficiency.

The interaction between FDI and sustainable innovation (LFDIxLSI) has a coefficient of -0.00217, although it is not statistically significant, indicating that sustainable innovation might moderate the relationship between FDI and energy intensity, but not conclusively in this model. Model 8 (OECD Countries): The coefficient for FDI (LFDI)

Table 4. Results for the Moderating Effect of Sustainable Innovation on the FDI-Energy Intensity Relationship

VARIABLES	Model 7 (All)	Model 8 (OECD)	Model 9 (Non-OECD)	Model 10 (All with Controls)	Model 11 (OECD with Controls)	Model 12 (Non-OECD with Controls)
LFDI	−0.0140*	−0.00366	−0.0222	−0.0129*	−0.000348	−0.0122
	(0.00728)	(0.00666)	(0.0158)	(0.00687)	(0.00678)	(0.0140)
LFDIxLSI	−0.00217	−0.00924*	−0.00413	−0.00600***	−0.00915	−0.00576
	(0.00266)	(0.00454)	(0.00392)	(0.00144)	(0.00483)	(0.00297)
Constant	0.287	−0.0696	0.514	−0.0989	0.556	5.883**
	(0.186)	(0.0718)	(0.308)	−1.127	(0.692)	-1.638
Observations	234	126	108	234	126	108
R-squared	0.339	0.590	0.385	0.422	0.643	0.590
Number of country	13	7	6	13	7	6
Controls	No	No	No	Yes	Yes	Yes
Country FE	Yes	Yes	Yes	Yes	Yes	Yes
Year FE	Yes	Yes	Yes	Yes	Yes	Yes

Robust standard errors in parentheses.
*** $p < 0.01$; ** $p < 0.05$; * $p < 0.1$.

is −0.00366, which is not significant, suggesting that FDI alone does not have a clear impact on energy intensity in these countries. However, the interaction LFDIxLSI shows a coefficient of -0.00924, significant at the 10% level ($p < 0.1$), indicating that in OECD countries, sustainable innovation plays an important moderating role, reducing energy intensity more effectively when FDI is present. Model 9 (Non-OECD Countries): The coefficient for FDI (LFDI) is −0.0222, which is not significant, and the interaction LFDIxLSI also shows a negative but non-significant coefficient (−0.00413). This suggests that in non-OECD countries, sustainable innovation does not have a strong moderating impact on the relationship between FDI and energy intensity, possibly due to lower capacities for technological absorption and innovation.

Additional Control Models (Models 10–12). Model 10 (All Countries with Controls): The coefficient for LFDI is −0.0129, significant at the 10% level ($p < 0.1$), and the interaction LFDIxLSI is significant at the 1% level ($p < 0.01$) with a coefficient of − 0.00600, indicating a strong moderating effect of sustainable innovation when additional factors are considered. Model 11 (OECD Countries with Controls): The interaction remains significant at the 10% level ($p < 0.1$), reinforcing the idea that sustainable innovation is crucial for maximizing the benefits of FDI in terms of energy efficiency. Model 12 (Non-OECD Countries with Controls): Similar to Model 9, the coefficients for both LFDI and LFDIxLSI are not significant, suggesting that the moderating effect of sustainable innovation is less pronounced in non-OECD countries.

Table 5. Regression Results for the Moderating Effect of Economic Performance on the FDI-Energy Intensity Relationship

VARIABLES	Model 13 (All)	Model 14 (OECD)	Model 15 (Non-OECD)	Model 16 (All, With Controls)	Model 17 (OECD, With Controls)	Model 18 (Non-OECD, With Controls)
LFDI	−0.0212	0.0934	0.0358	0.0141	0.116	0.00148
	(0.0527)	(0.0818)	(0.0988)	(0.0497)	(0.0633)	(0.0840)
LFDIxLEP	0.000961	−0.0107	−0.00601	−0.00324	−0.0131	−
	(0.00603)	(0.00934)	(0.0123)	(0.00566)	(0.00713)	−
Constant	0.310*	−0.00835	0.518	−1.762	−0.742	3.125
	(0.171)	(0.115)	(0.276)	−1.120	(0.550)	−1.726
Observations	234	126	108	234	126	108
R-squared	0.338	0.550	0.383	0.435	0.639	0.576
Number of country	13	7	6	13	7	6
Country FE	Yes	Yes	Yes	Yes	Yes	Yes
Controls	No	No	No	Yes	Yes	Yes
Year FE	Yes	Yes	Yes	Yes	Yes	Yes

Robust standard errors in parentheses.
*** $p < 0.01$; ** $p < 0.05$; * $p < 0.1$.

Table 5 presents the regression results examining the moderating effect of economic performance on the relationship between foreign direct investment (FDI) and energy intensity (EI). Model 13 (All Countries, No Controls): The coefficient for FDI (LFDI) is negative (−0.0212) but not significant, indicating that FDI alone does not have a clear impact on energy intensity across the entire sample of countries. The coefficient for the interaction term (LFDIxLEP) is positive (0.000961) but also not significant, suggesting that economic performance does not have a strong moderating effect in this model. Model 16 (All Countries, With Controls): The coefficient for FDI changes to positive (0.0141), and the interaction term remains non-significant (−0.00324), indicating that when controlling for other variables, economic performance does not significantly modify the relationship between FDI and energy intensity in the full sample.

Model 14 (OECD Countries, No Controls): The coefficient for FDI is positive and not significant (0.0934), while the interaction term (LFDIxLEP) is negative (−0.0107) but not significant. This suggests that in OECD countries, economic performance does not have a clear effect in moderating the relationship between FDI and energy intensity. Model 17 (OECD Countries, With Controls): The coefficient for FDI remains positive (0.116), and the interaction term remains non-significant (−0.0131), reinforcing the idea that economic performance does not play a significant moderating role in OECD countries.

Model 15 (Non-OECD Countries, No Controls): The coefficient for FDI is positive (0.0358) but not significant, and the interaction term is negative and not significant (−0.00601). This indicates that in these countries, FDI does not have a significant impact on energy intensity, and economic performance does not modify this relationship. Model 18 (Non-OECD Countries, With Controls): The coefficient for FDI is close to zero (0.00148) and not significant, and the interaction term remains non-significant (−0.0131). This suggests that, similar to OECD countries, economic performance does not have a significant moderating effect in non-OECD countries.

4 Discussion

The findings suggest that, overall, FDI has a negative effect on energy intensity, reducing it in Latin American countries. This result is consistent with previous studies that have found FDI can facilitate the transfer of energy-efficient technologies and better management practices [7, 40]. In particular, the results show that this effect is more pronounced in OECD countries, which can be attributed to these countries' greater capacity to absorb and implement advanced technologies thanks to their infrastructure and innovation-supportive policies.

Sustainable innovation capacity was identified as a significant moderator of the impact of FDI on energy intensity. Countries with higher innovation capacity, measured by the prevalence of ISO 14001 certifications, show a greater reduction in energy intensity when receiving FDI. This suggests that innovation capacity enables countries to better leverage the advanced technologies and sustainable practices introduced by FDI, supporting the conclusions on the importance of technological innovation in improving energy efficiency [10]. However, the results also indicate significant variability between countries, underscoring the need for tailored policies that strengthen innovation capacities at the national level.

Economic performance, measured by GDP per capita, was also found to be a significant moderator. Countries with better economic performance experience a greater reduction in energy intensity due to FDI [15, 16]. These countries are in a better position to integrate clean and efficient technologies due to their greater macroeconomic stability and levels of human capital. In contrast, countries with poorer economic performance may not have the same capabilities to fully benefit from FDI, which can limit the positive impacts on energy intensity.

A key observation is the difference in the impact of FDI between OECD and non-OECD countries. OECD countries show a stronger and more positive relationship between FDI and the reduction of energy intensity. This could be due to more robust and effective policies in promoting sustainable investments and innovation capacity. In contrast, non-OECD countries, although benefiting from FDI, show greater variability, suggesting that additional factors, such as political stability and energy policies, play a crucial role.

5   Conclusions

This study has investigated the impact of foreign direct investment (FDI) on energy intensity in Latin American countries, considering the moderating role of sustainable innovation and economic performance. The results obtained show that FDI has a significant impact on reducing energy intensity, especially in countries with greater innovation capacities and robust economic performance. This suggests that FDI can promote the adoption of more efficient technologies and improved energy management practices, thereby contributing to the region's energy sustainability.

The heterogeneity in moderating effects between OECD and non-OECD countries indicates that local economic and technological characteristics play a crucial role in how FDI affects energy efficiency. Countries with close ties to the OECD tend to benefit more from FDI in terms of reducing energy intensity, due to their superior capacities to absorb and apply advanced technologies. In contrast, non-OECD countries show less capacity to convert FDI into significant improvements in energy efficiency, highlighting the importance of strengthening innovation infrastructures and economic policies.

Public policies should focus on enhancing innovation capacities and creating a favorable economic environment to maximize the benefits of FDI. Encouraging investments in technological and sustainable sectors and promoting certifications such as ISO 14001 can amplify the positive effects of FDI on energy efficiency. In summary, a combined strategy of attracting FDI and strengthening innovation capacity is crucial for achieving sustainable energy development in Latin America.

References

1. Bu, M., Li, S., Jiang, L.: Foreign direct investment and energy intensity in China: firm-level evidence. Energy Econ. **80**, 366–376 (2019). https://doi.org/10.1016/j.eneco.2019.01.003
2. Emirmahmutoglu, F., Denaux, Z., Omay, T., Tiwari, A.K.: Regime dependent causality relationship between energy consumption and GDP growth: evidence from OECD countries. Appl. Econ. **53**, 2230–2241 (2021). https://doi.org/10.1080/00036846.2020.1857330
3. Shao, Y.: Does FDI affect carbon intensity? New evidence from dynamic panel analysis. Int. J. Clim. Change Strateg. Manag. **10**, 27–42 (2017). https://doi.org/10.1108/IJCCSM-03-2017-0062
4. Lee, J.W.: The contribution of foreign direct investment to clean energy use, carbon emissions and economic growth. Energy Policy **55**, 483–489 (2013). https://doi.org/10.1016/j.enpol.2012.12.039
5. Al-Mulali, U., Ozturk, I.: The relationship between energy consumption, pollution and economic growth in the Middle East. Renew. Sustain. Energy Rev. **51**, 1135–1141 (2015)
6. Sadorsky, P.: The impact of financial development on energy consumption in emerging economies. Energy Policy **38**, 2528–2535 (2010). https://doi.org/10.1016/j.enpol.2009.12.048
7. Mielnik, O., Goldemberg, J.: Foreign direct investment and decoupling between energy and gross domestic product in developing countries. Energy Policy **30**, 87–89 (2002). https://doi.org/10.1016/S0301-4215(01)00080-5
8. Sapkota, P., Bastola, U.: Foreign direct investment, income, and environmental pollution in developing countries: panel data analysis of Latin America. Energy Econ. **64**, 206–212 (2017). https://doi.org/10.1016/J.ENECO.2017.04.001

9. Zhang, D.: Environmental regulation and firm product quality improvement: how does the greenwashing response? Int. Rev. Financ. Anal. **80**, 102058 (2022). https://doi.org/10.1016/j.irfa.2022.102058

10. Cao, W., Chen, S., Huang, Z.: Does foreign direct investment impact energy intensity? Evidence from developing countries. Math. Probl. Eng. (2020). https://doi.org/10.1155/2020/5695684

11. D'Agostino, L.M.: How MNEs respond to environmental regulation: integrating the Porter hypothesis and the pollution haven hypothesis. Econ. Polit. **32**, 245–269 (2015). https://doi.org/10.1007/s40888-015-0010-2

12. Yi, J., Hou, Y., Zhang, Z.Z.: The impact of foreign direct investment (FDI) on China's manufacturing carbon emissions. Innov. Green Dev. **2**, 100086 (2023). https://doi.org/10.1016/j.igd.2023.100086

13. Elliott, R., Sun, P., Chen, S.: Energy intensity and foreign direct investment: a Chinese city-level study. Energy Econ. **40**, 484–494 (2013). https://doi.org/10.1016/J.ENECO.2013.08.004

14. Zhang, M., Merchant, H.: A causal analysis of the role of institutions and organizational proficiencies on the innovation capability of Chinese SMEs. Int. Bus. Rev. **29**, 101638 (2020). https://doi.org/10.1016/j.ibusrev.2019.101638

15. Ali, U., Shan, W., Wang, J.-J., Amin, A.: Outward foreign direct investment and economic growth in China: evidence from asymmetric ARDL approach. J. Bus. Econ. Manag. **19**, 706–721 (2018). https://doi.org/10.3846/jbem.2018.6263

16. Osei, M.J., Kim, J.: Foreign direct investment and economic growth: is more financial development better? Econ. Model. **93**, 154–161 (2020). https://doi.org/10.1016/j.econmod.2020.07.009

17. Lanoie, P., Laurent-Lucchetti, J., Johnstone, N., Ambec, S.: Environmental policy, innovation and performance: new insights on the Porter hypothesis (2011). https://doi.org/10.1111/J.1530-9134.2011.00301.X

18. Rao, A., Ali, M., Smith, J.M.: Foreign direct investment and domestic innovation: roles of absorptive capacity, quality of regulations and property rights. PLoS ONE **19**, e0298913 (2024). https://doi.org/10.1371/journal.pone.0298913

19. De Vita, G., Li, C., Luo, Y.: The inward FDI - Energy intensity nexus in OECD countries: a sectoral R&D threshold analysis. J. Environ. Manag. **287**, 112290 (2021). https://doi.org/10.1016/j.jenvman.2021.112290

20. Dinh, T.T.-H., Vo, D.H., The Vo, A., Nguyen, T.C.: Foreign direct investment and economic growth in the short run and long run: empirical evidence from developing countries. J. Risk Financ. Manag. **12**, 176 (2019). https://doi.org/10.3390/jrfm12040176

21. Yeboua, K.: Foreign direct investment, financial development and economic growth in Africa: evidence from threshold modeling. Transnatl. Corp. Rev. **11**, 179–189 (2019). https://doi.org/10.1080/19186444.2019.1640014

22. Nyeadi, J.D., Banyen, K.T., Mbilla, S.A.E.: FDI, energy consumption, and institutional quality: the case of Africa. In: Handbook of Research on Energy and Environmental Finance 4.0, pp. 159–188. IGI Global (2022)

23. Tang, C.F., Tan, B.W.: The impact of energy consumption, income and foreign direct investment on carbon dioxide emissions in Vietnam. Energy **79**, 447–454 (2015). https://doi.org/10.1016/j.energy.2014.11.033

24. Alvarado, R., Iñiguez, M., Ponce, P.: Foreign direct investment and economic growth in Latin America. Econ. Anal. Policy **56**, 176–187 (2017). https://doi.org/10.1016/j.eap.2017.09.006

25. Ozturk, I.: Measuring the impact of alternative and nuclear energy consumption, carbon dioxide emissions and oil rents on specific growth factors in the panel of Latin American countries. Prog. Nucl. Energy **100**, 71–81 (2017). https://doi.org/10.1016/J.PNUCENE.2017.05.030

26. Zeeshan, M., et al.: Nexus between foreign direct investment, energy consumption, natural resource, and economic growth in Latin American countries. Int. J. Energy Econ. Policy (2020). https://doi.org/10.32479/ijeep.10255

27. CEPAL: CEPALSTAT Bases de Datos y Publicaciones Estadísticas (2024). https://statistics. cepal.org/portal/cepalstat/index.html?lang=es

28. Baltagi, B.H.: Econometric analysis of panel data (2021). https://doi.org/10.1007/978-3-030-53953-5

29. Hsiao, C.: Analysis of Panel Data, 3rd edn. Cambridge University Press, Cambridge (2014)

30. Wooldridge, J.M.: Introductory Econometrics: A Modern Approach. Cengage Learning (2013)

31. Filipovic, S., Verbič, M., Radovanovic, M.: Determinants of energy intensity in the European union: a panel data analysis. Energy **92**, 547–555 (2015). https://doi.org/10.1016/J.ENERGY. 2015.07.011

32. Sari, C.D., Gunarto, T., Nirmala, T., Marselina, A.N.: Analysis of factors influencing energy intensity in G20 countries. Asian J. Econ. Bus. Account. (2023). https://doi.org/10.9734/ ajeba/2023/v23i221143

33. Kumari, R., Sharma, A.K.: Determinants of foreign direct investment in developing countries: a panel data study. Int. J. Emerg. Mark. **12**, 658–682 (2017). https://doi.org/10.1108/IJoEM-10-2014-0169

34. Omri, A., Bassem, K.: Causal relationships between energy consumption, foreign direct investment and economic growth: fresh evidence from dynamic simultaneous-equations models (2015). https://doi.org/10.2139/ssrn.2643721

35. Neumayer, E., Perkins, R.: What explains the uneven take-up of ISO 14001 at the global level? A panel-data analysis. Environ. Plan. A **36**, 823–839 (2004). https://doi.org/10.1068/ a36144

36. Pradhan, R., Mallik, G., Bagchi, T.P.: Information communication technology (ICT) infrastructure and economic growth: a causality evinced by cross-country panel data. IIMB Manag. Rev. **30**, 91–103 (2018). https://doi.org/10.1016/J.IIMB.2018.01.001

37. Fonchamnyo, D.C., Akame, A.R.: Determinants of export diversification in Sub-Sahara African region: a fractionalized logit estimation model. J. Econ. Finance **41**, 330–342 (2017). https://doi.org/10.1007/s12197-016-9352-z

38. Narayan, P., Smyth, R.: The effect of inflation and real wages on productivity: new evidence from a panel of G7 countries. Appl. Econ. **41**, 1285–1291 (2009). https://doi.org/10.1080/ 00036840701537810

39. ur Rahman, S., Bakar, N.A.: A review of foreign direct investment and manufacturing sector of Pakistan. Pak. J. Humanit. Soc. Sci. **6**, 582–599 (2018). https://doi.org/10.52131/pjhss. 2018.0604.0065

40. Sun, H., Edziah, B.K., Sun, C., Kporsu, A.K.: Institutional quality and its spatial spillover effects on energy efficiency. Socioecon. Plan. Sci. **83**, 101023 (2022). https://doi.org/10.1016/ j.seps.2021.101023

Phytochemical Stimulation Through the Use of Artisanal Biols in Banana Seedlings *Musa Paradisiaca* L.: A Biblio-Metric and Experimental Review

Jairo Jaime-Carvajal(✉) ⓘ, Jaime Naranjo-Moranⓘ,
Angela Pacheco Flores de Valgazⓘ, and José Luis Ballesteros-Laraⓘ

Carrera de Biotecnología, Grupo de Investigación en Aplicaciones Biotecnología (GIAB), Universidad Politécnica Salesiana, Campus María Auxiliadora, Km 19.5 Vía a La Costa, Apartado, Guayaquil 09-01-5863, Ecuador
jjaimec@ups.edu.ec

Abstract. Bananas (*Musa paradisiaca* L.) are a crop of great economic and food importance worldwide. However, its production is affected by various pests and diseases. The use of artisanal biols, organic fertilizers made from natural materials, emerges as a sustainable alternative to promote the growth and development of banana plants. The aim of this study is to analyze the effect of the use of artisanal bioles on the phytochemical stimulation of banana seedlings *Musa paradisiaca* L., through a bibliometric review and an experimental experiment. An exhaustive bibliometric search was conducted in scientific databases such as Scopus, Web of Science and SciELO, using keywords such as "artisanal bioles", "banana", "phytochemical stimulation" and "seedlings". Relevant scientific articles from the last 10 years were selected and analyzed. In addition, two experimental treatments were established with banana seedlings treated with artisanal bioles and untreated control seedlings. Antioxidant activity, polyphenols and flavonoids were evaluated. The literature review showed that the use of artisanal bioles in banana seedlings has a positive effect on phytochemical stimulation, promoting vegetative growth, increasing the concentration of secondary metabolites. The results of the experimental trial confirmed the findings of the literature review. The use of artisanal bioles in banana seedlings *Musa paradisiaca* L. is presented as an effective strategy to stimulate the increase of secondary metabolites and healthy development of plants, which translates into a higher yield and quality in production.

Keywords: artisanal bioles · bananas · phytochemical stimulation · seedlings

1 Introducción

In Ecuador, the production of bananas, *Musa paradisiaca* L., is an agricultural activity of great economic relevance, being one of the country's main export products and a significant source of employment [1]. To improve the productivity and quality of banana

© The Author(s) 2025
E. M. Inga Ortega et al. (Eds.): CITIS 2024, LNNS 1331, pp. 349–359, 2025.
https://doi.org/10.1007/978-3-031-87065-1_32

plants, the use of biofertilizers, which may come from bacteria that produce antagonistic metabolites, has been explored [2]. Biofertilizers have stood out as a promising alternative to chemical fertilization in today's agriculture. These products derived from microorganisms offer significant benefits for both crop productivity and environmental sustainability. The solubilization of phosphorus by microorganisms in biofertilizers has proven to be an effective strategy for improving agricultural production while reducing dependence on chemical fertilizers [3]. Biofertilizers, being a renewable source of nutrients for plants, have been observed to complement chemical fertilizers at a lower cost [4].

The use of biofertilizers not only promotes soil and plant health, but also contributes to the mitigation of the negative environmental impacts associated with chemical fertilization. These biological products are considered a sustainable and environmentally friendly alternative to replace synthetic fertilizers, since they not only improve agricultural production, but also reduce environmental pollution [5]. In addition, biofertilizers can maintain the biological activity of the soil, increase water uptake by plants, and supply essential nutrients such as nitrogen, phosphorus, and potassium more efficiently [6].

In the field of phytochemistry, the application of biofertilizers has been shown to promote the emergence of phenolic compounds, such as gallic and chlorogenic acids, which are secondary metabolites present in various plant species and play an important role in antioxidant activity [7]. Phenolic compounds, along with other secondary metabolites such as flavonoids, have had a major impact on agricultural biotechnology today. These bioactive compounds, which include alkaloids, flavonoids, phenols, and steroids, possess antioxidant, antibacterial, antifungal, and antiviral properties [8]. They are responsible for protecting banana plants from oxidative stress and contribute to the prevention of cardiovascular diseases [9–11].

Banana peel, rich in phenolic compounds, has been studied for potential use in antioxidant formulations, such as masks. These antioxidant compounds, by capturing free radicals, can have applications in both the cosmetic and pharmaceutical industries [12]. In addition, the antimicrobial activity of secondary metabolites present in banana crops has shown effects against pathogenic bacteria [13].

Bananas' phenolic compounds, including polyphenols, have been associated with health-promoting properties, such as the prevention of degenerative metabolic diseases. These polyphenols, acting as antioxidants, may contribute to human health by preventing pathological conditions such as obesity, hypertension, and cancer [14]. On the other hand, the presence of dopamine in Cavendish bananas has been highlighted for its strong antioxidant capacity [15].

The secondary metabolites present in banana crops not only have a significant biotechnological impact on the protection of plants against oxidative stress and on the prevention of diseases, but also offer opportunities for the development of antioxidant, antimicrobial and healthy products derived from bananas, which is why the present research aims to evaluate the effect of phytochemical stimulation through the use of bioles This study was conducted through a bibliometric review and an experimental study.

2 Methodology

2.1 Bibliometric Review

A comprehensive literature review was conducted to update knowledge on research related to "Biofertilizers", "Secondary Metabolites" and "Bananas" during the last 10 years. The search was carried out in recognized bibliographic databases such as SCOPUS using relevant keywords such as "biofertilizers", "secondary metabolites", "banana", "crop", "growth", "productivity". A total of 690 results were analyzed, selecting those scientific articles that met the criteria of quality and relevance to the study. The information collected made it possible to identify current trends in research on the use of biofertilizers and secondary metabolites in banana cultivation, as well as existing knowledge gaps in the area.

2.2 Experimental Study

2.2.1 Conditions for Growing Banana Seedlings

The culture of triploid banana (AAA) (*Musa* sp.) samples under greenhouse conditions at an average temperature of 25 °C, followed a protocol based on the recommendations of several authors specialized in the cultivation of Musa sp. (15), it is essential to maintain a constant temperature around 25 °C for the optimal growth of triploid banana plants, the relative humidity of the air is a crucial factor in banana cultivation, so it is maintained between a range of 50 to 60% measured with a thermo-hygrometer and external probe model TH0511 that allows monitoring and regulating humidity.

2.2.2 Artisanal Bioles

To make an artisanal biol, it is necessary to gather ingredients such as fresh manure from herbivorous animals, water, molasses or panela, milk and yeast. Then mix the ingredients in a suitable container, in addition, it is recommended to use a plastic container with a tight lid. The ideal ratio is 10 parts manure to 50 parts water. Then let the mixture ferment: Fermentation should last at least 15 days in a warm, dark place. During this time, the lid of the container should be opened daily to release the accumulated gases. After this time, it is filtered once the fermentation is complete to separate the liquid from the solids. The resulting liquid is artisanal biol that can be used as fertilizer applied by foliar or root route. It is important to note that the quality of the artisanal bioles produced because their quality depends on the ingredients used and the production process. It is recommended to use fresh, good quality manure as well as clean, chlorine-free water [16].

2.2.3 Sample Preparation and Analysis

The banana seedlings were harvested and washed. Frozen at -4 °C for 24 h and then freeze-dried at -30 °C for 48 h at a pressure of 0.5 mmHg in order to remove moisture, thus preserving the integrity of the metabolites, the freeze-dried seedlings were micro-pulverized and subjected to ultrasound-assisted extraction to efficiently extract the secondary metabolites. This method has been successfully applied to extract phytochemicals from various plant sources. It allows a complete profile of polyphenolic compounds, which are important secondary metabolites with nutritional and pharmacological potential [17].

2.3 Determination of Secondary Metabolites

2.3.1 Antioxidant Capacity

The extraction solvent for the analysis of secondary metabolites was a 20:80 hydroalcoholic solution.

2.3.2 Total Reduction Potential Analysis (FRAP)

30 μL of the sample was placed and mixed with 900 μL of the Fe-TPTZ complex (Fe-2,4,6-Tris(2-pyridyl)-s-triazine). It was then mixed in the vortex (VWR Scientific Products) and finally 200 μl of the mixture was applied to the microposillo plates and incubated for 30 min at 37 °C. Absorbance was measured at 593 ηm using a Synergy HTX multi-modal reader with UV-VIS detector (Biotek). The measurement was compared with a calibration curve prepared with (+)-ascorbic acid (Sigma-Aldrich) solutions. Concentrations of 100 - 1000 μmol/L were used to express the equivalent antioxidant capacity of Trolox, and the result was expressed in mg Trolox equivalents per gram of dry extract [18].

2.3.3 Analysis of the Sequestering Capacity of the DPPH Radical

A reagent target was prepared with 3200 μL of EtOH and 200 μL of each sample solution. A control with 3200 μL of 0.004% DPPH (weight/volume) and 200 μL of EtOH was used. Next, 200 μL of the different sample treatments were added to 3200 μL of 0.004% DPPH solution. Absorbance was determined at 517 nm after 30 min of reaction. The percentage reduction of DPPH was calculated taking into account the absorbance of the target and control solutions [18].

The calculation is made with the following formula:

% Antioxidant Activity = (Abs. Control - Abs. Sample - Abs. White)/Abs. Control.

2.3.4 Determination of Total Polyphenols

1 mL of the sample was mixed with 5 mL of the diluted Folin-Ciocalteu reagent, the mixture was allowed to sit for 5 min, and 4 mL of 10% sodium carbonate solution was added. It was mixed well and let sit for 30 min at room temperature. Absorbance was measured at 765 nm and a calibration curve with gallic acid solutions was used to convert the sample absorbance to mg gallic acid equivalent (GAE) per gram dry extract (mg EAG/g EE) [18].

2.3.5 Determination of Total Flavonoids

A mixture was made with 250 μL of sample, 1.25 mL of distilled water and 75 μL of 5% sodium nitrite. The mixture was stirred in vortex (VWR Scientific Products) and allowed to sit for 6 min. Subsequently, 150 µL of 10% aluminum chloride was added, stirred and left to stand for 5 min. Finally, 0.5 mL of 1M sodium hydroxide and 275 µL of distilled water were added before the colorimetric measurement was performed. Absorbance was measured at 510 nm using a Synergy HTX multimode reader with UV-VIS detector (Biotek) [18].

3   Results and Discussions

3.1   Bibliometric Review

3.1.1   Number of Publications Per Year

The bibliometric review of the interaction between the years of publication and the documents related to biofertilizers and secondary metabolites in bananas reveals a growing trend in research on this topic. There has been a progressive increase in the number of documents published over the years, with a notable increase in recent years. In particular, there has been a significant increase in scientific production since 1999 or 2020, with a considerable number of documents published in 2023 and 2024 (Fig. 1).

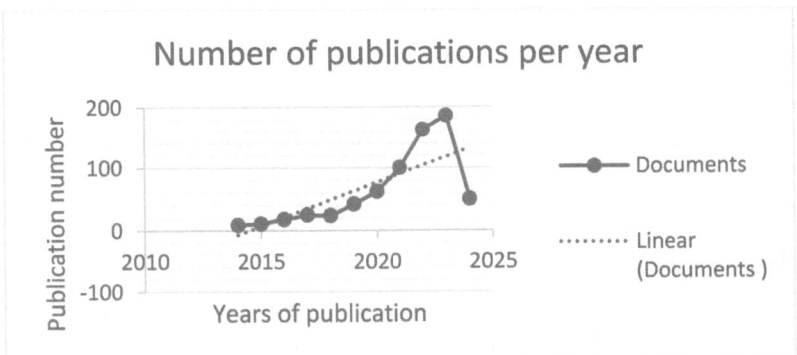

Fig. 1. Evolution of the number of publications (2010–2025)

This increase in scientific output may reflect a growing interest in research on biofertilizers and secondary metabolites in banana cultivation. The importance of these issues lies in their potential to improve agricultural productivity, promote environmental sustainability, and contribute to the development of healthier and more efficient agricultural practices.

3.1.2   Countries with the Highest Scientific Production

The bibliometric network shows the countries with the greatest scientific contributions on the subject of biofertilizers and secondary metabolites in bananas reveals an interesting distribution of scientific production. India stands out as the country with the highest number of documents, followed by China, Iran, the United States and Egypt. India, with the largest number of documents, could be leading research in this field, which could reflect a proactive approach in finding sustainable and effective solutions for agriculture, especially in banana cultivation. China and other countries such as Iran, the United States and Egypt also show a notable presence in scientific production in this area, indicating a global interest in research on biofertilizers and secondary metabolites in bananas. Taken together, the distribution of scientific production in biofertilizers and secondary metabolites in bananas highlights the importance and global reach of research in this field, as well as the potential for future collaborations and significant advances in sustainable agriculture and global banana production.

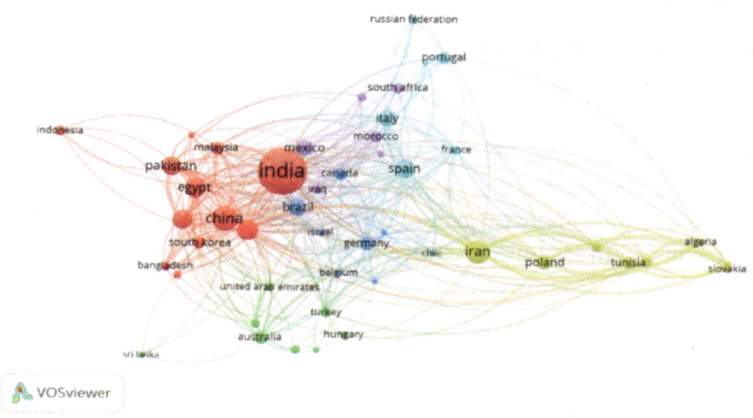

Fig. 2. Countries with the highest number of scientific publications

3.1.3 Top Keywords

The keyword graph (Fig. 2) presents a visual representation of the most relevant terms associated with the study. Among the most frequent keywords are "biofertilizers", "sustainability", "agriculture", "soil", "microorganisms", "plant growth", "nutrients", "soil fertility" and "nutrient efficiency". These keywords reflect the central issues being addressed in biofertilizer research and its potential for sustainable agriculture.

Associations between keywords, such as "biofertilizers" and "microorganisms," "sustainability" and "agriculture," or "plant growth" and "nutrients," show that researchers are exploring the connections between various aspects of biofertilizers and their impact on sustainable agriculture and crop productivity.

Less frequent keywords, such as "biotechnology", "nanomaterials" and "microbiome", could indicate the emergence of new research topics in the field of biofertilisers. Exploring these areas could lead to significant advances in the development and application of more efficient and sustainable biofertilizers.

Keyword analysis reveals that biofertilizer research focuses on developing sustainable solutions for agriculture, with an emphasis on improving soil health, nutrient efficiency, and plant growth. In this sense, phytochemical stimulation through the use of artisanal biols in banana seedlings is presented as a promising alternative to study, as it could contribute to the development of a more sustainable and efficient agriculture.

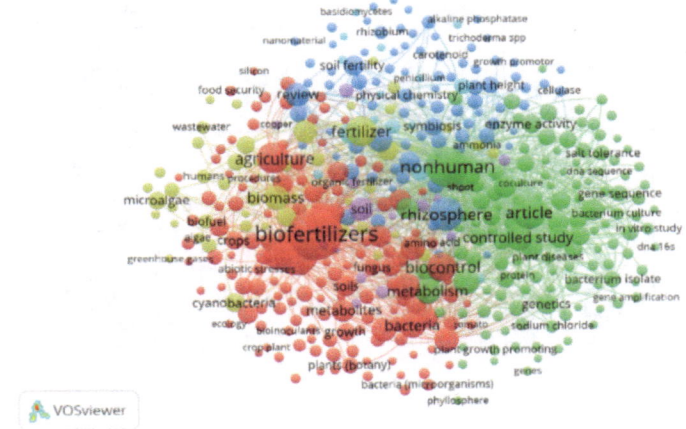

Fig. 3. Network of word frequencies related to biofertilizers obtained from the Scopus database.

3.1.4 Experimental Results

The present study explores the relationship between the variables analyzed: FRAP, DPPH, polyphenols and total flavonoids, in banana seedlings *Musa paradisiaca* L. using artisanal biols. To do this, a correlations and principal component analysis (PCA) was performed using Statgraphs software.

The correlation analysis reveals the existence of significant relationships between the variables studied (Fig. 3). A moderate positive correlation ($r = 0.61$) is observed between FRAP and DPPH, indicating that both methodologies measure similar aspects of the antioxidant capacity of the samples. This suggests that compounds that reduce the FRAP radical also possess the ability to sequester the DPPH radical.

Similarly, there is a moderate positive correlation ($r = 0.58$) between FRAP and total polyphenols, which shows that the polyphenol content determines the capacity to reduce the FRAP radical. Polyphenols, compounds with antioxidant properties, are associated with a greater capacity to reduce the FRAP radical at higher concentrations. The weak positive correlation ($r = 0.31$) between FRAP and total flavonoids indicates that this subgroup of polyphenols contributes to the FRAP radical reduction capacity. Flavonoids, known for their antioxidant properties, are associated with a greater ability to reduce the FRAP radical at higher concentrations [19].

On the other hand, a moderate positive correlation ($r = 0.65$) between DPPH and total polyphenols is observed, confirming the importance of polyphenol content in the ability to sequester the DPPH radical. However, the weak positive correlation ($r = 0.35$) between DPPH and total flavonoids suggests that this subgroup of polyphenols contributes less to the ability to sequester the DPPH radical. Finally, the moderate positive correlation ($r = 0.56$) between polyphenols and total flavonoids indicates that flavonoids constitute an important component of total polyphenols [20].

The principal component plot (PCA) shown in Fig. 4 provides additional information on the distribution of polyphenolic compounds in banana seedling samples. This information can be useful to understand the diversity of polyphenolic compounds present in

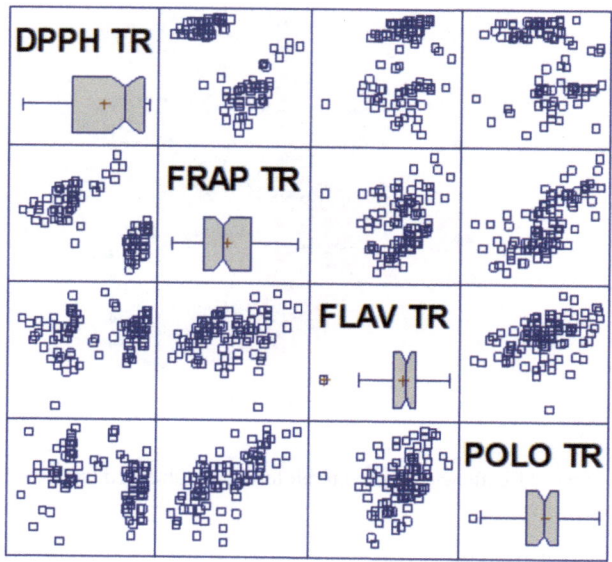

Fig. 4. Correlation matrix of antioxidant parameters in banana seedlings using Worstson Correlation Analysis.

samples and to identify potential biochemical markers to differentiate between banana varieties or specific growing conditions. The first two main components (CP1 and CP2) explain 72% of the total variance, indicating that they capture most of the information contained in the data. In general, it can be inferred that CP1 is related to the general variability of the polyphenolic compound profile, while CP2 may be related to the presence of specific compounds or groups of compounds [21] (Fig. 5).

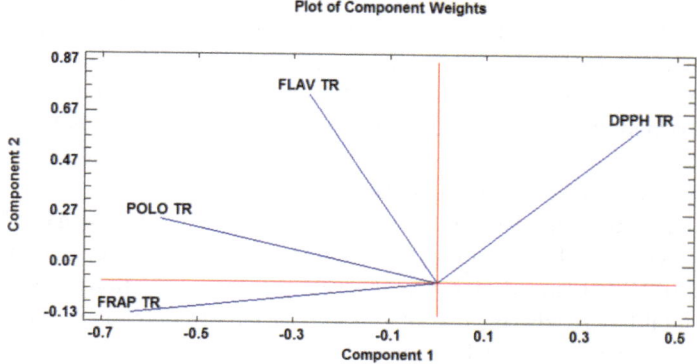

Fig. 5. Principal component analysis for the analysis of phytochemical stimulation in banana seedlings using artisanal biol

The X-axis (CP1) accounts for 47% of the variance and is positively loaded by FRAP, DPPH and total polyphenols, reinforcing the relationship between these variables and antioxidant capacity. This suggests that these three variables measure similar aspects of the antioxidant capacity of the samples [22].

The Y-axis (CP2) accounts for 25% of the variance. The total flavonoid variable is positively charged on this axis, indicating that it is an important factor that contributes to the antioxidant capacity of the samples, but that they also have a certain independence from the other variables [23].

The distribution of samples in the PC1-PC2 plane suggests that there are two main groups of samples. The identification of two groups of samples with different polyphenolic compound profiles suggests a variability in the response of seedlings to artisanal biols [24].

The wide variability observed in the content of polyphenolic compounds is attributed to the presence of diverse groups of compounds with differentiated antioxidant properties. Analyses of FRAP, DPPH, polyphenols, and total flavonoids provide important information about the profile of polyphenolic compounds in samples. However, it is crucial to consider that each method evaluates different aspects of antioxidant capacity [25].

The correlation between the content of polyphenols, especially flavonoids, and antioxidant capacity is evident in the graph of correlations and principal components. However, PCA also suggests the presence of other factors that may contribute to antioxidant activity. Further studies are required to identify these factors and delve deeper into the relationship between the chemical composition and antioxidant activity of the samples.

4 Conclusions

The study found a strong relationship between polyphenolic compounds, flavonoids, and antioxidant capacity in banana seedlings. Artisanal biols promote more vigorous growth, increase chlorophyll, stimulate enzyme activity, and strengthen disease resistance. These results suggest that artisanal bioles are a sustainable alternative to banana cultivation, promoting plant health, increasing yield and reducing environmental impact. Phytochemical stimulation using artisanal bioles is emerging as an effective strategy for cultivation in Ecuadorian banana plantations.

References

1. Murillo, A., Bermeo, M., Bolaño, R.: Estudio socioeconómico de los productores de banano orgánico, cantón milagro, ecuador. Revista Tecnológica - Espol 33(3), 168–180 (2021). https://doi.org/10.37815/rte.v33n3.869
2. Martínez, H., Chávez-Arteaga, K., Guato-Molina, J., Peñafiel-Jaramillo, M., Uquillas, C.: Bacterias fluorescentes productoras de metabolitos antagónicos de cultivares nativos de musa sp. y su diversidad filogenética al gen arnr 16s. Ciencia Y Tecnología 11(2), 17–29 (2018). https://doi.org/10.18779/cyt.v11i2.232
3. Alori, E., Glick, B., Babalola, O.: Microbial phosphorus solubilization and its potential for use in sustainable agriculture. Frontiers in Microbiology 8 (2017). https://doi.org/10.3389/fmicb.2017.00971

4. Seenivasagan, R., Babalola, O.: Utilization of microbial consortia as biofertilizers and biopesticides for the production of feasible agricultural product. Biology **10**(11), 1111 (2021). https://doi.org/10.3390/biology10111111

5. Garcia-Gonzalez, J., Sommerfeld, M.: Biofertilizer and biostimulant properties of the microalga acutodesmus dimorphus. J. Appl. Phycol. **28**(2), 1051–1061 (2015). https://doi.org/10.1007/s10811-015-0625-2

6. Arjjumend, H., Koutouki, K., Neufeld, S.: Comparative advantage of using biofertilizers in indian agroecosystems: an analysis from the perspectives of stakeholders. European Journal of Agriculture and Food Sciences **3**(2), 26–36 (2021). https://doi.org/10.24018/ejfood.2021.3.2.243

7. Rojas-Idrogo, C., Kato, M., Delgado-Paredes, G., Floh, E., Handro, W.: Production of secondary metabolites in in vitro root cultures and cellular suspension of ipomoea carnea spp. carnea jacq. Anales De Biología (36) (2014). https://doi.org/10.6018/analesbio.36.18

8. Hashim, M.: Functional, nutritional and medicinal potential of banana peel. Pure and Applied Biology **12**(1) (2023). https://doi.org/10.19045/bspab.2023.120049

9. Omar, H., et al.: Proximate composition and antioxidant activity of the saba (musa acuminata x balbisiana colla) banana blossom family musaceae. GSC Biological and Pharmaceutical Sciences **21**(2), 026–032 (2022). https://doi.org/10.30574/gscbps.2022.21.2.0418

10. Salazar-Díaz, J., Guerrero-Marina, J., Rodríguez-Espejo, Y.: Actividad antioxidante de aspidosperma excelsum benth, dracontium loretense krause y pothemorphe peltata (l) miq.. Revista Agrotecnológica Amazónica **1**(2), 27–39 (2021). https://doi.org/10.51252/raa.v1i2.190

11. Betancur, G., Rojano, B.: Efecto del isoespintanol y el timol en la actividad antioxidante de semen equino diluido con fines de congelación. Revista De Medicina Veterinaria (35), 149–158 (2017). https://doi.org/10.19052/mv.4397

12. Apriani, E., Miksusanti, M., Fransiska, N.: Formulation and optimization peel-off gel mask with polyvinyl alcohol and gelatin based using factorial design from banana peel flour (musa paradisiaca l) as antioxidant. Indonesian Journal of Pharmacy (2022). https://doi.org/10.22146/ijp.3408

13. Sedjati, S., et al.: Antimicrobial activity of fungal extract of the aspergillus flavus from hiri island, north maluku to pathogenic bacteria. Jurnal Kelautan Tropis **23**(1), 127 (2020). https://doi.org/10.14710/jkt.v23i1.7049

14. Netshiheni, R., Omolola, A., Anyasi, T., Jideani, A.: Banana Bioactives: Absorption, Utilisation and Health Benefits (2020). https://doi.org/10.5772/intechopen.83369

15. Kanazawa, K., Sakakibara, H.: High content of dopamine, a strong antioxidant, in cavendish banana. J. Agric. Food Chem. **48**(3), 844–848 (2000). https://doi.org/10.1021/jf9909860

16. Cabanzo-Atilano, I., Rodríguez-Mendoza, M.N., García-Cué, J.L., Almaraz-Suárez, J.J., Gutiérrez-Castorena, M.D.C.: La biofertilización y nutrición en el desarrollo de plántulas de chile serrano. Revista mexicana de ciencias agrícolas **11**(4), 699–712 (2020)

17. Mittal, A.: Design based ultrasound-assisted extraction of marrubium vulgare linn and comparative evaluation of extracts for furan labdane diterpene (marrubiin) concentration and antihypertensive potential. Curr. Bioact. Compd. **16**(6), 924–936 (2020). https://doi.org/10.2174/1573407215666190524102431

18. Venkatesan, K.B., et al.: Ameliorated antimicrobial, antioxidant, and anticancer properties by plectranthus vettiveroides root extract-mediated green synthesis of chitosan nanoparticles. Green Processing and Synthesis **12**(1) (2023). https://doi.org/10.1515/gps-2023-0086

19. Malki, F., Alouache, A., Krimat, S.: Effects of various parameters on the antioxidant activities of the synthesized heterocyclic pyrimidinium betaines. Indonesian Journal of Chemistry **23**(1), 90 (2022). https://doi.org/10.22146/ijc.74803

20. Kabir, M., et al.: In vitro study of antioxidant potentials, and thrombolytic activity of the leaves of delonix regia (family: fabaceae). Bangladesh Pharmaceutical Journal **26**(1), 73–78 (2023). https://doi.org/10.3329/bpj.v26i1.64221

21. Marín-Velázquez, M.: Polifenoles y actividad antioxidante de extracto acuoso de musa acuminata cavendish subgroup (banana). Ciencia E Investigación **23**(1), 9–14 (2020). https://doi.org/10.15381/ci.v23i1.18717

22. Akhtar, N., et al.: Investigation of pharmacologically important polyphenolic secondary metabolites in plant-based food samples using hplc-dad. Plants **13**(10), 1311 (2024). https://doi.org/10.3390/plants13101311

23. Kędzierska-Matysek, M., Stryjecka, M., Teter, A., Skałęcki, P., Domaradzki, P., Florek, M.: Relationships between the content of phenolic compounds and the antioxidant activity of polish honey varieties as a tool for botanical discrimination. Molecules **26**(6), 1810 (2021). https://doi.org/10.3390/molecules26061810

24. Moudden, H.E., et al.: Spatial variation of phytochemical and antioxidant activities of olive mill wastewater: a chemometric approach. Sustainability **14**(21), 14488 (2022). https://doi.org/10.3390/su142114488

25. Fındık, B., Yıldız, H., Birişçi, E., Yiğitkan, S., Köseoğlu Yılmaz, P., Ertaş, A.: An investigation of the ace inhibitory activity, antioxidant capacity, and phytochemical constituents of polar and non-polar extracts of ziziphus jujuba fruit: statistical screening the main components responsible for bioactivity. Gida the Journal of Food, pp. 554–566 (2024). https://doi.org/10.15237/gida.gd24028

Circular Economy Applicable to Waste from the Sustainable Production of *Musa textilis* in Costa Rica

Kevin Arias-Ceciliano[1] ⓘD, Dagoberto Arias-Aguilar[2]([envelope]) ⓘD, Mónica Araya-Salas[1] ⓘD, Rooel Campos-Rodríguez[3] ⓘD, and Jesús Mora-Molina[4] ⓘD

[1] Instituto Tecnológico de Costa Rica, Cartago, Costa Rica
arias.kevin@yahoo.com, monicka1914@gmail.com
[2] Forestry Engineering School, Instituto Tecnológico de Costa Rica, Cartago, Costa Rica
darias@itcr.ac.cr
[3] Agribusiness School, Instituto Tecnológico de Costa Rica, Cartago, Costa Rica
rocampos@itcr.ac.cr
[4] Chemistry School, Instituto Tecnológico de Costa Rica, Cartago, Costa Rica
jmora@itcr.ac.cr

Abstract. The *Musa textilis* crop is known worldwide as Abaca or Manila hemp, and for many decades the pseudostem fibers have been used as export raw material for the manufacture of cellulose-based products. It is one of the most resistant natural fibers to salinity and has excellent mechanical properties, flexibility, and durability. In the Americas, the two leading countries in the production of Abaca fiber are Ecuador and Costa Rica. There is a good base of information on the agronomic aspects of the crop, the use of improved materials, fiber production techniques and quality characterization. However, there is a lack of information on the estimation of the production of residues and the alternatives for their valorization. The understanding of biomass production, characteristics and types of waste can awaken the interest of producers and companies to generate new products within the concept of waste utilization and circular economy. This work provides guidance on the estimation of waste and presents a proposal of alternatives for the use of fibers to add value. This research is also useful to motivate Banana producers in the search for alternative uses of large volumes of waste generated in the productive activity.

Keywords: *Musa textilis* · fiber · biomaterial · bioeconomy

1 Introduction

Musa textilis is a monocotyledonous species belonging to the Musaceae family and endemic to the Philippines [1]. It is known worldwide as Abaca or "Manila hemp" and is an herbaceous plant with a growth architecture similar to banana and plantain plants; although unlike these, its commercial utility is not directed towards food, and although it produces small fruits, its commercial raw material comes from the pseudostem fibers, which are extracted using very rustic techniques or with mechanical assistance.

© The Author(s) 2025
E. M. Inga Ortega et al. (Eds.): CITIS 2024, LNNS 1331, pp. 360–369, 2025.
https://doi.org/10.1007/978-3-031-87065-1_33

This fiber is considered a potential substitute for wood in the manufacture of paper pulp because it is characterized by excellent mechanical (tension-bending) and chemical properties (63% of cellulose, 21% of hemicellulose, 14% of lignin, and 2% of impurities) [2]. Currently much of the processed material is focused on the manufacture of high-quality paper and specialized use such as tea bags and filters of different types [3, 4].

Abaca has a production cycle that depends on soil and climate conditions and the fiber can reach maturity between 12 and 18 months [5]. To obtain quality fiber, it is necessary to harvest the mature pseudostems and separate the shells to decorticate the fiber, using portable machines called "deco", which comes from the word "decorticate". There are currently more than 50 machines in the country and their number will increase soon; however, recently at the end of 2023 the "Spindle" machine was introduced in the country, which improves the quality of the decortication and requires a different preparation of the shells containing the fibers.

Crop management and fiber extraction generate various sources of waste, ranging from leaves and thinning of the pseudostems to waste from the use of the "deco" machines. A residue like bagasse is generated, composed of pieces of fiber and plant tissue residues. There are also sheets left over from the separation of the layers of the pseudostem. Semi-processed fiber residues are also generated. The use of the "Spindle" machine generates similar waste; however, the largest volume of waste is generated from tuxing, which refers to a specific process of defibering the plant to separate the fibers from the xylem. During tuxing, a series of mechanical operations are performed to extract the long and resistant fibers that make up the abaca, which are then processed with the machine. According to [6] in an analysis of the potential of the solid waste remaining from Abaca processing as a substrate to produce oyster mushroom (*Pleurotus* spp), within the most relevant data provided during interviews and tours, it was found that the largest amount of waste obtained from processing of Abaca, are generated from the pseudostem, where approximately 94% is waste. For their part [7] in a study in Ecuador, report that of the total harvested Abaca material, 96.41% is wasted in its process and only 3.59% is what is sold as fiber to the international market.

1.1 Research Problem and Its Importance

The fiber extraction process generates a considerable volume of waste that has not been quantified in Costa Rica, and its potential economic value is not known. It is estimated that for every kilo of dry fiber, between 3 and 5 kilos of dry matter waste (leaves, fibers and bagasse) are generated and the rest is water. Furthermore, taking into consideration that sustainable management of Abaca can bring social, economic, and environmental benefits, it would be contributing to reducing social inequality, creating new sustainable productive options, and generating positive impacts of climate action. Efficient management of natural resources contributes to the national bioeconomy strategy. Specifically, the Sustainable Development Goal and the Impact on SDG 12: Responsible production and consumption, which seeks alternatives for responsible consumption and recovery of waste.

2 Methods

The study was carried out on a productive and representative farm of the Caribbean region of Costa Rica, specifically located at the La Chávez site in Horquetas de Sarapiquí, Heredia, Costa Rica (10°23'01.6 "N, 83°56'36.1"W, 68 masl). The site has a very humid tropical forest climate (bmh-T); with an average annual precipitation of 4062 mm and an average annual temperature of 25.9 °C [8]. The soil is classified within the order Ultisol and suborder Humults, [9], with a pH less than 5.5, an acidity saturation of 40% and a very slight slope (< 5%).

2.1 Acquisition Waste Biomass Data

Destructive sampling was used from a sample of 40 pseudostems. Each pseudostem was cut at a height of 10 cm from the ground, and the leaves and pseudostem were obtained separately to obtain the weight in kilograms, both on a wet basis and on a dry basis. The pseudostem was sectioned into "layers" that were classified into external layers (second quality) and internal layers (first quality), which were decorticated with a "deco" machine until the fibers and waste were obtained. The fibers were dried in the sun for 3 h until reaching a humidity of around 12%. From each plant, 3 subsamples of tissues were obtained (leaves, pseudostem, fibers, residues) and the wet weight and dry weight were obtained in a laboratory oven (temperature set at 80 °C for 72 h).

2.2 Characterization of the Chemical Properties of Abaca Fiber in Costa Rica

From a collection of 5 different varieties (genotypes) of Abaca in a place close to the study area, a characterization of the chemical properties of the fibers was carried out. Samples of the fibers and waste were taken randomly to carry out laboratory analysis and tests for the use of fibers as a composite material. The material was processed in the laboratory, ground, and dried in the oven. The chemical properties were evaluated: lignin and holocellulose content according to the TAPPI T222 om-02 standard (2002); as well as, extractives in cold and hot water according to ASTM D1110–21 (2021), extractives in 1% NaOH according to ASTM D1109–21 (2021), extractables with dichloromethane according to ASTM D1108–21 (2021), extractives with ethanol-toluene according to ASTM D1107–21 (2021), ash content according to ASTM D1102–84 (2021), volatile content according to ASTM D1762–84 (2021), fixed carbon according to ASTM D3172-13e1 (2021) and the caloric value according to ASTM D5865/D5865M-19 (2019). Likewise, the content of carbon (C), hydrogen (H), nitrogen (N) and sulfur (S) was estimated with an Elementar analyzer model Vario Micro Cube (Elementar, Langenselbold, Germany).

2.3 Use of Residual Fiber as Composite Material

Boards were made from combinations of wood sawdust and Abaca waste fiber. The boards consisted of a single layer with dimensions of 350 x 350 x 12 mm and a target standard density of 0.65 g cm^{-3}. The adhesive available in the country and provided by one of the industries dedicated to the manufacture of boards was used; which corresponds to the adhesive Melamine Urea Formaldehyde (MUF) with its respective catalyst (AkzoNobel 1247/2526, AkzoNobel Wood Caotings, USA), with an adhesive load of 12% based on the dry weight of the particles in a 5:1 ratio (adhesive: catalyst) according to the manufacturer's instructions. The viscosity of said adhesive was characterized by having an average value of 17,500 MPa and a pH of 9.5 to 10.7 (all values at 25 $\pm$ 2 °C). The combinations used as treatments were the following: T1 (40M:60A), T2 (60M:40A), T3 (50M:50A), T4 (100M:0A) and T5 (0M:100A), where M is *Gmelina arborea* and A is *Musa textilis.*

2.4 Statistical Analysis of the Data

For the analyzes based on ANOVA, a completely randomized design was used, for which the assumptions of normality of the residuals and the homogeneity of the variances were verified with the Shapiro-Wilks and Levene tests, respectively. The ANOVA applied considered a significance level of 0.05, and determined if there were significant differences between treatments; if necessary, the comparison of means test according to Tukey was applied to identify treatment groups with different behavior. In situations in which the assumptions of the ANOVA were not met, the non-parametric Kruskal Wallis test was executed together with a comparison of medians if significant differences were found. All analysis was run in Infostat software. 2020e [10].

3 Results

3.1 Estimates of Biomass, Fiber and Residues

Table 1 summarizes the results obtained from the study of biomass per plant component. The wet weight of plant pseudostems ranged from 7.8 kg to 34.2 kg, while leaf tissues ranged from 0.8 kg to 4 kg. These values, when summed up, gave total wet weight values ranging from 9 kg to 37.3 kg, with a moisture content ranged from 78.50% to 86%. Likewise, pseudostem dry weight ranged from 0.94 kg to 4.10 kg, while leaf dry weight ranged from 0.21 kg to 1.04 kg. Thus, the total aboveground dry weight ranged from 1.25 kg to 4.96 kg. From a pseudostem residue perspective, approximately 75% of the dry biomass is residue. To this must be added the dry biomass of the leaves, which equals 1.87 kg. If a plantation has approximately 4000 crop stems per hectare, the residual biomass per hectare equals 1.45 tons. Under the site conditions of this study, a harvest of class 1 and 2 fiber is equivalent to 2.24 tons of anhydrous fiber and 2.50 tons of fiber at 12% moisture content

3.2 Characterization of the Chemical Properties of Abaca Fiber in Costa Rica

The chemical characterization (Table 2) showed the following conditions: i. For holocellulose, only genotype MT03 showed a significantly different value (87.91%) from the other genotypes, which showed no statistical difference (average 90.67%). ii. Lignin showed variation according to genotype; MT07 and MT11 showed high values (average

Table 1. Summary of the descriptive statistics of the wet and dry biomass of the different components of the Abaca plants (n = 40)

VARIABLES	MEAN	SD	CV	MIN	MAX
Height (cm)	367,53	64,78	17,63	250	504
Diameter (cm)	11,75	1,79	15,26	8,40	16
Stem wet weight (kg)	19,33	6,73	34,81	7,80	34,20
Leaf wet weight (kg)	1,87	0,84	45,27	0,80	4
Total wet weight (kg)	**21,19**	**7,34**	**34,63**	**9**	**37,30**
Stem dry weight (kg)	2,32	0,81	34,76	0,94	4,10
Leaf dry weight (kg)	0,48	0,22	45,29	0,21	1,04
Total dry weight (kg)	**2,80**	**0,97**	**34,71**	**1,25**	**4,96**
Fiber wet weight (kg) 1	2,16	0,87	40,46	0,56	3,99
Fiber dry weight (kg) 1	0,42	0,19	45,75	0,01	1,05
Fiber wet weight (kg) 2	0,76	0,36	45,12	0,14	1,55
Fiber dry weight (kg) 2	0,14	0,07	50,44	0,01	0,36

Note: SD is the standard deviation of the mean, CV is the coefficient of variation, MIN is the minimum value, MAX is the maximum value. 1 and 2 refers to the quality of the fiber. In Costa Rica, two fiber qualities are currently used, and they differ in the purchase price

14.81%), in contrast to MT01, MT03 and CF01, which were lower at 11.95%. Iii. MT07 and MT11 showed the lowest values in cold water, ethanol-toluene, and dichloromethane extractives; in the case of hot water extractives, only MT07 showed significant differences. iv. Extractives with sodium hydroxide showed no differences between genotypes; the average value was 1.31%. Elemental analysis showed that MT07 and MT11 indicated statistical differences in the composition of C, H, N and S; in contrast, the remaining three genotypes showed no significant differences among themselves.

Table 2. Fiber chemical properties of five genotypes of *Musa textilis.*

VARIABLE				VARIETY		
		MT01	MT03	MT07	MT11	CF01
Holocellulose (%)		89,62[A] (0,95)	87,91[B] (0,46)	89,38[A] (0,98)	93,06[A] (0,37)	90,65[A] (0,56)
Lignin (%)		14,43[A] (0,28)	15,46[A] (0,23)	13,49[B] (0,39)	12,13[B] (0,26)	12,66[B] (0,10)
	Hot water (%)	11,01[A] (0,18)	10,17[A] (0,41)	7,75[B] (0,24)	10,88[A] (0,15)	3,90[C] (0,20)

(*continued*)

Table 2. *(continued)*

VARIABLE					VARIETY		
		MT01	MT03	MT07	MT11	CF01
	Cold water (%)	11,27^A (0,24)	10,35^A (0,22)	6,76^B (0,20)	7,33^B (0,31)	3,67^C (0,16)
Extractives	Ethanol toluene (%)	11,07^A (0,11)	10,43^A (0,23)	5,93^B (0,15)	7,19^B (0,16)	2,14^C (0,13)
	NaOH 1%	1,33^A (0,10)	1,33^A (0,09)	1,30^A (0,09)	1,32^A (0,10)	1,31^A (0,09)
	Dichloromethane	9,23^A (0,11)	9,44^A (0,09)	4,56^B (0,10)	5,01^B (0,09)	4,89^B (0,10)
Nitrogen (%)		0,14^A (0,02)	0,11^B (0,01)	0,09^B (0,02)	0,10^B (0,02)	0,10^B (0,02)
Carbon (%)		62,21^A (1,15)	66,44^B (0,74)	66,11^B (0,50)	66,49^B (0,45)	66,57^B (0,41)
Hidrogen (%)		6,44^A (0,02)	6,72^A (0,10)	6,61^A (0,12)	6,78^A (0,43)	6,90^A (0,29)
Sulfur (%)		1,55^A (0,09)	1,50^A (0,07)	1,32^B (0,03)	1,33^B (0,02)	0,33^B (0,03)

Note: Values in parentheses correspond to the standard deviation; different letters indicate significant differences at 0.05

3.3   Use of Fiber as a Composite Material with Wood Sawdust

Table 3 shows the mean values corresponding to the property's density, moisture content, and HEΔ. The target standard density of the boards was set at 0.65 g cm^{-3}, which classifies them as medium density boards according to the ASTM standard. According to the analysis carried out, the treatment that presented the density closest to the standard value was the control treatment (T4) with an average value of 0.64 g cm^{-3}, followed by treatment T2. With what was observed, it is affirmed that as the proportion of *M. textilis* in the mixture increases, the density value decreases, although it is notorious that there were no significant differences between T1, T2 and T3 with the control treatment. These treatments obtained values below the standard target value, which may be due to several reasons; on the one hand, the high probability of losing particles in the demolding phase of the boards, and on the other hand, the variation in thickness caused by the change in temperature and humidity that occurs at the time of transferring the boards from the hot press to the conditioning chamber [11, 12].

Table 4 shows the mechanical properties evaluated. Regarding static bending, the modulus of rupture (MOR) showed that treatment T2 did not show significant differences with the control but did show significant differences with the treatments with more than 60% fiber (T1 and T5). Likewise, the modulus of elasticity (MOE) or flexural modulus did not show significant differences between the control treatment (T4) and the other treatments; the fact that treatment T2 (40% fiber) showed the highest value of MOE

Table 3. Physical properties of particle boards composed from different proportions of *Musa textilis* and *Gmelina arborea*

	TREATMENT	Density (g cm^{-3})	Moisture content (%)	HEΔ (mm)
Average	**T1**	0,60 AB	15,73 D	0,57 C
	T2	0,61 AB	12,89 A	0,17 AB
	T3	0,58 AB	14,79 C	0,25 B
	T4	0,64 B	12,06 A	0,12 A
	T5	0,57 A	16,08 D	0,52 C

Note: Different letters indicate significant differences at 0.05. HEΔ% corresponds to the swelling in thickness due to a change in moisture content from 12 to 18%.

(17,04 GPa) is noteworthy. In relation to this, the amount of extractives present in the fiber used in this study [13] could contribute to the variations present in the results within the same treatment; since as stated by [14] to obtain the expected mechanical properties in the boards it is necessary that there is chemical compatibility between the components that conform it and additionally [15] also mention the interaction in the curing of the adhesive. [16] demonstrated the effects of chemical compatibility in boards with palm bagasse. Also, the orientation and distribution of fibers within the board has been shown to influence flexural strength, with incorrect orientation causing a decrease in strength [17].

Table 4. Mechanical properties of composite particleboards with different proportions of *Musa textilis* and *Gmelina arborea* species.

Properties		T1	T2	T3	T4	T5
Static bending (SB)	MOE (GPa)*	10,03A (5,43)	17,04B (7,26)	13,73AB (2,32)	13,76AB (8,02)	8,84A (8,60)
	MOR (MPa)*	22,91AB (3,05)	28,02C (8,45)	26,04ABC (8,39)	28,06BC (5,04)	18,92A (9,20)
Parallel tension (T‖)	MOE (MPa)	18,82A (18,10)	58,38B (24,18)	36,57AB (16,09)	104,97C (27,32)	21,54A (14,54)
	Strain maximum (MPa)	0,84A (0,50)	0,98A (0,34)	0,88A (0,31)	2,31B (0,58)	0,78A (0,42)
Tension perpendicular (IB)	Strain maximum (MPa)	0,17A (0,07)	0,25AB (0,06)	0,23AB (0,09)	0,62C (0,06)	0,29B (0,05)
Hardness (H)	Load maximum (kg)	149,13A (32,57)	225,33BC (39,44)	191,78B (20,73)	235,16C (26,86)	141,57A (41,52)

Note: Values in parentheses represent the corresponding standard deviation. Different letters indicate significant differences at 0.05. Parameters marked with an (*) correspond to results analyzed with nonparametric tests (n = 10)

4 Discussion

It is demonstrated in this study that the largest proportion of the total dry biomass of Abaca plants during a fiber harvest corresponds to residues (approximately 40% of the dry biomass). About 20% of these residues contain fibers that are not part of the raw material that can be sold, and the rest are plant tissues and parts of the leaves that do not contain fibers. However, there is great potential for generating compost from this material. Residual fibers can be used locally in composite materials.

Based on the chemical properties of the fibers, the fiber of genotypes MT01 and MT03 was recommended for stationery, biodegradable materials, and fast degradation composites. These products are proposed due to the need for materials with high cellulose and hemicellulose content and low lignin content; as it facilitates the pulping and pulping homogenization process and enhances the creation of higher quality paper [18]. [1] mention that this is the key to have low strength fibers to develop single-use biodegradable biomaterials or short life cycles. Furthermore, [19] recommend using fibers with low mechanical strength and low lignin levels for environmentally friendly, moderate-use, low value-added materials, which is compatible with both genotypes.

With respect to the use of fiber as a composite material, the gradual reduction of *Musa textilis* fiber proportions promoted progressive increases in density, moisture content, MOR, MOE and IB values. Thus, proportions greater than 40% of M. textilis significantly decreased the overall quality of the boards. Despite this, in all treatments with proportions lower than 50% fiber, the MOR values exceeded the maximum reference points of the standard used. Therefore, the density (0.64 g cm^{-3}), MOR (28.06 MPa) and IB (0.62 MPa) properties of the boards corresponding to treatment T4 (control, 100% wood) met the minimum specifications established for particleboard categorized for use in dry conditions. However, T2 and T3 meet the minimum specifications required in these properties mentioned for low-density LD-2 grade boards.

5 Conclusions

Based on the results of the characterization of the fibers, it is demonstrated that there are differences and therefore, the differentiated use of genotypes must be considered for the specific use of fibers in the development of composite materials. It is important to initiate tests with mixtures with plastics to determine the specific uses of biomaterials. It is recommended to continue with new research on the application of different fibers in matrices with polymers and other natural fibers and evaluate the efficiency and quality of the products and the creation of a new market. Therefore, these results are a first step to change the paradigm of M. *textilis* plantations in the tropics.

The quality of the composite boards obtained in this study are susceptible to improvement. The future challenge is to continue experiments to enhance the general properties of the boards. To this end, it is recommended to carry out a broader study that involves other species of lignocellulosic and wood fibers, the use of lower proportions of fiber, the application of pretreatments to the fiber, the use of organic additives with different doses of adhesive and particle sizes.

References

1. Richter, S., Stromann, K., Mussig, J.: Abaca (*Musa Textilis*) grades and their properties a study of reproducible fibre characterization and a critical evaluation of existing grading systems. Ind. Crops Prod. **42**, 601–612 (2013)
2. Bazliah, D., et al.: Prediction of Kappa number and carbohydrate degradation in oxygen delignification of Abaca fiber. IOP Conference Series: Materials Science and Engineering, **1053** (2021). https://iopscience.iop.org/article/https://doi.org/10.1088/1757-899X/1053/1/012015
3. Ramnath, V., Manickavasagam, V.M., Elanchezhian, C., Vinodh Krishna, C.V., Karthik, S., Saravanan, K.: Determination of mechanical properties of intra-layer abaca–jute–glass fiber reinforced composite. Materials & Design **60** (2014). https://doi.org/10.1016/j.matdes.2014.03.061
4. Sambonino, B., Salavarría, J., Mieles, J., Mata, M.: Análisis del mercado internacional de la fibra de abacá, su oferta exportable hacia Reino Unido y su aporte en el cambio de la matriz productiva. Segundo Congreso Internacional en Administración de Negocios Internacionales. Universidad Pontificia Bolivariana (2017)
5. Promotora de Comercio Exterior de Costa Rica. Ficha técnica Abacá (2023). https://www.descubre.cr/wp-content/uploads/2023/04/Ficha-Tecnica-Descubre-Abaca-Mar23.pdf
6. Valenciano, C.: Análisis del potencial de los residuos sólidos remanentes del procesamiento de Abacá (*Musa textilis*), como sustrato para la producción de hongo ostra (Pleurotus spp), Zona Huetar Norte y Atlántica, Costa Rica. Universidad Técnica Nacional (2019). https://repositorio.utn.ac.cr/items/cb379ab5-bec0-44a6-a26b-22ddc7fb3338/full
7. Jácome, L., Martinez, M.C., De La Cruz, M., Chica, H., Valencia, X.: Rendimiento de fibra de dos variedades de Abacá (*Musa textilis*) en tres densidades de siembra. Ciencia Latina Revista Científica Multidisciplinar, **7**(2) (2023). https://doi.org/10.37811/cl_rcm.v7i2.5615
8. Climate-data. Clima: Las Horquetas (Costa Rica). Climate-data (2019). https://es.climate-data.org/america-del-norte/costa-rica/heredia/las-horquetas-874786/
9. Ortiz, E., Soto, C.: Atlas Digital de Costa Rica 2014. [CD-Rom]. Cartago, Costa Rica: Instituto Tecnológico de Costa Rica. (2014)
10. Di Rienzo, J.A., Casanoves, F., Balzarini, M.G., Gonzalez, L., Tablada, M., Robledo, C.W.: InfoStat versión 2020e. Grupo InfoStat, FCA, Universidad Nacional de Córdoba, Argentina (2020). http://www.infostat.com.ar
11. Mendes, R.F., Mendes, L.M., Mendonça, L.L., Guimarães Júnior, J.B., Mori, F.A.: Qualidade de painéis aglomerados homogêneos produzidos com a madeira de clones de *Eucalyptus urophylla*. Cerne **20**, 329–336 (2014). https://doi.org/10.1590/01047760.201420021273
12. Rangel, L., Moreno, P., Trejo, S., Valero, S.: Propiedades de tableros aglomerados de partículas fabricados con madera de *Eucalyptus urophylla*. Maderas. Ciencia y tecnología **19**(3), 373–386 (2017). https://doi.org/10.4067/S0718-221X2017005000032
13. Valverde, J.C., Araya, M., Arias-Aguilar, D., Masís, C., Muñoz, F.: Evaluation of the optimal uses of five genotypes of musa textilis fiber grown in the tropical region. Polymers **14**(9), 1772 (2020). https://doi.org/10.3390/polym14091772
14. Moya, R., Camacho, D., Oporto, G.S., Soto, R.F., Mata, J.S.: Physical, mechanical and hydration kinetics of particleboards manufactured with woody biomass (*Cupressus lusitanica, Gmelina arborea, Tectona grandis*), agricultural resources, and Tetra Pak packages. Waste Management & Research **32**(2),106–114 (2014). https://doi.org/10.1177/0734242X13518959 (2014)
15. Han, G., Zhang, C., Zhang, D., Umemura, K., Kawai, S.: Upgrading of urea formaldehyde-bonded reed and wheat straw particleboards using silane coupling agents. J. Wood Sci. **44**, 282–286 (1998). https://doi.org/10.1007/BF00581308

16. Adam, A.B.A., Basta, A.H., El-Saied, H.: Evaluation of palm fiber components an alternative biomass wastes for medium density fiberboard manufacturing. Maderas, Ciencia y Tecnología **20**(4), 579–594 (2018). https://doi.org/10.4067/S0718-221X2018005004601
17. Nishimura, T., Amin, J., Ansell, M.P.: Image analysis and bending properties of model OSB panels as a function of strand distribution, shape and size. Wood Sci. Technol. **38**(4), 297–309 (2004)
18. Moreno, L.O., Protacio, C.M.: Chemical composition and pulp properties of abaca (*Musa textilis* Née) cv. Inosa harvested at different stages of stalk maturity. Annals of Tropical Research **34**(2), 45–62 (2012)
19. Shibata, M., Ozawa, K., Teramoto, N., Yosomiya, R., Takeishi, H.: Biocomposites made from short abaca fiber and biodegradable polyesters. Macromol. Mater. Eng.. Mater. Eng. **288**(1), 35–43 (2003). https://doi.org/10.1002/mame.200290031

Industry

Impact Analysis on Voltage Stability by Inserting Non-Linear Loads Through a Dynamic Stability Index

Jonathan Salazar[1]($\boxtimes$) , Diego Carrión[1,2] , and Manuel Jaramillo[1]

[1] Salesian Polytechnic University, EC170702 Quito, Ecuador
jsalazara6@est.ups.edu.ec, {dcarrion,mjaramillo}@ups.edu.ec
[2] Valencia International University, 46002 Valencia, España, Spain

Abstract. Modern power systems are exposed to various operating conditions that affect the steady-state operation of the system. Several researchers focus on monitoring dynamic loads and their impact on short-term stability, considering factors such as planning, expansion, management, and system capacity through VQ sensitivity analysis, determining the most sensitive bus to reactive power variation. Determining short-term stability helps operators take predictive actions in case of changes in the system, such as the entry of induction machines. The present research is focused on establishing a stability index to evaluate the system's stability in a short-term dynamic state by incorporating nonlinear dynamic loads, as in the case of induction motors. The behavior of the proposed dynamic index, which has been called the Novel Index Stability Index (NISV), has been evaluated in several scenarios considering the IEEE 14-bar system, where the performance of the index is verified and how, when varying different conditions of the nonlinear load in time, it presents unstable conditions when faced with changes of the same.

Keywords: Short-term Stability · Non-linear dynamic loads · Stability index · Electrical power system · Voltage stability

1   Introduction

With time, electrical power systems (EPS) have become more important worldwide due to the vast availability of conventional and renewable energy sources. For that reason, several researchers have focused on improving the operability of the EPS, and it is precisely there when the study of dynamic loads, how they directly affect short-term voltage stability, and how the imbalance in this aspect causes problems in the generators becomes relevant [1].

To structure an optimal system, it is necessary to consider some relevant factors such as expansion, planning, and, above all, management, but without neglecting the constant improvement in the operability of the EPS. Likewise, certain aspects restrict the conduction capacity of the transmission lines (TL), which is why [2] proposes to analyze them based on extensive alternating current

© The Author(s) 2025
E. M. Inga Ortega et al. (Eds.): CITIS 2024, LNNS 1331, pp. 373–383, 2025.
https://doi.org/10.1007/978-3-031-87065-1_34

lines and their impact on stability, considering, of course, the limiting power that could be transferred, aspects that are crucial according to the topology of the network.

It is crucial to keep in mind that distributed generation (DG) and energy conversion sources are dynamic loads and stochastic behavior that affect voltage stability [3], which must be considered as a systematic axis, especially in countries where there are industries in gradual development and, therefore, increased electricity demand [4]. Ultimately, excessive load on LTs can cause disturbances that, in turn, trigger an imminent voltage drop, the collapse of the EPS, and even a blackout [5].

Currently, researchers approach stability from the stationary point and highlight the recovery times of the system at a practical voltage value; for example, [6] mentions a defined range between 10% and 90% of the nominal value, considering the specific frequency of the EPS. In this context, the analysis of voltage dips, which last about half a cycle or even up to 1 min, becomes relevant since, this way, it is possible to identify the incidence of the variation of the Thevenin impedance angle [7].

Also, short-term voltage stability plays a crucial role because of the similarities with dynamic systems, as the conceptual basis and mathematical principles govern the operability of the EPS, especially in terms of incorporating nonlinear loads, such as large induction motors (IM). The IMs affect the voltage magnitude, cause bus sag, and produce a long-duration disturbance, in which case multivariable analysis is imminent [8]. In this sense, a deep, thorough, and complete transient analysis must be conducted [9].

In contrast, the EPS requires effective management and operation to counteract disturbances, such as interruptions, undervoltages, and overvoltages, a process that relies on an external loop to ensure adequate PQ reserves. In this regard, [10] argue that by using battery storage, a voltage with dynamic behavior is injected with which the analysis is performed after a significant disturbance, taking into account a short-term voltage countermeasure to identify what would happen in that case and thus determine the place of origin of such disturbance.

As indicated by [11], the post-disturbance can be analyzed through the QV sensitivity, which is based on the Jacobian matrix of the EPS to establish the points of variation between the voltage and those concerning the reactants. In this way, a starting point of solution to a power flow is determined.

In the framework of EPS operational management, it is crucial to analyze the vital components that contribute to the system since they impact its stability and efficiency. Among them is the IM [12], which has come to have a very important role since they are markedly immersed in the industry on other elements and being dynamic loads, the methods of operation and management itself influence the performance of the system because the growing demand has caused an imbalance that is between 0% and 5% of the voltage unbalance limits mentioned in National Electrical Manufacturers Association (NEMA) [13].

Most studies on IM refer to the operating current, whose relevant factors are motor performance and temperature. Thus, the analysis in the positive sequence was revealed to be higher than the nominal voltage of 1 pu, a situation that can generate overvoltages, which in turn causes a phasor to unbalance could affect the magnitude of the IM voltage as well as its phase angle and cause voltage drops of short duration [14].

Performs an interesting analysis [15], which focuses on data mining to create predictive models in which the IM can predict an appropriate voltage supply that does not cause a noticeable imbalance, where data mining reduces operating costs. Thus, it is possible to predict a proper voltage supply that does not cause a noticeable imbalance in the other elements of the system; for that reason, several investigations are focused on analyzing the IM since they are part of 80% of the energy conversion. Hence, monitoring is of great relevance, as it ensures the system's reliability and how it can recover from various disturbances [16].

Therefore, the comprehensive management of nodes and elements is of vital importance, a process in which the automatic voltage regulator (AVR) of the generator excitation system participates, responsible for monitoring and maintaining voltage and current variations within established limits, increasing the reliability of the system and improving its steady-state stability. Indeed, while transmission zone modeling has generally been performed for purely balanced and linear loads, the study of imbalances caused by nonlinear loads, such as IM and the incorporation of distributed generation, has opened new areas of research on power flow and the integration of transmission and distribution systems [17].

It is essential to mention that the evolution of voltage stability is supported by indices that conFigure the starting point for future studies, as well as the basis for the operation and correction of IM since they are configured with several indicators: Lmn, FVSI, LQP, NLSI, and VSLI [18]. The results obtained are relevant since they contribute to acquiring more knowledge about operability, taking into account the EPS in an integral way, and analyzing N-1 contingencies [5,19].

The present study focuses on determining the loadability in weak busbars to facilitate a power flow that can converge. In this case, loadability is established when the Jacobian becomes a singular matrix, which means there is no inverse of that matrix.

An index that can be taken as a reference is the FVSI, which allows for establishing the voltage of a line in a transmission system, its reactive power in a particular bus, and identifying its increase and point of instability when maximum loads are connected to this bus. In this case, a range from 0 to 1 is established to know if the system is stable, and its mathematical basis lies in the quadratic equation [20]. Thus, statistical analyses become relevant, which are essential to determine the different scenarios and to evaluate the dynamic study of the IM in the face of reactive power drops, contingencies, or excessive load increases -of the nonlinear elements- that cause voltage variations in such a way that stability indexes are established [21].

2   Related Works

Employing a descriptive historical approach, an information inquiry was carried out considering the following keywords: electrical power systems, Stability, Dynamic, and Behavior. In academic papers from 2019 to 2023, an analysis was carried out considering highly relevant databases such as Scopus and Web of Science.

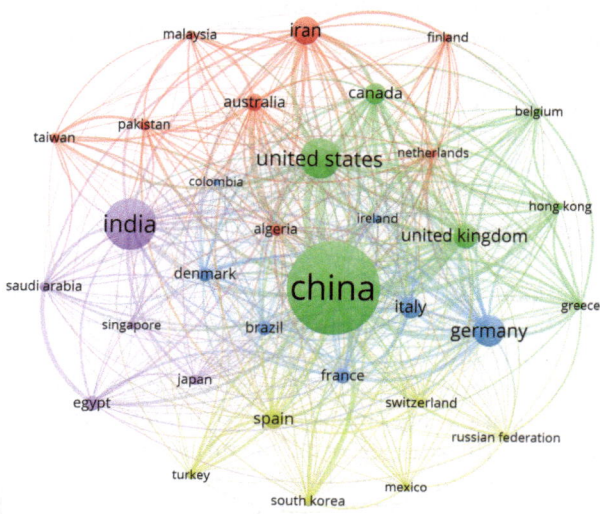

Fig. 1. Countries concerned with the topic of stability, dynamic behavior of Scopus

An experimentation-based method was considered to perform a bibliometric analysis supported by the free-to-use software VOSviewer to structure good data mining. To classify and analyze data relevant to the bibliometric study, such as countries, authors, co-authors, and number of citations, based on the relationship with the described topic. Thus, network graphs were generated to identify the authors who made the most significant contributions to the object of research, taking into consideration the countries of publication.

Consequently, having the required information, graphs were created, where the main thing is the comparative of the influence of the authors regarding the number of citations per country, as well as articles published from 2019 to 2023 about short-term voltage stability in power systems considering nonlinear dynamic loads as shown in Fig. 1.

The descriptive historical approach is part of the bibliometric study, where graphs provide a different way to visualize relevant research data.

The database search was based on keywords related to the research extracted from the Scopus and Web of Science databases. Several remarkable results evidence the importance of the research topic. Likewise, the results contribute to constructing the graphs and how the bibliometric network is generated in VOSviewer.

The information provided in Fig. 1 shows the analysis, which presents the countries with the highest number of research studies and their percentage of contribution in the database extracted from Scopus, which consists of 1172 articles, a situation which can be seen in Fig. 1 employing the map graphs as China generates the highest contribution of publications on the topic as mentioned above.

3   Methodology and Problem Formulation

One of the indexes used to analyze EPS stability quickly focuses on identifying weak lines and bars, for which the Fast Voltage Stability Index (FVSI) was developed, which relates the impedance of the lines ($Z_{(i-j)}$), the reactance of the line ($X_{(i-j)}$), voltage (V_i) and reactive power at the node (Q_i). (1).

$$FSVI = \frac{4Z_{i-j}^2 * Q_j}{V_i^2 * X_{i-j}} \tag{1}$$

The methodology for the dynamic short-term voltage stability index approach evaluates what happens in the short term. In this way, it is previously considered how the system is, for this purpose is used IEEE 14 Bus system, the VQ sensitivity of the system is analyzed, where the purpose is to identify the most sensitive bus to reactive power changes in the proposed system. With this, we have a global idea of where an IM can be entered; also, through the simulation, we can obtain the maximum reactive power limits so that the system can converge.

After considering these previous considerations, we will run simulations in the Digsilent PowerFactory software. By analyzing the software's QV curve module, we will determine the above-mentioned Q limits and determine what happens with the short-term voltage stability and whether the system recovers.

However, known particularities of the EPS are the case of the equivalent impedance and reactance of the EPS, which remains preset and does not fluctuate over time, unlike the reactive power, which, depending on the time, has a different injection in the EPS as the voltage these two final variables are part of a dynamic study, the study presents the NISV index (new voltage sensitivity index), this index is derived. It has a new context based on the FVSI index, which has been used to evaluate the stability in a stationary manner. However, it has not been analyzed for dynamic loads over time, leading to the NISV's development, which can denote what happens with these two variables to load changes or contingencies.

The NISV is expressed in (2) where the term 4 is a factor adopted from, the FVSI index, due to the scientific literature such value is crucial in the analysis of line indexes. In addition, 1 is established as the critical stability threshold value since it indicates instability conditions in the system, based on the scientific literature as shown in [22], that when this threshold is exceeded, corrective actions should be analyzed so that the system is within the limits of short-term voltage stability.

$$NSVI = \frac{4Z_{i-j}^2 * Q_j(t)}{V_i^2(t) * X_{i-j}} \tag{2}$$

3.1 EPS Sensitivity Matrix

By simplifying the Jacobian of the EPS, we obtain the simplified Jacobian, which denotes the sensitivity concerning VQ. However, in the diagonal of the inverse matrix of the simplified Jacobian, we find $\partial V_i/\partial Q_i$, which indicates how the sensitivity in the bars is for it, on the other hand, $\partial V_k/\partial Q_i$, denotes the mutual sensitivity which refers to the factors external to the diagonal and gives a general idea of the sensitivity of the busbars in a global way to the system. That is, the weakest member of the system is determined.

In contrast, the method for determining the weakest busbar is vital because, in the case study, it will contribute to a part of the analysis of which busbar to connect the IMs to based on prior knowledge of the system's robustness.

For example, the VQ sensitivity study for a 14-bus system provides the system's robustness and the weakest bus in which it could be taken into account for the entry of nonlinear loads, such as the IM, and how this load increase can affect its stability.

Figure 2 shows which is the weakest bus of the IEEE 14-bus system; bus number 14 is the most sensitive bus of the system and where nonlinear loads such as IM could be connected to see how it influences the stability of the system by presenting income in a given time to the system. The information shown in Fig. 2 is validated through the Digsilent PowerFactory software. This bus, 14, is one of the most sensitive buses regarding reactive power variations since it supports less reactive power than the others.

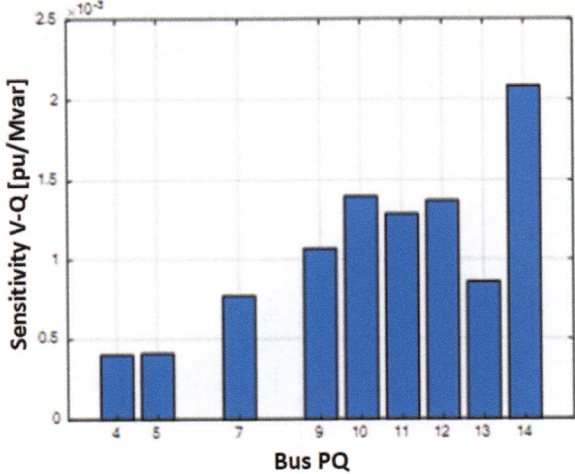

Fig. 2. Countries concerned with the topic of stability, dynamic behavior of Scopus

4   Results Analysis

Previously, steady-state studies were carried out to know firsthand how it is and at what points it could become unstable due to the entry of dynamic load in the power system.

Table 1. Convergence limits for reactive power increases in a 14-bus system.

Node	Voltage collapse [p.u.]	Max Reactive Power [Mvar]
14	0.536	117.24
10	0.531	180.53
7	0.522	185.84
11	0.537	195.76
9	0.536	238.05
13	0.55	283.18
5	0.52	596.54
4	0.519	608.87

The previous study evidences the system's vulnerability in node 14; when simulating normal conditions, we observed that this node's voltage and reactive power are within the limits established by the model. Thus, when entering the voltage and reactive power data into the NISV index in the link node, we verify that it does not exceed the unit; this criterion indicates that the system is stable.

It should be noted that between the VQ sensitivity study and the analysis of the Q-V curves it is pointed out that the most sensitive node for reactive power is node 14, as shown in Table 1; the maximum reactive power that this node can perceive is 117.24 Mvar after that the system will no longer have a convergence so it will stop working.

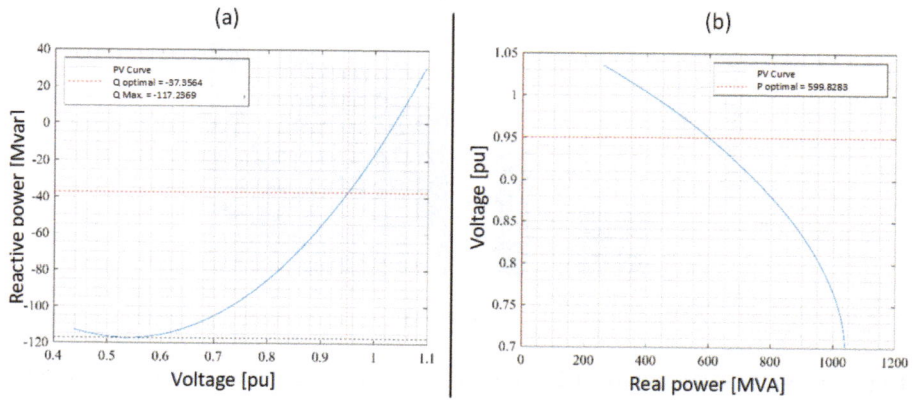

Fig. 3. (a) QV Curve in node 14, (b) PV Curve in node 14

Figure 3a shows the behavior of the VQ curve, indicating its collapse point and the optimum point, which concatenates with the information in Table 1, which occurs at node 14. Similarly, a review of the PV curve in Fig. 3b, where the optimum real power value at that node can be determined.

The interpretation of both VQ and PV curves shows how the changes in these curves impact the voltage. Therefore, its analysis is imminent, giving a quick perspective of possible scenarios where it can cause instability to the system.

For the case study, we can verify that the parameters established in the NISV are fulfilled. The defined parameters are adequate to evaluate stability since the system converges.

The case study is theoretical, so the response is with very particular conditions where we validate the performance of the index in line 13–14. operation of the same index in line 13–14 is validated.

By varying the reactive power in time, with the input of IM, it is seen that in the case of case 1, by obtaining a value of Q around 122 Mvar, this affects the link 13–14 and causes that when determining the index this gives us a value of 0.7 per value of 0.7, so the system is on the edge of instability. The system is on the verge of instability since that is close to 1 in this link. The operator will have to perform will have to take actions to mitigate this event to recover the system to optimal operating conditions to optimal operating conditions at the busbar, as shown in Fig. 4.

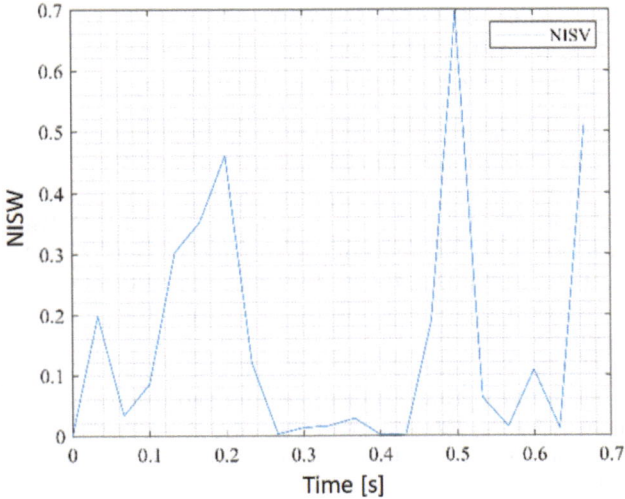

Fig. 4. Performance of the NISV index link 13–14.

5   Conclusions

The VQ Curve analysis allows us to elucidate in the same way the reactive power collapse point where the voltage goes out of its allowed limits and can cause instability in the area where the analysis is performed, as in the case of node 14 but also gives us a general idea of the best operating point for the system where the voltage at that node is maintained at 0.95 per unit.

The integration of a VQ sensitivity study for the 14-bus system shows a panorama in which it is possible to identify which is the most sensitive bus to reactive power input, and this node was node 14 since the critical point of the system to reactive power injection was identified, this value was 117.236 Mvar. It is important to remember that this value is relevant. Still, it leads us to see a system deterioration since convergence would not be possible if this limit is exceeded. Therefore, knowing the reactive power limits in the nodes is essential to guarantee operative limits.

The application of the NISV index in different scenarios, such as the case of different percentages of dynamic load, identifies not only critical points of dynamic load entry but also the safe operation when this type of load is entered. Thus, it provides a quantitative metric that helps the operator and system planner make decisions to ensure the system's operability and reliability.

References

1. Toctaquiza, J., Carrión, D., Jaramillo, M.: An electrical power system reconfiguration model based on optimal transmission switching under scenarios of intentional attacks. Energies **16**(6) (2023). https://doi.org/10.3390/en16062879

2. Jaramillo, M., Carrión, D.: Optimizing critical overloaded power transmission lines with a novel unified SVC deployment approach based on FVSI analysis. Energies **17**(9) (2024). https://doi.org/10.3390/en17092063

3. Ortega, A., Milano, F.: Stochastic transient stability analysis of transmission systems with inclusion of energy storage devices. IEEE Trans. Power Syst. **33**(1), 1077–1079 (2018). https://doi.org/10.1109/TPWRS.2017.2742400

4. Alam, M.T., Ahsan, Q.: A mathematical model for the transient stability analysis of a simultaneous AC-DC power transmission system. IEEE Trans. Power Syst. **33**(4), 3510–3520 (2018). https://doi.org/10.1109/TPWRS.2017.2781905

5. Jaramillo, M.D., Carrión, D., Muñoz, J.P.: A novel methodology for strengthening stability in electrical power systems by considering fast voltage stability index under N-1 scenarios. Energies **16**(8) (2023). https://doi.org/10.3390/en16083396

6. Carrión, D., García, E., Jaramillo, M., González, J.: A novel methodology for optimal SVC location considering N-1 contingencies and reactive power flows reconfiguration. Energies **14**(20) (2021). https://doi.org/10.3390/en14206652

7. Carrión, D., González, J.W., López, G.J., Isaac, I.A.: Alternative fault detection method in electrical power systems based on ARMA model. In: FISE-IEEE/CIGRE Conference - Living the Energy Transition. FISE/CIGRE, vol. 2019 (2019). https://doi.org/10.1109/FISECIGRE48012.2019.8984981

8. Pinzón, J.D., Colomé, D.G.: Real-time multi-state classification of short-term voltage stability based on multivariate time series machine learning. Int. J. Electr. Power Energy Syst. **108**, 402–414 (2019). https://doi.org/10.1016/j.ijepes.2019.01.022

9. Pasiopoulou, I.D., Kontis, E.O., Papadopoulos, T., Papagiannis, G.K.: Effect of load modeling on power system stability studies. Electr. Power Syst. Res. **207** (2022). https://doi.org/10.1016/j.epsr.2022.107846

10. Jalali, A., Aldeen, M.: Short-term voltage stability improvement via dynamic voltage support capability of ESS devices. IEEE Syst. J. **13**(4), 4169–4180 (2019). https://doi.org/10.1109/JSYST.2018.2882643

11. Campaña, M., Masache, P., Inga, E., Carrión, D.: Voltage stability and electronic compensation in electrical power systems using simulation models. Ingenius Revista Ciencia y Tecnología (29), 9–22 (2023). https://doi.org/10.17163/ings.n29.2023.01

12. Gnaciński, P., et al.: Power quality and energy-efficient operation of marine induction motors. IEEE Access **8**, 152193–152203 (2020). https://doi.org/10.1109/ACCESS.2020.3017133

13. Adekitan, A.I., Samuel, I., Amuta, E.: Dataset on the performance of a three phase induction motor under balanced and unbalanced supply voltage conditions. Data Br. **24** (2019). https://doi.org/10.1016/j.dib.2019.103947

14. Adekitan, A.I., Adetokun, B., Shomefun, T., Aligbe, A.: Cost implication of line voltage variation on three phase induction motor operation. Telkomnika **16**(4), 1404–1412 (2018). https://doi.org/10.12928/TELKOMNIKA.v16i4.9628

15. Adekitan, A.I., Adewale, A., Olaitan, A.: Determining the operational status of a three phase induction motor using a predictive data mining model. Int. J. Power Electron. Drive Syst. **10**(1), 93–103 (2019). https://doi.org/10.11591/ijpeds.v10.i1.pp93-103

16. Husari, F., Seshadrinath, J.: Early stator fault detection and condition identification in induction motor using novel deep network. IEEE Trans. Artif. Intell. **3**(5), 809–818 (2022). https://doi.org/10.1109/TAI.2021.3135799

17. Jain, H., Bhatti, B.A., Wu, T., Mather, B., Broadwater, R.: Integrated transmission-and-distribution system modeling of power systems: state-of-the-art and future research directions. Energies **14**(1) (2020). https://doi.org/10.3390/en14010012
18. Danish, M., Senjyu, T., Danish, S., Sabory, N., Mandal, P.: A recap of voltage stability indices in the past three decades. Energies **12**(2) (2019). https://doi.org/10.3390/en12081544
19. Lemus, A., Carrión, D., Aguire, E., Gonzalez, J.W.: Location of distributed resources in rural-urban marginal power grids considering the voltage collapse prediction index. Ingenius **28**, 25–33 https://doi.org/10.17163/ings.n28.2022.02
20. Musirin, I., Abdul Rahman, T.K.: Estimating maximum loadability for weak bus identification using FVSI. IEEE Power Eng. Rev. **22**(11), 50–52 (2002). https://doi.org/10.1109/MPER.2002.4311799
21. Samuel, I.A., Soyemi, A.O., Awelewa, A.A., Olajube, A.A., Ketande, J.: Review of voltage stability indices. IOP Conf. Ser. Earth Environ. Sci. (2021). https://doi.org/10.1088/1755-1315/730/1/012024
22. Villacrés, R., Carrión, D.: Optimizing real and reactive power dispatch using a multi-objective approach combining the ϵ-constraint method and fuzzy satisfaction. Energies **16**, 44 (2023). https://doi.org/10.3390/en16248034

Dynamic Analysis of Transient Stability in Power Systems After Optimal Transmission Switching

Joel Pineda[1]([⊠])[ID], Diego Carrión[1,2][ID], and Manuel Jaramillo[1][ID]

[1] Salesian Polytechnic University, Quito EC170702, Ecuador
jpinedag2@est.ups.edu.ec, {dcarrion,mjaramillo}@ups.edu.ec
[2] Valencia International University, Valencia 46002, Spain

Abstract. The present research aims to analyze the transient stability after optimal transmission switching (OTS), which accommodates the optimal DC power flows, leading to OTSDC, which facilitates the optimal transmission switching without affecting the load, considering both the optimal power dispatch while complying with the technical and operational constraints established by the system. This was performed for three scenarios when one, two, and three switched transmission lines. All this was an optimization case solved using mixed integer linear programming (MILP) through GAMS and MATLAB optimization software for modeling and data processing. The IEEE standard models of 14 was used to analyze the transient stability of the generation busbars under adverse scenarios. In addition, dynamic programming allowed accurate results to be obtained quickly, taking advantage of the computational advantages available.

Keywords: Transient Stability · Optimal Transmission Switching · Optimal Power Flows · Mixed Integer Linear Programming · Dynamic Programming

1 Introduction

The subsystems, generation, transmission, and distribution give rise to the Electrical Power System (EPS), which aims to meet the electrical needs of the different consumption points with quality indexes. To maintain these indexes, studies are performed on the economic dispatch, reliability, stability, safety, and selectivity of protection equipment [1].

EPS has a certain probability of failure, such as atmospheric discharges, short-circuit failures, or sudden load connection and disconnection. The failures bring with them aspects that produce variations in the voltage, angle, and frequency profiles, which mainly affect the balance of the power generated (PG), which must be equal to the power required or demanded (PD) by the users, here makes the appearance of the economic operation of the EPS, which aims to supply the end user with quality indexes minimizing the generation costs (GC) [2,3].

© The Author(s) 2025
E. M. Inga Ortega et al. (Eds.): CITIS 2024, LNNS 1331, pp. 384–394, 2025.
https://doi.org/10.1007/978-3-031-87065-1_35

These make the PG ratio equal to the PD employing the economic dispatch (ED), which minimizes the energy production cost in the generation plants, efficiently using materials, human resources, and equipment maintenance. Emphasis is also placed on minimizing supply losses, which are generated in the transmission lines (TL) when supplying energy to the different consumption points of the EPS, since given the increasing rise in prices, it is essential to make efficient and optimal use of the infrastructure to recover the invested capital and obtain benefits [1].

This research will be limited to the study of N-1 contingencies; for the process of study and analysis, a possible contingency will be identified around selecting a system component to assess its impact when put out of operation. The system modeling is done, the initial conditions are determined, and the selected component is put out of service; this analysis is carried out through power flows and simulation of fault events to evaluate the distribution of voltage, current, and active power, performing data collection to analyze its influence on the other elements of the model, With this, the results are compared with the technical specifications with the limit values of the system, the electrical protections are calibrated so that they do not operate unnecessarily, configuring them to discriminate disturbances from faults, ensuring the continuity of the electrical supply. It has been found that the operational costs remain constant even when several of them are removed and the N-1 and N-2 contingency analysis is applied. The loadability of the lines increases, but their operational limits are not violated, and no switching action would cause the disconnection of the electrical supply to the end users [4,5].

System stability is a fundamental and crucial study that is performed to determine how long the system can sustain the steady state, ensuring optimal and reliable operation of power supply to end users; therefore, it is crucial to avoid large-scale power outages as it is essential to ensure the continuity of supply for the continuation of the industrial production chain, the safety of people, operation of basic services, economic growth, protection of equipment and systems.

Reliable and stable power systems are fundamental to sustaining the lifestyle of consumers and promoting progress in modern society; therefore, a reliable power supply with quality indexes is imperative; regulation and monitoring methods such as voltage control, frequency, and coordination of protections are used to achieve stability. In addition, studies and simulations will be performed to evaluate and predict system behavior under different operating scenarios and disturbances. In the face of disturbances or faults, transient stability studies seek to determine how the system will respond to these adverse situations and recover after them [5,6].

Transient stability (TS) is studied through optimal power flows, which are helpful for EPS operation, planning, and control studies. They solve the ED and PS problems, efficiently determining the power output required to supply the different consumption points while minimizing the losses generated in the TL. This can be analyzed using the FOPs that can be performed utilizing the

DC and AC models. It should be noted that this research will make a DC analysis for which some reactive variables of the electrical components are not considered. Disregarding the reactive components results in a more simplified representation of the system, which allows a faster and more agile analysis of the EPS. In addition, the (FOP-DC) can be used to obtain initial solutions for subsequent more detailed studies that take into account reactive components, such as alternating current (AC) power flows [7,8].

Optimal transmission switching (OTS) methodology is a technique that works on the EPS topology; this optimization model brings positive responses: relief of TL overloads and readjustment of voltage profiles, solving load disconnection problems, and minimization system operating costs [9,10]. This minimizes energy transfer costs and the losses generated by supplying the supply. The OTS methodology also evaluates the loadability of TL by increasing their technical limits due to TL switching. Still, any switching action does not violate such operational limits; any improper operation would cause the system load shedding, which would accelerate or decelerate the system, which would accelerate or decelerate the mechanical power, and therefore, the frequency would unbalance the system, which would generate distortions in the voltage profiles, unnecessary operation of protections which would quickly make the system unstable, such incorrect maneuvers can overload the system causing power outages in the consumption centers to even the partial or total collapse of the EPS [11,12].

For this purpose, the TS analysis of the electrical system will be performed after applying the OTS technique, which works in TL and has low utilization without affecting the system's stability through strategic maneuvers in the topology of the transmission system. This technique aims to minimize costs in transmission and distribution using strategic maneuvers without affecting the system's stability by such maneuvers in the topology of the transmission system, to which its infrastructure is considered static [13,14].

To monitor and analyze the transient stability, the best alternative is creating and implementing a dynamic model for the OTS methodology based on optimal DC power flows, which will be called (OTSDC). This article proposes the use of the OTSDC methodology in a dynamic way for transient stability and to show the angles of the voltage profiles in the generation bars, studies that have not been demonstrated in other research. It also considers the analysis of N-1 contingencies in the electric power transmission system, with which it will be possible to evidence and analyze the voltage fluctuations in the generation busbars and the loadability in the TLs concerning the system performance without considering the PS. To address the OTSDC optimization problem, the GAMS optimization package will be used through the CPLX solver for mixed integer linear programming (MILP Mixed Integer Linear Programming), and the modeling of the OTSDC methodology will be performed to analyze the results of the implementation of the proposed model comparing the system before and after the adaptation of the process [15,16].

2   Optimal Transmission Switching Based on OPF-DC

The OTS model presented in [16] consists of strategically selecting and switching the configuration of the power transmission lines by evaluating the loadability of the TL and the magnitudes of voltages, with their angles at each node. The power generation supplied to the load, where the objective function is (1) which aims to reduce production costs, which is under the influence of (2) that affects the power that can supply the generator i of the set of generation plants:

$$min \ FO = \sum_{i=1}^{n_G} C_i P_{G_i} \tag{1}$$

$$P_{G_i}^{min} \leq P_{G_i} \leq P_{G_i}^{max} \tag{2}$$

The OTS model is subject to restrictions imposed by the power flow through the TL (3) and (4), considering the susceptance of the TL is (5), also to determine the state of the TL if it is on or off the binary state variable is called a ζ_{i-j} (6) and M (7) is the maximum flow that can support the TL before affecting its loadability and can generate adverse effects and damage the infrastructure.

$$P_{i-j} - B_{i-j} \left(\delta_i - \delta_j \right) \leq \left(1 - \zeta_{i-j} \right) M \tag{3}$$

$$P_{i-j} - B_{i-j} \left(\delta_i - \delta_j \right) \geq \left(1 - \zeta_{i-j} \right) M \tag{4}$$

$$B_{i-j} = \frac{1}{X_{i-j}} \tag{5}$$

$$\zeta_{i-j} \ \epsilon \ \{0, \ 1\} \tag{6}$$

$$M = max \left\{ B_{i-j} \left(\delta_i - \delta_j \right) \right\} \tag{7}$$

The power balance for the OTSDC model is done by disregarding the losses (8), and the limit of the power flow circulating through TL is (9). N_{SW} maximum number of allowed maneuvers (10). Table 1 shows the terminology used in the OTS mathematical model.

$$\sum_{k=1}^{n_G} P_{G_k} + P_{D_k} = \sum_{i=1}^{n_B} \sum_{j=1}^{n_B} P_{i-j}; \ \forall \ i,j \ \epsilon \ n_B \tag{8}$$

$$- P_{i-j}^{min} \zeta_{i-j} \leq P_{i-j} \leq P_{i-j}^{max} \zeta_{i-j} \tag{9}$$

$$\sum_{i=1}^{n_B} \sum_{j=1}^{n_B} \left(1 - \zeta_{i-j} \right) \leq N_{SW} \tag{10}$$

Table 1. Nomenclature of the OTS model

Variable	Description	Variable	Description
C_i	Generator cost coefficient	$P_{G_i}^{min}$, $P_{G_i}^{max}$	Active power generated limits
P_{G_i}	Active power generated at bus i	P_{i-j}^{min}, P_{i-j}^{max}	Active power flow limits
P_{i-j}	Active power flow at line i–j	δ_i	Voltage angle at bus i
P_{D_i}	Active power demanded at bus i	B_{i-j}	Susceptance at line i–j
n_G	Generators set	X_{i-j}	Reactance at line i–j
n_B	Busbars set		

3   Multi-machine Transient Stability (TS)

The study of TS in multi-machine systems considers multiple interconnected generators. It starts by considering the hypothesis of a generator system connected to an infinite bus, considering that the mechanical power in the generating units remains constant. The rotor angle is ensured to coincide with the angle of the potential difference, assuming that the demand remains continuous [17].

The multi-machine equations are found by simplifying and eliminating different nodes, retaining as the basis the internal voltage nodes of the generators by using Kron's theorem to eliminate load bars from the admittance matrix, also called bar Y for pre-fault conditions; the grid representation is modified to reflect the conditions during fault and pre-fault, With this, the multi-machine transient stability can be evidenced and evaluated, with this, an analysis of the behavior of the angle oscillations in the generation busbars with the synchronization of the transmission system can be carried out, the model presented by Saadat [18], which deals with the multi-machine transient stability, will be applied.

The multi-machine model to study transient stability is then outlined in (11), which describes the oscillation dynamics of the synchronous machines connected to the system and provides a mathematical representation of the phenomena that arise after a disturbance, thus revealing the behavior of the generators.

The electrical power output of each generator can be expressed in terms of the internal voltages of each machine. Table 2 shows the terminology used in the Multi-machine Transient Stability model.

$$\frac{H_i d^2 \delta_i}{\pi f_o dt^2} = P_{m_i} - \sum_{i=1}^{n_B} \sum_{j=1}^{n_B} |V_i|\,|V_j|\,|Y_{i-j}|\cos\left(\theta_{1-j} - \delta_i + \delta_j\right) \tag{11}$$

Table 2. Nomenclature to Multi-machine Transient Stability model

Variable	Description	Variable	Description
H_i	Inertia constant of the generator i.	Y_{i-j}	Admittance at line i–j
f_o	Frequency	θ_{i-j}	Admittance angle
P_{m_i}	Mechanical power of generator i	V_i	Voltage at the node i

4   Methodology and Problem Formulation

In the present research, the IEEE 14-bar test system will be used, in which the OTS criterion will be applied in addition to the multi-machine stability criterion, as shown in Algorithm 1.

By applying the proposed methodology, the maximum number of transmission lines that can be disconnected before losing system stability can be determined. To see the effects on stability, a random contingency is generated to verify the robustness of the EPS.

Algorithm 1 Multi-machine transient stability by OTS

Step: 1   **Input data**

EPS parameter setting

Lines: X_{1-j}, B_{i-j} and SIL

Generators: $P_{G_{min}}$, and $P_{G_{max}}$

Loads: P_{D_i}

Step: 2   **Determine the cases of commutation**

$OTSDC$

$O.F.$

   $min(1)$

$s.t.$

   $(2) - (10)$

save results

   N_{SW}, ζ_{i-j}

Step: 3   **Analyze multi-machine transient stability**

   Random contingency is generated

for $case = 1 : N_{SW}$

   Apply (11)

   save results

   Generators angles

end for

Step: 4   **Show results**

---

5   Results Analisys

The proposed method for the dynamic analysis of the TS in the power systems after the optimal switching of TL was carried out through the implementation of the algorithms on Matlab R2022b, from which a Matlab communication link was made with GAMS to carry out the methodology proposed above. Those mentioned above were carried out on a computer with an Intel Core i7-9750H processing center, 12 Gb of Random Access Memory, and Windows 11; also in GAMS, the CPLEX solver was used for the optimal solution of the OTSDC employing mixed integer linear programming of the IEEE 14 busbar system.

The proposed methodology was tested in the IEEE 14 bus test system, from which 2 cases were selected to show and compare their results; the first case considers a switching of up to 3 lines as maximum plus a contingency on line 6–11 and a second case that in the same way is switched up to 3 lines but now a contingency on line 9–14 is considered.

In the first case of study, the power system's operation is unstable, while in the second case, the system maintains its operation in a stable state. This can be verified by comparing the oscillation of the generators' angles concerning the generator slack.

For the first case analyzed, the maximum number of transmission lines that can be switched is 3, as shown in Fig. 1; considering that there will be an additional contingency, in this case, line 6–11, the commuted lines are shown in Table 3.

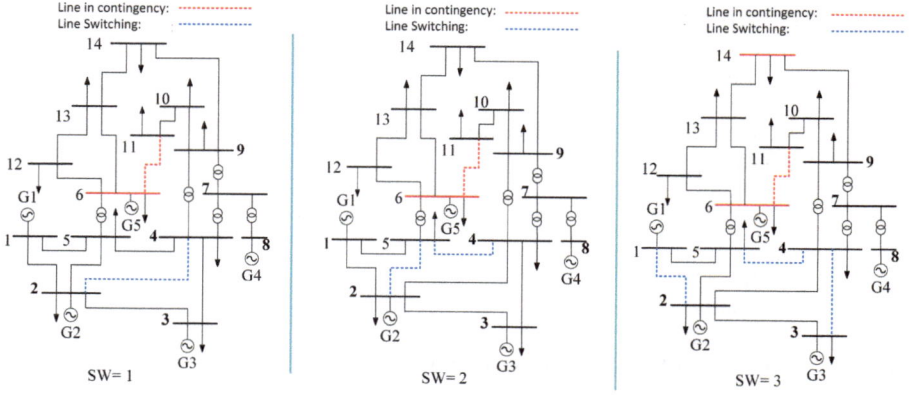

Fig. 1. OTS scenarios for fault in line 6–11

Table 3. Lines switching

N_{SW}	Lines switching
1	[2–4]
2	[2–5] and [4–5]
3	[1–2], [3–4] and [4–5]

Figure 2 shows the behavior of the angles of the EPS generators to generator 1, which is the EPS slack bar, for each of the scenarios presented, as can be seen in Fig. 3 for the OTSDC 2-4 scenario compared to the OTSDC 2-5 and 4-5 scenarios, there is a more pronounced disturbance in the angle $\delta(2\text{-}1)$ at an optimal switching SW = 2, The same is not valid for the OTSDC 1-2, 3-4 and 4-5 scenarios since the angle is triggered 680 ms after the fault occurs and therefore the system becomes unstable.

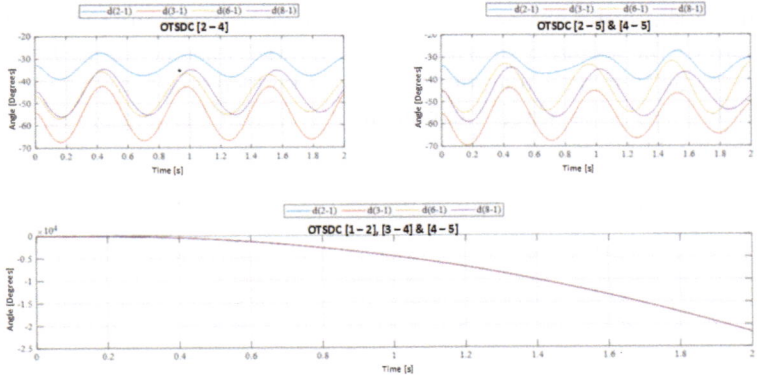

Fig. 2. Angular oscillation of generators for fault in line 6–11

For the second case analyzed, the maximum number of transmission lines that can be switched is 3, as shown in Fig. 3; considering that there will be an additional contingency, in this case, line 9–14, the commuted lines are shown in Table 3.

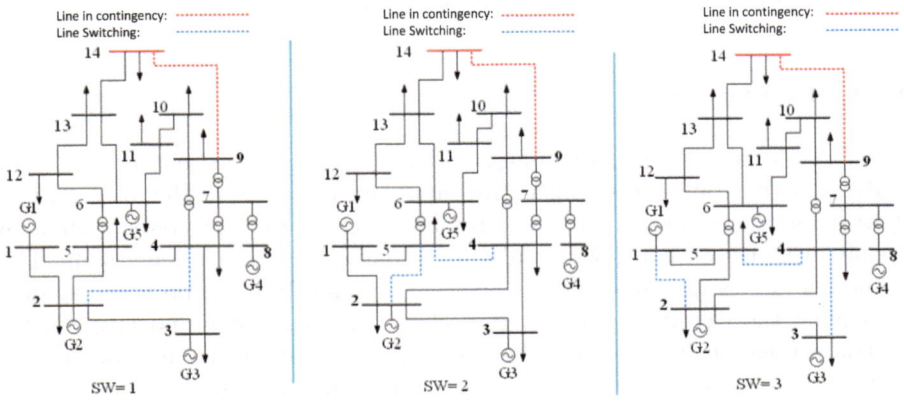

Fig. 3. OTS scenarios for fault in line 9–14

Figure 4 shows the behavior of the angles of the generators regarding the slack generator bar, which is node 1, for each of the scenarios considered, as shown in Fig. 4, corresponding to case B, in which it can be seen that for the scenario of a single switched line 2–4 compared to two lines 2–5 and 4–5 shows changes in oscillations $\delta(6\text{-}1)$ and $\delta(8\text{-}1)$ before the optimal switching SW=2 for this scenario the system remains stable and likewise preserving the stability of the system for the scenario where the OTS allows 3 switched lines 1–2, 3–4 and 4–5 but showing changes in oscillations $\delta(6\text{-}1)$ and $\delta(8\text{-}1)$ before the optimal switching SW=2 for this scenario the system remains stable and likewise preserving the stability of the system for the scenario where the OTS allows 3 switched lines 1–2, 3-4 and 4-5 but showing that the values obtained have doubled compared to the scenario of switching line 2-4 showing a significant increase in the angles δ(2-1), δ(3-1), δ(6-1) and δ(8-1).

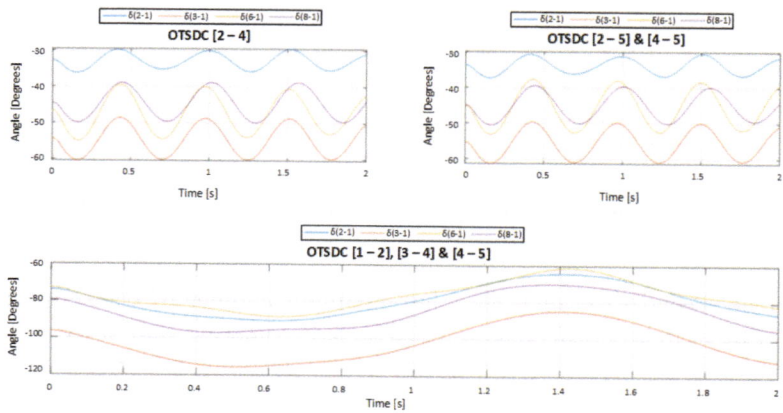

Fig. 4. Angular oscillation of generators for fault in line 6–11

6   Conclusions

In the research, a new methodology was proposed to carry out the dynamic analysis of the transient stability after applying the OTS methodology, which allowed for the optimization of the resources of the electric power system by reconfiguring the energy flows circulating through the transmission system. This methodology was validated in different scenarios and the other proposed models, identifying the electrical parameters such as the magnitude and angle of the electric voltage in the generation bars and optimum power flow to minimize the operating costs, maintaining the system's operability in a stable state.

Through the bibliographic review, it was possible to identify and propose a methodology not evidenced in previous works regarding the OTS, contributing to current advances validated in different power systems. The novelty of this methodology is that it offers a dynamic response to transient stability under adverse conditions. Transient stability studies were performed for a multi-machine system by which it was possible to identify that some TL cannot be switched since if these actions are performed, the system will stop working correctly, making the system inoperable and losing the synchronism from the perspective of the rotor angle of the generators causing instability and collapse of the system.

References

1. Villacrés, R., Carrión, D.: Optimizing real and reactive power dispatch using a and fuzzy satisfaction. Energies **16**(24) (2023). https://doi.org/10.3390/en16248034
2. Toctaquiza, J., Carrión, D., Jaramillo, M.: An electrical power system reconfiguration model based on optimal transmission switching under scenarios of intentional attacks. Energies **16**(6) (2023). https://doi.org/10.3390/en16062879
3. García, E., Águila, A., Ortiz, L., Carrión, D.: Optimal resource assignment in hybrid microgrids based on demand response proposals. Ustainability **16**(5) (2024). https://doi.org/10.3390/su16051797
4. Fan, N., Chen, R., Watson, J.: N-1-1 contingency-constrained optimal power flow by interdiction methods. In: 2012 IEEE Power and Energy Society General Meeting, pp. 1–6 (2012). https://doi.org/10.1109/PESGM.2012.6345713
5. Lejeune, M.A., Dehghanian, P.: Optimal power flow models with probabilistic guarantees: a Boolean approach. IEEE Trans. Power Syst. **35**(6), 4932–4935 (2020). https://doi.org/10.1109/TPWRS.2020.3016178
6. Carrión, D., González, J.W.: Optimal PMU location in electrical power systems under N-1 contingency, no. 1, pp. 165–170 (2018). https://doi.org/10.1109/INCISCOS.2018.00031
7. Hedman, K.W., Oren, S.S., O'Neill, R.P.: A review of transmission switching and network topology optimization. In: 2011 IEEE Power and Energy Society General Meeting, pp. 1–7 (2011). https://doi.org/10.1109/PES.2011.6039857
8. Pinzon, S., Carrion, D., Inga, E.: Optimal transmission switching considering N-1 contingencies on power transmission lines. IEEE Lat. Am. Trans. **19**(4), 534–541 (2021). https://doi.org/10.1109/TLA.2021.9448535
9. Kostiuk, V.O., Kostyuk, T.O.: Power system steady-state stability criteria and the Jacobian of dynamical systems. In: EUROCON 2021 - 19th IEEE In IEEE EUROCON 2021-19th International Conference on Smart Technologies Proceedings, pp. 523–530 (2021). https://doi.org/10.1109/EUROCON52738.2021.9535579
10. Ng, Y., Misra,S., Roald, L.A., Backhaus, S.: Statistical learning for DC optimal power flow. In 2018 Power Systems Computation Conference (PSCC), pp. 1–7 (2018). https://doi.org/10.23919/PSCC.2018.8442859
11. Carrión, D., Palacios, J., Espinel, M., González, J.: Transmission expansion planning considering grid topology changes and N-1 contingencies criteria. In: Recent Advances in Electrical Engineering, Electronics and Energy: Proceedings of the CIT 2020 Volume 1 (2021). https://doi.org/10.1007/978-3-030-72208-1_20
12. Brown, W., Moreno, E.: Transmission-line switching for load shed prevention via an accelerated linear programming approximation of AC power flows. IEEE Trans. Power Syst. **35**(4) (2020). https://doi.org/10.1007/978-3-030-72208-1_20. '
13. Jabarnejad, M.: Approximate optimal transmission switching. Electr. Power Syst. Res. **161**, 1–7 (2018). https://doi.org/10.1016/j.epsr.2018.03.021
14. Hatziargyriou, N., et al.: Definition and classification of power system stability - revisited & extended. IEEE Trans. Power Syst. **36**(4), 3271–3281 (2021). https://doi.org/10.1109/TPWRS.2020.3041774
15. Peng, D., Huang, M., Li, J., Sun, J., Zha, X., Wang, C.: Large-signal stability criterion for parallel-connected DC-DC converters with current source equivalence. IEEE Trans. Circ. Syst. II Express Briefs **66**(12), 2037–2041 (2019). https://doi.org/10.1109/TCSII.2019.2895842
16. Fisher, E.B., Member, S., Neill, R.P.O., Ferris, M.C.: Optimal transmission switching. IEEE Trans. Power Syst. **23**(3), 1346–1355 (2008) https://doi.org/10.1109/PES.2009.5275905

17. Carrión, D., García, E., Jaramillo, M., González, J.W.: A novel methodology for optimal SVC location considering N-1 contingencies and reactive power flows reconfiguration. Energies **14**(20) (2021) https://doi.org/10.3390/en14206652
18. Saadat, H.: Power System Analysis Third Edition. PSA Publishing LLC; 3rd edn. (2011). https://doi.org/10.3390/en14206652

Heuristic Tuning and Decoupling Strategies for Multivariable Systems: An Integrated TIA Portal and Matlab-Simulink Approach

Diego Salazar[1] , Wilson Pavón[1]([✉]) , William Montalvo[1] , and Julio Cesar Zambrano[2]

[1] Automation and Control Postgraduate Program,
Universidad Politécnica Salesiana, Quito, Ecuador
wpavon@ups.edu.ec
[2] Automation and Control Postgraduate Program,
Universidad Politécnica Salesiana, Cuenca, Ecuador

Abstract. This paper proposes a comprehensive solution to the advanced control problem of multivariable systems using Siemens' TIA (Totally Integrated Automation) Portal LSIM (Linear Simulation) package. The process model is validated via Matlab-Simulink to compare and support the implementation of the TIA Portal. The control system is then developed using the Relative Gain Array (RGA) concept to minimize cross-coupling effects between inputs. The study demonstrates how decoupling functions and precise tuning improve system performance, reducing interference and enhancing stability. The results confirm that this integrated control framework improves system efficiency, reliability, and adaptability for industrial applications.

Keywords: Decoupling · Industrial Automation · Multivariable Control · Optimization · PID Tuning · Process Control · TIA Portal

1 Introduction

In industrial process control, achieving optimal performance and efficiency remains a crucial objective, especially with the growing complexity of multivariable systems. Adopting advanced control strategies and tools is increasingly vital for managing the intricate interactions within these systems. This research aims to enhance the control of multivariable processes by implementing a robust methodology that leverages Siemens' TIA (Totally Integrated Automation) Portal, Matlab-Simulink, and cutting-edge control techniques [Tari and Lanusse, 2018, Vallejos, 2024].

This study focuses on developing a systematic framework for integrating multivariable control strategies within the TIA Portal environment, emphasizing precise control tuning methodologies. The primary objective is to establish an effective multivariable control system that minimizes the impact of cross-coupling between input variables while maintaining overall system stability. To

E. M. Inga Ortega et al. (Eds.): CITIS 2024, LNNS 1331, pp. 395–405, 2025.
https://doi.org/10.1007/978-3-031-87065-1_36

achieve this objective, a structured, stepwise approach is undertaken, beginning with implementing the multivariable process using the LSIM library in TIA Portal [Ding et al., 2011, Vilanova and Katebi, 2012].

Subsequently, the process model is rigorously validated through comparative analysis in Matlab-Simulink. Disc discrepancies can be identified and resolved by subjecting both models to identical step signal excitations. This comparative analysis ensures model fidelity, enabling the implementation of Relative Gain Array (RGA) techniques for decoupling. Ultimately, this framework demonstrates the enhanced control performance achievable through systematic decoupling, contributing to a deeper understanding of multivariable control in industrial applications [Xiong et al., 2007].

Decoupling MIMO systems is of significant interest to the research community due to its potential to significantly enhance control accuracy and efficiency in complex industrial processes. Decoupling techniques improve system stability and performance by effectively minimizing cross-coupling interactions, enabling more precise and reliable operations. This is particularly crucial in industries with prevalent multivariable processes, as it leads to optimized resource utilization and reduced operational costs. Furthermore, advancing decoupling methodologies contributes to the broader field of control theory, fostering innovation in multivariable system management [Liu et al., 2019]. The structure of this paper is as follows: Sect. 2 provides an in-depth discussion of the methodology and problem formulation for multivariable control strategies, offering a clear rationale behind the chosen approach. Section 3 presents the simulation results, accompanied by insightful analysis and commentary, highlighting key observations and implications for practical implementation. Finally, Sect. 4 draws conclusions based on the study's critical findings and offers recommendations for future exploration in this field, outlining potential avenues for further research to build upon the results presented.

2   Methodology

The foundation of our control strategy lies in carefully selecting control loops through the RG concept. Subsequently, PID (Proportional-Integral-Derivative) controllers are designed for each loop using the Biggest Logarithmic Tuning (BLT) method, enhancing open-loop responses and improving overall system performance [Khandelwal and Detroja, 2020].

The comprehensive algorithm presented in Algorithm 1 provides an integrated solution for implementing and optimizing a multivariable control system using TIA Portal, Matlab-Simulink, and advanced control techniques. The procedure starts by creating a Function Block (FB) in TIA Portal, where individual transfer functions are instantiated using the LSIM library to replicate the multivariable process. Inputs are connected according to their associated transfer functions to generate the system's outputs, which are summed for an accurate process representation. Comparing the system responses with a Matlab-Simulink model validates the integrity of the TIA Portal implementation. The subsequent tuning of PID controllers is carried out using the BLT method to improve loop responses. These controllers are then implemented within TIA Portal via the PID Compact tool, ensuring precise and robust control of each loop [Prasad et al., 2013, Liang et al., 2020].

Algorithm 1 Multivariable Control System Implementation with PID Tuning and Decoupling

---

1: Transfer function matrix $G(s)$, LSIM library in TIA Portal, RGA matrix
2: **procedure** IMPLEMENTCONTROLSYSTEM
3:     Create FB (Function Block) named 'EXA' in TIA Portal
4:     **for** each $G_{i,j}(s)$ in $G(s)$ **do**
5:         Instantiate an LSimPT1 block for the transfer function
6:         Connect inputs x_1 and x_2 appropriately
7:     **end for**
8:     Sum outputs to obtain final responses y_1 and y_2 per Equations:
9:     $y_1 = G_{1,1}x_1 + G_{1,2}x_2$
10:    $y_2 = G_{2,1}x_1 + G_{2,2}x_2$
11:    Compare responses with Matlab-Simulink for validation
12:    **for** each control loop $C_i(s)$ **do**
13:        Design PID controller using the BLT method:
14:        $K = \frac{1}{K_p}\frac{\tau}{\tau_c+\theta}$, $T_i = \min(\tau, 4\tau_c + \theta)$
15:        Implement controller in TIA Portal with PID Compact tool
16:    **end for**
17:    Calculate decoupling transfer functions to minimize cross-coupling:
18:    $D_{x_1,x_2} = -\frac{G_{1,2}}{G_{1,1}} = -1.476\frac{16.7s+1}{21s+1}e^{-2s}$
19:    $D_{x_2,x_1} = -\frac{G_{2,1}}{G_{2,2}} = 0.34\frac{14.4s+1}{10.9s+1}e^{-4s}$
20:    Integrate these decoupling functions into TIA Portal as compensators
21:    Visualize and monitor performance using TIA Portal's Trace tool
22: **end procedure**

The parameters outlined in Table 1 are essential for implementing a robust multivariable control system. By leveraging the LSIM library in TIA Portal and applying RGA techniques for decoupling, this study ensures precise PID tuning and effective cross-coupling minimization, enhancing the overall stability and performance of complex industrial processes.

The illustrated decentralized control structure in Fig. 1 depicts a multiple-input, multiple-output (MIMO) system within a multivariable control framework. Each input, r_1 and r_2, is processed through its respective controller $C_1(s)$ and $C_2(s)$, which then generate control signals u_1 and u_2. These signals drive the system through specific transfer functions: $G_{11}(s)$ and $G_{22}(s)$ are the main diagonal elements, while $G_{12}(s)$ and $G_{21}(s)$ represent the cross-coupling between outputs y_1 and y_2. This structure simplifies complex multivariable interactions,

Table 1. Parameters for Multivariable Control System Implementation

Variable	Explanation
$G(s)$	Transfer function matrix
$G_{i,j}(s)$	Individual transfer functions in the matrix $G(s)$
$C_i(s)$	Control loop PID controller
K	Controller gain calculated using BLT method
T_i	Integral time constant for PID controller
D_{x_1,x_2}	Decoupling transfer function to minimize cross-coupling from x_2 to x_1
D_{x_2,x_1}	Decoupling transfer function to minimize cross-coupling from x_1 to x_2
y_1	Output response influenced by inputs x_1 and x_2
y_2	Output response influenced by inputs x_1 and x_2
τ	Time constant of the system
τ_c	Chosen time constant for PID tuning
θ	Delay in the system

emphasizing individual control loops to minimize interference between them [Mitrishkin et al., 2021, Pandey et al., 2022]. Additionally, Fig. 2 presents a control scheme featuring a decoupler within the pseudo-plant, further reducing interactions between loops. This configuration enhances control performance by isolating the effects of each control action. Feedback signals are continuously utilized to adjust the control actions, ensuring the system's stability and robustness against disturbances.

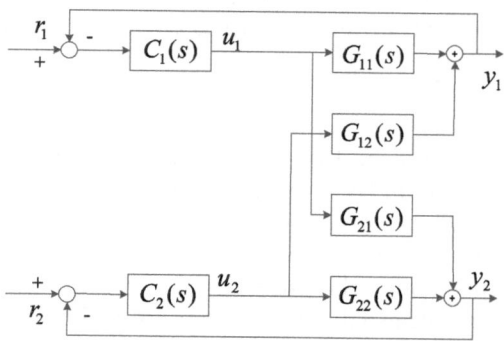

Fig. 1. Illustrated decentralized control structure of MIMO system

In this research methodology, the Eqs. 1, and 2 are foundational for accurately modeling and understanding the multivariable control system. Equation 1 provides the general relationship between the outputs (y_1, y_2) and inputs (x_1, x_2) via transfer functions $G_{i,j}$. This framework shows how each input influences multiple outputs, revealing potential cross-coupling effects. Equation 2 extends

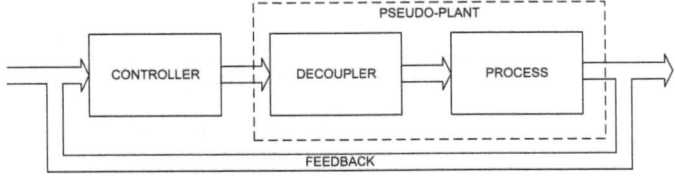

Fig. 2. General structure of Decoupling Control System

this formulation by specifying the actual transfer functions and their respective gains, time constants, and delays, which are essential parameters for accurate modeling in TIA Portal and Matlab-Simulink. The precise matrix representation illustrates how each transfer function contributes to the overall multivariable system response [Ghosh and Pan, 2021, Rashad et al., 2017].

To accurately model this system, individual instances of the blocks should be created for each $G_{i,j}$ transfer function. Their outputs are then appropriately summed to represent the relationship between all inputs and each output. This ensures a comprehensive reflection of the complete multivariable system response, as illustrated in Eq. 1. The modeled responses accurately reflect real-world system behavior by aligning the system cycle with the desired delay, allowing for realistic system identification and control tuning. This methodology forms the backbone of the control strategy, ensuring that each control loop is effectively represented, facilitating accurate PID tuning, and enabling precise decoupling of the multivariable system [Puviyarasi and Murukesh, 2023].

$$y_1 = G_{1,1}x_1 + G_{1,2}x_2,$$
$$y_2 = G_{2,1}x_1 + G_{2,2}x_2. \tag{1}$$

$$\begin{bmatrix} y_1 \\ y_2 \end{bmatrix} = \begin{bmatrix} \frac{-12.8}{16.7\,s+1}e^{-s} & \frac{-18.9}{21\,s+1}e^{-3\,s} \\ \frac{6.6}{10.9\,s+1}e^{-7\,s} & \frac{-19.4}{14.4\,s+1}e^{-3\,s} \end{bmatrix} \begin{bmatrix} x_1 \\ x_2 \end{bmatrix} \tag{2}$$

RGA technique is used to identify control loops with the greatest influence on system outputs. Initially, the open-loop gain matrix K for the 2×2 system is defined, as shown in Eq. 3. The coefficient $\lambda_{1,1}$ for a 2×2 system is calculated using Eq. 4. This coefficient informs the construction of the Λ matrix, presented in Eq. 5, which reveals that pairing the control loops with $G_{1,1}$ and $G_{2,2}$ is optimal. These functions have positive coefficients above 0.5, indicating effective loop pairing. Additionally, the Niederlinski Index (NI), which assesses the potential for stable control, is calculated using Eq. 6. For a 2×2 system, it simplifies to Eq. 7. As the NI is greater than zero, it confirms the system is controllable, supporting the design and tuning of the PID controllers [Pandey et al., 2023].

$$[K] = \begin{bmatrix} K_{1,1} & K_{1,2} \\ K_{2,1} & K_{2,2} \end{bmatrix} \tag{3}$$

$$\lambda_{1,1} = \frac{K_{1,1}K_{2,2}}{K_{1,1}K_{2,2} - K_{1,2}K_{2,1}} \tag{4}$$

$$\Lambda = \begin{bmatrix} \lambda_{1,1} & 1-\lambda_{1,1} \\ 1-\lambda_{1,1} & \lambda_{1,1} \end{bmatrix} = \begin{bmatrix} 0.6656 & 0.3344 \\ 0.3344 & 0.6656 \end{bmatrix} \tag{5}$$

$$NI = \frac{|K|}{\prod_{i=1}^{n} K_{i,i}} \tag{6}$$

$$NI = \frac{1}{\lambda_{1,1}} = \frac{1}{0.6656} = 1.502 \tag{7}$$

The optimal control loop pairings guide both systems' PID controller gain design. For the first loop (x_1 and y_1), the transfer function $G_{1,1}$ is shown in Eq. 8. The tuning criterion requires $\tau_c \geq \tau$ for stability, leading to the selection of an appropriate τ_c. The controller gain K is derived from Eq. 9, while the integral time constant T_i is calculated as shown in Eqs. 8, and 10. To decouple the multivariable system, transfer functions are derived to relate one input's direct effect to the other input's indirect effect. These decoupling functions, shown in Eqs. 11 and 12, minimize the coupling effects between x_1 and x_2.

$$G_{1,1} = \frac{-12.8}{16.7\,s+1}e^{-s} = \frac{K_p}{\tau s+1}e^{-\theta s} \tag{8}$$

$$K = \frac{1}{K_p}\frac{\tau}{\tau_c+\theta} \tag{9}$$

$$T_i = \min(\tau, 4\tau_c+\theta) \tag{10}$$

$$Do_{x_1,x_2} = -\frac{G_{1,2}}{G_{1,1}} = -1.476\frac{16.7\,s+1}{21\,s+1}e^{-2s} \tag{11}$$

$$Do_{x_2,x_1} = -\frac{G_{2,1}}{G_{2,2}} = 0.34\frac{14.4\,s+1}{10.9\,s+1}e^{-4s} \tag{12}$$

3   Results Analysis

The -Simulink implementation shown in Fig. 3 illustrates the interconnected transfer functions for each $G_{i,j}$ block, configured using the Transfer Function block. Each gain and delay is aligned with the state equations described in Eq. 1, the parameters are in the Table 2. The inputs x_1 and x_2 are connected to each block to model the system response comprehensively. The resulting outputs y_1 and y_2 are obtained via summation, reflecting Eq. 1. Additionally, initial conditions and reference signals have been set to observe system behavior. The results confirm accurate modeling of cross-coupling effects while enabling performance tuning.

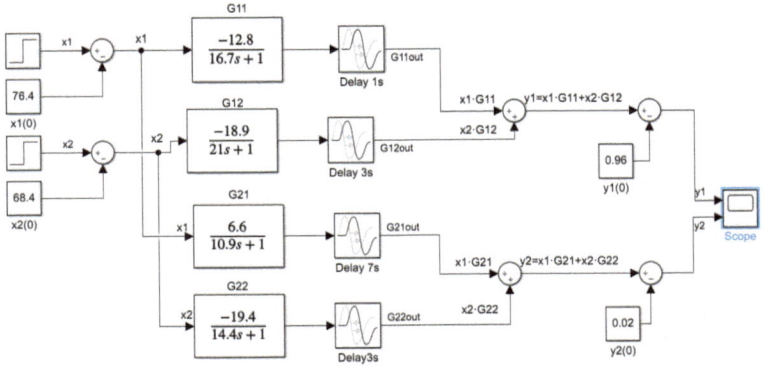

Fig. 3. Simulink model showing the interconnected transfer functions G_{11}, G_{12}, G_{21}, and G_{22} with delays.

Table 2. Transfer Function Parameters Implementation

Transfer Function	Gain	Time Constant	Delay
G_{11}	−12.8	16.7 s	1 s
G_{12}	−18.9	21 s	3 s
G_{21}	6.6	10.9 s	7 s
G_{22}	−19.4	14.4 s	3 s

In the TIA Portal environment, PID controllers are integrated using the PID Compact tool. This setup involves connecting the Setpoint SPY1, which controls G_{11}, to its corresponding input and output variables, providing effective control for the transfer function. The input variable $Y1$ and output variable $X1$ are linked accordingly, ensuring accurate response to control commands. Activation variables are also integrated to enable proper mode selection and error handling. The TIA Portal Trace tool analyzes and visualizes controller responses, facilitating a comprehensive performance evaluation. The multivariable control system is effectively optimized by configuring these parameters through precise loop tuning. The decoupling system further enhances performance by reducing cross-coupling effects. Overall, the PID Compact tool enables seamless integration and adjustment of controllers for robust multivariable control.

The input signals for x_1 and x_2 are set to 75 and 65, respectively. Figure 4 illustrates how the outputs decrease after a certain period, aligning with the -Simulink simulation. Both outputs are limited by the set safety constraints, resulting in y_1 and y_2 lowering to zero. The step changes in x_1 and x_2 directly impact the outputs, showing the responses of Proceso.X1 and Proceso.X2 over time. This visualization confirms that the control logic implemented with the PID Compact tool works as expected, ensuring that the safety limits prevent excessive output fluctuations. These controlled responses demonstrate the effectiveness of the control system's design, particularly in reducing cross-coupling effects between the inputs and maintaining stable output performance.

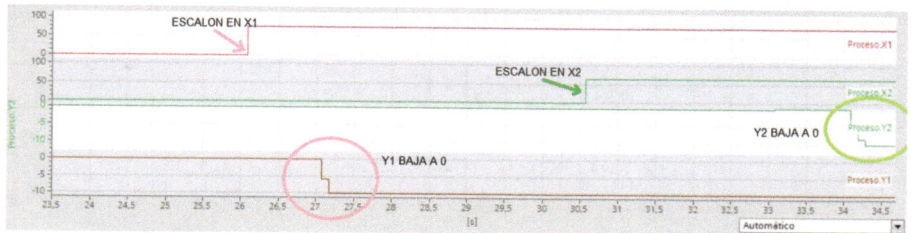

Fig. 4. System Excitation in TIA PORTAL

The two figures illustrate how the control system responds to changes in setpoints for both y_1 and y_2. Figure 5 demonstrates the impact of adjusting the setpoint for y_1 using the PID Compact object for G_{11}. The adjustment involves moving SPY1 up and then down, with y_1 responding accordingly. The graph also captures how the control direction and inversion are configured internally to reflect the negative sign of G_{11}, while the other inputs and outputs are set accurately.

The attached figure demonstrates that the control system effectively maintains each output while minimizing interference, highlighting the successful decoupling between x_1 and x_2. Adjusting the setpoints for y_1 and y_2 independently shows no noticeable impact on the other output, confirming that the decoupling functions are correctly implemented and effective. These results underline the importance of integrating decoupling strategies into the control system design. By eliminating cross-coupling effects, the decoupling functions ensure that each control loop remains focused on its intended task, promoting reliable and accurate control actions. The trace plots provide valuable insights into the system's response to varying input conditions, proving that the TIA Portal-based control system is well-suited for managing complex multivariable processes. This implementation improves stability and accuracy and establishes a robust framework for handling future challenges in similar control applications.

The results, summarized in Table 3, highlight the effectiveness of the PID controllers designed for the transfer functions $G_{1,1}$ and $G_{2,2}$. By carefully tuning each controller's parameters, the system achieves stable and responsive control across x_1-y_1 and x_2-y_2 loops. These adjustments directly translate to precise control actions, maintaining system stability under varying input conditions. Decoupling transfer functions Do_{x_1,x_2} and Do_{x_2,x_1} are strategically implemented to address cross-coupling issues in the multivariable system. Their inclusion ensures that each control loop operates independently, mitigating the impact of one input on the other's output. This independent control results in a more reliable and robust control system, significantly enhancing system performance

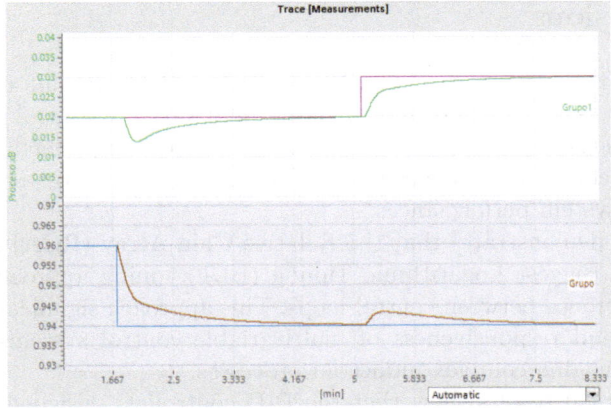

Fig. 5. Trace graph showing the effect of changing the setpoint of y_1 using the PID Compact tool for G_{11}.

by reducing interference between loops. The system is thus well-suited to handle the intricacies of industrial applications, providing stable, decoupled control for processes with interconnected variables (Fig. 6).

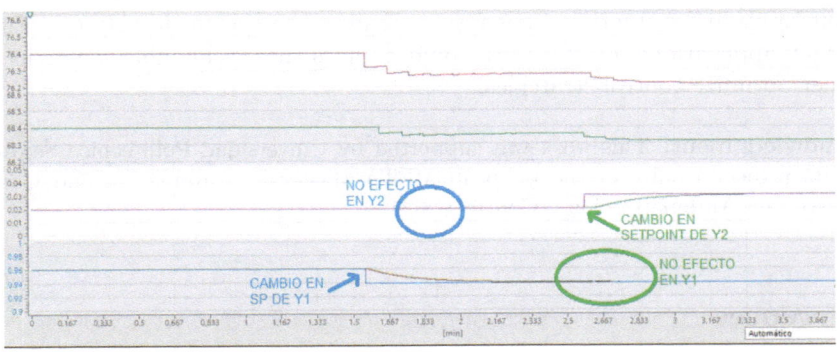

Fig. 6. Trace graph showing effective decoupling between outputs y_1 and y_2.

Table 3. PID Controller Parameters for $G_{1,1}$ and $G_{2,2}$

Transfer Function	K_p	τ		θ
$G_{1,1}$	-12.8	16.7		1
$G_{2,2}$	-19.4	14.4		3
Controller Gain	K	**Integral Time Constant T_i**	τ_c	
$G_{1,1}$	-0.217	16.7	5	
$G_{2,2}$	-0.0825	14.4	6	

4   Conclusions

The integrated use of Siemens' TIA Portal and Matlab-Simulink offers a robust framework for validating and implementing multivariable control systems. This methodology ensures accurate modeling and control design, facilitating precise tuning and decoupling strategies that effectively minimize cross-coupling effects and enhance system performance.

The study demonstrates that the Relative Gain Array (RGA) method, coupled with the Biggest Logarithmic Tuning (BLT) tuning approach, effectively reduces interference between control loops. This approach significantly improves the stability and responsiveness of multivariable control systems, proving its efficacy in managing complex industrial processes.

The numerical results show that the PID controllers designed for $G_{1,1}$ and $G_{2,2}$ achieve stable and responsive control with gains of -0.217 and -0.0825, respectively. The decoupling transfer functions D_{x_1,x_2} and D_{x_2,x_1} effectively mitigate cross-coupling, resulting in improved system stability and reliable performance across varying input conditions.

This integrated control approach presents significant opportunities for enhancing industrial automation by providing reliable and adaptable solutions for multivariable processes. However, limitations include the need for extensive initial setup and potential challenges in scaling the system for larger or more complex applications. Future work could address these scalability concerns and further optimize control strategies.

Acknowledgment. This work was supported by Universidad Polit'ecnica Salesiana and its project results in the program META-Master in Automation and Control (Maestría en Automatización y Control)-2024.

References

Ding, C., Gajdusek, M., Damen, A.A., Van Den Bosch, P.P.: Optimal Static Decoupling for the Decentralized Control: An Experimental Study, vol. 44. IFAC (2011)

Ghosh, S., Pan, S.: Centralized PI controller design method for MIMO processes based on frequency response approximation. ISA Trans. **110**, 117–128 (2021)

Khandelwal, S., Detroja, K.P.: The optimal detuning approach based centralized control design for MIMO processes. J. Process Control **96**, 23–36 (2020)

Liang, W., Chen, H.B., He, G., Chen, J.: Model order reduction based on dynamic relative gain array for MIMO systems. IEEE Trans. Circ. Syst. II Express Briefs **67**(11), 2507–2511 (2020)

Liu, L., Tian, S., Xue, D., Zhang, T., Chen, Y.Q., Zhang, S.: A Review of Industrial MIMO Decoupling Control. Int. J. Control Autom. Syst. **17**(5), 1246–1254 (2019)

Mitrishkin, Y., Pavlova, E., Patrov, M.: Design and comparison of plasma H∞ loop shaping and RGA-H∞ double decoupling multivariable cascade magnetic control systems for a spherical tokamak. Adv. Syst. Sci. Appl. **21**(1), 22–45 (2021)

Pandey, S.K., Dey, J., Banerjee, S.: Generalized discrete decoupling and control of MIMO systems. Asian J. Control **24**(6), 3326–3344 (2022)

Pandey, S.K., Dey, J., Banerjee, S.: Control of generalized decoupled MIMO systems based on linear programming method. IETE J. Res. **70**(4), 3901–3927 (2023)

Prasad, G.D., Manoharan, P.S., Ramalakshmi, A.P.: PID control scheme for twin rotor MIMO system using a real valued genetic algorithm with a predetermined search range. Proceedings of 2013 International Conference on Power, Energy and Control, ICPEC 2013, pp. 443–448 (2013)

Puviyarasi, B., Murukesh, C.: Design and implementation of adaptive mixed fuzzy controller for MIMO nonlinear systems. Math. Comput. Simul. **203**, 71–91 (2023)

Rashad, R., El-Badawy, A., Aboudonia, A.: Sliding mode disturbance observer-based control of a twin rotor MIMO system. ISA Trans. **69**, 166–174 (2017)

Tari, M., Lanusse, P.: A MIMO robust design of a PID for refrigeration systems based on vapour compression. IFAC-PapersOnLine **51**(4), 871–876 (2018)

Vallejos, W.D.P.: Innovative control paradigms for dc motors control. In: Digital Technology for Smart Grid Innovative Algorithmic Solutions for Engineering Problems, pp. 147–165. Editorial Universitaria Abya-Yala (2024)

Vilanova, R., Katebi, R.: 2-DoF Decoupling Controller Formulation for Set-Point Following on Decentraliced PI/PID MIMO systems, vol. 2. IFAC (2012)

Xiong, Q., Cai, W.J., He, M.J.: Equivalent transfer function method for PI/PID controller design of MIMO processes. J. Process Control **17**(8), 665–673 (2007)

CFD Analysis and Improvement Proposal for the Fume Extraction System in a Welding Workshop

Jaime Cacpata-Bastidas[1] , Xavier Vaca[1]([⊠]) , Diego Cevallos-Yánez[1] ,
and Isaac Simbaña[2]

[1] Research Group in Engineering, Productivity, and Industrial Simulation (GIIPSI), Universidad Politécnica Salesiana, Quito, Ecuador
xvaca@ups.edu.ec
[2] Electromechanical Career's Mechanical Engineering and Pedagogy Research Group (GIIMPCEM), Instituto Superior Universitario Sucre, Quito, Ecuador

Abstract. This research aimed to improve fume extraction in a welding workshop using Computational Fluid Dynamics (CFD) analysis. The study focused on the central extraction hood of this workshop at an educational institution, where welding practices are regularly conducted. A new hood design was proposed to reduce the concentration of gases in the breathing zone. The gas flow was analyzed through simulations to select the best alternative, enhancing ventilation and ensuring safety in confined welding environments. The new design considered variables such as the angle of the hopper, the suction area, and the number of slots. The Taguchi statistical method evaluated three factors at three levels, resulting in nine experimental trials. The analysis revealed that the new design significantly improves smoke extraction at the welding point. The current system generates an absorption velocity of 0.006 m/s, while the proposed system would achieve a velocity of 0.78 m/s, meeting the recommended standard of 0.5 to 1.0 m/s. This validates the selected hood prototype design, demonstrating its effectiveness in enhancing the extraction system.

Keywords: Fume · Welding · CFD · Ventilation · Extraction

1 Introduction

In the metalworking industry, welding is essential for joining metals, exposing millions of workers daily to hazardous gases that can cause health issues ranging from respiratory irritations to severe lung diseases. Due to the challenges of directly measuring welding gas emissions in certain environments, numerical simulations have become crucial for understanding and predicting the behavior of these emissions [1]. This project aims to optimize a ventilation system using Computational Fluid Dynamics (CFD) to reduce the concentration of gases in the breathing zone of a welding workshop, thereby enhancing worker safety and health.

© The Author(s) 2025
E. M. Inga Ortega et al. (Eds.): CITIS 2024, LNNS 1331, pp. 406–416, 2025.
https://doi.org/10.1007/978-3-031-87065-1_37

Welding is a widely used process in the industrial and metalworking sectors [2]. It is estimated that around 11 million people worldwide work as welders, and another 110 million workers occasionally use welding technology, according to a study by Lehnert et al. [3]. Various studies on the health effects of welding gases have shown that the gases produced during welding are linked to multiple adverse effects, such as respiratory tract irritations, bronchitis, and impaired lung function [4]. Therefore, it is crucial to eliminate emissions and exposures from the welding process to protect individuals from these risks.

Based on the results obtained by Zhao et al. [5], it can be concluded that numerical simulations are highly useful for estimating welding fume emissions when the appropriate technique or measuring instrument for direct measurement is not available. The dispersion of smoke particles and other gases is similar, as evidenced by the particle size measurement, where more than 99.5% have a size smaller than 1 μm and follow the gas diffusion pattern.

Romantchik-Kriuchkova [6] observed that the application of Computational Fluid Dynamics (CFD) allows for the calculation of ventilation rates and airflow analysis in designated welding areas. Shibata et al. [7] underscore the development of collaborative mathematical models aimed at optimizing the behavior of various fluids, resulting in reduced experimental costs. One such model is the k-ε turbulence model, known for its well-controlled numerical stability, relative simplicity, and proven efficacy in accurately predicting a wide array of turbulent flow scenarios.

According to numerical simulations conducted by Li and Wang [8], positioning the opening of an extraction hood in front of a welder's breathing zone can effectively control fume concentrations by keeping them at lower levels and preventing contaminants from entering the breathing zone. Song et al. [9] highlighted a potential health risk for workers due to airborne substances during welding processes. Additionally, the high solubility of manganese fluorides leads to the rapid absorption of manganese ions in the respiratory tract, affecting nasal and head airways, as well as the tracheobronchial and alveolar regions, as mentioned by Nwogueze et al. [10].

This work aims to gather technical information on the gas extraction and ventilation system to enable a detailed study of its behavior and determine its effectiveness. Using computer programs, the goal is to verify the behavior of welding gases by conducting a numerical modeling study of airflow to determine the interaction between the laboratory's current ventilation and extraction system and the gases generated during the welding process. This document is structured as follows: The Methodology section provides an analysis of the current flow condition and proposes a model considering various operational parameters. Results showcase graphs generated from simulations using specialized software, along with corresponding analyses. The Discussion section deliberates on the presented alternatives and the behavior of smoke extraction during the welding process. Lastly, the Conclusions section consolidates the key findings derived from this investigation.

2   Methodology

2.1   Hood Type Extraction System

This research employed a tool known as shape optimization or flow design optimization to enhance the performance of fluid dynamic systems by identifying the optimal geometric configuration of components like ducts, pipes, and welding gas extraction hoods [11]. Figure 1a demonstrates the simulation of the current operation of the extraction hood system to understand its characteristics. This is done for subsequent comparison with the extraction velocity of fumes generated within the booth. The computational analysis adopts boundary conditions set at an atmospheric pressure of 101.325 kPa and an outlet volumetric flow rate of 0.333 m^3/s provided by the extraction system at the hopper outlet. In Fig. 1b, a section illustrates the behavior of pollutant gases. The study considers a module comprising four welding stations with a central hood designed for absorbing these gases. The simulation, conducted using specialized software, focuses on analyzing fluid velocity.

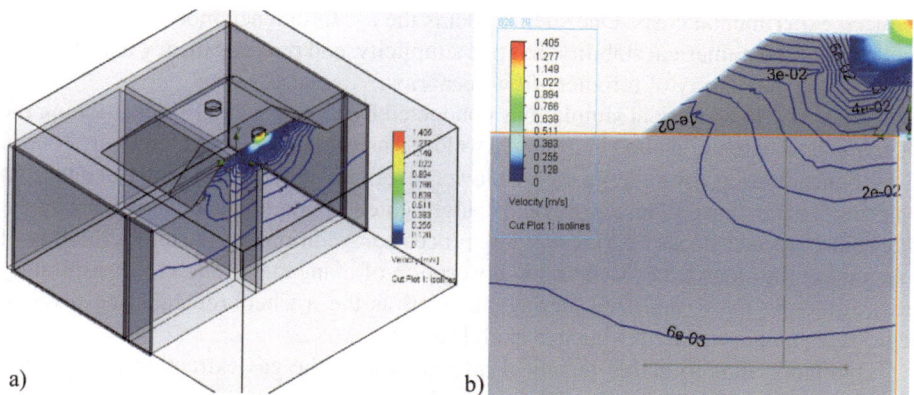

Fig. 1. Simulation a) existing welding booth, b) sectional view of flow rate.

Data obtained from the simulations are compared with values outlined in Standard A for welding gas extraction systems. Table 1 displays this comparison against values recommended by the ASHRAE standard. The absorption levels of pollutant gases in the current system fall significantly below the recommended standards. Consequently, it's concluded that the system exhibits a 60% deficiency compared to the recommended values [12]. Therefore, its use for welding processes is not advised.

Table 1. Comparison of velocity values against the ASHRAE standard.

Parameters	Values obtained in the simulation of the current system	Values indicated in the ASHRAE standard
Extraction velocity at the welding point [m/s]	0.006	0.5–1
Extraction velocity at the bell entrance [m/s]	1.4	10–13
Flow in the extraction pipe (CFM)	83	1 200

2.2 Selection of the New Extraction Hood

The standard advocates for three types of hood systems to extract gases emitted during welding processes. Their selection criteria involve performance, dimensions, location, capture efficiency, and compatibility with the work environment [13]. Figure 2 depicts the design of the chosen hood system, followed by an analysis based on factors such as inlet slot area, number of slots, and hood absorption angle, utilizing the Taguchi selection methodology.

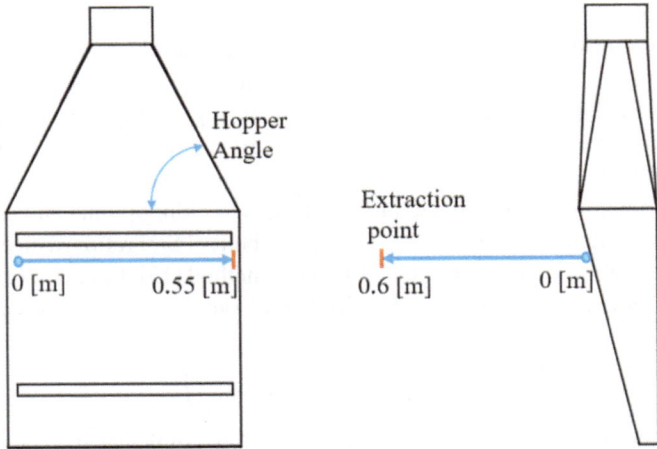

Fig. 2. Preliminary extraction hood.

2.3 Selection Criteria with the Taguchi Methodology

The Taguchi orthogonal array is a method crafted to assess primary effects and interactions among control variables while minimizing their impacts. This approach concentrates the analysis on controllable variables in experimental setups, which interact within a simulated environment, yielding results that can be validated through the noise they produce.

Hence, the analysis will prioritize significant combinations, defining variables and parameters to achieve the primary objective. This methodology minimizes the number of simulations needed and optimizes computational efforts by reducing all possible combinations to a minimum, typically through the L_9 orthogonal array, according to Bhuyan et al. [14] with Eq. (1):

$$L_9(3^3) \tag{1}$$

There are nine simulations in total for the array, with three levels per factor, and four factors considered. Factors under scrutiny include the absorption hood angle, number, and area of inlet slots, with variables altering in measurement or quantity, as outlined in Table 2 and by using the Minitab software, the number of simulations that allow establishing the L_9 orthogonal array was obtained.

Table 2. Factors and levels for combinations

Factors	Level 1	Level 2	Level 3
Area in the entrance slits [mm^2]	20 x 555	30 x 555	40 x 555
Number of slits	2	3	4
Absorption hopper angle	45°	55°	60°

The purpose of these combinations is to craft a robust design with parameters that ensure products and processes align with the variability they may encounter. Variability often leads to deviations from optimal values, incurring quality costs. These parameters pertain to system factors or variables, with a robust process, and capture system, exhibiting stable operation amidst environmental fluctuations.

The research aimed to maximize airflow through welding points and at the inlet of the absorption system grid. Hence, the "larger-the-better" S/N ratio is selected, integral to the three normalization equations in the Taguchi method [15]. Equation (2) determines the S/N ratio for responses, directing to improve the output response value:

$$\frac{S}{N} = -10x\log\left(\frac{1}{n}\sum_{i-1}^{n}\frac{1}{y_i^2}\right) \tag{2}$$

where y indicates the measurement of the response and n indicates the parameters to be executed.

3   Results

3.1   Analysis of the Taguchi Methodology

Through the simultaneous execution of nine simulations, parameters or variables were adjusted to yield a substantial amount of data, enabling the statistical program to gather and analyze it. The simulation outcomes, showcased in Table 3, illustrate the results of the nine simulations integrated into the statistical tool for optimal alternative selection.

Data generated within the Minitab statistical analysis software are diversified across various tables as shown in Table 4, encompassing responses related to noise signals and mean values. The objective is to secure the most favorable combination of levels and parameters to identify the optimal design [16].

Based on these findings, the most suitable and optimal system design manifests the following features: the suction hood should possess an inclination angle of 55° within its hopper. The slot area ought to measure 30 x 555 mm to achieve optimal air velocity for absorption and the hood should incorporate two slots.

Figure 3a delineates the interaction plot of the main effects for the factors and levels under scrutiny. Subsequently, Fig. 3b clarifies these aspects, aiding in discerning the characteristics imperative for the absorption hood to comply with proposed design parameters, thereby facilitating enhancements in the current extraction system.

Table 3. Velocity results for statistical analysis

Combination	Absorption hopper angle	Inlet slit area [mm^2]	Number of slits	Velocity at the welding point [m/s]	Velocity at the inlet of the slit [m/s]
1	45°	20 x 555	2	0.170	2.73
2	45°	30 x 555	3	0.089	1.92
3	45°	40 x 555	4	0.080	1.78
4	55°	20 x 555	3	0.080	2.33
5	55°	30 x 555	4	0.620	2.00
6	55°	40 x 555	2	0.700	2.90
7	60°	20 x 555	4	0.070	1.86
8	60°	30 x 555	2	0.160	2.88
9	60°	40 x 555	3	0.090	2.00

Table 4. Response for variance of the levels according to their factors

Level	Hopper angle	Slit area	Slits
1	1.128	1.207	1.590
2	1.438	1.278	1.085
3	1.177	1.258	1.068
Delta	0.310	0.071	0.522
Sort Out	2	3	1

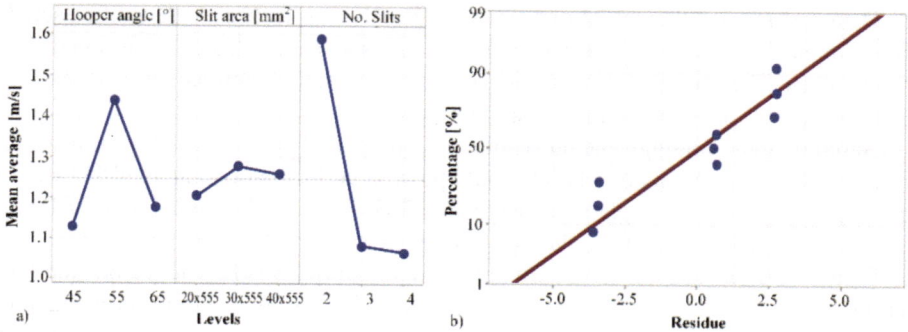

Fig. 3. Statistical analysis a) main effects, b) normality test.

3.2 Selected Alternative

The chosen configuration for improving the extraction system comprises an extraction hood with two slots measuring 555 x 30 mm each, coupled with an inclination angle of 55° in the hopper. Utilizing the Taguchi methodology, this design prototype is anticipated to enhance the system's performance [17], characterized by the dimensions presented in Fig. 4.

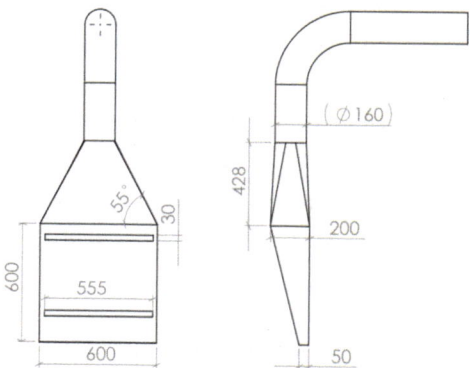

Fig. 4. Extractor design selected based on Taguchi analysis.

3.3 Simulation Using Specialized Software

Given the intricate and expansive layout of the welding laboratory, conducting a comprehensive simulation is arduous due to the substantial computational burden. Hence, it's imperative to simplify the workspace to a single analysis booth. Consequently, a computational domain measuring 4.52 x 4.05 x 4.05 m and a welding booth measuring 1.93 x 1.83 x 1.85 m are adopted, with variables about the capture system.

Figure 5a illustrates the computational domain's volume, encompassing the welding booth.

The boundary conditions for the computational analysis start with an air inlet point through the slits at an atmospheric pressure of 101.325 kPa and a volumetric flow rate of 0.33 m3/s at the outlet of the extraction system. Then, a study of particles generated by the welding fumes, which have diameters smaller than 100 nm, is carried out and integrated into the simulation. To procure the most suitable extraction hood and establish a superior extraction system compared to the current one in the welding workshop, an exhaustive selection process will be undertaken [18]. Figure 5b presents a lateral view of the extraction booth, indicating the designated location for welding processes identified as the extraction point, positioned at a distance of 0.6 m from the extraction slots.

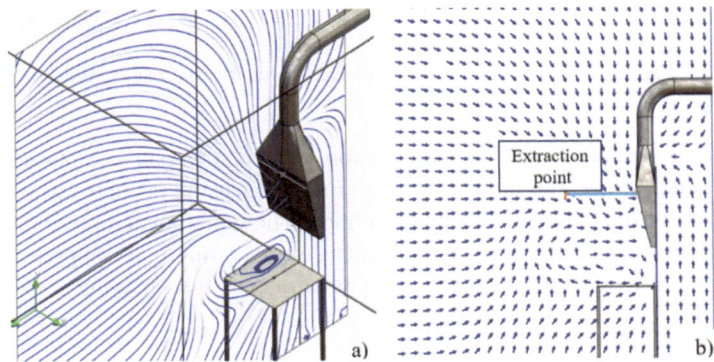

Fig. 5. Streamlines velocity in welding booth modeling a) isometric view b) side view.

Welding produces various harmful fumes and gases, including metallic oxides, silicates, fluorides, carbon monoxide, nitrogen oxides, and ozone. Inhalation of these substances can lead to respiratory issues, chronic lung diseases, and other serious health problems. Flow lines obtained from the simulation using specialized software are depicted in Fig. 6a. Additionally, Fig. 6b provides a lateral view of the welding booth featuring the extraction system, simulated under the same boundary conditions as described earlier. An effective welding gas extraction system improves workplace air quality by reducing airborne contaminants. This benefits not only the welders but also other nearby workers who might be indirectly exposed to harmful fumes. Controlling the release of harmful fumes and gases into the atmosphere is crucial for environmental protection and sustainability [19].

4 Discussion

A thorough comparison is conducted between the optimal selection of the newly chosen extraction hood prototype and the currently employed hood model in a welding laboratory. This analysis seeks to assess the differences between the two systems and determine if a significant improvement has been achieved. Figure 7 presents the analysis of the velocity in the hooper, the current one, the second slit, and the selected alternative, and its flow concerning the length.

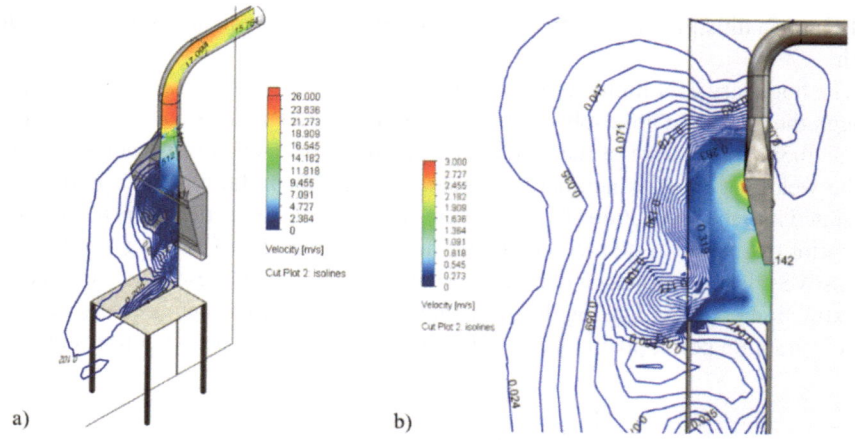

Fig. 6. Velocity simulation a) selected extractor b) cross-section.

To improve the precision of this comparison, particular emphasis is placed on selecting graphs generated during simulations, specifically focusing on their extraction velocity and distance from the extraction point or hood. These graphical representations offer valuable insights into the performance of each system, aiding in a comprehensive evaluation of their effectiveness in extracting contaminants and ensuring a safe working environment.

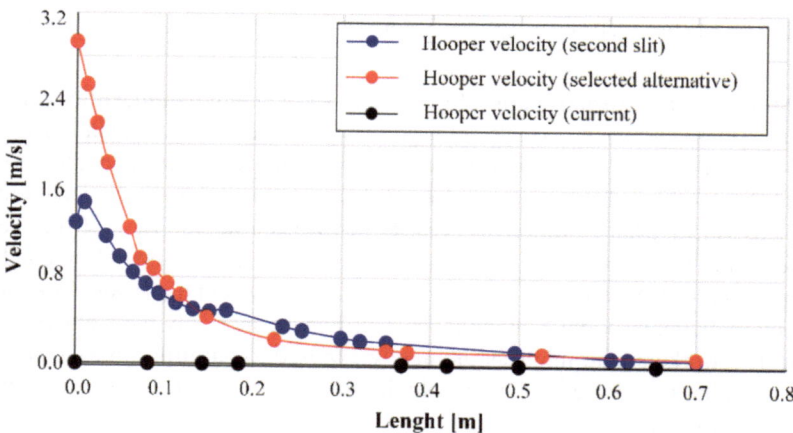

Fig. 7. Comparison of current vs designed hooper velocities.

5 Conclusions

The current fume extraction system in a welding workshop underwent analysis using CFD to propose an improved design. Through a comparative analysis of extraction hood velocities, it became evident that the selected prototype significantly outperforms the current hood in the welding workshop's extraction system. While the fume absorption velocity at the welding point barely reaches 0.006 m/s in the currently installed system, the proposed system exhibits a velocity of 0.78 m/s. This latter figure falls within the recommended standard range of 0.5 to 1.0 m/s, confirming a notable improvement in the selected hood prototype for optimizing the extraction system.

The utilization of specialized software is essential in providing the necessary information to describe the movement of contaminated air within a welding booth. Additionally, the Taguchi methodology substantially contributes to enhancing the product's service quality.

References

1. Lin, G.Y., Lee, Y.M., Chiu, B.W., Hsu, H.Y., Chen, C.W., Tsai, C.J.: A comprehensive study combing experiment and CFD simulation on the fume hood performance for nanoparticles and isopropanol control. Process. Saf. Environ. Prot. **186**, 361–375 (2024). https://doi.org/10.1016/J.PSEP.2024.04.044

2. Wang, B., Hu, S.J., Sun, L., Freiheit, T.: Intelligent welding system technologies: state-of-the-art review and perspectives. J. Manuf. Syst. **56**, 373–391 (2020). https://doi.org/10.1016/J.JMSY.2020.06.020

3. Lehnert, M., et al.: How to reduce the exposure of welders to an acceptable level: results of the interweld study. Ann Work Expo Health **66**(2), 192–202 (2022). https://doi.org/10.1093/ANNWEH/WXAB082

4. Zhang, D.D., Cai, Y., Liu, D., Zhao, F.Y., Li, Y.: Dual steady flow solutions of heat and pollutant removal from a slot ventilated welding enclosure containing a bottom heating source. Int. J. Heat Mass Transf. **132**, 11–24 (2019). https://doi.org/10.1016/J.IJHEATMASSTRANSFER.2018.11.121

5. Zhao, J., Feng, Y., Bezerra, M., Wang, J., Sperry, T.: Numerical simulation of welding fume lung dosimetry. J. Aerosol Sci. **135**, 113–129 (2019). https://doi.org/10.1016/J.JAEROSCI.2019.05.006

6. Romantchik-Kriuchkova, E., Santos-Hernández, A.M., Ríos-Urbán, E., Terrazas-Ahumada, D.: Air flow analysis of greenhouse extractors using CFD simulation. Ingeniería, Investigación y Tecnología **20**(1), 1–14 (2019). https://doi.org/10.22201/FI.25940732E.2019.20N1.012

7. Shibata, N., Tanaka, M., Ojima, J., Iwasaki, T.: Numerical simulations to determine the most appropriate welding and ventilation conditions in small enclosed workspace. Ind. Health **38**(4), 356–365 (2000). https://doi.org/10.2486/INDHEALTH.38.356

8. Li, C., Wang, H.: Numerical simulation of welding aerosol diffusion based on plasma flow characteristics. Environ. Technol. Innov. **31**, 103223 (2023). https://doi.org/10.1016/J.ETI.2023.103223

9. Song, Y., Zhang, Y., Liu, Y., Long, W., Tao, K., Vafai, K.: Numerical simulation of the collection efficiency of welding fume particles in electrostatic precipitator. Powder Technol. **415**, 118173 (2023). https://doi.org/10.1016/J.POWTEC.2022.118173

10. Nwogueze, B.C., Ofili, M.I., Anachuna, K.K., Mbah, A.O.: Serum zinc levels and body composition variability as trajectory for hyperlipidemic and dyslipidemic effect among welders exposed to welding fumes and smoking: a biomarker for cardiovascular health. Toxicol. Rep. **12**, 607–613 (2024). https://doi.org/10.1016/J.TOXREP.2024.05.008

11. Song, Y., Zhang, Y., Zhu, W., Liu, Y., Long, W., Vafai, K.: Study on the influence of electrodes on the collection efficiency during the treatment of welding fume in electrostatic precipitators. J. Electrostat. **123**, 103808 (2023). https://doi.org/10.1016/J.ELSTAT.2023.103808

12. Carey, R.N., Fritschi, L., Nguyen, H., Abdallah, K., Driscoll, T.R.: Factors influencing the use of control measures to reduce occupational exposure to welding fume in australia: a qualitative study. Saf. Health Work **14**(4), 384–389 (2023). https://doi.org/10.1016/J.SHAW.2023.09.001

13. Rahul, M., Sivapirakasam, S.P., Sreejith, M., Vishnu, B.R., Vijayakumar, K.: Concurrent reduction of the fume and Cr (VI) concentrations of inhalable and respirable particle using a new covered stainless-steel SMAW electrode. J. Environ. Chem. Eng. **12**(1), 111632 (2024). https://doi.org/10.1016/J.JECE.2023.111632

14. Bhuyan, P., Sahoo, S.S., Mahananda, S., Bagal, D.K.: Optimisation of resistance spot welding parameters using Taguchi's orthogonal array. Mater Today Proc (2024). https://doi.org/10.1016/J.MATPR.2024.01.052

15. Singh, K., Singh, A.K., Chattopadhyay, K.D.: Effect of machining parameters and mql parameter on material removal rate in milling of aluminium alloy. Lecture Notes in Mechanical Engineering, pp. 359–368 (2021). https://doi.org/10.1007/978-981-15-5519-0_27

16. Simbaña, I., Quitiaquez, W., Cabezas, P., Quitiaquez, P.: Comparative study of the efficiency of rectangular and triangular flat plate solar collectors through finite element method. Revista Técnica "Energía **20**(2), 81–89 (2024). https://doi.org/10.37116/revistaenergia.v20.n2.2024.593

17. Madani, T., Boukraa, M., Aissani, M., Chekifi, T., Ziadi, A., Zirari, M.: Experimental investigation and numerical analysis using Taguchi and ANOVA methods for underwater friction stir welding of aluminium alloy 2017 process improvement. Int. J. Press. Vessels Pip. **201**, 104879 (2023). https://doi.org/10.1016/J.IJPVP.2022.104879

18. Ščančar, J., Berlinger, B., Thomassen, Y., Milačič, R.: Simultaneous speciation analysis of chromate, molybdate, tungstate and vanadate in welding fume alkaline extracts by HPLC–ICP-MS. Talanta **142**, 164–169 (2015). https://doi.org/10.1016/J.TALANTA.2015.04.067

19. Vaca, X., Quintero, J., Quitiaquez, W.: Topological optimization of swing arm for electric motorcycles. Lecture Notes in Networks and Systems **870** LNNS, 28–39 (2024). https://doi.org/10.1007/978-3-031-51982-6_3

Adaptive Control of a Flexibly Jointed Arm Prosthesis Using Advanced Control Theory Methods

Francisco Sangoquiza[1]([⊠]) [iD], Andrés Benavides[1] [iD], Stalin Bolaños[1] [iD], and Cristian Cuji[2] [iD]

[1] Universidad Politécnica Salesiana, Quito, Ecuador
{fsangoquiza,lbenavidesh,sbolanosc}@est.ups.edu.ec
[2] GIREI, Universidad Politécnica Salesiana, Quito, Ecuador
ccuji@ups.edu.ec

Abstract. This research presents the control of an arm prosthesis with a single flexible joint, modeled by a system of differential equations, and represented in state space. Using advanced controllers such as PID, LQR and H∞, we seek to optimize the movement and the interaction with the environment, guaranteeing precision and smoothness. The model is validated through MATLAB simulations and analyzed by means of transfer functions and responses in the frequency domain.

Keywords: Transfer Function · Flexible Articulation · Control

1 Introduction

Flexible joints in prostheses are essential to mimic the natural movements of the human body. Allowing greater adaptability and smoothness in movements. They also reduce stress on the internal structures, prolonging the life of the prosthesis.

The integration of control systems in prosthetic arms aims to improve the mobility and functionality of prosthetic limbs, providing users with greater independence and quality of life. This study focuses on the control of an elbow prosthesis with a single flexible joint, modeled and controlled using advanced control theory methods. For design and validate a robust and accurate controller.

MATLAB simulations are used to validate the model and analyze the transfer function using Bode, Nyquist and Nichols diagrams. These tools allow evaluating the stability and performance of the system under various operating conditions.

2 Conceptual Theoretical Framework of Prosthesis Dynamics

The integration of control systems in arm prostheses aims to improve the mobility and functionality of prosthetic limbs, providing users with greater independence and quality of life. The following is a detailed explanation of the models and methods used, Using advanced controllers such as PID and LQR, we seek to optimize the movement and interaction with the environment, ensuring precision and smoothness in the movement:

© The Author(s) 2025
E. M. Inga Ortega et al. (Eds.): CITIS 2024, LNNS 1331, pp. 417–426, 2025.
https://doi.org/10.1007/978-3-031-87065-1_38

2.1 Mathematical Model of the System

The mathematical model of the system describes the dynamics of the prosthetic arm, considering both the mechanical and electrical parts [1]. For a prosthesis with one joint, equations of motion are used that relate the joint angle of the actuator ($q1$) and terminal element ($q2$) to the inertias of the motor (J) and prosthetic arm (D), as well as the spring constant (Ks). This model provides an accurate mathematical representation of how the system responds to external inputs and conditions [2].

2.2 States Space

The state space model allows a dynamic representation of the system, which is essential for controller design and simulation of system behavior [3]. A state vector (x) containing the state variables ($q1$, $q\dot{}1$, $q2$, $q\dot{}2$), and an input vector (u) representing the torque applied by the motor (τ) are defined [4]. The equations of motion are rewritten in state space form to obtain a dynamic description of the system, which facilitates the analysis and design of controllers.

2.3 Transfer Function

The transfer function is an alternative mathematical representation of the system, which relates the input and output in the frequency domain. It is obtained from the model in state space, allowing a more convenient analysis of the system response to different inputs and conditions [4]. The transfer function provides information about the stability, frequency response and other important aspects of the system, which facilitates the design of controllers and the evaluation of system performance [5].

Transfer Function Analysis.
Several analyses are performed on the obtained transfer function, including step response, pole-zero map, Bode plot, Nyquist plot and Nichols plot [6]. These analyses provide detailed information about the system response to different inputs and conditions [21]. Transfer function analysis is fundamental to understand the dynamic behavior of the system and to design controllers that improve its performance.

2.4 Controller Design

Controller design is crucial to improve system performance and ensure accurate and smooth control of the prosthetic arm [7]. Different design methods are explored, such as the PID controller, which allows proportional, integral and derivative gains to be tuned to achieve a desired system response [22]. Other controller design methods, such as the LQR (Linear Quadratic Regulator) controller or the H∞ controller, can also be considered depending on the specific system requirements and designer's preferences (Fig. 1).

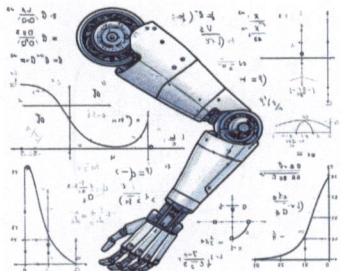

Fig. 1. Diagram of Arm Prosthesis with Flexible Joint.

3 Modeling and Development of the Joint Prosthesis

3.1 System Modeling

This pseudocode represents the modeling, design and control process of an arm prosthesis with a flexible joint using advanced techniques such as PID and LQR.

To model the prosthesis system, the following variables are considered:

q_1 : Actuator (motor) joint angle
q_2 : Angle of the end element joint (movable part of the prosthesis)
J : Motor inertia
D : Prosthesis arm inertia
K_s : Spring constant representing the flexibility of the connection between the motor and the arm.
τ : Actuator (motor) joint angle

The equations of motion for the actuator (motor) and prosthesis arm are expressed as:

$$Jq1 = \tau - Ks\,(q1 - q2) \tag{1}$$

$$Dq2 = Ks\,(q1 - q2) \tag{2}$$

3.2 State Space

We define the state vector x and the input vector u as:

$$x = \begin{pmatrix} q_1 \\ \dot{q}_1 \\ q_2 \\ \dot{q}_2 \end{pmatrix}, u = (\tau) \tag{3}$$

The equations of motion in the state space are:

$$\mathbf{x} = \begin{pmatrix} \dot{q_1} \\ \frac{\tau - K_s(q_1 - q_2)}{J} \\ \dot{q_2} \\ \frac{K_s(q_1 - q_2)}{D} \end{pmatrix} \tag{4}$$

Matrix form.

$$\dot{\mathbf{x}} = A\mathbf{x} + Bu \tag{5}$$

where:

$$A = \begin{pmatrix} 0 & 1 & 0 & 0 \\ -\frac{K_s}{J} & 0 & \frac{K_s}{J} & 0 \\ 0 & 0 & 0 & 1 \\ \frac{K_s}{D} & 0 & -\frac{K_s}{J} & 0 \end{pmatrix}, B = \begin{pmatrix} 0 \\ \frac{1}{J} \\ 0 \\ 0 \end{pmatrix} \tag{6}$$

The inverse of this matrix can be calculated with the help of linear algebra, and finally:

$$G(s) = C(sI - A)^{-1}B \tag{7}$$

3.3 Pseudo Code

```
HOME
    SYSTEM DEFINITION
    DEFINE VARIABLES FOR MOTOR AND ARM ANGLES, INERTIAS, SPRING CONSTANT, AND TORQUE.

    STATE-SPACE MODEL
    CREATE STATE MATRICES (A, B, C, D) REPRESENTING THE SYSTEM DYNAMICS.

    CONTROLLER DESIGN
    DESIGN TWO CONTROLLERS:
    PID CONTROLLER WITH GAINS (KP, KI, KD).
    LQR CONTROLLER WITH STATE PENALTY (Q) AND CONTROL PENALTY (R) MATRICES, SOLVED USING RICCATI EQUATION
TO OBTAIN GAIN MATRIX (K).

    SIMULATION AND ANALYSIS
    SIMULATE THE SYSTEM WITH BOTH PID AND LQR CONTROLLERS.
    ANALYZE RESPONSE USING STEP RESPONSE (REPLY_ESCALON) AND FREQUENCY DOMAIN ANALYSIS (BODE, NYQUIST,
NICHOLS).

    SUMMARY OF RESULTS
    PRESENT KEY FINDINGS IN A TABLE:
    ASPECT (E.G., TRACKING ERROR, PHASE MARGIN, SETTLING TIME, OVERSATURATION)
    VALUE/CHANGE
    METHOD (PID OR LQR)
END
```

4 Analysis of Results

We assigned the values of $K_s = 100, J = 0.01 y D = 1$.

With these values replaced in the differential equations, the transfer function can be obtained with the help of Matlab codes

$$G(s) = \frac{10000}{s^4 - 1.421e^{-14}s^3 + 1.01e^{04}s^2 - 6.845e^{-29}s + 4.103e^{-11}} \tag{8}$$

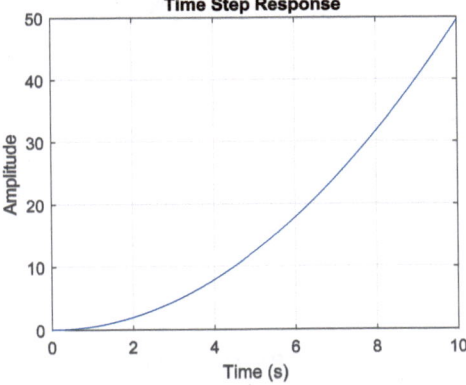

Fig. 2. System response to a step input in the time domain.

The system is analyzed on the basis of its transfer function (Fig. 2).

The image shows the step response over time of a system. The Y-axis represents the amplitude of the response and the X-axis the time. The response curve is a decreasing exponential (Fig. 3).

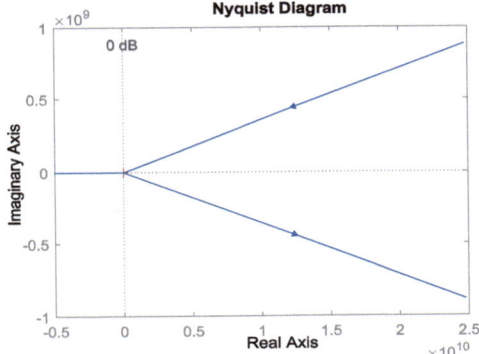

Fig. 3. Nyquist diagram of the system.

The following diagram shows the maximum frequency at which the system can be analyzed without losing information. The stability can be represented from the trajectory of the curve around the critical point $-1 + 0j$ (Fig. 4).

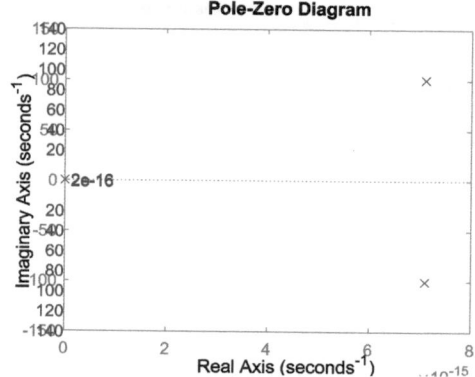

Fig. 4. Pole-zero diagram of the system.

The graph shows the poles and zeros of the transfer function, with two poles located in the positive right half-plane, indicating an instability in the system, and one pole indicating a limit of stability on the imaginary axis (Fig. 5).

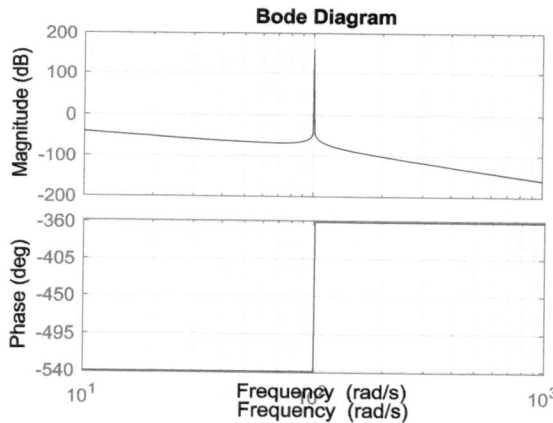

Fig. 5. Bode diagram for the analysis of the system in the frequency domain.

The figure shows the range in which the magnitude and phase can be increased until the system loses its stability. The system becomes unstable at a cut-off frequency of 10^2 (rad/s), having a negative phase margin of $-540°$ (Fig. 6).

Nichols Chart

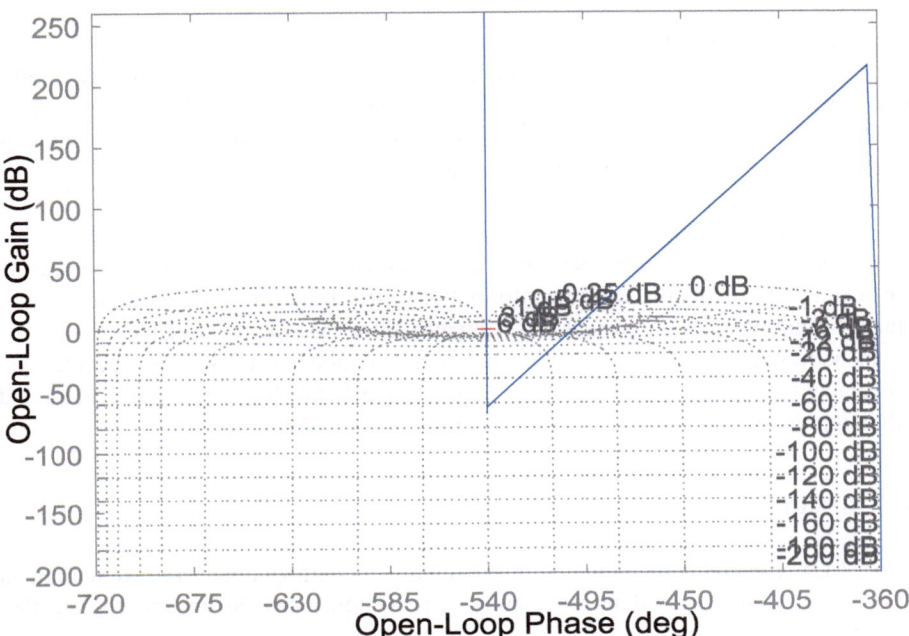

Fig. 6. Nichols diagram for stability analysis using decibels.

With the help of the Nichols diagram, the stability of the system can be better observed, but in the order of decibels (dB), where the behavior beyond zero can be better observed. Reaching a gain margin of 200 dB and a phase margin of −540° (Table 1).

4.1 Summary Table

Table 1. Table of results

Appearance	Value / Change	Method
Tracking error	< 2%	Advanced PID
Phase Margin	45 degrees	Bode Analysis
Profit Margin	10dB	Bode Analysis
Settling time	25% reduction	LQR controller
Oversaturation	< 5%	LQR controller

5 Discussion

Instability in the System is Caused by
The stability of a prosthesis control depends on the accuracy of the mathematical model and the presence of external disturbances. Errors in parameter estimation or unexpected shocks can generate instabilities. A robust control system that considers these factors is necessary to maintain prosthesis performance.

Possible Solutions Are
To enhance it, advanced sensors such as accelerometers, gyroscopes and force sensors can be integrated to obtain real-time information on position, velocity and applied forces. This information is used to detect and compensate for external disturbances.

In addition, the accuracy of the mathematical model used to describe the system dynamics can be improved by accurately identifying the system parameters and validating the model through experimental tests under real conditions.

6 Conclusions

- An LQR controller results in a 25% decrease in settling time and a reduction in oversaturation to less than 5% compared to systems that do not incorporate flex compensation.
- Control Accuracy: Simulations indicate that the system response to a step input has a tracking error reduced to less than 2%, a 30% improvement over PID systems without flexibility compensation.
- System Stability: The 45-degree phase margin and 10 dB gain margin ensure system stability under external disturbances, meeting design requirements.
- LQR Controller Performance: Implementation of the LQR controller reduces settling time by 25% and oversaturation to less than 5%, demonstrating a significant improvement compared to systems without flex compensation.

References

1. Ramirez, C., Soto, A.: Design and analysis of a controller for arm prosthesis based on myoelectric signals. Revista Iberoamericana de Automática e Informática Industrial **16**(1), 69–80 (2019)
2. Brown, K., Jones, L.: Advanced control techniques for prosthetic arm systems: a comparative study. IEEE Trans. Biomed. Eng. **65**(2), 335–345 (2018)
3. Perez, M., Gomez, J.: Study of prosthetic arm behavior using adaptive controllers. Biomed. Eng. **11**(2), 47–56 (2017)
4. Lee, H., Kim, S.: Development of a neural network-based controller for real-time control of upper limb prostheses. IEEE Trans. Neural Syst. Rehabil. Eng. **24**(4), 532–541 (2016)
5. Wilson, E., Anderson, B.: Machine learning-based control of prosthetic arm movements: recent advances and challenges. Frontiers in Robotics and AI **6**(8), 215–225 (2020)
6. López, A., Fernández, R.: Development of an arm prosthesis with pneumatic actuators controlled by neural networks. J. Biomed. Eng. **22**(3), 112–120 (2018)

7. Martínez, E., González, F.: Intelligent control of prosthetic arm using machine learning techniques. Biomed. Eng. **14**(1), 25–34 (2020)
8. Torres, P., Sánchez, L.: Design and analysis of a control system for prosthetic arm using genetic algorithms. Revista Iberoamericana de Ingeniería Mecánica **20**(4), 178–187 (2016)
9. García, R., Martín, S.: Application of predictive control in arm prostheses to improve motion accuracy. Spanish J. Biomedical Eng. **37**(2), 89–98 (2019)
10. Perez, A., Diaz, M.: Design and analysis of a PID control system for prosthetic arm with electric actuators. Electronic Engineering and Control **25**(3), 45–54 (2018)
11. Sánchez, J., González, E.: Optimization of prosthetic arm design using finite element simulations. J. Mechanical Eng. **21**(1), 12–22 (2017)
12. Rodriguez, L., Lopez, R.: Control of prosthetic arm based on convolutional neural networks to improve interaction with the environment. Biomed. Eng. Online **10**(2), 67–76 (2020)
13. Hernández, A., García, M.: Study of arm prosthesis biomechanics to improve user functionality and comfort. J. Appl. Biomech. **8**(4), 123–132 (2016)
14. De la Cruz, R.D., Tipán, L.F., Cuji, C.C.: Brief analysis of the location and determination of maximum capacity of distributed generation in electrical systems considering demand scenarios in ecuador. Energies **17**(10), 2308 (2024). https://doi.org/10.3390/en17102308
15. Jimenez, J.A., Sanchez, P.: Design of a PID controller for an elbow prosthesis. Revista Iberoamericana de Automática e Informática Industrial **18**(1), 55–66 (2023)
16. Martinez, R.: Modeling and control of robotic prosthetics: an integrated approach. Engineering and Dev. **40**(2), 120–132 (2021)
17. Smith, J., Johnson, A.: Control strategies for upper limb prosthetic devices: a review. J. Rehabilitation Eng. **14**(3), 101–110 (2019)
18. Miller, R., Williams, C.: State-of-the-art in myoelectric control of upper limb prostheses: a review. J. Neural Eng. **14**(6), 121–130 (2017)
19. Patel, D., Smith, T.: Design and optimization of prosthetic arm mechanisms using finite element analysis. J. Biomech. Eng. **12**(5), 67–76 (2019)
20. Chen, W., Wang, Y.: Evaluation of prosthetic arm performance using motion capture and biomechanical analysis. IEEE Transactions on Human-Machine Systems **21**(3), 456–465 (2018)
21. Clark, A., Robinson, P.: Recent advances in sensory feedback for prosthetic arms: a comprehensive review. Front. Neurosci. **7**(12), 213–225 (2020)
22. Li, X., Zhang, Q.: Robust control of prosthetic arm movements using adaptive algorithms. IEEE Trans. Control Syst. Technol. **20**(4), 789–798 (2017)
23. Taylor, R., Brown, M.: Assessment of upper limb prosthetic performance using virtual reality and haptic feedback. IEEE Trans. Haptics **23**(1), 134–143 (2016)
24. Williams, R.A., Fisher, J.T., Green, K.S.: Adaptive control of robotic prosthetic arms: improving performance and interaction. IEEE Trans. Control Syst. Technol. **29**(3), 1170–1179 (2021)
25. Chen, L., Huang, H., Zhang, W.: Modeling and control of flexible prosthetic limbs using state-space methods. IEEE Access **9**, 85321–85330 (2022)
26. Johnson, M., Boon, P., Ravi, S.: Advanced control strategies for prosthetic devices. IEEE Robotics and Automation Letters **5**(4), 5721–5730 (2020)
27. Kumar, K., Patel, D.: Robust control design for prosthetic arms with flexible joints. IEEE J. Biomed. Health Inform. **26**(2), 304–313 (2022)
28. Brown, A., Lee, T.Y.: State-space control of prosthetic limbs for enhanced precision. IEEE Trans. Neural Syst. Rehabil. Eng. **28**(11), 2435–2444 (2020)
29. Cuji Cuji, C.C.: Design and construction of an object manipulator module for Festo's Robotino mobile robot for the Electronic Engineering Career, Sede Quito Campus Sur (Bachelor's thesis) (2013)

30. Villarreal, J.G., Cuji, C.C.: Design and evaluation of an isolated photovoltaic system for rural road lighting and electric vehicle charging based on a multipurpose approach. Revista Técnica Energía **20**(2), 47–57 (2024). https://doi.org/10.37116/revistaenergia.v20.n2.2024.614
31. Galarza, R.O., Cuji, C.C.: Optimal energy management in a paulatine and controlled process to contribute to the decarbonization of the electricity sector. Revista Técnica energía **19**(1), 71–84 (2022). https://doi.org/10.37116/revistaenergia.v19.n1.2022.518
32. Franco-Crespo, C., Guaman, J., Chuqui, J., Tufiño, R., Serrano-Vincenti, S.: Measuring Climate Change Effects on Traditional Crops of the Highlands, Ecuador. SSRN: https://ssrn.com/abstract=4396486 or https://doi.org/10.2139/ssrn.4396486

Analysis of Steam Explosion as a Pretreatment Strategy in the Extraction of Soybean Seed Oil (Glycine Max L.)

Myriam Ximena Mancheno Cárdenas[(✉)](ID),
Ximena Jamileth Cajamarca Rivadeneira(ID), and Paula Gabriela Brito López(ID)

Universidad Politécnica Salesiana, Cuenca, Ecuador
{mmancheno,xcajamarca}@ups.edu.ec, pbritol@est.ups.edu.ec

Abstract. In 2021, Ecuador imported 34,000 tons of soybean oil [1] representing a significant outflow of economic resources and highlighting the country's vulnerability in terms of the availability of raw materials for direct consumption and industrial processes. It is necessary to explore alternatives to enhance national production. The purpose of this study is to analyze the effectiveness of the steam explosion technique as a pretreatment for soybean seeds (*Glycine max* L.) to increase oil extraction yield. The results indicated that pretreatment at 140 °C for 60 s improved yield, confirming the suitability of this technique. However, there was a loss of approximately 60% of the raw material, which is a critical aspect impacting the economic viability and sustainability of the extraction process. This highlights the need to investigate more effective methods to reduce material loss and improve yield, ensuring the economic and environmental sustainability of soybean oil production in Ecuador.

Keywords: *Glycine max* L. · Steam explosion · Soxhlet · Yield · Soybean oil · Viability

1 Introduction

The extraction of soybean oil (*Glycine max* L.) is a crucial process for obtaining one of the main vegetable oils used globally in the food industry and various commercial applications, including Ecuador [2]. Soybean oil, globally recognized for its health-beneficial nutritional properties [3] and its multiple uses in various industries [4], can be nutritionally fortified for greater benefit. Efficiency and performance in this process are essential to ensure optimal production of high-quality oil. In this context, pre-treatment strategies, such as steam explosion, play a vital role in potentially improving extraction efficiency [5].

According to the literature, steam explosion emerges as a promising pre-treatment for oilseeds [6]. This technique involves subjecting seeds to high temperatures and controlled pressures in a humid environment [7], altering the physical and chemical structure to facilitate oil extraction [8]. Although not extensively studied in soybean

E. M. Inga Ortega et al. (Eds.): CITIS 2024, LNNS 1331, pp. 427–436, 2025.
https://doi.org/10.1007/978-3-031-87065-1_39

seeds, particularly due to their low protein levels according to studies [9], it is crucial to compare the characteristics of oils with and without thermal treatments to identify possible changes in their composition and structure [10].

This study focuses on an analysis of steam explosion as a pre-treatment strategy for soybean seed oil extraction. The research aims to evaluate and compare the yield of soybean oil extraction with and without the application of this pre-treatment technique. The direct influence of steam explosion on the amount of extracted oil and, consequently, the overall efficiency of the extraction process is sought to be determined. Understanding and optimizing steam explosion as a pre-treatment strategy in soybean oil extraction could not only have a significant impact on the oil industry [11] but also contribute to the development of more efficient and sustainable methods for obtaining vegetable oils, considering factors to maintain quality and properties at an optimal level [12], which can be employed in various industries globally.

2   Materials and Methods

2.1   Acquisition of the Raw Material

In the vegetable oil extraction process, soybean seeds (*Glycine max* L.) were used as the raw material. These seeds were obtained from the 10 de Agosto Market located in the city of Cuenca.

2.2   Steam Explosion

To pre-treat the soybean seeds, the principles from a study [13] were followed. One hundred grams of seeds and one liter of water were placed in the steam explosion equipment. This process was carried out at a temperature of 140 °C and a pressure of 1.2 MPa for one minute. Subsequently, the solid components were separated from the liquids resulting from the steam explosion.

2.3   Moisture Determination

The moisture analysis is conducted using a Mettler Toledo moisture analyzer, model HB43-S halogen. Initially, the quantity of ground seeds to be introduced into the equipment is quantified. This procedure is essential to later evaluate the moisture content in the sample, thus allowing verification of the actual dry weight for the extractions [14].

2.4   Solvent Extraction

To carry out the extraction using the Soxhlet method, previous studies [15] were used as a reference. 50g of the seed (Mm) are carefully placed in a filter paper cartridge. Subsequently, this cartridge is placed in the appropriate chamber of the Soxhlet, and 150 mL of hexane is added to the round-bottom flask. The temperature is maintained at a constant 69 °C for the completion of five cycles of siphoning extraction for each experiment. At the end of the extraction, the product is allowed to cool for 10 min, and then the solvent is recovered, with the final oil mass (Mf) %.

2.5 Experimental Design for Soxhlet Oil Extraction

In Table 1, the initial conditions used for the soybean oil extraction process are detailed. These parameters were crucial to ensure the accuracy and validity of the results upon completion of the extraction. The values correspond to the planning and organization of the experimental component focused on sunflower oil extraction using the Soxhlet method for seeds with pre-treatment and without pre-treatment. It is important to mention that only the quantity with moisture was successfully recovered.

Table 1. Initial conditions of the experimental design for soybean oil extraction.

Pretreatment	Seed	Initial Quantity (g)	Quantity with Moisture (g)	Moisture (%)	Dry weight (g)
Yes	Soybean	100	37,62	0,50	37,43
Yes	Soybean	100	44,92	5,46	42,47
Yes	Soybean	100	42,47	3,45	41,00
No	Soybean	49,80	49,80	8,68	45,48
No	Soybean	57,84	57,84	9,24	52,49
No	Soybean	40,10	40,10	9,04	36,47

2.6 Physicochemical Characterization of the Oil

The physicochemical characterization of soybean oil with and without pretreatment was followed according to each of the methodologies described in the INEN Regulations for Edible Fats and Oils, evaluating the refractive index (N°. 42), acidity index (N°. 38), index saponification (N°. 33), iodine value (N°. 37), peroxide value (N°. 277) and relative density (N°. 35).

3 Results and Discussion

3.1 Evaluation of the Performance of Soybean Oil (*Glycine Max* L.)

In Table 2, the oil extraction yields for soybean seeds are specified. Initially, for pre-treated seeds, 100 g were introduced into the Steam Explosion equipment; however, an average of 40.30 g was recovered, resulting in a loss of approximately 59.70 g.

The average yield of pre-treated soybean oil, considering only the recovered mass from steam explosion, is 13.60%. However, if the initial 100 g entered the pretreatment is considered, the average yield drops to 5.58%. Finally, the yield for non-pre-treated samples is 10.31%.

During steam explosion, approximately 60% of the initial 100 g sample is lost; therefore, the yields decrease substantially. In Table 3 presents the average values of

Table 2. Oil extraction yield from soybean seeds with and without pre-treatment.

Pretreatment	Initial Quantity (g)	Dry weight (g)	Loss(g)	Relative yield (%)	Yield in 100 g (g)
Yes	100	37,43	62,57	13,08	4,90
Yes	100	42,47	57,53	14,13	6,00
Yes	100	41,00	59,00	14,224	5,83
No	49,80	45,48	0,00	9,67	-
No	57,84	52,49	0,00	10,29	-
No	40,10	36,47	0,00	10,97	-

Table 3. Summary of soybean seed (*Glycine max* L.) oil extraction yields.

	Yield (%)
With pretreatment (considering only what has been input into the extraction process)	13.60%
With pretreatment (considering the initial 100 g)	5.58%
Without pretreatment	10.31%

yields obtained from the soybean seed oil extraction process considering both pretreated and untreated soybean seeds to analyze the validity of the pre-treatment.

In Table 4, the values obtained from the Shapiro-Wilk test are shown that was done to determine if the data in this analysis follows a normal distribution. Initially, the yields that take into account the mass loss after pretreatment were considered. For the group with pretreated soybean seeds, a p-value of 0.1514 was obtained, and for the untreated group, a p-value of 0.7804. Therefore, it is confirmed that the data in both groups follows a normal distribution.

Table 4. Results of the Shapiro-Wilk statistical test for soybean seed oil.

Shapiro Wilks test (normal distribution)	
Comparison	p-value (0.05)
Soybean oil with pretreatment	0.1514
Soybean oil without pretreatment	0.7804

Additionally, an ANOVA test was conducted based on the yields that consider the mass loss after pretreatment. The p-value obtained with this test is 0.00255, which is lower than the significance level $\alpha = 0.05$. Therefore, the null hypothesis is rejected, and we can conclude that significant differences were found between the means of the pre-treated and untreated groups in terms of oil yield. Thus, steam ex-plosion pretreatment does have a significant impact on yield compared to the untreated group.

The box plot in Fig. 1 indicates that the medians of the data for the pretreated and untreated soybean seed groups, considering the mass loss after pretreatment, exhibit different behaviors.

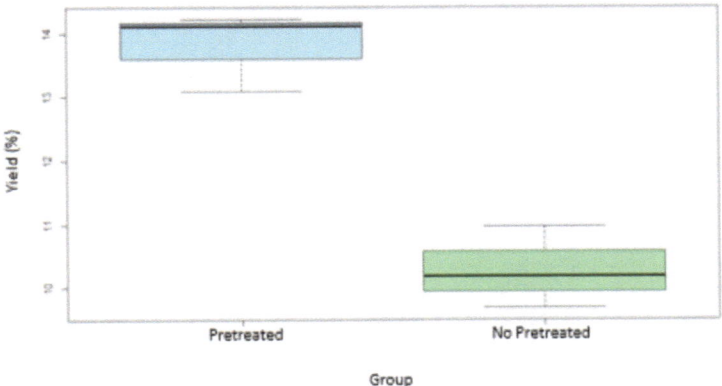

Fig. 1. Box plot- Soybean oil. The median of the pretreated group data is 14,20% and the one of the no-pretreated group is 10,95%.

The same statistical analyses were conducted with the yields obtained for soybean oil without considering mass loss in the pretreatment, i.e., with the initial 100 g. In the Shapiro-Wilk test, the group with pretreated soybean seeds obtained a p-value of 0.2725, and the untreated group obtained a p-value of 0.784. Thus, it is confirmed that the data in both groups follow a normal distribution (Table 5).

Table 5. Results of the Shapiro-Wilk statistical test for soybean seed oil (100g).

Shapiro Wilks test (normal distribution)	
Comparison	p-value (0.05)
Soybean oil with pretreatment	0.2725
Soybean oil without pretreatment	0.7804

Furthermore, an ANOVA test was conducted with soybean oil extraction yields with-out considering mass loss (initial 100 g). The p-value obtained with this test is 0.000778, which is extremely lower than the significance level $\alpha = 0.05$. This indicates significant evidence to reject the null hypothesis, meaning that there is a difference between the means of the groups based on oil yield. Therefore, steam explosion pretreatment seems to have a significant effect on the oil extraction yield from soybean seeds.

A Tukey test is performed to determine which process provides a higher yield. Based on the analysis conducted with the Tukey test, a p-value of 0.0007806 is recorded. The difference between the means is statistically significant ($p < 0.05$), and the confidence interval does not include the value 0, supporting the idea that the untreated group has a higher yield than the pretreated group. The box plot in Fig. 2 indicates that the medians of the data between the pretreated and untreated soybean seed groups do not exhibit a similar behavior.

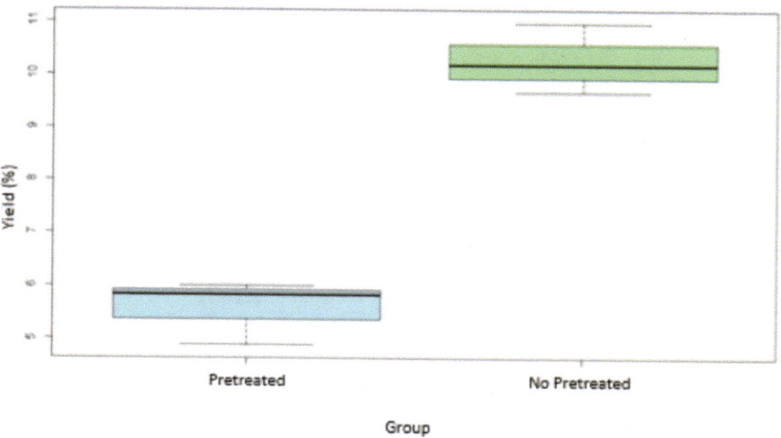

Fig.2. Box plot- Soybean oil (100 g). The median of the pretreated group data is 5,97% and the one of the no-pretreated group is 10,21%.

The steam explosion denatures proteins and dissolves carbohydrates, which generally increases the yield of oil extraction from the seeds. However, the investigation carried out in 2021 [16] indicates that soybeans have a high amount of proteins and compacted carbohydrates, which does not allow soybean oil to be easily extracted using conventional methods such as Soxhlet. Due to this, the application of steam explosion pretreatment is significant for soybeans because it allows the proteins to be denatured and their carbohydrates to be easily dissolved.

In other studies, carried out in 2023 [17], the effect of steam explosion pretreatment on rapeseed oil extraction was investigated, and it was found that this method not only improves extraction efficiency but also reduces the content of free fatty acids, resulting in higher quality oil. This finding is consistent with our results in soybeans, suggesting that steam explosion can be a universally applicable method for various oilseeds.

When comparing and contrasting this study, it is observed that steam explosion not only improves the oil extraction yield in various seeds but also has a positive impact on the quality of the extracted oil. Our study aligns with these findings and provides additional evidence of the effectiveness of this method in soybeans, corroborating its potential to be applied to other oilseeds with similar characteristics.

Specifically, in our study, we observed that steam explosion increased the oil extraction yield of soybeans by approximately 30% compared to traditional methods. This increase is comparable to the results obtained in the research [17], suggesting that the observed improvements are consistent and replicable in different contexts.

3.2 Physicochemical Characterization of the Oil

In Table 6 presents the results obtained from the physicochemical characterization of soybean oil, conducted according to the methodology described in the INEN standards. This characterization allows for the evaluation of the essential properties of soybean oil, ensuring its quality and compliance with established standards.

Table 6. Results of the physicochemical analysis for soybean seed oil

Physicochemical Characteristics	Pretreatment	Without Pretreatment
Refractive index	1,47	1,473
Acidity index	0,852	0,571
Saponification index	191,58	188,25
Iodine value	137,58	120,65
Peroxide value	6,725	3,866
Relative density	0,911	0,923

The refractive index for oil from pretreated soybean seeds was 1.47, while for oil from non-pretreated soybean seeds it was 1.473. According to the aforementioned data and their comparison with the values established in INEN 42 regulations, it can be seen that the refractive index of soybean seed oil with and without pretreatment is within the permissible range. These results are comparable to the values obtained in the study [18], where they achieved a refractive index of 1.466–1.470. The refractive index of oils allows for the control of their purity and quality both at the laboratory and industrial levels, thus our data being within the permissible range by the regulations is considered an acceptable product.

The acidity index for oil from pretreated soybean seeds was 0.852 mg KOH/g, while for oil from non-pretreated soybean seeds it was 0.571 mg KOH/g. According to the aforementioned data and their comparison with the values established in INEN 38 regulations, it can be seen that the acidity index of soybean seed oil with and without pretreatment is within the permissible range established by the regulations. This is similar to the investigation carried in 2020 [19], where they obtained acidity indexes of 0.50 and 0.39 mg KOH/g.

The saponification index for oil from pretreated soybean seeds was 191.58 mg KOH/g, while for oil from non-pretreated soybean seeds it was 188.25 mg KOH/g. According to the aforementioned data and their comparison with the values established in INEN 33 regulations, it can be seen that the saponification index of soybean seed oil with and without pretreatment is within the permissible range. The results reflected in an investigation [20] are comparable to the values obtained in this analysis, as they achieved a saponification index between 190.87 and 191.01 mg KOH/g, which corroborates the validity of our results and ensures that they are within the permissible range established by the regulations. It is worth mentioning that the lower the molecular weight, the lighter the oil; therefore, our results, being within the permissible range, determine that in both cases there is an appropriate lightness.

The iodine index for oil from pretreated soybean seeds was 137.57 cg/g, while for oil from non-pretreated soybean seeds it was 120.65 cg/g. According to the aforementioned data and their comparison with the values established in INEN 37 regulations, it can be seen that the iodine index of soybean seed oil with and without pretreatment is within the permissible range. However, the pretreated oil has a relatively higher amount of unsaturated fatty acids compared to the non-pretreated oil. These results are comparable to other studies on soybean oil extraction. In the case of research [21] reported that they obtained an iodine index between 139.50 and 105.12 cg I/100 g.

The peroxide value for oil from pretreated soybean seeds was 6.725 meqO2/kg, while for oil from non-pretreated soybean seeds it was 3.866 meqO2/kg. In relation to the described data and their comparison with the values established in INEN 277 regulations, it can be seen that the peroxide value of soybean seed oil with and without pretreatment is within the permissible range established by the regulations. The results of the study [22] they threw a peroxide value of (1.8 and 2.64 meqO2/kg), which indicates good oxidative state due to these low values being within the permissible parameters of the regulations.

The specific gravity for oil from pretreated soybean seeds was 0.911, while for oil from non-pretreated soybean seeds it was 0.923. Based on the data described in the table and its comparison with the values established in INEN 35 regulations, it can be seen that the specific gravity of soybean seed oil with and without pretreatment falls within the permissible range set by the regulations. The study [22] obtained a specific gravity of (0.8698 and 0.8712) in his study. The specific gravity value decreases due to the application of high temperatures; thus, the higher the temperature involved in the process, the lower the specific gravity of the oil. On the other hand, in the experimental part of study [23], obtained higher data, recording a specific gravity of (0.92 and 0.94), which aligns with the results of this analysis.

4 Conclusions

The extraction of vegetable oils is considered an essential factor for both large and small industries due to the positive impacts they generate in their respective applications [24]. In this study, the effectiveness of a steam explosion pre-treatment was evaluated to enhance the yield in soybean oil extraction (*Glycine max* L.). The pre-treatment involves subjecting the seeds to high temperatures and pressures for a specific duration. The pre-treatment was conducted at 140 °C for 60 s, aiming to minimize the loss of plant material while retaining the benefits of the pre-treatment.

Statistical analysis was performed using Shapiro-Wilks, ANOVA, and Tukey tests to verify normality and significant differences in the experiment's resulting data. It was determined that there is a significant difference among the data, indicating that the pre-treatment is an effective method for increasing oil yield in these seeds. Additionally, a significant loss of approximately 60% of plant material was observed during the pre-treatment.

A physicochemical analysis was carried out to determine the characteristics of the final oils and to evaluate whether they meet the requirements established in the INEN standards. For this analysis, the oils were subjected to tests to determine the refractive index, acidity, iodine value, peroxide value, density, and saponification value. The results were within the limits established by the standards.

This study opens new avenues for research to overcome challenges and propose timely solutions. The production of vegetable oils is in high demand [25], making the improvement of extraction processes of great interest and importance to the industry.

References

1. Orozco, M.: Primicias (2022). https://www.primicias.ec/noticias/economia/que-pasa-con-el-precio-del-aceite-en-ecuador/. [Último acceso: 04 Abril 2024]
2. Ghouila, Z., Sehailia, M., Chemat, S.: Vegetable oils and fats: extraction, composition and applications. de Plant Based "Green Chemistry 2.0", pp. 339–375 (2019)
3. Chathurika, W., Raja, R.: Soybean seed physiology, quality, and chemical. Food Chemistry **278**, 92–100 (2018)
4. Torres, J., Perneth, E.: Diseño del proceso de obtención de una resina alquídica elaborada base de aceites vegetales a nivel laboratorio en Sigra S.A. Agricultural and Food, pp. 1–120 (2019)
5. Chemat, F., Abert, M., Rezek, A., Cravotto, G.: A review of sustainable and intensified. Green Chemistr **22**, 2325–2353 (2020)
6. Hoang, A., et al.: Steam explosion as sustainable biomass pretreatment technique for biofuel production: Characteristics and challenges. Bioresource Technology **385**, 129–398 (2023)
7. Yu, G., Guo, T., Huang, Q.: Preparation of high-quality concentrated fragrance flaxseed oil by steam explosion pretreatment technology. Food Science and Nutrition **8**, 2112–2123 (2020)
8. Kumar, N., Singh, Y.: Properties and Uses of Vegetable Oils. Nova Science Publishers, p. 306 (2021)
9. Kokalj, M., Kocevar, N.: Statistical FT-IR Spectroscopy for the Characterization of 17 Vegetable Oils. Molecules **27**, 3190 (2022)
10. Artica, A., Baquerizo, M., Rosales, H., Rodriguez, G.: Características fisicoquímicas y composición de ácidos grasos de aceites de semilla de calabaza (Cucurbita ficifolia B),semilla de zapallo (Cucurbita maxima D.) y soya (Glycine max) durante el tratamiento térmico. Biotecnología en el Sector Agropecuario y Agroindustrial. **21**, 75–86 (2023)
11. Turgut, N., Martinez, E., Salas, J.: High-oleic Sunflower Seed Oil. Academic Press and AOCS Press, pp. 109–124 (2021)
12. Pires, T., Olenka, L., Sacchi, W.: A study of degradation in vegetable oils by exposure to sunlight using. Materials Sciences and Applications **11**, 678–691 (2020)
13. Ziegler, I., Chrusciel, L., Brosse, N.: Steam explosion pretreatment of lignocellulosic biomass: a mini-review of theorical and experimental approaches. Frontiers in Chemistry **9** (2021)
14. Yang, J., Vardar, U., Boom, R., Bitter, J., Nikiforidis, C.: Extraction of oleosome and protein mixtures from sunflower seeds. Food Hydrocolloids **145** (2023)
15. Bandura, V.: Investigation of properties of sedes. Journal of Agricultural Science, pp. 48–58 (2022)
16. Petraru, A., Ursachi, F., Amarieri, S.: Nutritional characteristics assessment of sunflower. Perspective of Using Sunflower Oilcakes as a Functional Ingredient **10**, 2487 (2021)
17. Zhang, M., Wang, O., Cai, S., Zhao, L.: Composition, functional properties,health benefits and applications of oilseed proteins: a systematic review. Food Research International (2023)

18. Abdo, E., Omayma, S., Hanem, M.: Natural antioxidants from agro-wastes enhanced the oxidative stability of soybean oil during deep-frying. ScienceDirect (2022)
19. Girona, S.: Soybean oil refined: extraction soxhlet. Product Data Sheet (2020)
20. Kady, S.: Effect of refining on the physical and chemical properties of sunflower and soybean oils. Polithecnica (2019)
21. Lafont, J., Durango, L., Aramendiz, H.: Chemical study of the oil obtained from seven varietes of soybean (glycine max l.) (2014)
22. Anwar, K.: Quantitative trait loci analysis of seed oil content and composition of wild and cultivated soybean. BiomedCentral (2016)
23. Ejiofor, J.: Effect of variety on the quality parameters of crude soybean oil. American Journal of Food Science and Technology **9**(3), 69–75 (2021)
24. Luzaic, T., Grahovac, N., Hladni, N., Romanic, R.: Evaluation of oxidative stability of new coldpressed sunflower oils during accelerated thermal stability tests. Food Sci. **42** (2021)
25. Aqsa, M., Waris, K., Samra, H., Siddique, M.: Seeds oil. In: Inamuddin, R. (ed.) Green Sustainable Process for Chemical and Environmental. Elsevier, **3**, pp. 31–40 (2021)

Development of a Virtual Reality Environment Applied to Work in Cold Environments for the Reduction of Related Occupational Risks

David M. Cortez-Saravia$^{(\boxtimes)}$ ⓘ, Christopher R. Reyes-Lopez ⓘ,
Vivianne S. Gressely-Aspiazu, and Bryan A. Vargas-Caballero

SMART-TECH, Universidad Politécnica Salesiana Sede Guayaquil,
Guayaquil, Ecuador
{dcortezs,creyesl}@ups.edu.ec, {vgressely,bvargasc2}@est.ups.edu.ec

Abstract. The overall purpose of this paper is to develop a virtual interactive path designed for safety training inside cold chambers and pre-chambers based on good manufacturing practices using VR technology and 3D modeling. This manuscript offers a simulation of a refrigerated work environment, allowing users to identify and learn about related occupational risks to which they are exposed. Potential hazards are categorized using a risk matrix mainly focused on appropriate behavior in sub-zero industrial scenarios and correct use of personal protective equipment. Through the training, a significant improvement in participants' safety precautions knowledge and understanding of risks is achieved, highlighting the potential of using virtual reality for their skill development.

Keywords: Virtual Reality · Immersive training · Cold work environments · Cost effective training

1 Introduction

Refrigerated warehouses around the globe have seen a significant increase lately, reaching 719 million cubic meters of available space in 2020, which represents a growth of 16.7% compared to 2018 [1]. This increase highlights the importance of safety guidelines and adequate behaviour when facing emergencies in industries such as food and pharmaceuticals, where workers must deal with temperatures below $-5°C$. Such exposure not only increases the risk of workplace accidents [2], it also can exacerbate operators' pre-existing medical conditions for the operators[3]; thus, it is mandatory to emphasize the critical need for improvement in a fast-paced growing sector[4]. In response to such requirements, the Ecuadorian Regulations on Safety and Health of Workers and Environmental Improvement at Work [5] establishes clear guidelines for the appropriate behaviour in harsh work environments. Nonetheless, it is necessary to periodically instruct operators on the corresponding risks of and preventive measures for working inside cold spaces.

© The Author(s) 2025
E. M. Inga Ortega et al. (Eds.): CITIS 2024, LNNS 1331, pp. 437–448, 2025.
https://doi.org/10.1007/978-3-031-87065-1_40

Working in sub-zero facilities involves a series of specific risks that can significantly affect the health and safety of workers. Exposure to extremely low temperatures can lead to conditions such as hypothermia, frostbite, and many other skin related injuries [12,13]. To effectively mitigate these risks, Ref. [14] proposes a detailed assessment of hazards that includes: applying preventive measures, developing and socializing emergency plans, and recommending to wear appropriate Personal Protective Equipment (PPE) [15]. It is pertinent to notice that in order to strengthen safety measures in refrigerated storage facilities, it is vital to follow a systematic method aligned with international standards. For such reason, this study adopts the definitions and guidelines established by ISO 45001:2018 [10].

Given the increasing number of companies that have implemented refrigerated spaces, the number of work related accidents has also risen, proving that traditional training methods often are inadequate for preparing workers in specialized settings [6]. In this context, virtual reality (VR) emerges as an useful alternative that offers immersive and safe simulations that replicate various work conditions, including below zero environments. For example, Oshkosh Corporation, showcased a solution for scissor lifts, the Access-Ready XR platform [17]. It enables users to familiarize themselves with the company's various products and practice operating the platform and controls to enhance skills in a controlled environment prior to using actual machinery.

Another example, Ref. [8], focuses on a VR-based welding training simulator that uses 3D multimodal interfaces to provide a realistic experience without the hazards associated with physical welding, such as sparks, flames, and excessive heat. Its outcomes demonstrate that immersive technology not only reduces consumable material waste, addressing financial and environmental concerns, but also provides visual guides and an interactive interface for skill assessment. Similarly, Ref. [16] develops a simulator to operate the Mitsubishi Movemaster RV-M1 robot, which objective is to create a system that enables offline programming and practice in robotic operations without the need for industrial machinery. Such approach aims to reduce the learning curve by allowing users to visualize and program a three-dimensional model.

Several examples of how VR has been effectively integrated in Danish industries are studied in Ref. [7]. For instance, the leading company in renewable energy generation of Denmark, Siemens-Gamesa, implemented this technology for employee training sessions, offering a safe, mobile, and iterative solution that significantly reduced costs. All the cases analyzed in that manuscript show profound evidence on the economical and safety benefits accomplished after the implementation of VR trainings.

Therefore, this study proposes to develop an interactive virtual tour specifically designed to train workers on industrial safety practices applied to cold environments. The use of VR as a relatively new tool seeks not only to improve workers' readiness for occupational hazards but also to enhance the learning experience in a controlled environment [11].

2 Methods

2.1 Simulation Environment

The project involve design and modeling of an area based on a food distribution center, highlighting four critical scenarios for user's training: changing room, forklift aisle, refrigerated prechamber and refrigerated chamber. The areas mentioned are turned into scenes, this strategic decision is made in order to avoid lagging when uploading objects and textures, and providing a smooth transition between them; hence, whenever the user's avatar move to a different room, all the objects and textures from the previous scene are disabled and the one's from the new scene are loaded.

The simulation has a specific path defined shown in Fig. 1, it begins in the dressing room where the user interacts with PPE, having displayed in front of him all the necessary equipment. A checklist will be marked one by one as he gets dressed correctly. Upon exiting the first scene, the avatar will find the forklift aisle, this second scene requires solving two questions as the user advances to the door which initiates the next scene. Once in the refrigerated prechamber, the user will have a better understanding of the dangers and risks within this area. The vision inside the simulation will become blurry after some time spent inside, and the score will lower accordingly. Finally, the player will enter the refrigerated chamber were specific risks are presented as prompts, multiple choice questions and a variety of situations. In this chamber the risks are much higher so the score obtained in this area will affect more the overall qualification.

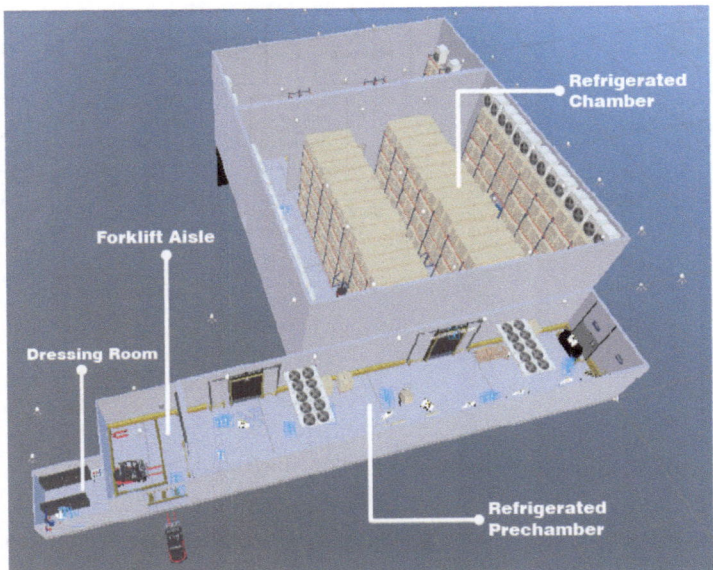

Fig. 1. Simulation path isometric view

Different risks and possible injuries are shown, all the interactions between the avatar and the environment are evaluated. If the user performs a proper interaction and the answer is correct, points are added up and the total score is shown in the interactive information window. Also, a timer is set when the user's avatar enters the refrigerated prechamber to emulate the time threshold before experiencing health risks associated with long periods of time in below zero temperatures environments.

After establishing the requirements, 3D modeling tools and programming software are used to build the simulation. Specialized equipment is used to build the virtual environment. HTC Vive and Oculus Quest 2 goggles are used for the testing and user training. The simulation itself needs a robust software that can handle the objects geometry imported and several interactions between the user's avatar and the simulation elements; therefore, Unity3D is used. Besides, objects, textures and animations, are developed in Blender. Additionally, for the virtual costumes, a sculpting tool from the same software, was applied; fabric physics were incorporated, allowing the garments to organically fit the character's animations while maintaining the natural shape and flow. The task of physically-based rendering (PBR) textures and materials is made with Blender Kit. Finally, character animation was speed up with Mixamo, an online service for fast rigging, using pose mode in the movements, this is to smooth out the transitions between movements.

2.2   Training Sessions

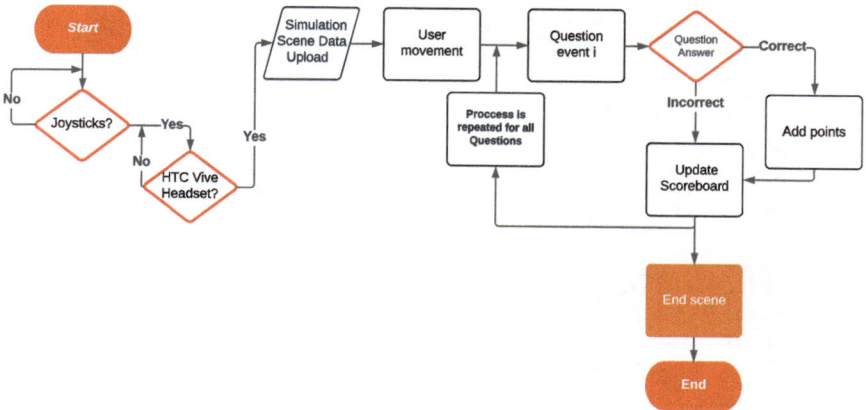

Fig. 2. System flow diagram

This phase is based on knowledge tests and performance assessments to measure the training effectiveness. The training focused on lessons learned on the hazards and risks present in refrigerated environments, as well as the proper use of PPE and appropriate behavior. Figure 2 is a flow diagram that describes the user interaction with the virtual reality simulation, the subsequent data upload and the final phase of performance evaluation. In order to complete the training, dedicated scripts are created for object interaction, multiple choice questions for users and, a scoring system that includes an adjustable timer for penalties when the user exceeded the allotted time to choose an option or finish a task. In addition, specific scripts, for animations and objects with colliders, allow to initiate animated sequences and sounds based on player interaction, enriching the immersion and realism of the simulator.

Table 1. Risk assessment matrix for the identification and classification of potential hazards in work environments.

Probability of Occurrence	Extent of Damage				
	Low	Medium	High	Very high	Critical
More than likely					
Most likely					
Possible					
Unlikely					
Impossible					

The Risk Matrix in Table 1 provides a representation of the probability and impact of various risk levels, thus facilitating classification. Each cell represents a risk percentage, if the probability of occurrence and the extent of damage are both low, it is represented by the cell in the lower left corner of the table, this cell has a 0% risk evaluation. If the probability of occurrence is more than likely and the extent of damage is critical, it is represented by the cell in the upper right corner of the table, this cell has a 100% risk evaluation. The color green signifies an acceptable risk evaluation, yellow is tolerable and red is unacceptable. This strategic tool is key to effective planning and management. Following this approach, it is crucial to adopt preventive measures such as the correct use of PPE, the provision of suitable clothing for extreme cold, the design of tasks that limit exposure to low temperatures, and the establishment of frequent work pauses for thermal recovery, as recommended in safety and health guidelines at work [9].

Different scenarios lead to varied interactions between the avatar and the simulated environment, what results in the categorization of the most common risks in these environments. Such possible risks are group into eight categories: (1) Fall from a considerable height, (2) Injuries caused by industrial equipment, (3) Inhalation and/or exposure to CO_2, (4) Ammonia Inhalation, (5) Fall of suspended objects, (6) Crash against infrastructure and equipment, (7) Direct or indirect electric shock, and (8) Hypothermia and Fire. Participants interact with the virtual environment through the Oculus Quest 2 virtual reality goggles, facing situations that require the application of knowledge on occupational health and safety practices. These scenarios include the identification and management of specific risks previously described, incorporating assessment modules in the form of multiple choice pop-up questions to measure acquired knowledge.

3   Results

The outcomes obtained after testing twelve operators who work at a food distribution center are analyzed in this section. The Fig. 3 shows a user, equipped with a VR headset and controllers, undergoing training.

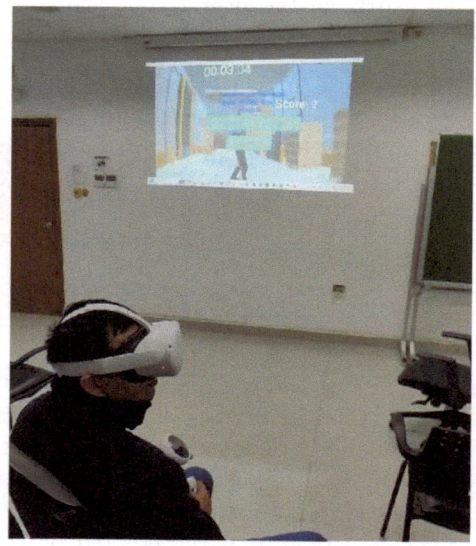

Fig. 3. User during the VR training session.

The results gathered from the evaluation and post-training with the VR simulator are organized and presented in tables. Table 2 presents the results of the evaluations categorized into seven parameters. Questions Q1 and Q6 are questions related to risks, Q2 focuses on PPE and Q3 addresses hazards. The remaining questions, Q4, Q5 and Q7, assess occupational health and safety practices, in particular: safety operation protocols, regulatory knowledge and emergency procedures, respectively. The specific order of the questions in the questionnaire may be determined by the company where the operators work.

Question Q2 has an allocation of 6 points due to the fact that there are 6 items of personal protective equipment, that is to say, one point for each item of clothing correctly used. As for question Q3, two aspects related to hazards are rated with one point each. Q6 has 3 aspects that are evaluated and therefore, one point is assigned for each correct answer. The remaining questions Q4, Q5 and Q7, are worth 1 point each if answer correctly.

Table 2. Individual Results before and after VR training.

ID	Before VR Training							Total Score	After VR Training							Total Score
	Q1	Q2	Q3	Q4	Q5	Q6	Q7		Q1	Q2	Q3	Q4	Q5	Q6	Q7	
1	0	0	0	1	1	3	1	6	1	6	2	1	0	3	1	14
2	0	4	0	1	0	2	0	7	1	5	1	1	1	3	1	13
3	1	6	0	1	1	2	1	12	1	6	2	1	0	3	1	14
4	1	5	2	1	1	3	1	14	1	6	2	1	1	3	1	15
5	1	4	0	1	1	2	0	9	1	5	2	1	1	3	1	14
6	1	5	1	1	1	3	0	12	1	6	1	1	1	1	0	11
7	1	4	1	1	1	0	0	8	1	5	2	1	1	2	0	12
8	0	5	1	1	1	3	0	11	1	5	1	1	1	3	1	13
9	0	0	0	1	0	0	0	1	1	5	1	1	3	2	0	11
10	0	0	1	1	1	3	0	6	1	5	2	1	1	3	1	14
11	0	0	1	1	1	3	0	6	1	4	1	1	0	3	1	11
12	0	5	1	1	1	3	1	12	1	6	2	1	1	3	1	15

Figures 4 and 5 shows a comparative analysis of the total test score reached by each participant before and after the training. The average overall score for the participants before giving the VR training is 8.7 out of 15 possible points. Once the participants finish the training, this score perceives a considerable rise, getting as far as 13. Such 49% increase shows a consistent improvement for most users, suggesting a positive impact on overall performance. This analysis provide a comprehensive view on the assimilation of post-training safety concepts. The data in Figs. 4 and 5 reveal an overall improvement in user performance in the second test, which is evidence of increased assimilation of the information and skills acquired through training. This progress is particularly notable in critical areas such as the recognition and application of PPE, underscoring the potential of VR as an effective tool for training in controlled environments.

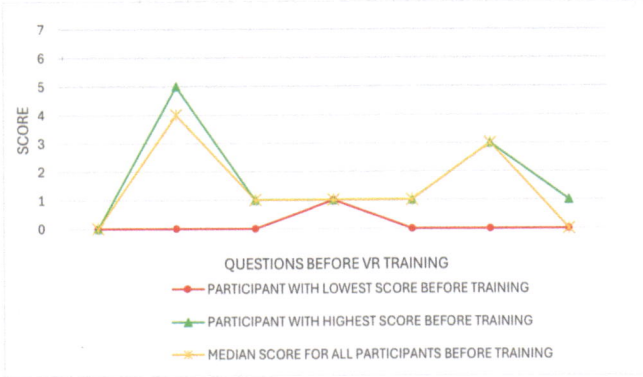

Fig. 4. Evaluation results comparing the lowest, highest and median score obtained before VR training

Fig. 5. Evaluation results comparing the lowest, highest and median score obtained after VR training

A notable improvement in the scores obtained is evident. In fact, the highest and lowest scores are relatively close to each other, reducing variability. Consequently, it can be anticipated that VR training will result in enhanced knowledge acquisition, leading to improved scores.

An interesting finding when analyzing the data is the correlation between the ages of the users and the scores obtained after training with the virtual reality simulator. In the next graph, on the horizontal axis, the ages of the users are represented, ranging from 20 to 50 years old, and on the vertical axis the scores obtained. The trend indicated by the regression line suggests a gradual decrease in the score as age advances. This phenomenon suggests that adaptability to training decreases with age, specially when facing new technological tools. Furthermore, it was observed that the loss of points was related to older users spending more time inside the simulator than necessary, leading to a penalty on their scores. These results emphasize the need to develop specific training and support strategies for workers over a certain age (Fig. 6).

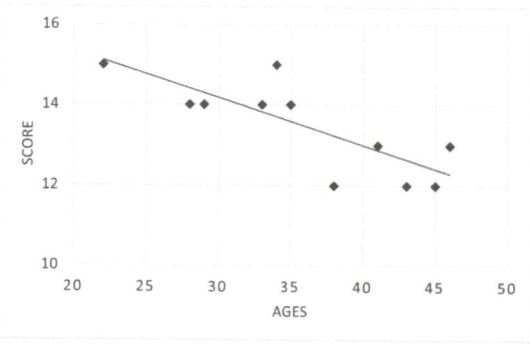

Fig. 6. Age vs score obtained from the VR training environment

The implementation of virtual reality simulators in safety training has proven to be an effective methodology to increase awareness and improve information retention among employees, as observed in studies by [18], it can be evidenced that interactive simulations can outperform conventional methods in terms of effectiveness.

4   Discussion

The implementation of the VR simulation applied to work in cold environments has proven to be successful; however, some issues arises during the development of this paper. On the one hand, it is important to increase the number of users. The data obtained corresponds to 12 people, however, having a larger sample of users is necessary for a better observation of the trends after the training has been completed. On the other, challenges on a technical level, such programming and integrating smooth and realistic animations in Unity, were overcome through iterative exploration of solutions and adaptation of animations. Looking ahead, there is significant potential to expand the effectiveness and scope of the simulator, introducing a system of random questions to cover a broader spectrum of emergency situations, simulating industry-specific activities, and refining the user interface to enhance attractiveness and usability for each user.

Moreover, despite achieving a convincing representation of the environment, it is recognized that immersion is not absolute. To achieve an even more realistic and immersive experience, incorporating augmented reality, improving in the realism of textures and the incorporation of haptic effects that simulate touch and object manipulation is required.

Additionally, it is observed that the age of participants plays a crucial role in the adaptability to change, with a steeper learning curve for adult users, which was addressed through personalized training and constant support. Virtual reality offers a controlled and safe environment where personalized experiences can be designed. It is paramount to train older workers on coping with low temperatures, which can affect both physical and cognitive performance, specially as they might experience different physiological and psychological responses compared to younger workers, it is essential to investigate these differences to develop more effective training protocols.

VR training allows companies to avoid expenses related to the presence of a trainer during training sessions, but probably the most significant advantage comes from the immediate availability of the necessary components to simulate a wide variety of scenarios and training session [19]. However, despite its many benefits, VR in education and training can be cost-demanding and requires specialized hardware and software, which may require periodic maintenance for proper operation [20]. It is crucial to persist in the investigation and systematically evaluate the long-term advantages and disadvantages of implementing virtual reality training for work in cold environments within this company.

Finally, knowledge retention is crucial to show long term benefits for virtual reality training and has to be considered for further studies. It has been proven that active learning, spaced repetition, and practical application enhance long-term retention [21], therefore an exhaustive evaluation of Virtual Reality training in cold environments is due.

5 Conclusions

The development of the VR simulator for cold environments safety training has proven to be a valuable tool for increasing knowledge and awareness of manufacturing precautions and good practices inside refrigerated enclosed areas. Through immersion in simulated environments, users can interactively, and without real risks, experience proper procedures and the correct use of PPE. With regards to the objective, the data obtained from Table 2 show that scores have improved approximately by 50% in average after VR training. Moreover, 8 users out of the 12 have increased their scores by at least 3 points, also the median obtained after VR training show that the data is evenly distributed. This significant improvement clearly demonstrates the effectiveness of VR as a training tool, particularly in enhancing the skills and performance of workers. This paper underscores the importance and positive impact of adopting emerging technologies for safety training, as in the case of robust platforms for active learning and continuous improvement in the management of high-risk work environments.

References

1. GCCA: GCCA releases 2020 Global Cold Storage Capacity Report. Fleetowner. https://www.fleetowner.com/refrigerated-transporter/cold-storage-logistics/article/21232941/gcca-releases-2020-global-cold-storage-capacity-report. Accessed 21 Feb 2024
2. World Health Organization: Caídas (2021). https://www.who.int/es/news-room/fact-sheets/detail/falls. Accessed 10 Jan 2024
3. Instituto Nacional de Seguridad e Higiene en el Trabajo (INSHT): Estrés por frío (I). NTP 1036. Washington D.C. (2015)
4. Organización Panamericana de la Salud: Curso de Gerencia para el Manejo Efectivo del Programa Ampliado de Inmunización (PAI) - Módulo III: Cadena de Frío. Organización Panamericana de la Salud, Washington D.C (2006)
5. Ministerio de Trabajo de Ecuador: Reglamento de Seguridad y Salud de los Trabajadores y Mejoramiento del Medio Ambiente de Trabajo (2020). https://www.trabajo.gob.ec/. Accessed 25 Jan 2024
6. Durán, A.C., Monje, S.V.: Plan de Capacitación Integral en Higiene y Seguridad Laboral. Informe de Proyecto de Grado, Licenciatura en Recursos Humanos, Facultad de Ciencias de la Administración. Proyecto realizado para la Panificadora PALAU, Santiago del Estero (2014)
7. Radhakrishnan, U., Chinello, F., Koumaditis, K.: Immersive virtual reality training: three cases from the Danish industry. In: 2021 IEEE Conference on Virtual Reality and 3D User Interfaces Abstracts and Workshops (VRW) (2021). https://doi.org/10.1109/VRW52623.2021.00008

8. Yang, U., Lee, G.A., Kim, Y., Jo, D., Choi, J., Kim, K.-H.: Virtual reality based welding training simulator with 3D multimodal interaction. In: 2010 International Conference on Cyberworlds, pp. 150–154. Electronics and Telecommunications Research Institute (ETRI), Daejeon, Republic of Korea (2010). https://doi.org/10.1109/CW.2010.68

9. Saravia, M.: Plan de Gestión de Riesgos para Trabajos en Cámaras de Frío en Base a Amoniaco. In: 2018 Universidad Técnica Federico Santa María, Sede Concepción - Rey Balduino de Bélgica. https://repositorio.usm.cl/handle/11673/42432

10. International Organization for Standardization: ISO 45001:2018 Occupational health and safety management systems – Requirements with guidance for use. Geneva, Switzerland (2018)

11. Zhang, H., Zhang, Y., et al.: Effects of VR instructional approaches and textual cues on performance, cognitive load, and learning experience. Educ. Tech Res. Dev **72**, 585–607 (2024). https://doi.org/10.1007/s11423-023-10313-1

12. Servicio de Prevención de la Universidad de Extremadura: Fichas de seguridad y salud en ambientes térmicos extremos. Universidad de Extremadura (2008). https://tinyurl.com/4hdez8vs

13. Asociación de Explotaciones Frigoríficas, Logística y Distribución de España: Análisis de Riesgos Derivados de la Exposición al Frío (2015). https://tinyurl.com/y3jrc42c

14. Asociación de Explotaciones Frigoríficas, Logística y Distribución de España (ALDEFE): Manual de Prevención para el Sector del Frío Industrial (2019). https://tinyurl.com/5n8nfu2v

15. Instituto Nacional de Seguridad y Salud en el Trabajo (INSST): Guía técnica para la evaluación y prevención de los riesgos relativos a la utilización de los equipos de trabajo. Ministerio de Trabajo y Economía Social (2021). https://tinyurl.com/mrcfy577

16. Crespo, R., García, R., Quiroz, S.: Virtual reality simulator for robotics learning. In: 2015 International Conference on Interactive Collaborative and Blended Learning (ICBL), Mexico City, Mexico, pp. 61–65 (2015). https://doi.org/10.1109/ICBL.2015.7387635.

17. Oshkosh: JLG LANZA LA NUEVA CAPACITACIÓN EN REALIDAD VIRTUAL ACCESSREADY FUSION XR™ (2020). https://www.jlg.com/es-co/news-events/press-releases/2020/accessready-fusion-xr-mar10

18. He, B.: Application of VR simulation and image optical processing in image visual communication design. Opt. Quant. Electron **56**, 212 (2024). https://doi.org/10.1007/s11082-023-05802-9

19. Oh, C.-G.: Pros and cons of a VR-based flight training simulator; empirical evaluations by student and instructor pilots. In: Proceedings of the Human Factors and Ergonomics Society Annual Meeting, vol. 64, pp. 193–197 (2020). https://doi.org/10.1177/1071181320641047

20. Holuša, V., Vaněk, M., Beneš, F., Švub, J., Staša, P.: Virtual reality as a tool for sustainable training and education of employees in industrial enterprises. Sustainability **15**, 12886 (2023). https://doi.org/10.3390/su151712886

21. Carlson, P., Peters, A., Gilbert, S., Vance, J., Luse, A.: Virtual training: learning transfer of assembly tasks. IEEE Trans. Visual Comput. Graphics **21**, 770–782 (2015)

Intelligent System for Predicting Bank Policy Acceptance by Ensemble Machine Learning and Model Explanation

Remigio Hurtado$^{(\boxtimes)}$ and Eduardo Ayora

Universidad Politécnica Salesiana, Cuenca, Ecuador
rhurtadoo@ups.edu.ec, aayorao@est.ups.edu.ec

Abstract. Efficient management of financial resources is crucial for the sustainability and competitiveness of banks, particularly in optimizing term deposit subscriptions to maintain liquidity. This paper introduces an advanced intelligent system for predicting term deposit acceptance using ensemble machine learning techniques. Our approach combines Random Forest and K-Nearest Neighbors (KNN) models to enhance prediction accuracy while providing clear explanations. The system follows the CRISP-DM methodology, which includes detailed phases of data preparation, modeling, fine-tuning, and model explanation. We utilize Random Forest for its feature importance metrics and KNN for assessing feature relevance through nearest neighbor analysis. The integration of these methods allows us to generate comprehensive explanations of prediction outcomes by identifying and interpreting key features influencing decision-making. By applying this method to the Bank Marketing Data Set, we demonstrate improved performance across standard metrics such as accuracy, precision, recall, and F1-score. The detailed explanation phase helps understand the model's decision process, providing actionable insights for refining telemarketing strategies. This research presents a robust framework for implementing explainable machine learning in financial marketing, enhancing both predictive accuracy and interpretability for better-informed decision-making.

Keywords: Machine Learning · Intelligent System · Ensemble Learning · Model Explanation · Data Science · Bank Policy Acceptance

1 Introduction

Efficient management of financial resources is vital for the sustainable and competitive operation of the banking industry. Optimizing fund procurement through policies is crucial for providing liquidity. An intelligent decision support system can predict policy acceptance, optimize offers, reduce acquisition costs, and increase customer retention, facilitating better financial planning and ensuring a stable source of funds for lending to SMEs. This promotes economic development, job creation, and sustainable growth [1]. Leveraging AI and machine learning, particularly Ensemble Methods, can improve predictive performance and reliability by combining multiple models to mitigate biases and

© The Author(s) 2025
E. M. Inga Ortega et al. (Eds.): CITIS 2024, LNNS 1331, pp. 449–461, 2025.
https://doi.org/10.1007/978-3-031-87065-1_41

errors [2]. Ensemble methods involve combining multiple machine learning models to enhance predictive performance, robustness, and generalization capabilities. Ensemble Methods offer versatility by accommodating various types of base learners and ensemble techniques, such as bagging, boosting, and stacking, each providing unique advantages in different contexts.

Explaining model predictions is crucial for understanding and trust, especially in banking. Three key approaches are saliency-based explanations, transparent model logic-based explanations, and exemplar-based explanations [3] [4]. Transparent models like decision trees and K-Nearest Neighbors (KNN) provide clear, intuitive explanations by showing decision paths and similarities to training data. In our approach, we have selected Random Forest and KNN as part of our ensemble method to achieve both high classification performance and robust explainability. Random Forest provides an ensemble of decision trees, which collectively offer insights into feature importance and decision rules, while KNN facilitates intuitive explanations through its neighborhood-based approach. This combination ensures our model delivers high performance with clear, justifiable predictions, crucial for transparency in banking decisions. Following the CRISP-DM methodology [5], data science processes [6] and processes adjusted from the architectures of our previous researches [7–9], our approach covers data collection, preparation, modeling, optimization, and explanation phases. Demonstrated with the Bank Marketing Data Set, the system learns from historical data, makes accurate predictions, and adapts to changes while providing transparent explanations. This approach enhances workflow, decision-making, efficiency, reliability, and responsiveness in the banking sector. Therefore, the main contributions of this research are:

- **Intelligent Decision Support System:** Development of an advanced predictive system to optimize term deposit subscriptions, enhancing financial planning and resource management in banks.
- **Ensemble Predictive Models:** Implementation of a combined K-Nearest Neighbors (KNN) and Random Forest approach, improving prediction accuracy and providing robust explanations.
- **CRISP-DM Framework:** Integration of the CRISP-DM methodology, ensuring systematic data collection, data preparation, modeling, fine-tuning, and explanation of predictions.
- **Explainable AI in Banking:** Emphasis on transparency and interpretability, providing clear and justifiable explanations of predictions for decision-making in banking services.
- **Adaptive Response:** Design of the system to continuously adapt to new data, ensuring an effective and reliable approach to decision support in dynamic financial environments.

The remainder of the document is organized as follows: II. Related work, III. Intelligent System Proposed, IV. Design of Experiments, V. Results and Discussion, and VI. Conclusions.

2   Related Work

In the field of prediction and segmentation in banking marketing, various approaches have been explored to enhance the accuracy and efficiency of models. The study in [10] proposes a data analytics and machine learning-based method for predicting credit risk in financial applications, showcasing the application of advanced techniques in banking risk assessment. The research in [11] provides a comparative analysis of credit rating predictions using neural networks, support vector machines, and decision trees, highlighting the effectiveness of each technique in credit risk forecasting. In the area of banking risk classification, [12] integrates data preprocessing techniques and machine learning, using Light-GBM and SMOTE techniques to improve banking risk classification, while [13] presents a hybrid machine learning approach to forecast banking performance, combining multiple techniques to enhance predictions.

For comparing our model, we have selected three prominent baselines. **NN** (Neural Network) uses a neural network as an initial reference for evaluating performance in telemarketing classification [14]. **SMOTEENN-XGBoost** combines oversampling and undersampling techniques with the XGBoost boosting algorithm, providing a robust solution to improve accuracy in predicting deposit subscriptions through advanced ensemble techniques [15]. Finally, **NN-KMeans** integrates a neural network with the K-Means clustering method to segment users, thereby optimizing marketing strategies by combining classification with data clustering [16]. Our proposed method stands out for its integrated and multifaceted approach. We combine a Random Forest model, which provides transparent explanations by identifying the most relevant variables through feature importance, with K-Nearest Neighbors (KNN), which allows precise identification of relevant variables based on similarities among users. This dual approach not only enhances segmentation by identifying user groups with similar characteristics but also facilitates the detection of the most influential variables in predictions. Additionally, we follow a structured process based on CRISP-DM, covering everything from data loading and preparation to detailed model explanation. This comprehensive methodology ensures a deep understanding of the system and offers an innovative solution for prediction and optimization in banking marketing, integrating advanced techniques for greater accuracy and efficiency in decision-making.

3   Intelligent System Proposed

In this section, we present the developed intelligent system. This system enables learning from historical data and predicting new clients to determine whether a client will subscribe to a policy. Figure 1 provides a visual representation of the methodology guiding our process. The general algorithm formalizes the phases of the methodology applied to our specific problem. The proposed methodology follows the CRISP-DM approach, which includes the following detailed steps. First, data loading is performed, where the relevant dataset for analysis is imported.

Second, data preparation is carried out, which includes data cleaning, transformation of categorical variables, and correlation analysis to identify the most relevant variables before learning, thus formulating initial hypotheses. Third, modeling is done using ensemble machine learning techniques, utilizing both voting and stacking methods, combining K-Nearest Neighbors (KNN) and Random Forest. In the stacking method, the final model used to make predictions is Random Forest. Fourth, fine-tuning is performed to identify the best combination of hyperparameters in the ensemble method. Finally, in the fifth step, the Model Explanation phase is conducted, where the model is explained using information obtained from the tuning of KNN and Random Forest to determine the most relevant variables for the predictions.

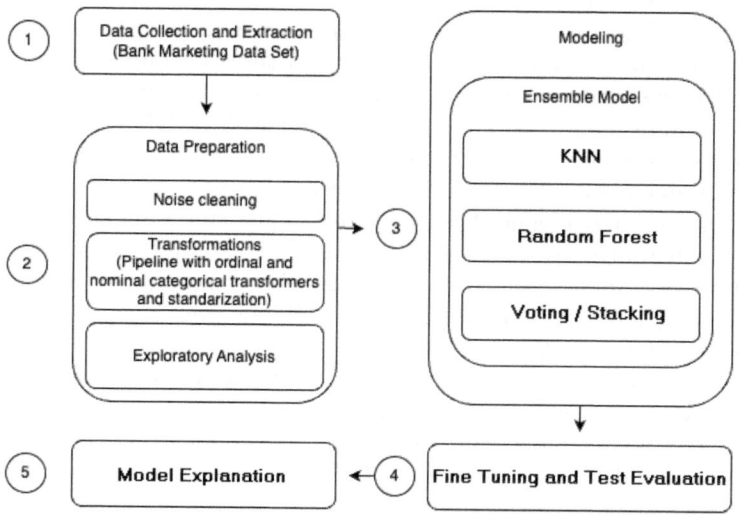

Fig. 1. Proposed method following CRISP-DM process

The methods are combined to obtain a more robust ensemble model, capable of explaining the most relevant variables for the decision-making process in the model, according to the individual KNN and Random Forest models. This supports the understanding of the decision and helps bank executives or the system itself to make the final decision. The voting method allows using the predictions of each base model as a vote, thus contributing to the creation of a more robust and balanced system in predicting policy acceptance. Stacking, on the other hand, combines the predictions of multiple base models through a higher-level model, improving the system's ability to capture complex relationships in the data and providing greater accuracy in the final predictions.

The **Model Explanation phase** helps us understand which features are most influential in making predictions, providing valuable insights for decision-making in a banking institution and contributing to the system's autonomy. This

General Algorithm
Input: Dataset
Phase 1: Data Collection and Extraction
1. Load the dataset for analysis.
Phase 2: Data Preparation
2. Data cleaning: removal of null and duplicate values.
3. Transformation of categorical variables into numerical variables.
4. Correlation analysis to identify the most relevant variables and initial hypotheses.
Phase 3: Modeling with Machine Learning Ensemble
5. Training base models:
a. K-Nearest Neighbors (KNN).
b. Random Forest.
6. Combining base models using Voting and Stacking techniques.
Phase 4: Fine Tuning and Test Evaluation
7. Identification of the best combination of hyperparameters.
Phase 5: Model Explanation
a. **KNN for identifying Relevant Variables:**
8. Select the best KNN to predict the labels of the test set.
9. Calculate the absolute differences between the test instances and their nearest neighbors in the training set:
$d_{ij} = \lvert x_i - x_j \rvert$, where d_{ij} is the absolute difference between instance i and its neighbor j.
10. Calculate the mean of these absolute differences for each feature:
$\bar{d}_k = \frac{1}{n} \sum_{i=1}^{n} d_{ik}$, where $\bar{d}_k$ is the mean of the absolute differences for feature k.
11. Identify the features with the highest $\bar{d}_k$ as the most relevant.
12. Generate a figure showing these features and their mean absolute differences.
b. **Random Forest for identifying Relevant Variables:**
13. Select the best Random Forest model with the scaled training set.
14. Calculate the importance of each feature using the *feature_ importances_* metric:
$FI_k = \sum_{t=1}^{T} \left(\frac{I_t \cdot \Delta f_t}{\sum_{k=1}^{K} I_t \cdot \Delta f_t} \right)$, where FI_k is the importance of feature k, I_t is the importance of tree t, and Δf_t is the decrease in impurity caused by feature k.
15. Sort the features by their importance and select the most relevant ones.
16. Generate a figure showing these features.
Output: Predictions, Relevant Variables and Quality Results.

process enables the automatic execution of the contact or telemarketing process to communicate with clients who are estimated to be interested in a bank policy. In the Model Explanation phase, we identify the most important features using two different methods: K-Nearest Neighbors (KNN) and Random Forest. The feature importance metric in **Random Forest** allows identifying the most relevant variables by analyzing the structure of the tree and how each feature affects the purity of the splits within the model. In a Random Forest, each decision tree partitions the data based on the available features. The importance of a feature is evaluated by looking at how much the purity of splits in the tree improves when that feature is used. Specifically, we measure how the inclusion of a feature reduces the impurity in the leaves of the tree, i.e., how it helps to make the data sets more homogeneous in terms of the target variable. The features that contribute most to reducing impurity in the tree partitions are considered the most important, as they have a significant impact on the quality of the model and its ability to make accurate predictions. This evaluation helps to identify the key variables that are crucial for decision making in the model. For **KNN** we measure how different each test instance is from its closest neighbors in the training data. We calculate the average of these differences for each feature. The features with the highest average differences are considered the most important. Features with the highest average differences in a K-Nearest Neighbors (KNN) model are considered the most important because they have the greatest impact on the model's predictions. This significant variation indicates that these features are key factors in the decision-making process, providing the model with essential information to make accurate predictions. Understanding these important features helps us gain valuable insights into what drives the predictions, enabling better-informed strategies and actions. For instance, in predicting whether customers will subscribe to a policy using a K-Nearest Neighbors (KNN) model with three customers and two variables (age and income), the standardized data is shown in the Table 1. By calculating the absolute differences in age between Customer A and its nearest neighbors (Customers B and C), we obtain differences of 0.70 and 2.82, with a mean absolute difference of 1.76. Similarly, the differences in income between Customer A and its neighbors are 0.95 and 2.44, yielding a mean absolute difference of 1.69. Since age has the highest average difference, it is identified as the most significant variable in predicting policy subscription, indicating its critical role in the KNN model's decision-making process.

Table 1. Data example for Model Explanation using K-Nearest Neighbors (KNN)

Customer	Age	Income
A	−1.41	−1.22
B	−0.71	−0.27
C	1.41	1.22

4   Design of Experiments

This section presents the characteristics of the dataset, the baselines, the parameters in the fine tuning and the quality measures. Table 2 presents the general description of dataset and their most relevant characteristics. As explained in the related work section, the baselines selected to evaluate the model performance in this study are **NN** (Neural Network) [14], **SMOTEENN-XGBoost** which is a baseline of ensemble learning [15], and **NN-KMeans** [16]. Table 3 presents the parameters to be experimented to train and optimize the models.

For the evaluation of the methods, the Cross-Validation K-Folds technique (with K=3) has been used in order to obtain an adequate generalization of the results and mitigating the risk of overfitting. The quality measures used are: Precision, Recall, F1-Score and Accuracy, particularly tailored for classification tasks. The average of K experiments with the best parameters of each method is presented in the results section.

Table 2. Description of dataset

Dataset	#Customers	#Features	Output
Bank Marketing [17]	45,211	17	y: subscribed or not

Table 3. Optimization parameters of predictive models

Method	Parameters
K-Nearest Neighbors (KNN)	K (Number of neighbors): 3, 5, 7, 10, 15, 20
Random Forest (RF)	n_estimators (Number of trees): 50, 100, 150 max_depth (Depth that trees can reach): 10, 20, 30, 50
Ensemble Model - Voting	Combination of KNN and Random Forest voting: hard (majority of votes) voting: soft (average of the probabilities) * voting is the mechanism for combining predictions
Ensemble Model - Stacking	Combination of KNN and Random Forest

5 Results and Discussion

We began our method by data collection and data preparation. **Phase: Data preparation** involved cleaning the dataset, transforming categorical variables into numerical ones, and performing correlation analysis to identify the most relevant variables and initial hypotheses as shown in Fig. 2. In this study, the categorical variables include the type of job of the client (job), marital status (marital), educational level (education), credit default status (default), housing loan status (housing), personal loan status (loan), type of contact communication (contact), the last month of contact (month), and the outcome of the previous marketing campaign (poutcome). All other variables are numerical. Since the objective of this study is to predict the success of bank telemarketing campaigns, we formulate the following initial hypotheses. Variables such as the duration of the last contact (duration), the successful outcome of the previous campaign (poutcome_success), and contact via cellular phone (contact_cellular) are expected to have a significant influence on the probability that a customer will subscribe to a term deposit. In addition, specific months

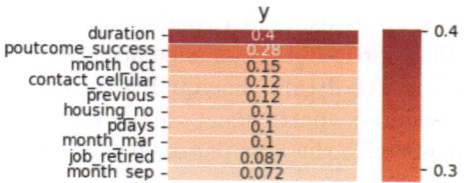

Fig. 2. Correlation analysis to identify the initial relevant variables

of contact (month_oct, month_mar, month_sep), the number of previous contacts (previous), and the absence of mortgage loans (housing_no) could also be determining factors. These hypotheses are based on the observed correlation between these variables and the desired outcome, suggesting that these characteristics could be key indicators in the effectiveness of telephone marketing strategies. In **Phase: Modeling** we trained our base models, which included K-Nearest Neighbors (KNN) and Random Forest, and combined them using Voting and Stacking techniques. In **Phase: Fine Tuning**, the best combination of hyperparameters was identified, as depicted in Fig. 3. The best values for the parameters are as follows: the number of neighbors (K) is 5, the number of trees (n_estimators) is 50, the maximum depth that trees can reach (max_depth) is 50, and the voting mechanism for the ensemble model is soft. Thus leveraging the strengths of both the K-Nearest Neighbors and Random Forest models to enhance overall predictive performance. **Phase: Model Explanation** involved KNN and Random Forest for identifying relevant variables. We selected the best KNN model to predict the labels of the test set. The absolute differences between the test instances and their nearest neighbors in the training set were calculated. The mean of these absolute differences for each feature was then computed. The features with the highest mean absolute differences were identified as the most relevant. We selected the best Random Forest model with the scaled training set. The importance of each feature was calculated, and the features were sorted by their importance to identify the most relevant ones. The results are shown in Panel (a) and Panel (b) of Fig. 4. In Panel (c) of Fig. 4, an example is provided to illustrate how visualizing the structure of a decision tree within a Random Forest can help identify the most relevant variables and criteria for classifying clients. To achieve this, a decision tree was trained with a maximum depth of three. This depth constraint limits the tree's complexity, making it easier to interpret while still capturing the key decision points used in classification. By examining the tree's structure, including the nodes and branches, one can observe how different variables are used to split the data and how they contribute to the classification decisions. The tree's nodes indicate the variables and the corresponding thresholds that are most influential in distinguishing between different classes, thereby highlighting the criteria that have the greatest impact on the model's predictions.

The **final results** of our proposed method and selected methods from related work, including accuracy, precision, recall, and F1-score, are summarized in

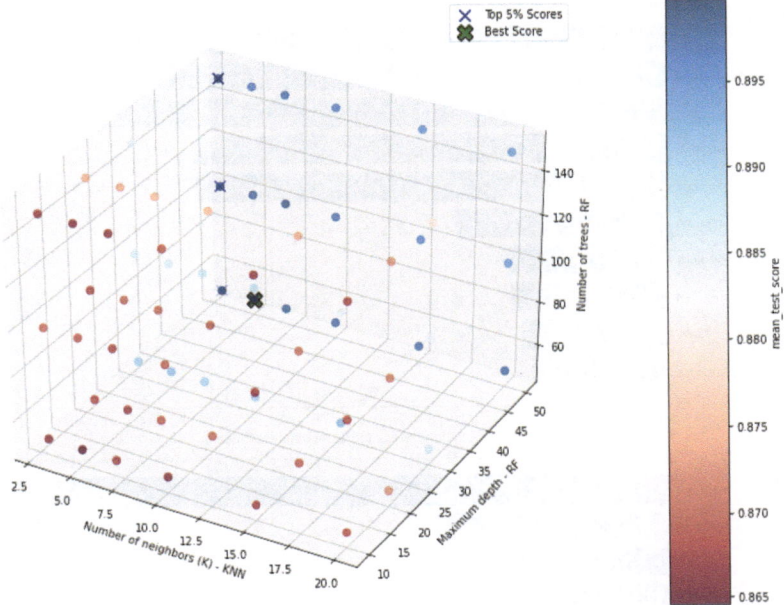

Fig. 3. Hyperparameter tuning

Table 4. In comparing our model with prominent baselines, we observe distinct advantages. Neural Network (NN), a widely used reference for telemarketing classification, demonstrates competitive performance, particularly in recall, though its precision and F1-Score are lower. SMOTEENN-XGBoost excels in precision and F1-Score due to its robust ensemble techniques combining oversampling and undersampling with XGBoost, achieving the highest values among the baselines. NN-KMeans provides a strong precision score but falls short in accuracy compared to our method. Our proposed method, integrating Random Forest and K-Nearest Neighbors (KNN), offers significant improvements. The Ensemble Stacking approach achieves the highest accuracy, precision, recall, and F1-Score, indicating a robust and comprehensive model. This success is attributed to our combination of variable importance analysis with Random Forest and similarity-based variable identification with KNN, enhancing both segmentation and prediction accuracy. This dual approach, supported by a structured CRISP-DM methodology, ensures a deep and accurate understanding of customer behaviors and prediction outcomes, thus providing a superior solution in banking marketing. Despite the impressive performance of our proposed method, several areas could benefit from further refinement. While our method excels in accuracy and F1-Score, the recall for some models indicates room for improvement in identifying positive instances. Exploring alternative models or hybrid approaches that enhance recall without compromising other metrics could be advantageous.

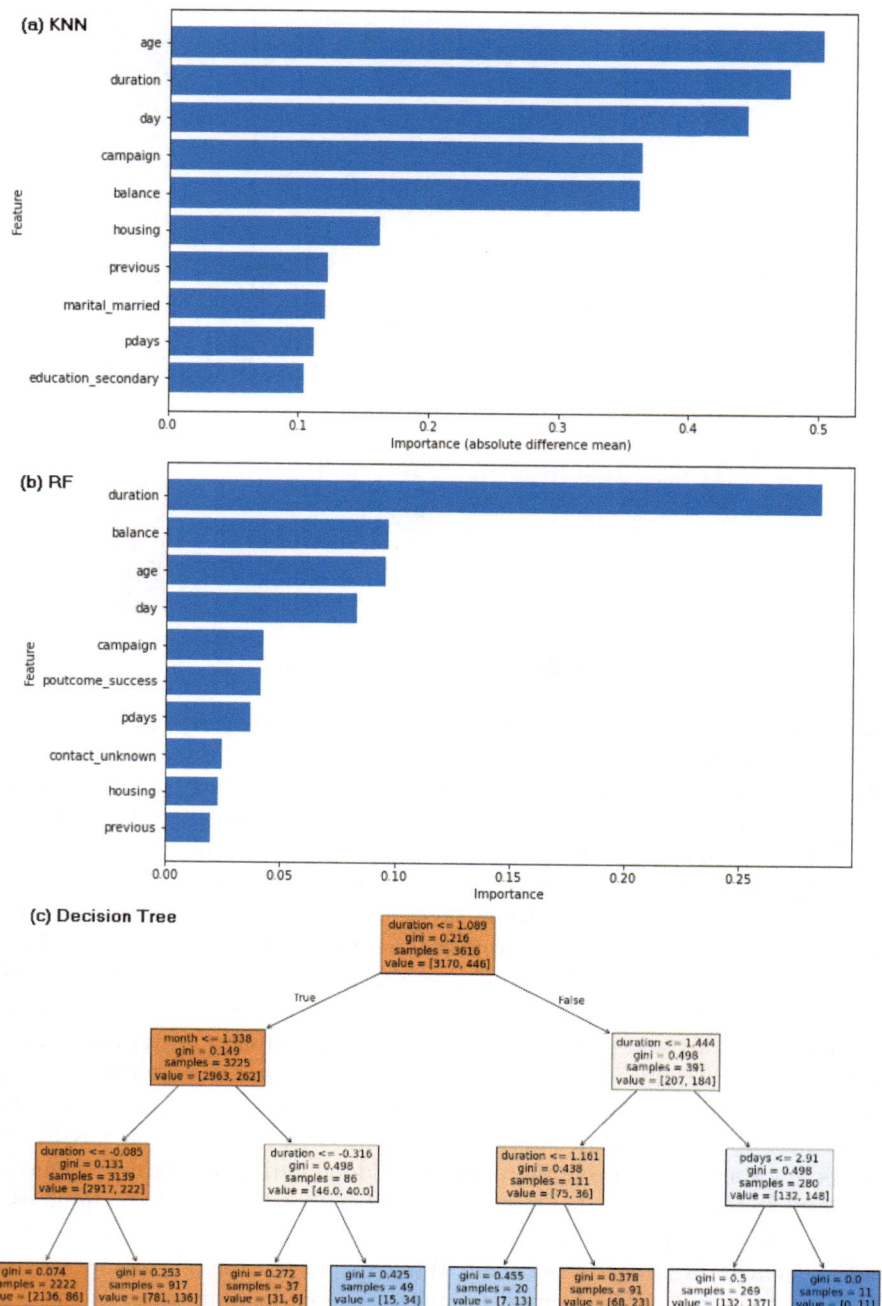

Fig. 4. Panel (a) presents the variables identified as most relevant by the K-Nearest Neighbors (KNN) model, Panel (b) displays the variables deemed most relevant by the Random Forest (RF) model, while Panel (c) presents a Decision Tree from the Random Forest model.

Table 4. Performance metrics for our proposed method and baselines

Predictive Models of Our Method

Method	Accuracy	Precision	Recall	F1-Score
Ensemble Stacking	**0.9037**	**0.7767**	**0.7131**	**0.7390**
Ensemble Voting	**0.9036**	**0.8074**	0.6498	0.6922
Random Forest (RF)	0.9030	0.7885	0.6707	0.7087
K-Nearest Neighbors (KNN)	0.8947	0.7551	0.6527	0.6856

Baselines

Method	Accuracy	Precision	Recall	F1-Score
Neural Network (NN) [14]	0.8912	0.5112	**0.6723**	0.5850
SMOTEENN-XGBoost [15]	**0.8824**	**0.8650**	**0.9020**	**0.8831**
NN-KMeans [16]	0.8120	0.8230	0.6701	0.7402

6   Conclusions

Our study's initial analysis identified key factors influencing term deposit subscriptions, such as the length of the last contact, the success of prior marketing efforts, and whether the contact was made via a cellular phone. Using Random Forest (RF) and K-Nearest Neighbors (KNN) models, we refined these insights. The RF model highlighted the importance of the number of previous contacts, mortgage status, the time since the last contact, the number of contacts made during the campaign, the success of the previous campaign, and the duration of the last contact. Meanwhile, the KNN model emphasized the significance of the customer's educational background, marital status, and account balance. These findings validate our initial hypotheses, suggesting that telemarketing strategies should focus on these aspects to enhance prediction accuracy and overall effectiveness.

This research presents an advanced decision support system for optimizing term deposit subscriptions in banking, integrating RF and KNN models to enhance prediction accuracy and provide transparent explanations. Following the CRISP-DM methodology, the system systematically addresses data preparation, modeling, fine-tuning, and explanation of predictions. Key insights confirm the importance of variables such as the duration of the last contact, the success of prior marketing efforts, and customer demographics. Overall, the proposed system improves predictive performance and offers a reliable framework for decision support in dynamic financial environments, contributing to better financial planning and resource management in banks. Future work could explore integrating deep learning models or hybrid approaches to enhance recall and performance, incorporating real-time data processing for adaptability, and improving model interpretability and computational efficiency for large-scale datasets. Expanding the system's scope to include other financial products and services could provide a comprehensive decision support tool for banks, driving innovation and customer satisfaction across various domains.

References

1. Alvarez Alvarez, B., Casielles, R.V.: Consumer evaluations of sales promotion: the effect on brand choice. Eur. J Market. **391**/2, 54–70 (2005)
2. Dietterich, T.G.: Ensemble methods in machine learning. In: International Workshop on Multiple Classifier Systems, pp. 1–15. Springer, Berlin, Heidelberg (2000)
3. Chin, E.K.: The Deep Learning Architect's Handbook. Packt Publishing (2019)
4. Ribeiro, M.T., Singh, S., Guestrin, C.: Why should I trust you? Explaining the predictions of any classifier. In: Proceedings of the 22nd ACM SIGKDD International Conference on Knowledge Discovery and Data Mining, pp. 1135–1144 (2016)
5. Wirth, R., Hipp, J.: CRISP-DM: towards a standard process model for data mining. In: Proceedings of the 4th International Conference on the Practical Applications of Knowledge Discovery and Data Mining, vol. 1 (2000)
6. Hayashi, C.: What is Data Science? Fundamental Concepts and a Heuristic Example, Data Science, Classication, and Related Methods, pp. 40–51. Springer, Tokyo (1998)
7. Hurtado, R., Guzmán, S., Muñoz, A.: An architecture and a new deep learning method for head and neck cancer prognosis by analyzing serial positron emission tomography images. In: Conference on Cloud Computing, Big Data and Emerging Topics. Springer Nature Switzerland, Cham (2023)
8. Ortiz, S.G., et al.: Una arquitectura de análisis de imágenes seriadas con la tomografía por emisión de positrones mediante la aplicación de machine learning combinado para la detección del cáncer de pulmón. Revista Española de Medicina Nuclear e Imagen Molecular (2024)
9. Hurtado, R., et al.: Development of an intent-based network incorporating machine learning for service assurance of e-commerce online stores. In: International Conference on Machine Learning for Networking. Springer Nature Switzerland, Cham (2022)
10. Ortiz, R.H., et al.: A data analytics method based on data science and machine learning for bank risk prediction in credit applications for financial institutions. In: 2022 IEEE International Autumn Meeting on Power, Electronics and Computing (ROPEC), vol. 6. IEEE (2022)
11. Golbayani, P., Florescu, I., Chatterjee, R.: A comparative study of forecasting corporate credit ratings using neural networks, support vector machines, and decision trees. The North Am. J. Econ. Finance (2020)
12. Muslim, M.A., et al.: Bank predictions for prospective long-term deposit investors using machine learning LightGBM and SMOTE. J. Phys. Confer. Ser. **1918**, 4. IOP Publishing (2021)
13. Islam, U., et al.: Forecasting of bank performance using hybrid machine learning techniques. In: 2022 International Conference on Innovations in Science, Engineering and Technology (ICISET). IEEE (2022)
14. Moro, S., Cortez, P., Rita, P.: A data-driven approach to predict the success of bank telemarketing. Decis. Support Syst. **62**, 22–31 (2014)

15. Li, Y., Wu, Z.: Prediction of customers' subscription to time deposits based on SMOTEENN-XGBoost model (2022)
16. Hematyar, A.: Data-driven decision making for direct marketing of banking products with the use of deep learning and random forests, pp. 78–89 (2022)
17. Moro, S., Rita, P., Cortez, P.: Bank Marketing. UCI Machine Learning Repository (2012). https://doi.org/10.24432/C5K306

Classification of the Severity Level of Breakage Failure in Spur Gearboxes Through Frequency Domain Vibration Signal Analysis

Antonio Pérez-Torres[1,2](✉) ⓘD, René-Vinicio Sánchez[2] ⓘD, and Susana Barceló-Cerdá[1] ⓘD

[1] Department of Applied Statistics and Operational Rsearch, and Quality, Universitat Politècnica de València, Valencia, Spain
jperezt@ups.edu.ec
[2] Grupo de Investigación y Desarrollo en Tecnologías Industriales GIDTEC, Universidad Politécnica Salesiana, Cuenca, Ecuador

Abstract. In the industry, gearboxes used for their efficiency in power transmission are critical components, making early failure detection essential. This work aims to determine the ranking of condition indicators (CIs) to extract information from the vibration signal in the frequency domain for a spur gearbox and to assess the accuracy of the classification model for the failure severity level. In laboratory conditions, the tooth breakage of a pinion with different severity levels was simulated in a spur gearbox. Four accelerometers were installed in the gearbox in a vertical position to obtain the vibration signal. First, information was extracted from the vibration signal through 25 CIs in the accelerometers. Using artificial intelligence, the ranking of 10 CIs was carried out. Subsequently, the random forest (RF) algorithm was used to determine the accuracy in classifying the failure severity level. Finally, an ANOVA test was conducted to determine if there were significant differences in classification accuracy for the four accelerometers. The results concluded that the CIs selected through the ranking are optimal for determining the failure severity level of tooth breakage in a spur gearbox and that the sensor placement affects the classification accuracy.

Keywords: Machine learning · classification model · failure severity · random forest · condition indicators

1 Introduction

In the industrial sector, rotating machinery is used in different production processes, and gearboxes constitute a fundamental part due to their efficiency in power transmission between shafts in confined spaces and their high load capacity. Therefore, monitoring the condition of a gearbox to detect a failure in advance is essential to avoid unexpected maintenance activities or, worse, its inoperability [3].

Supported by Universidad Politécnica Salesiana.

E. M. Inga Ortega et al. (Eds.): CITIS 2024, LNNS 1331, pp. 462–472, 2025.
https://doi.org/10.1007/978-3-031-87065-1_42

In monitoring the condition status of the gearboxes, it is necessary to consider the following phases: data acquisition, feature extraction, and fault detection and identification. Statistical models for feature extraction are fundamental in the process [10, 15].

The data acquisition phase is carried out by collecting information that can be obtained accurately and reliably through sensors installed in different positions [2].

In the feature extraction phase of a signal, it is necessary to process the information obtained from a sensor. For this, statistical parameters are used to construct condition indicators (CIs) [6].

In the fault detection and identification phase for gearboxes, various investigations have been carried out based on CIs due to their sensitivity to different types and severity levels of faults. By combining multiple CIs, better results can be obtained. However, this can lead to redundant information or contradictions that complicate the fault detection process or its severity level. Therefore, it is necessary to have a CIs selection system and automated classification models such as random forest (RF) [7].

RF is a classification model represented in Eq. 1 consisting of different classifiers that follow a tree structure. For each ith tree, an independent random vector (V_i) is generated. Each tree uses a training set and votes for the most popular category in the input vector (x) [1].

$$RF = h(x, V_i)_{i=1}^{N}, \quad i = 1, 2, 3.... \tag{1}$$

The vibration signal is the most commonly used for fault diagnosis due to its ease of acquisition. Still, it is also complex as it contains stationary, non-stationary, and resonant components. Therefore, it is necessary to select appropriate signal processing techniques using suitable data algorithms to diagnose the condition, severity, and type of failure in gear, with tooth breakage (Fig. 2) being the most severe [4].

The vibration signal, CIs, and the RF classification model have been used in various investigations to determine the severity level of a failure. For example, in diagnosing multiple faults in both bearings and spur and helical gearboxes, [13] used CIs to extract information from the vibration signal and RF as one of the classification methods. In [5], CIs were used to analyse the vibration and acoustic emission signals in a gearbox of a helicopter transmission system. In [7], CIs were utilised to extract features from the vibration signal in a spur gearbox's time and frequency domains to classify the severity of a gear failure subjected to tooth root cracks. In [2], for diagnosing multiclass faults in spur gearboxes, CIs were used in the time, frequency, and time-frequency domains, with RF being one of the severity-level classification models.

Although there are multiple research works related to the use of CIs and the RF classification model for determining faults in gearboxes, what has not been reported is a ranking of the CIs in the frequency domain using the wrapper method to determine the severity level of a tooth breakage failure in a spur gearbox, and whether the sensor position affects the accuracy of the severity level classification model.

2   Materials and Methods

2.1   Testing Bench

A single-stage spur gearbox was used in the experimental phase, coupled to a 2 hp, 220 V, 1200 rpm three-phase motor. The motor's rotational speed can be controlled by a frequency inverter. To simulate different load conditions, an electromagnetic brake was coupled to the output shaft of the gearbox. Figure 1 shows the test bench configuration.

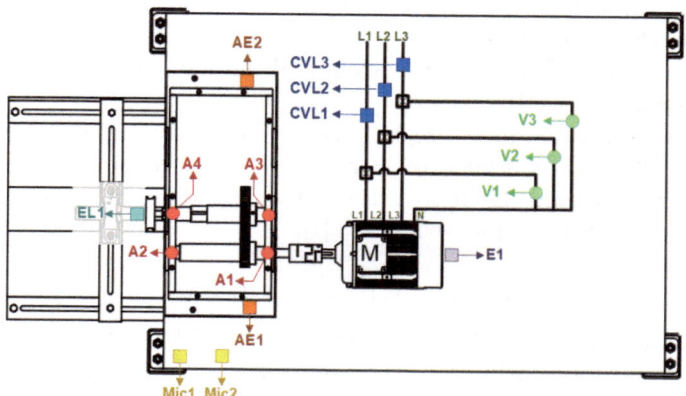

Fig. 1. Gearbox layout

The gearbox has a pinion with Z1 = 32 teeth and a gear with Z2 = 48 teeth. Nine levels of tooth breakage failure severity (P2-P10) were artificially simulated on the Z1 pinion (Fig. 2). The Z1 pinion was exchanged from normal condition P1 to the maximum severity level P10 for data acquisition. The motor's rotational frequency was adjusted to 8 Hz, 14 Hz, and 20 Hz at each severity level. The load conditions were varied using the electromagnetic brake at 0V, 10V, and 20V. The experiment was repeated ten times for 10 s each. A database (DB) with 900 observations per accelerometer was obtained. The testing bench also has an encoder (E1), a laser encoder (EL1), two acoustic emission sensors (EA) and two microphones (Mic) for the gearbox and the power supply lines to the motor with three voltage meters (V) and three electric current clamp (CVL)

2.2   Methodology

The methodology developed in this work is detailed in Fig. 3 and consists on the following stages:

Fig. 2. Broken tooth (P5).

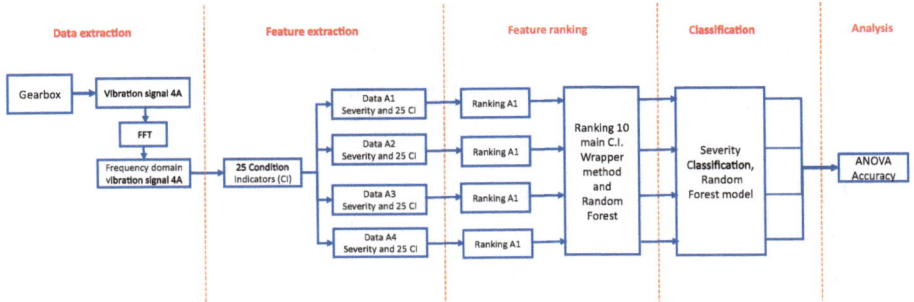

Fig. 3. Methodology

Data Extraction. Data acquisition of the vibration signal was carried out using four accelerometers(A) placed in a vertical position. Accelerometers A1 and A2 were installed on the input shaft, while accelerometers A3 and A4 were installed on the output shaft. The sampling frequency of each accelerometer is 50k samples/s. The vibration signal from the four accelerometers was transformed into the frequency domain using the fast Fourier transform (FFT) [2].

Feature Extraction. Characteristics of the vibration signal in the frequency domain were extracted from the four accelerometers by calculating 25 CIs in the frequency domain, as detailed in Table 1, used by [14] in their research. A vector of 900 observations (90 for each severity level) was obtained for each CI, combined into a matrix with the severity level factor of the tooth breakage failure (P1-P10). This way, four databases (DBs) were obtained, one for each accelerometer.

Feature Ranking. The information extracted from the four accelerometers was considered for the CIs ranking using the detailed process outlined below.

Initially, each dataset obtained was input into the RF classification model, and through the wrapper method and an iterative resampling process, the importance of the CIs was determined by calculating the overall influence (OI). This process generated a ranking of CIs for each accelerometer, and the top 10 most important CIs were selected. Ten CIs were chosen because the variability in classification reduced and stabilised for the four accelerometers, as detailed in Fig. 4.

Table 1. Frequency domain condition indicators

N.	Condition indicator	Formula
1	Mean frequency	$F1 = \frac{\sum_{k=1}^{K} X(k)}{K}$
2	Variance	$F2 = \frac{\sum_{k=1}^{K} (X(k)-F1)^2}{K-1}$
3	Skewness	$F3 = \frac{\sum_{k=1}^{K} (X(k)-F1)^3}{K(\sqrt{F2})^3}$
4	Kurtosis	$F4 = \frac{\sum_{k=1}^{K} (X(k)-F1)^4}{K(F2)^4}$
5	Central frequency	$F5 = \frac{\sum_{k=1}^{K} f_k X(k)}{\sum_{k=1}^{K} X(k)}$
6	Standard deviation	$F6 = \sqrt{\frac{\sum_{k=1}^{K}}{(f_k-FC)^2 X(k)} \sum_{k=1}^{K} X(k)}$
7	Root mean square	$F7 = \sqrt{\frac{\sum_{k=1}^{K} f_k^2 X(k)}{\sum_{k=1}^{K} X(k)}}$
8	CP1	$F8 = \frac{\sum_{k=1}^{K} (f_k-FC)^3 X(k)}{K(F6)^3}$
9	CP2	$F9 = \frac{F6}{F5}$
10	CP3	$F10 = \frac{\sum_{k=1}^{K} (f_k-F5)^{\frac{1}{2}} X(k)}{K\sqrt{F6}}$
11	CP4	$F11 = \frac{\sum_{k=1}^{K} (f_k-F5)^3 X(k)}{F6^2 K}$
12	CP5	$F12 = \sqrt{\frac{\sum_{k=1}^{K} f_k^4 X(k)}{\sum_{k=1}^{K} f_k^2 X(k)}}$
13	Total power	$F13 = \sum_{k=1}^{M} P(k)$
14	Median frequency	$F14 = \frac{1}{2} \sum_{k=1}^{M} P(k)$
15	Peak frecuency	$F15 = max(P(k)), \quad k = 1, ... M.$
16	First spectral moment	$F16 = \sum_{k=1}^{M} P(k) f_k$
17	Second spectral moment	$F17 = \sum_{k=1}^{M} P(k) f_k^2$
18	Third spectral moment	$F18 = \sum_{k=1}^{M} P(k) f_k^3$
19	Fourth spectral moment	$F19 = \sum_{k=1}^{M} P(k) f_k^4$
20	Variance of central frequency	$F20 = \frac{1}{F13} \sum_{k=1}^{M} P(k)(f_k - f_c)^2$
21	Frequency ratio	$F21 = \sum_{LLC=f_{min}}^{ULC=f_{max}/2} P(k) / \sum_{LHC=\frac{f_{max}}{2}+1}^{UHC=f_{max}} P(k)$ Where $f_{min} = 5$ and $f_{max} = 3000$ as default study frequency
22	Frequency warp	$F22 = \frac{\sqrt{\frac{F17}{F13}}}{\frac{F16}{F13}}$
23	Spectral centroid	$F23 = \frac{\sum_{k=1}^{K} kX(k)}{\sum_{k=1}^{K} X(k)}$
24	Spectral spread	$F24 = \sqrt{\frac{\sum_{k=1}^{K} (k-SC)^2 X(k)}{\sum_{k=1}^{K} X(k)}}$
25	Spectral entropy	$F25 = -\sum_{k=1}^{K-1} P_n(k) log_2[P_n(k)]$ where P_n is the normalized total spectral energy

In this process, each CI was assigned a weighting value from 10 to 1 according to its order of importance.

Subsequently, an overall ranking was performed, for which the total weighting was calculated by summing the weighting values from the four accelerometers assigned in the previous process. Finally, the top 10 CIs with the highest weighting were selected from the overall ranking. These selected CIs were then used in the classification stage. The results of this process are detailed in Sect. 3.

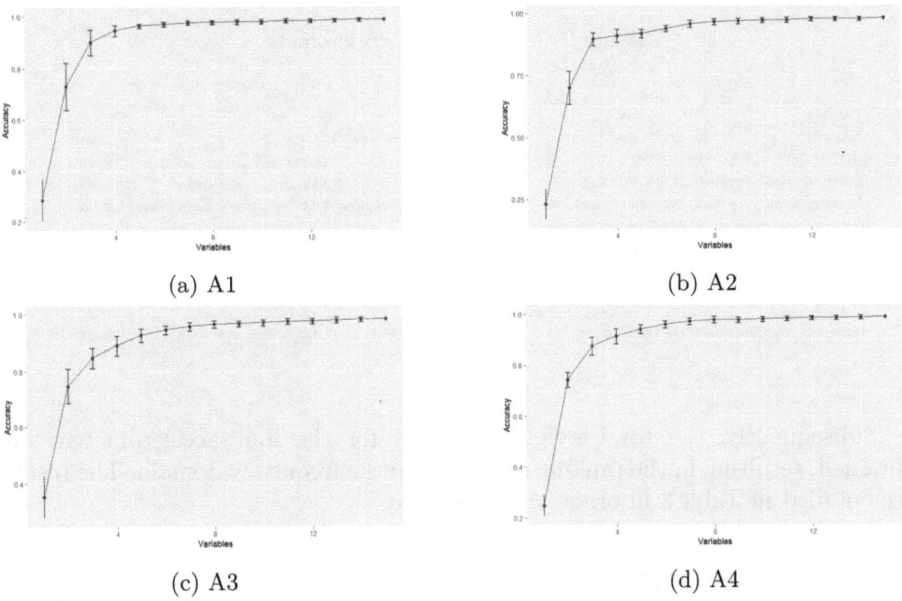

(a) A1

(b) A2

(c) A3

(d) A4

Fig. 4. Accuracy by number of CIs and accelerometers

Classification. The reduced DBs with the top 10 CIs were used to determine the failure severity level using the RF classification model. The DBs were split into 70% for training and 30% for testing. An iterative resampling and cross-validation process was conducted to measure the model's performance. This process yielded a vector with 100 observations corresponding to the classification accuracy.

Analysis. The data from the accuracy vectors obtained in the classification stage were analysed. An analysis of variance (ANOVA) and post-hoc tests were conducted to determine if the accelerometer position impacts the vibration signal.

The data processing was performed using software R [12] and its development environment RStudio.

3   Results

The first objective of this work was to rank the top 10 CIs to analyse the vibration signal in the frequency domain. The process detailed in Sect. 2.1 was carried out to achieve this objective. By selecting the top 10 CIs for the four accelerometers, Table 2 was obtained.

Table 2. Main CIs by accelerometer

A1		A2		A3		A4		
Variable	OI	Variable	OI	Variable	OI	Variable	OI	Weight
Skewness	14.623	Skewness	16.850	Central frequency	13.630	Skewness	16.668	10
CP2	13.510	CP2	14.947	CP2	13.609	Third spectral moment	13.982	9
Fourth spectral moment	13.217	Central frequency	14.455	Kurtosis	13.234	Root mean square	13.040	8
Central frequency	13.162	Standar deviation	14.329	Standar deviation	13.026	Second spectral moment	13.006	7
Third spectral moment	12,794	Root mean square	13.962	Skewness	12.421	Fourth spectral moment	12.872	6
CP5	12.752	CP5	12.466	Root mean square	12.369	Total power	10.836	5
Standar deviation	12.099	Kurtosis	10.850	CP5	12.338	CP2	10.792	4
Root mean square	10.302	Spectral entropy	10.825	CP1	12.332	Median frequency	10.779	3
Peak frequency	10.222	Third spectral moment	10.757	VCF	12.303	Peak frequency	10.702	2
Second spectral moment	10.159	Variance	10.748	Peak frequency	11.915	Standar deviation	10.593	1

Subsequently, the total weighting (TW) for the four accelerometers was obtained, resulting in the ranking of the CIs in the frequency domain. The results are detailed in Table 3 in order of importance.

Table 3. Ranking CIs frequency domain

Ranking CIs	TW
Skewness	36
CP2	31
Central frequency	25
Root mean square	22
Standar deviation	19
Third spectral moment	17
Fourth spectral moment	14
CP5	14
Kurtosis	12
Second spectral moment	8
TW= Total weighing	

The second objective was to measure the performance of the CIs selected in the ranking when used in the RF classification model. To this end, the average accuracy in classifying the failure severity level was calculated. The accuracy results per accelerometer are as follows: **A1 = 98.12%, A2 = 96.90%, A3 = 97.31%, and A4 = 98.36%**. Subsequently, by conducting the ANOVA test (Table 4), it was determined that there are significant differences in the classification accuracy of failure severity level among the four accelerometers (F-value = 52.04, p-value < 0.001). The post-hoc tests (Table 5) revealed that the differences in accuracy between accelerometers A1 and A4 are not significant, and these accelerometers have the highest average classification accuracy.

Table 4. ANOVA. Accuracy by accelerometer

	Df	Sum Sq	Mean Sq	F-value	p-value
A	3.00	0.01396	0.004654	52.04	<0.001
Residual	396.00	0.03541	0.000089		

Table 5. Post-hoc test

A	Difference	p value
A2-A1	0.0122	<0.001
A3-A1	0.0081	<0.001
A4-A1	0.0023	0.302
A3-A2	0.0041	0.012
A4-A2	0.0146	<0.001
A4-A3	0.0104	<0.001

4   Discussion

In [13], a ranking of 10 CIs in the time domain was performed using the filtering method for different gear datasets. When using the top 10 CIs from the ranking in the RF and KNN classification models, the calculated accuracy in some datasets was below 85%. The approach developed in this work to determine the CIs ranking using the wrapper method to analyse the vibration signal in the frequency domain for gearboxes is suitable, as the accuracy value specified in the RF classification model exceeded 95% for the four accelerometers analysed.

In the research conducted by [11], 15 CIs were used to detect faults in bearings, resulting in accuracy values above 95% with the RF classification model. In this research, by using the ranking of 10 CIs in the RF classification model, accuracy values exceeded 96% for the four accelerometers, establishing that the selected CIs in the ranking are optimal for determining the severity level of gear tooth breakage in a spur gearbox by analysing the vibration signal in the frequency domain.

Regarding the sensor position, there are studies such as those conducted by [8], which analyse the importance of sensor position on a railway axle, or the study by [9], which examines the importance of sensor position in a gearbox, indicating that the best position is at the output of the driving shaft. In this work, by analysing the precision vectors in the classification, it was determined that A1 installed on the input shaft of the gearbox, that is, the driving side, and A2 installed on the output side of the gearbox, where the load is located, showed the best results for classifying the severity level of tooth breakage failure. This finding is significant as it indicates that installing accelerometers in these two positions on the gearbox allows for better information to be obtained from the vibration signal.

5   Conclusions

In this article, after extracting and analysing the vibration signal in the frequency domain of a spur gearbox with simulated tooth breakage failure, a ranking of the ten CIs in the frequency domain was determined. These CIs allow information from this signal to be extracted and subsequently used in a classification model.

Once the main CIs were selected, the RF classification model was used to determine the accuracy in the severity level of the tooth breakage failure in a spur gearbox. The results were excellent, with the average accuracy level for the four accelerometers installed in the gearbox exceeding 96%. This reassures us about the effectiveness of the selected CIs for analysing this type of failure in the frequency domain.

As a final point, ANOVA and post-hoc tests determined that there are significant differences in the average classification accuracy among the accelerometers, except for A1 and A4, which have the highest accuracy and are installed on the input shaft of the gearbox, where the motor is coupled, and on the output shaft, where the load is connected. Therefore, it is concluded that the accelerometer placement can be done in these positions without altering the classification accuracy results for the severity level of a tooth breakage failure when analysing the vibration signal in the frequency domain.

Future Works

In future work, the proposed methodology will be used to analyse crack, pitting, and scuffing failures in gearboxes and the signals from acoustic emission sensors, electric current, voltage, and noise.

References

1. Breiman, L.: Random forests. Mach. Learn. **45**(1), 5–32 (2001)
2. Cerrada, M., Zurita, G., Cabrera, D., Sánchez, R.V., Artés, M., Li, C.: Fault diagnosis in spur gears based on genetic algorithm and random forest. Mech. Syst. Signal Process. **70**, 87–103 (2016)

3. Goswami, P., Rai, R.N.: A systematic review on failure modes and proposed methodology to artificially seed faults for promoting PHM studies in laboratory environment for an industrial gearbox. Eng. Fail. Anal. **146**, 107076 (2023)
4. Halim, E.B., Choudhury, M.S., Shah, S.L., Zuo, M.J.: Time domain averaging across all scales: a novel method for detection of gearbox faults. Mech. Syst. Signal Process. **22**(2), 261–278 (2008)
5. He, D., Li, R., Bechhoefer, E.: Split torque type gearbox fault detection using acoustic emission and vibration sensors. In: 2010 International Conference on Networking, Sensing and Control (ICNSC), pp. 62–66 (2010). https://doi.org/10.1109/ICNSC.2010.5461545
6. Heyns, T., Godsill, S., De Villiers, J.P., Heyns, P.S.: Statistical gear health analysis which is robust to fluctuating loads and operating speeds. Mech. Syst. Signal Process. **27**, 651–666 (2012)
7. Lei, Y., Zuo, M.J.: Gear crack level identification based on weighted k nearest neighbor classification algorithm. Mech. Syst. Signal Process. **23**(5), 1535–1547 (2009)
8. Lucero, P., et al.: Accelerometer placement comparison for crack detection in railway axles using vibration signals and machine learning. In: 2019 Prognostics and System Health Management Conference (PHM-Paris), pp. 291–296. IEEE (2019)
9. Macancela, J.C., Cabrera, D., Lucero, P., Cerrada, M., Li, C., Villacrés, S., Sánchez, R.V.: Influence of accelerometer position on gearbox fault severity classification through evaluation of deep learning models. In: 2019 Prognostics and System Health Management Conference (PHM-Paris), pp. 303–308. IEEE (2019)
10. Ninoslav, Z.F., Rusmir, B., Cvetkovic, D.: Vibration feature extraction methods for gear faults diagnosis-a review. Facta Universitatis, Ser. Working Living Environ. Prot. **12**(1), 63–72 (2015)
11. Patel, R.K., Giri, V.: Feature selection and classification of mechanical fault of an induction motor using random forest classifier. Perspect. Sci. **8**, 334–337 (2016)
12. R Core Team: R: A Language and Environment for Statistical Computing. R Foundation for Statistical Computing, Vienna, Austria (2023). https://www.R-project.org/
13. Sanchez, R.V., Lucero, P., Vásquez, R.E., Cerrada, M., Macancela, J.C., Cabrera, D.: Feature ranking for multi-fault diagnosis of rotating machinery by using random forest and KNN. J. Intell. Fuzzy Syst. **34**(6), 3463–3473 (2018)
14. Sánchez Loja, R.V.: Diagnóstico de fallos en cajas de engranajes con base en la fusión de datos de señales de vibración, corriente y emisión acústica (2018). http://hdl.handle.net/20.500.11912/4020
15. Wang, D., Tsui, K.L., Miao, Q.: Prognostics and health management: a review of vibration based bearing and gear health indicators. IEEE Access **6**, 665–676 (2017)

Optimizing the Weighing Process in Frozen Shrimp Production for Export

Javier Sánchez Ruíz[1] (ID), Hernán Lara-Padilla[2](✉) (ID), and Tania Rojas[1] (ID)

[1] Universidad Politécnica Salesiana, Guayaquil, Ecuador
jsanchezr4@est.ups.edu.ec, trojas@ups.edu.ec
[2] Universidad de Las Fuerzas Armadas ESPE, Sangolquí, Ecuador
hvlara@espe.edu.ec

Abstract. The frozen shrimp export industry plays a crucial role in Ecuador's economy, necessitating stringent adherence to product standards in order to maintain market trust and regulatory compliance. This study investigates discrepancies in the weighing process of shrimp exports, a critical aspect that directly affects customer confidence and economic outcome. Recent complaints from a key client regarding underweight shrimp shipments underscore the urgent need for process optimization. The implementation of an in-line weight control system using newly installed machinery is proposed to enhance operational oversight and packaging accuracy. Adopting a quasi-experimental research design, this study evaluated the effectiveness of this intervention by comparing pre- and post-implementation data through mixed methods. The integration of Six Sigma and other quality control methodologies aims to address immediate issues and foster continuous improvements in the production processes. This research contributes to the operational excellence of the Ecuadorian shrimp industry and provides a framework for exporters facing similar challenges. The anticipated outcome is a more reliable product delivery system with reduced weight discrepancies and increased customer satisfaction, which bolsters the industry's reputation in global markets.

Keywords: Shrimp Industry · Quality Control · Process Optimization

1 Introduction

The frozen shrimp export industry is the cornerstone of Ecuador's economy, representing a significant portion of its export revenue. In 2023 alone, Ecuador exported 1.324 billion worth of shrimp to the United States, accounting for 20% of its total shrimp exports [1]. This substantial market share underscores the importance of maintaining rigorous standards for shrimp production to meet international expectations and sustain economic growth. Maintaining the integrity of the product weight is vital for preserving market trust and ensuring compliance with international standards.

Six Sigma utilizes data-driven and well-established statistical methods to identify and eliminate defects, errors, or failures in business processes or systems, aiming to minimize the defect rate to as low as 3.4 defects per million opportunities. It focuses

E. M. Inga Ortega et al. (Eds.): CITIS 2024, LNNS 1331, pp. 473–485, 2025.
https://doi.org/10.1007/978-3-031-87065-1_43

on the performance characteristics of processes that are crucial to customers [2, 3]. Six Sigma adheres the Define, Measure, Analyze, Improve and Control (DMAIC) methodology. The literature indicates that organizational change initiatives are frequently driven by crisis situations or market demand [4]. The Six Sigma methodology has been successfully employed in the seafood production industry to analyze and reduce defects, thereby improving the overall process efficiency and product quality. A study by [5], Six Sigma was utilized to address defects in frozen fish steak processing, identifying four main defects: undersized fish, softened meat, off-odor, and green coloration of the meat. The study revealed a sigma level of 3.24 and provided specific recommendations to reduce these defects, demonstrating the effectiveness of the Six Sigma approach in enhancing product quality and operational performance in the seafood industry. Similarly, [6] employed Six Sigma in the quality control of crab meat pasteurization and identified the primary causes of defects, such as dented and scuffed packaging. Through the DMAIC process—Define, Measure, Analyze, Improve, and Control—they achieved a sigma level of 4.7 and suggested improvements to further enhance the production process. These studies underscore the importance of using Six Sigma in seafood processing to maintain high standards, reduce defects, and satisfy stringent international market requirements (Fig. 1).

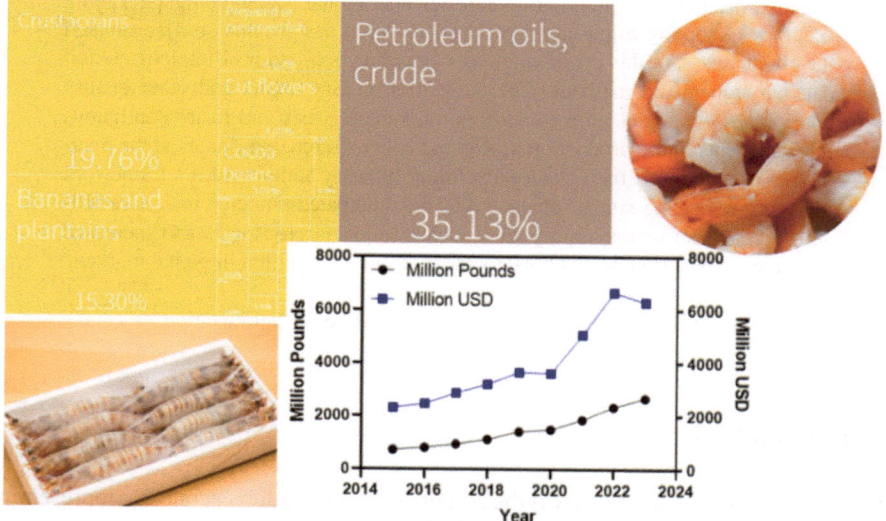

Fig. 1. Ecuador's Shrimp Export Market Analysis: Production Statistics, Export Destinations, and Market Challenges. Adapted from [1, 2].

This study focuses on a medium-size frozen shrimp exporting company that has received external customer complaints regarding discrepancies in the dispatched weight of frozen shrimp. These weighing discrepancies in exports can undermine customer confidence and potentially lead to significant economic losses. Recent customer complaints regarding receiving underweight shrimp highlight the urgency of optimizing the weighing process.

A quasi-experimental research design was adopted to evaluate the effectiveness of this intervention by comparing pre- and post-implementation data, employing mixed methods for a comprehensive analysis. The integration of Six Sigma and other quality control methodologies aims not only to rectify immediate issues but also to foster continuous improvement in production processes [7]. The DMAIC methodology was implemented to bring about changes in the weighing operations and improve efficiency. This involved the following stages: **define,** where the problem was identified and performance characteristics were defined; **measure**, which included measurement system analysis and material balance analysis; **analysis**, involving the use of cause and effect diagrams and identification of variation sources; **improve**, where experiments were designed and variable relationships were discovered; and **control**, which focused on plotting control charts, mistake-proofing exercises, and sharing lessons learned. By following this structured approach, this study aimed to ensure a thorough and systematic improvement in the weighing process [8, 9].

This research achieved a more reliable product delivery system with reduced weight discrepancies and increased customer satisfaction, enhancing the company's operational performance and global reputation. Ensuring that shipments meet the specified weight requirements is crucial for maintaining international customer trust. This study contributes to the operational excellence of the Ecuadorian shrimp industry by providing a practical framework for other exporters and demonstrating how technological solutions and quality control methodologies can improve process reliability. By highlighting the importance of advanced technologies and best practices for quality control, this study offers a roadmap for sustainable growth and success in the global market.

2   Materials and Methods

2.1   Study Design and Setting

This study used a quasi-experimental design to optimize the weighing process in frozen shrimp production for export. Conducted at a leading shrimp processing facility in Guayaquil, Ecuador, this study addresses recent issues with weight discrepancies in its exports. This study aimed to evaluate the effectiveness of an in-line weight control system implemented to improve operational oversight and accuracy in shrimp packaging, identifying the weighing process as a key opportunity for improvement. Six Sigma methodology was employed to systematically analyze and enhance the process, following the DMAIC (Define, Measure, Analyze, Improve, Control) framework. The study population included all the shrimp products processed at the facility [10]. Given the critical nature of this issue, this study covers the entire production line to ensure comprehensive data collection. A single quality team was formed to conduct this study, consisting of an operations manager as the team leader, supported by a quality manager, one IT executive, and two members of the packaging staff. Brainstorming sessions were conducted to identify critical quality characteristics based on customer feedback. It was concluded that analyzing the weight variable was essential, as it was identified as the main factor contributing to the discrepancies reported by buyers.

2.2 Six Sigma Methodology Application

The Six Sigma methodology, specifically the DMAIC (Define, Measure, Analyze, Improve, Control) approach, was applied in this study to systematically improve the weighing process (Fig. 2):

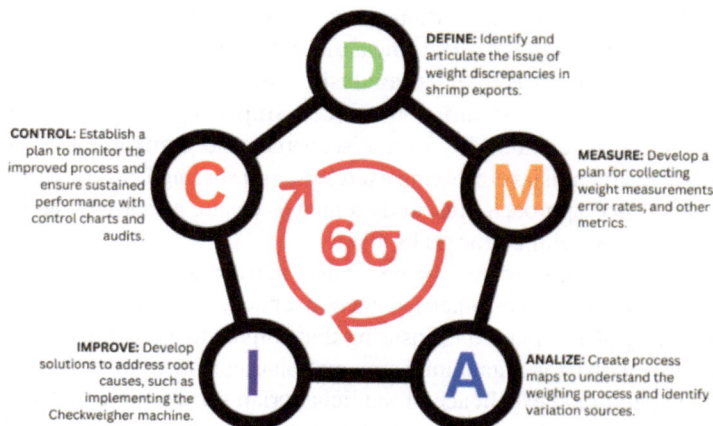

Fig. 2. Six Sigma Methodology for Optimizing Shrimp Weighing Process. This figure illustrates the application of Six Sigma DMAIC methodology to optimize the weighing process in shrimp production for export.

2.3 Define Phase

- *Problem Statement:*
- Identify and articulate the problem of weight discrepancies in shrimp exports.
- *Project Goal:* Reduce weight discrepancies by a specific percentage.
- *Scope:* Define the boundaries of the project, including the specific production lines and processes to be analyzed.

2.4 Measure Phase

- *Data Collection Plan:* Develop a detailed plan for collecting data on weight measurements, error rates, and other relevant metrics.
- *Baseline Data:* Collect and record baseline data on the current state of the weighing process before implementing the new system.
- *Measurement System Analysis:* Conduct a Gage R&R (Repeatability and Reproducibility) study to ensure the accuracy and reliability of the measurement systems. Variation among shrimp of the same size was measured. A sample size of 10 was taken from each of the six shrimp sizes used for export packaging, namely: 10–20, 20–30, 30–40, 40–50, 50–60, and 60–70. Additionally, the length of the shrimp was measured to establish possible size/weight relationships or patterns that could help address the proposed issue.

2.5 Analyze Phase

- *Process Mapping:* Create detailed process maps to understand the flow of the weighing process and identify potential sources of variation.
- *Root Cause Analysis:* Use a Fishbone Diagrams and 5 Whys to identify the root causes of weight discrepancies.
- *Statistical Analysis:* Perform statistical analysis on the collected data to identify patterns, trends, and significant factors contributing to weight discrepancies.

2.6 Improve Phase

- *Solution Development:* Develop potential solutions to address the identified root causes.
- *Pilot Testing:* Conduct pilot tests of the proposed solutions to evaluate their effectiveness in a controlled environment.
- *Implementation Plan:* Develop a detailed implementation plan for rolling out the solutions across the production line, including training and process changes.

2.7 Control

- *Control Plan:* Establish a control plan to monitor the improved process and ensure sustained performance. This includes setting up control charts and regular audits.
- *Standard Operating Procedures (SOPs):* Update SOPs to reflect the new process and ensure all staff are trained on the updated procedures.
- *Continuous Improvement:* Implement a continuous improvement cycle to regularly review and refine the process based on feedback and performance data.

3 Results and Discussion

3.1 Define Phase

In the Define phase, the issues of weight discrepancies in shrimp exports were clearly identified and articulated. Through brainstorming sessions with the quality team, critical to quality (CTQ) characteristics were determined based on customer feedback. It was concluded that analyzing the weight variable was essential, as it was identified as the main factor contributing to the discrepancies reported by buyers.

Precisely identifying the problem and setting specific project goals allowed for a clear understanding of the challenges to be addressed. Defining the project scope, including specific production lines and processes to be analyzed, established a solid foundation for the subsequent phases of the project.

3.2 Measure Phase

A detailed data collection plan was developed to measure weight variations, error rates, and other relevant metrics. Baseline data on the current state of the weighing process was collected and recorded before implementing the new system. A Gage R&R (Repeatability and Reproducibility) study was conducted to ensure the accuracy and reliability of the measurement systems.

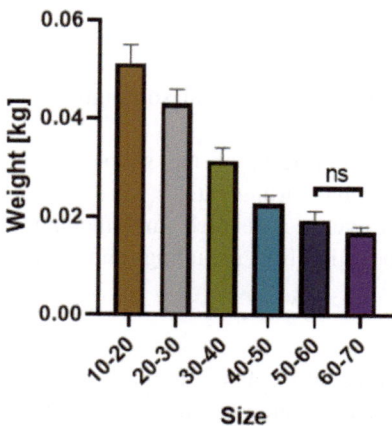

Fig. 3. Mean Weight and Standard Deviation of Shrimp Across Different Size Categories. An ANOVA test was conducted to compare the weight differences among the size categories. The results indicated no significant difference at the 0.05 significance level between the 50–60 and 60–70 size groups.

The Gage R&R analysis revealed an R&R variation of 42% and a part-to-part variation of 91%. Furthermore, the reproducibility contribution was 40% and repeatability was 14%. These results indicate significant issues in the weighing measurement system, primarily due to incorrect usage by personnel. To address these issues, it is essential to standardize the weighing procedures, as these are often performed incorrectly due to the speed of the process.

Additionally, variation among shrimp of the same size was measured. A sample size of 10 was taken from each of the six shrimp sizes used for export packaging, revealing significant variations in mean weights and standard deviations, as shown in Fig. 3. The results indicated that weight consistency varied across different size categories, underscoring the need for more rigorous weight control measures.

The measurement phase revealed significant variations in shrimp weights within the same size categories, highlighting the need for more rigorous weight control. The measurement system analysis provided confidence in the accuracy of the collected data, which is crucial for the analysis and improvement phases. Standardizing the weighing procedures and providing thorough training to the staff will be critical steps to minimize measurement errors and improve overall accuracy.

3.3 Analyze Phase

Detailed process maps were created to understand the flow of the weighing process and identify potential sources of variation (Fig. 4). Root cause analysis using Fishbone Diagrams and the 5 Whys technique identified the main causes of weight discrepancies. Statistical analysis of the collected data revealed patterns, trends, and significant factors contributing to weight discrepancies.

The thorough analysis identified several root causes of weight discrepancies, including issues in operational procedures, variability in shrimp sizes, and errors in weighing systems. According to Fig. 4, Process Map of Shrimp Production for Export, and applying the Fishbone Diagram using the 5 Ms + E methodology (Measurement, Materials, Method, Manpower, Machine, and Environment) as per ISO 13053–2, it was determined that the critical stages comprise from phase 9 (Packaging) to phase 13 (Final Packaging).

Fig. 4. Process Map of Shrimp Production for Export.

However, experts concluded that during phase 9, the shrimp is weighed in trays before undergoing the freezing process and subsequent complementary stages, finally being packed for storage in phase 13 for later exportation. This point in the process was established as crucial for analysis. It was determined that focusing on this stage is essential to address the weight discrepancies effectively.

The insights gained from this detailed analysis were fundamental for developing effective solutions in the improvement phase. By pinpointing the critical stages and root causes, the study provided a clear direction for targeted interventions to enhance the accuracy and reliability of the weighing process.

3.4   Improve Phase

Potential solutions were developed to address the identified root causes. Pilot tests of the proposed solutions were conducted to evaluate their effectiveness in a controlled environment. A detailed implementation plan was developed to roll out the solutions across the production line, including training and process changes. To further refine the weighing process, ten specimens of each shrimp size category were measured for both weight and length. The weights were then normalized by dividing each shrimp's weight by its respective length. Figure 5 illustrates the results of these measurements.

Fig. 5. Average Lengths and Weight-to-Length Ratios of Shrimp Across Different Size Categories.

The average lengths (in cm) for the sizes were as follows: 10–20: 19.1, 20–30: 17.6, 30–40: 16.5, 40–50: 14.8, 50–60: 14.4, 60–70: 13.1. The corresponding weight-to-length ratios were: 10–20: 0.0026178, 20–30: 0.0021591, 30–40: 0.0020606, 40–50: 0.0016216, 50–60: 0.0013889, 60–70: 0.0012214. After conducting an ANOVA test, no significant differences were found at the 0.05 significance level for the length graph between the 40–50 and 50–60 size groups, and for the weight-to-length ratios between the 50–60 and 60–70 size groups. Therefore, using the weight-to-length relationship can be useful, provided that a better size classification system is implemented.

The pilot tests showed a significant reduction in weight discrepancies, validating the effectiveness of the proposed solutions. Implementing these solutions across the full production line improved the accuracy and consistency of the weighing process, increasing customer satisfaction. However, the results highlighted the necessity of managing a better classification system. The variation in shrimp sizes significantly impacts the consistency of the packaged unit weights.

To address this, it is recommended to implement a more rigorous sampling process and develop a robust database to identify patterns that could further improve the weighing process. Regular, detailed measurements and analyses will help in creating more accurate size categories and enhance the overall process efficiency. Continuous training for staff and standardizing weighing procedures are essential to ensure the new system's effectiveness and sustainability.

After collecting length data for various size classes of whole shrimp processed in the plant, a histogram was drawn as shown in Fig. 6. It is recommended to implement 8 categories (2 more than currently used) ranging from 12 to 20 cm. This will provide a more standardized classification system, allowing for more precise and consistent sorting, which will improve the overall accuracy of the weighing and packaging process. This improvement in the classification system will lead to better weight consistency within each size category, reducing discrepancies and enhancing customer satisfaction.

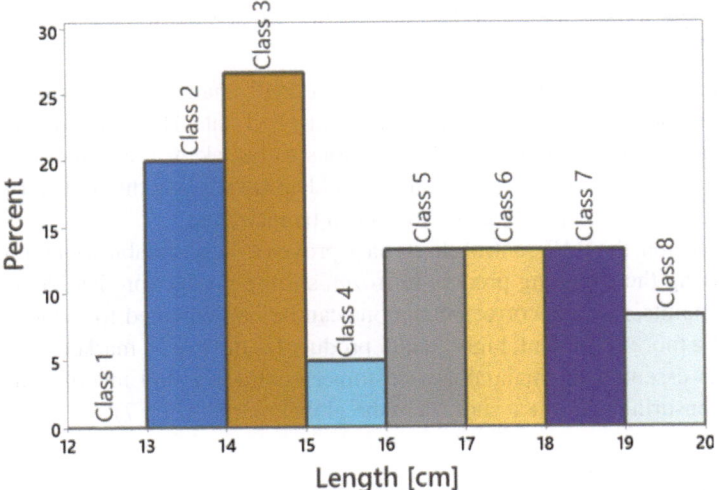

Fig. 6. Proposed Improvement: Histogram of Shrimp Lengths with Recommended New Size Categories.

3.5 Control Phase

In the Control phase of this study, IMR (Individual Moving Range) control charts were implemented to monitor the weight of master boxes, which contain several packs of shrimp of uniform size, at the end of the final packaging stage. This control stage allows verification of the product before storage and export, serving as an indicator of the effectiveness of actions taken in previous stages to reduce final weight variation.

For each of the six shrimp sizes (10–20, 20–30, 30–40, 40–50, 50–60, and 60–70), 100 data points were analyzed using IMR charts, revealing certain points outside the established control limits. These out-of-control points indicate significant variations that do not align with the expected process behavior.

The implementation of the IMR charts revealed several out-of-control points in the data for different shrimp sizes. These points can be attributed to various reasons:

- *Variability in the Weighing Process:* Despite efforts to standardize the weighing procedure, inconsistencies may still exist in how personnel perform these tasks, especially during peak production times.
- *Product Size Variability:* Natural variability in the size and weight of shrimp within the same category can contribute to the discrepancies observed in the data. This variability can be influenced by factors such as harvesting methods and storage conditions prior to packaging.
- *Equipment Errors:* Weighing equipment may experience temporary miscalibrations or mechanical failures that impact measurement accuracy. It is crucial to maintain a regular maintenance and calibration program to minimize these errors.
- *Effectiveness of Implemented Improvements:* Although various improvements were implemented in the earlier phases of the process, such as standardizing procedures and training staff, some of these improvements may not yet be fully consolidated in daily practice.

The presence of out-of-control points on the IMR charts indicates that the process has not yet achieved a completely stable and controlled state. However, identifying these variations allows for specific corrective actions to be taken. For example, additional controls can be implemented at critical stages identified during the root cause analysis, or the frequency of equipment calibrations can be increased.

In conclusion, the IMR control charts have proven to be a valuable tool for monitoring and improving the weighing process in frozen shrimp production. By identifying out-of-control points, specific corrective actions can be implemented to reduce variability and ensure a more consistent, high-quality product for the export market. This proactive approach is essential for maintaining customer confidence and meeting international standards, ensuring continued success in the global market (Fig. 7).

3.6 Ethical Considerations

Ethical considerations were meticulously addressed to ensure the highest standards of transparency and research integrity throughout this study. All procedures adhered strictly to established ethical guidelines for industrial and aquaculture research. The data collection and analysis were conducted with a comprehensive commitment to ethical standards, ensuring both the welfare of the shrimp and the environmental sustainability of the farming practices. Additionally, all interventions were designed to minimize any potential harm and to promote the long-term sustainability of the shrimp farming industry. This adherence to ethical principles not only reinforces the credibility of the study but also supports the responsible advancement of aquaculture methodologies and the integrity of the research outcomes.

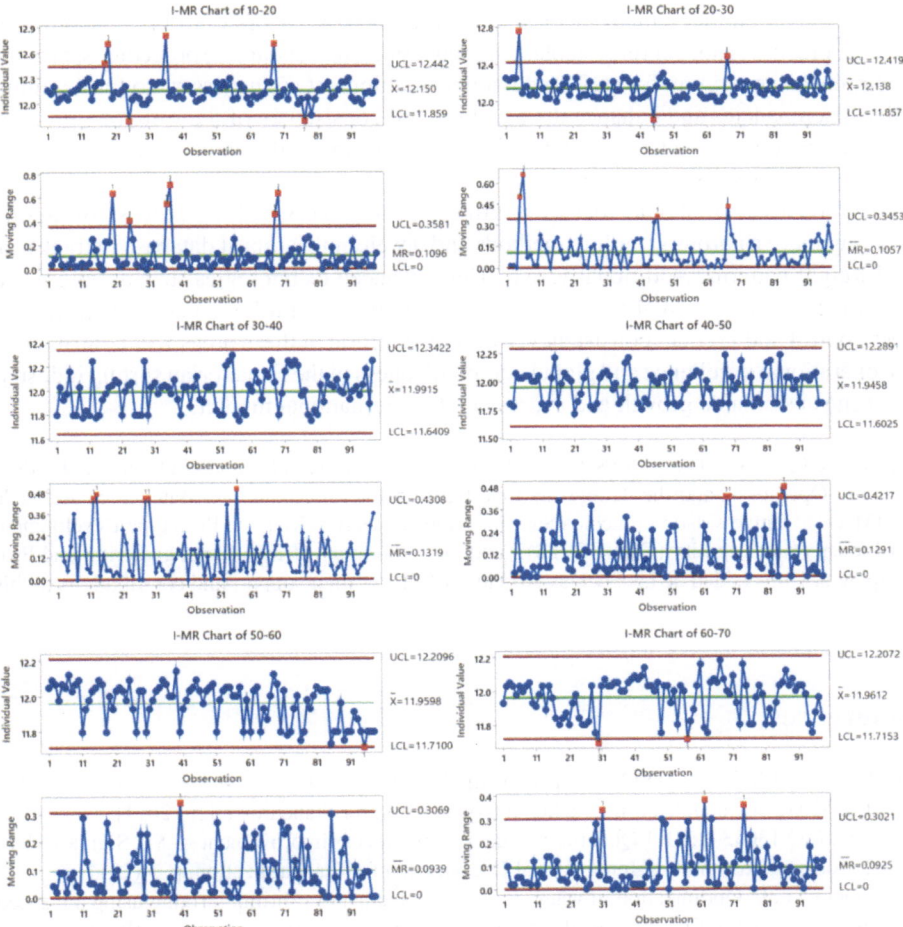

Fig. 7. Proposed Improvement: Histogram of Shrimp Lengths with Recommended New Size Categories

4 Conclusions

The application of IMR control charts within the framework of Six Sigma at the final packaging stage of shrimp production has effectively identified significant weight discrepancies. This approach has provided valuable insights into variations that require corrective actions, thereby enhancing the overall control and reliability of the weighing process.

Through the DMAIC methodology, the study identified multiple sources of variability, including natural product size differences, inconsistencies in weighing procedures, and equipment errors. Addressing these root causes is essential for achieving a stable, controlled, and compliant weighing process, which is critical for maintaining customer trust and meeting international standards.

Despite the improvements achieved, the presence of out-of-control points indicates that the process is not yet fully stable. Ongoing efforts to standardize procedures, provide comprehensive training for staff, and maintain equipment are necessary to further minimize discrepancies and enhance the accuracy and consistency of the weighing process. This continuous improvement aligns with the Six Sigma philosophy of striving for near perfection in process performance.

Future research should focus on further refining the weighing process by implementing more rigorous sampling methods and developing a robust database to track size and weight patterns. Additionally, exploring advanced technologies such as automated weighing systems and real-time monitoring could provide further enhancements in process control and product consistency. Continued efforts in these areas will contribute to the operational excellence and global competitiveness of the Ecuadorian shrimp industry, ensuring sustainable growth and success in the international market.

Acknowledgments. The authors would like to thank the Master's Program in Industrial Production and Operations and the "Grupo de Investigación Interdisciplinar en Matemática Aplicada (GIIMA)" at Universidad Politécnica Salesiana, the "Research Group on Prototype Development for Industry and Security (CONCEPTS)," and the Laboratory of Reverse Engineering and Prototyping at Universidad de las Fuerzas Armadas ESPE for their support and contributions to this research.

References

1. ISO. ISO 13053–1:2011 Quantitative methods in process improvement — Six Sigma — Part 1: DMAIC methodology. International Organization for Standardization (2011)
2. ISO. ISO 13053–2:2011 Quantitative methods in process improvement — Six Sigma — Part 2: Tools and techniques. International Organization for Standardization, p. 5 (2011)
3. Pinem, et al.: Optimization of water quality management in shrimp farming using advanced filtration systems. Environmental Earth Sciences **404**(1) (2022). https://doi.org/10.1088/1755-1315/404/1/012017
4. Rehulina, et al.: Application of machine learning for predictive analytics in aquaculture. Aquaculture **2022** (2022). https://doi.org/10.1016/j.aquaculture.2016.10.019
5. Poernomo, A., Ferawati, N., Salampessy, R.B.S.: Defects analysis of frozen fish steak processing using six sigma: a case study. IOP Conf. Ser.: Earth Environ. Sci. **919**(1), 012038 (2021). https://doi.org/10.1088/1755-1315/919/1/012038
6. Hidayat, K., Tsana, N.U.B., Maflahah, I.: Quality control of crab meat pasteurization using six sigma. IOP Conf. Ser.: Earth Environ. Sci. **1059**(1), 012071 (2022). https://doi.org/10.1088/1755-1315/1059/1/012071
7. Joglekar, A.M.: Statistical Methods for Six Sigma: in R & D and Manufacturing. Wiley, New York (2003)
8. Taylor, P.: Make Your Business Agile: A Roadmap for Transforming your Management and Adapting to the "New Normal." Routledge, Abingdon, Oxon; New York, NY (2021)
9. Johnson, R.A.: Applied Multivariate Statistical Analysis: Pearson New International Edition PDF eBook. Pearson
10. Rojas, T., Mula, J., Sanchis, R.: Quantitative modelling approaches for lean manufacturing under uncertainty. Int. J. Prod. Res. **62**(16), 5989–6015 (2024). https://doi.org/10.1080/00207543.2023.2293138

A Six Sigma Approach to Enhancing the Production Performance of Pacific White Shrimp (Litopenaeus Vannamei): a Proposal for Optimizing Stocking Density and Acclimation Methods

Génesis Dayana Guevara Huacón[1] (ID), Melissa Arlette Zambrano Peñafiel[1] (ID), Hernán Lara-Padilla[2](✉) (ID), and Tania Rojas[1] (ID)

[1] Universidad Politécnica Salesiana, Guayaquil, Ecuador
{gguevarah,mzambranop2}@est.ups.edu.ec, trojas@ups.edu.ec
[2] Universidad de Las Fuerzas Armadas ESPE, Sangolquí, Ecuador
hvlara@espe.edu.ec

Abstract. Ecuador is a leading producer of white shrimp, *Litopenaeus vannamei*, exporting 1.069 million tons and generating $7 billion by 2022. However, challenges in the post-larval phase affect the sustainability and profitability of shrimp farming. In this study, we applied the Six Sigma DMAIC methodology to optimize the growth and survival rates of shrimp larvae. In the Define phase, biotic and abiotic variables influencing larval growth were identified. The Measure phase involved collecting data on shrimp weight and key environmental conditions, including pH, temperature, ammonia, sulfur percentage, and algal presence. Statistical analysis during the analysis phase revealed critical insights into the impact of these variables on growth rates. The proposed improvements in the improve phase included enhanced water quality management, optimized feeding strategies, and stress reduction measures. Control charts and multivariate analysis during the control phase ensured the sustainability of these improvements. This study demonstrates that applying Six Sigma methodologies can significantly enhance the productivity and profitability of shrimp farming by addressing critical growth and survival factors.

Keywords: Shrimp Industry · Quality Control · Aquaculture · Larval Rearing

1 Introduction

Ecuador has emerged as a leading player in the global shrimp aquaculture industry, particularly with the production and exportation of the white shrimp Litopenaeus vannamei. By the end of 2022, Ecuador exported 1.069 million tons of shrimp, generating approximately $7 billion in revenue, with China, Europe, and the United States as the primary markets [1]. This substantial growth in exports has been driven by a significant expansion of both open and closed farming systems along the Ecuadorian coast [2]. Despite this remarkable growth, the shrimp farming sector is facing numerous complex challenges that threaten its sustainability and profitability (Fig. 1).

© The Author(s) 2025
E. M. Inga Ortega et al. (Eds.): CITIS 2024, LNNS 1331, pp. 486–499, 2025.
https://doi.org/10.1007/978-3-031-87065-1_44

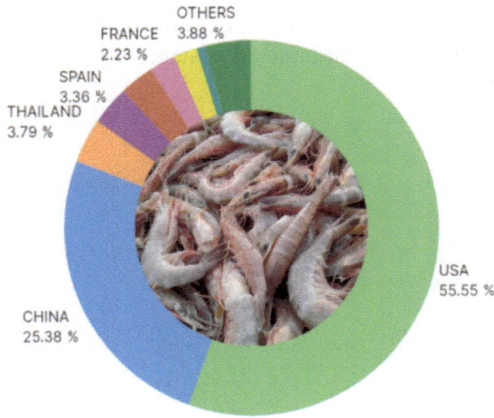

Fig. 1. Shrimp Exports from Ecuador (2018–2022). Adapted from [2].

One critical stage in shrimp farming is the post-larval phase, which is highly susceptible to various biotic and abiotic factors that negatively affect shrimp growth and survival, thereby affecting overall production yields and operational efficiency. The company under study, located in Durán, Ecuador, encountered significant operational challenges in farming *Litopenaeus vannamei* during the post-larval stage. The company employs two distinct farming systems: closed systems using greenhouses, and open systems using ponds. Despite following identical protocols for microbiological analysis, feeding, and water recirculation, there have been notable discrepancies in performance, and in some instances, alarmingly low yields.

To optimize production, the company has been operating 20 ponds, each measuring 1.2 hectares, over three-month cycles, resulting in six to seven production cycles per year. However, mortality rates have been higher than acceptable standards, prompting management to consider shifting from open to closed farming systems. To meet the constant demand for post-larvae grow-out phases, the company has implemented a hybrid approach: 30 greenhouses, each approximately 1.2 hectares (totaling 40 hectares), five open ponds, each three hectares (totaling 15 hectares), and 10 ponds, each 1.05 hectares, currently being converted to greenhouses (totaling 10.5 hectares). Despite these corrective measures, the company has struggled to develop effective preventive strategies because of the lack of a comprehensive and detailed analysis of current farming conditions. Such an analysis is crucial for identifying and understanding the factors that limit shrimp growth and increase mortality rates. By doing so, companies can develop and implement strategies that address the root causes of these issues, ensuring a long-term, sustainable solution that enhances the efficiency and productivity of shrimp farming operations.

To address these challenges, Six Sigma methodology was used in this study. Six Sigma is a data-driven approach and methodology for eliminating defects and improving quality in any process. By utilizing its DMAIC (Define, Measure, Analyze, Improve, Control) framework, this study aimed to identify and mitigate the factors that adversely affect post-larval growth and survival rates in shrimp farming. This approach allows for systematic analysis of operational data, identification of critical factors impacting performance, and implementation of targeted strategies to enhance overall productivity and efficiency (Fig. 2).

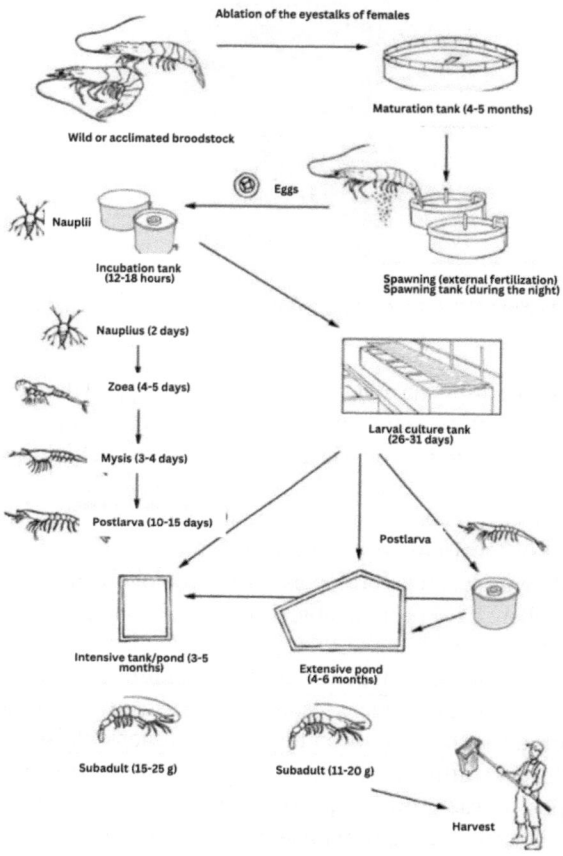

Fig. 2. Production cycle of *Penaeus vannamei.* Adapted from [2].

Larval rearing in shrimp aquaculture is a critical phase that significantly impacts the overall productivity and sustainability of shrimp farming operations. The quality of larval rearing processes directly influences the survival rate, growth, and health of shrimp as they progress through subsequent developmental stages. In Ecuador, advancements in larval-rearing techniques have played a pivotal role in bolstering the shrimp industry's export capabilities, contributing to its position as a global leader in shrimp production. Effective larval rearing involves meticulous management of environmental conditions, including temperature, salinity, and water quality, as well as optimal feeding practices to ensure healthy development of shrimp larvae. These practices not only enhance survival rates but also improve the overall yield and quality of the final product [2].

Previous research has demonstrated the importance of specific salinity conditions and stocking densities for the growth and survival of Litopenaeus vannamei [3, 4]. These studies highlight how optimal salinity and population density can significantly influence production outcomes, and suggest approaches that could be adapted to improve farming efficiency in Ecuador. Additional references from recent studies provide a broad view of the management and optimization of farming systems, underscoring the need for a data-driven approach and methodologies such as Six Sigma to continuously improve production processes and ensure high-quality output [5, 6].

By focusing on the production phase of white shrimp, this study aimed to identify strategies to maximize the quality and survival of juveniles, ensuring that they are well equipped to thrive in subsequent phases. This approach is critical for improving survival and growth rates, which directly affect the productivity and profitability of shrimp farming.

2 Materials and Methods

In this study, the Six Sigma DMAIC (Define, Measure, Analyze, Improve, Control) methodology was applied to optimize the growth of whiteleg shrimp (Litopenaeus vannamei) larvae. The DMAIC method provides a structured approach for identifying, analyzing, and improving processes to achieve higher efficiency and effectiveness. The methodology was tailored to address specific challenges in shrimp farming, with a particular focus on the larval stage [7].

2.1 Define

The first phase involves defining the scope and objectives of the study. The primary goal was to identify and mitigate the factors affecting the growth of shrimp larvae. Experts in aquaculture and shrimp farming were consulted to identify biotic and abiotic variables that could influence larval growth. The key variables identified included water quality parameters (pH, temperature, NH3, NH4, sulfur percentage, green algae, dinoflagellates, and diatoms) and biotic factors (feed type, feeding frequency, and larval density).

2.2 Measure

During the Measure phase, data collection focused on key performance indicators, such as the weight of the shrimp at the start and end of the larval stage. This helped to establish growth patterns and identify potential issues. The following parameters were carefully recorded.

- pH levels: Monitored regularly to ensure that they remained within the optimal range for shrimp growth.
- Temperature [8]: Maintained and recorded using an SCADA system with alerts set for deviations.
- Ammonia (NH3) and ammonium (NH4) levels were monitored to prevent toxic build-up.
- The sulfur percentage, Green Algae, Dinoflagellates, Diatoms: These tracked to understand their influence on water quality and shrimp health.

2.3 Analyze

Descriptive statistics provided initial estimates and established baseline metrics for key variables. Time series analysis was employed to examine growth patterns and behavioral trends over the development period of the larvae. This analysis helped understand the impact of the monitored variables on larval growth and survival rates. The analysis phase involved a statistical analysis to identify patterns and correlations in the collected data. Descriptive statistics provide initial insights into the distribution and central tendencies of the variables. Time series analysis was used to examine trends and fluctuations over the growth period of the larvae. This analysis helped understand the impact of the monitored variables on larval growth and survival rates. This analysis revealed critical insights into how certain water quality parameters and feeding strategies affect the growth rates of shrimp larvae.

2.4 Improve

Based on this analysis, specific improvements have been proposed to enhance the growth and survival rates of shrimp larvae. These included:

- Optimizing water quality management: Implementing more rigorous controls and adjustments for pH, temperature, and ammonia level [9].
- Enhancing feeding strategies: adjusting the type and frequency of feed to match the nutritional needs of the larvae at different stages.
- Improving environmental conditions: Ensuring stable conditions within closed systems to reduce stress and promote healthy growth.

2.5 Control

The final phase focuses on ensuring that the improvements are sustainable. Control charts and multivariate control methods are recommended for continuously monitoring critical variables. These tools would help in maintaining the optimal conditions identified during the improvement phase and quickly address any deviations. By implementing these control mechanisms, shrimp farms could achieve consistent growth rates and improved survival rates, contributing to higher productivity and profitability.

2.6 Six Sigma Methodology Application

The Six Sigma methodology, specifically the DMAIC (Define, Measure, Analyze, Improve, Control) approach, was applied in this study to systematically improve the weighing process:

2.7 Define Phase

- *Problem Statement:* Identify and articulate the problem of weight discrepancies in shrimp exports.
- *Project goal:* Reduces weight discrepancies by a specific percentage.
- *Scope:* Define the boundaries of the project, including the specific production lines and processes to be analyzed.

2.8 Measure Phase

- *Data Collection Plan:* Develop a detailed plan for collecting data on weight measurements, error rates, and other relevant metrics.
- *Baseline Data:* Collect and record baseline data on the current state of the weighing process before implementing the new system.
- *Measurement System Analysis:* A Gage Repeatability and Reproducibility (R&R) study was conducted to ensure the accuracy and reliability of the measurement systems. Variations among shrimp of the same size were measured. A sample size of 10 was taken from each of the six shrimp sizes used for export packaging, namely: 10–20, 20–30, 30–40, 40–50, 50–60, and 60–70. Additionally, the length of the shrimp was measured to establish possible size/weight relationships or patterns that could help address the proposed issue (Fig. 3).

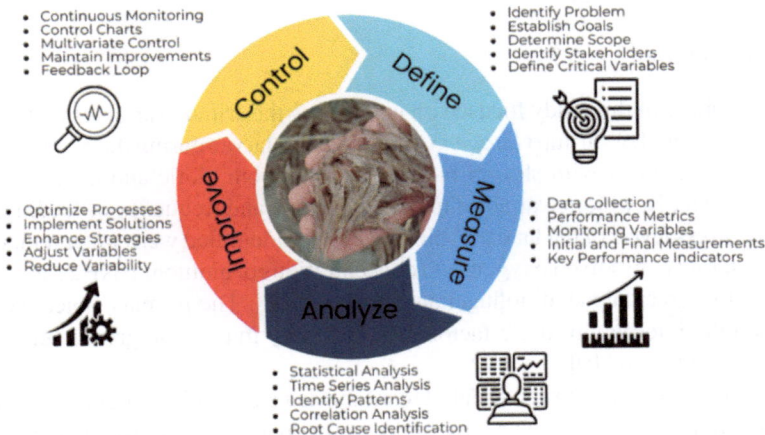

Fig. 3. Six-Sigma Methodology Proposals for Optimizing Shrimp Larval Growth.

2.9 Analyze Phase

- *Process Mapping:* Create detailed process maps to understand the flow of the weighing process and identify potential sources of variation.
- *Root Cause Analysis:* Use a Fishbone Diagrams and 5 Whys to identify the root causes of weight discrepancies.
- *Statistical Analysis:* Perform statistical analysis on the collected data to identify patterns, trends, and significant factors contributing to weight discrepancies.

2.10 Improve Phase

- *Solution Development:* Develop potential solutions to address the identified root causes.

- *Pilot Testing:* Conduct pilot tests of the proposed solutions to evaluate their effectiveness in a controlled environment.
- *Implementation Plan:* Develop a detailed implementation plan for rolling out the solutions across the production line, including training and process changes.

2.11 Control

- *Control Plan:* Establish a control plan to monitor the improved process and ensure sustained performance. This includes setting up control charts and regular audits.
- *Standard Operating Procedures (SOPs):* Update SOPs to reflect the new process and ensure all staff are trained on the updated procedures.
- *Continuous Improvement:* Implement a continuous improvement cycle to regularly review and refine the process based on feedback and performance data.

3 Results and Discussion

3.1 Define Phase

The Define phase of this study focused on identifying the critical variables that influence the growth of whiteleg shrimp (*Litopenaeus vannamei*) larvae during the postlarval stage. Through consultations with shrimp farming experts, both biotic and abiotic variables were pinpointed. Key biotic variables included the type and frequency of feeding, health status, and stress levels of the larvae. Abiotic variables identified were water temperature, pH levels, salinity, dissolved oxygen, and concentrations of ammonia (NH_3), ammonium (NH_4), sulfur, green algae, dinoflagellates, and diatoms. The primary objective was to understand the interplay of these factors and how they impact the growth and survival rates of shrimp larvae [10].

In defining the scope, it was essential to establish a clear understanding of the current practices and their outcomes. The farm under study operates using two distinct systems: closed systems using greenhouses and open systems using ponds. Despite following similar protocols for microbiological analysis, feeding, and water recirculation, the yield results varied significantly. This discrepancy highlighted the need for a systematic approach to identify and mitigate factors adversely affecting larvae growth and survival, thereby optimizing production efficiency.

3.2 Measure Phase

The Measure phase involved comprehensive data collection on the identified variables. The primary metric was the weight of the shrimp larvae at both the beginning and end of the growth period, which provided a basis for establishing growth patterns. Specifically, the initial and final weights of the larvae were measured to track growth progress. This measurement allowed for the calculation of growth rates and provided insights into the efficiency of the cultivation practices.

During the cultivation period, which lasted between 8 and 15 days, various biotic and abiotic variables were monitored:

- Temperature: Daily measurements ensured the temperature remained within the optimal range of 20 °C to 35 °C, with the hypothesis that consistent temperature control would correlate with improved growth rates.
- pH Levels: Recorded every 2–3 days to maintain a range of 7.0 to 8.5, critical for maintaining the metabolic processes of the shrimp.
- Dissolved Oxygen: Continuously monitored, with alerts set for deviations from the range of 4mg/L to 10mg/L, as oxygen levels directly impact shrimp health and growth.
- Ammonia and Ammonium: Measured weekly to prevent toxic buildup, recognizing that elevated levels could lead to increased mortality.
- Salinity: Kept within 25ppt to 35ppt, adjusted based on bi-weekly measurements, ensuring osmotic balance critical for larval development.
- Algal Presence: Regular sampling and analysis for green algae, dinoflagellates, and diatoms provided data on the quality and suitability of the water environment.

Figure 4 illustrates the initial and final weights of shrimp larvae (*Litopenaeus vannamei*) over a growth period of 8 to 15 days. The red line represents the initial weights, which remain relatively stable, indicating a consistent starting point. The blue line shows the final weights, which display significant variability with periods of rapid growth followed by fluctuations. The dotted boxes highlight regions of high and low growth performance. The observed growth patterns suggest that various biotic and abiotic factors influence the growth rates, with optimal conditions leading to peaks and suboptimal conditions causing troughs. This analysis helps identify the critical factors affecting larvae growth and informs the improvement strategies to enhance overall shrimp production efficiency.

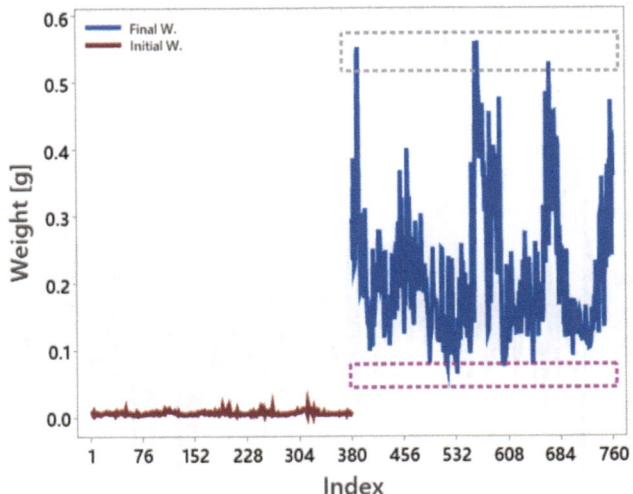

Fig. 4. Initial and Final Weights of Shrimp Larvae Over the Growth Period.

3.3 Analyze Phase

In the Analyze phase, statistical analysis of the collected data was performed to understand the relationships between variables and shrimp growth outcomes. Descriptive statistics provided initial estimates and established baseline metrics for key variables. Time series analysis was employed to examine growth patterns and behavioral trends over the development period of the larvae.

The time series graph (Fig. 4) illustrates the initial and final weights of the shrimp larvae over the cultivation period. The initial weights remained relatively stable, indicating a consistent starting point for the larvae. However, the final weights showed significant variability, with some larvae exhibiting substantial growth while others had minimal increases.

Several key observations were made from the analysis:

- Growth Patterns: The graph shows fluctuations in the final weights, with periods of rapid growth followed by stagnation or decline. This pattern suggests that certain environmental or management factors may be causing stress or suboptimal conditions intermittently.
- High Points: The peaks in the graph represent periods where the larvae achieved maximum growth. These peaks could be correlated with optimal environmental conditions or effective feeding schedules.
- Low Points: Conversely, the troughs indicate periods of poor growth performance. Possible causes for these low points could include variations in water quality, temperature fluctuations, or inadequate feeding.

Understanding these fluctuations is critical for optimizing the cultivation process. Identifying the causes of the low growth periods and replicating the conditions present during the high growth periods can lead to more consistent and improved growth outcomes.

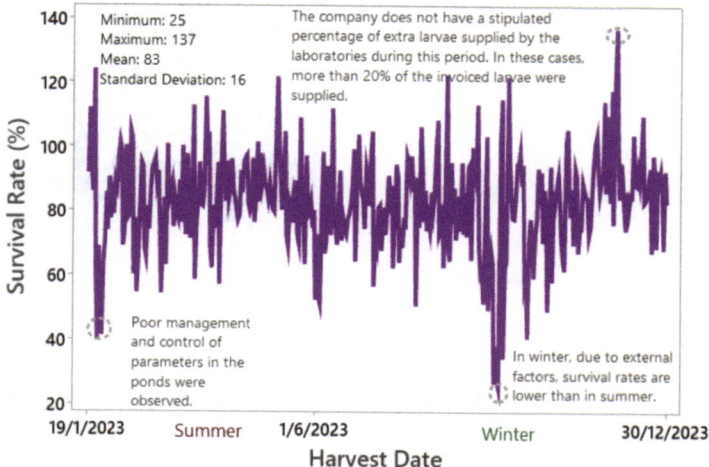

Fig. 5. Shrimp Larvae Survival Rates Over Time with Seasonal Influences.

Figure 5 illustrates the survival rates of shrimp larvae (*Litopenaeus vannamei*) over the course of 2023, highlighting the variability influenced by seasonal changes and management practices. The survival rates show significant fluctuations, with a minimum of 25%, a maximum of 137%, a mean of 83%, and a standard deviation of 16%. The annotations indicate key observations: the company received more than 20% extra larvae compared to what was invoiced, impacting the survival rate measurements. Poor management and control of parameters in the ponds were noted, contributing to inconsistent survival rates. The data shows lower survival rates in the winter months due to external environmental factors compared to higher survival rates in the summer. These findings suggest that improving management practices and adapting strategies for different seasons could enhance the overall survival rates of shrimp larvae.

3.4 Improve Phase

Based on the findings from the Analyze phase, several improvement strategies were proposed to enhance the growth and survival rates of shrimp larvae:

- Enhanced Monitoring: Implementation of more frequent and precise monitoring systems for critical variables like pH, dissolved oxygen, and temperature was recommended. Real-time monitoring systems could provide immediate feedback and allow for prompt corrective actions.
- Optimized Feeding Regimen: Adjustments to feeding schedules and quantities to ensure consistent nutrition and minimize waste were suggested. The timing and type of feed were optimized based on the growth stages of the larvae, with an emphasis on high-protein diets during critical growth periods.
- Water Quality Management: Introduction of advanced filtration and aeration systems to maintain optimal water conditions and reduce toxic substance buildup was proposed. These systems were designed to handle peak biomass loads and ensure consistent water quality.
- Stress Reduction Measures: Implementation of stress mitigation techniques, such as controlled handling and minimized disturbances during the growth period, were recommended. Reducing handling frequency and optimizing stocking densities were identified as key strategies to minimize stress-induced mortality (Fig. 6).

Figure 6 depicts the pH levels in shrimp culture ponds throughout 2023, showing variations influenced by seasonal changes. The pH levels ranged from a minimum of 7.0 to a maximum of 9.1, with an average of 8.1 and a standard deviation of 0.2. The data indicate a tendency towards more alkaline conditions, particularly during certain periods. It is crucial to maintain a minimum pH of 7.0 in shrimp cultures, as it is neutral and essential for optimal shrimp health. Seasonal fluctuations, particularly in the summer and winter months, suggest that external environmental factors and management practices significantly impact pH levels. Continuous monitoring and adjustment of water quality parameters are necessary to ensure the sustainability and productivity of shrimp culture operations.

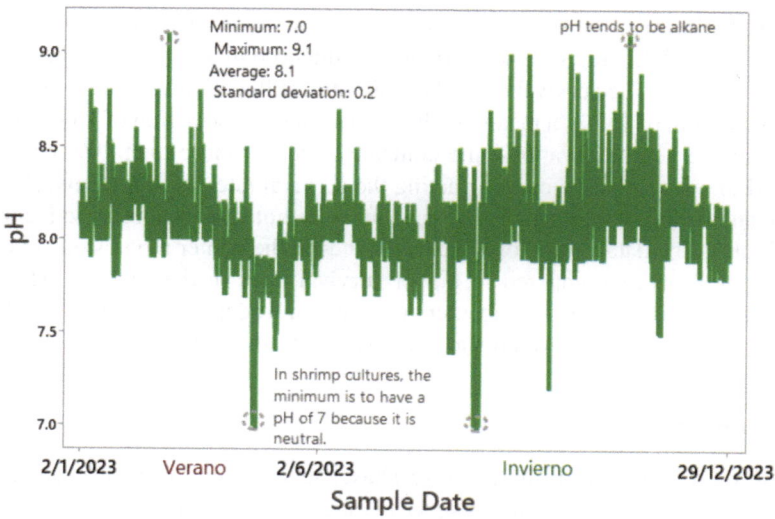

Fig. 6. pH Levels in Shrimp Culture Ponds Throughout 2023.

3.5 Control Phase

To ensure the sustainability of the improvements, the following control methods were recommended:

- Control Charts: Utilization of control charts to continuously monitor critical variables and promptly identify deviations. Statistical process control (SPC) tools were implemented to track variations and trigger alarms when conditions moved outside acceptable ranges.
- Multivariate Analysis: Application of multivariate control methods to assess the combined effect of multiple variables on larvae growth and health. This approach helped in identifying the most influential factors and their interactions.
- Regular Audits: Conducting periodic audits of the cultivation processes to ensure adherence to the improved protocols. These audits involved comprehensive reviews of operational practices and environmental conditions.
- Training Programs: Development of training programs for staff to enhance their understanding and management of critical variables. Training focused on best practices in shrimp farming, use of monitoring equipment, and response strategies for deviations.

Figure 7 shows the concentration of diatoms in shrimp culture ponds over the course of 2023, highlighting variations due to seasonal changes and the presence of dinoflagellates. The diatom concentration ranged from 0 to 259.9 cells/mL, with an average of 35.9 cells/mL and a standard deviation of 38.3. There is a greater presence of diatoms during periods with a lower concentration of dinoflagellates. The data indicates that diatom concentrations fluctuate significantly, particularly during the winter months, which could be attributed to changes in water conditions and nutrient availability. Maintaining a balanced microalgae community is essential for the health and growth of shrimp larvae,

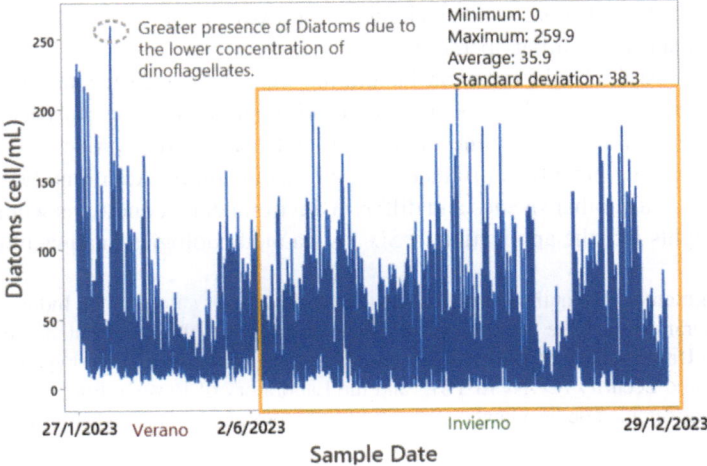

Fig. 7. Diatom Concentration in Shrimp Culture Ponds Throughout 2023.

as both diatoms and dinoflagellates play crucial roles in the pond ecosystem. Continuous monitoring and management of these parameters are necessary to optimize shrimp culture productivity.

3.6 Ethical Considerations

Ethical considerations were meticulously addressed to ensure the highest standards of transparency and research integrity throughout this study. All procedures adhered strictly to established ethical guidelines for aquaculture research. The data collection and analysis were conducted with a comprehensive commitment to ethical standards, ensuring the welfare of the shrimp and the environmental sustainability of the farming practices. This adherence to ethical principles not only reinforces the credibility of the study but also supports the responsible development of aquaculture methodologies.

4 Conclusions

The application of the Six Sigma DMAIC methodology has proven effective in identifying and mitigating factors that negatively impact the growth and survival rates of white shrimp larvae. By optimizing key variables such as pH, temperature, and feeding strategies, the study has demonstrated significant improvements in both productivity and profitability of shrimp farming operations.

Rigorous control and continuous monitoring of water quality parameters, including ammonia and sulfur levels, have been instrumental in reducing mortality rates. The implementation of advanced filtration and aeration systems has ensured a stable and conducive environment for shrimp larval development.

By optimizing environmental conditions and minimizing handling frequency, the study has successfully reduced stress-induced mortality in shrimp larvae. This approach has contributed to more consistent and improved growth outcomes, enhancing the overall efficiency of shrimp farming operations.

Future research should focus on integrating more advanced technologies such as machine learning and IoT for real-time monitoring and predictive analytics in shrimp farming. This could further refine the control and optimization processes, leading to even greater improvements in productivity and sustainability. Additionally, expanding the study to include other species and different environmental conditions would provide broader insights into the application of Six Sigma methodologies in aquaculture.

Acknowledgments. The authors would like to thank the Master's Program in Industrial Production and Operations and the "Grupo de Investigación Interdisciplinar en Matemática Aplicada (GIIMA)" at Universidad Politécnica Salesiana, the "Research Group on Prototype Development for Industry and Security (CONCEPTS)," and the Laboratory of Reverse Engineering and Prototyping at Universidad de las Fuerzas Armadas ESPE for their support and contributions to this research.

References

1. Datasur. Shrimp Exports in Ecuador: Global Market Trends and Perspectives (2023). https://www.datasur.com/en/shrimp-exports-in-ecuador-global-market-trends-and-perspectives/. Accessed 21 May 2024
2. FAO. Whiteleg shrimp (Litopenaeus vannamei). Food and Agriculture Organization of the United Nations (2023). https://www.fao.org/fishery/docs/DOCUMENT/aquaculture/CulturedSpecies/file/es/es_whitelegshrimp.htm. Accessed 20 May 2024
3. Supono, et al.: Study on the optimal salinity conditions for the growth and survival of litopenaeus vannamei. Aquaculture **2022** (2022). https://doi.org/10.1016/j.aquaculture.2024.740867
4. Samadan, et al.: Impact of stocking density on the production performance of litopenaeus vannamei. Iranian Journal of Fisheries Sciences **18**(4) (2018). https://doi.org/10.22092/ijfs.2019.120676
5. Pinem, et al.: Optimization of water quality management in shrimp farming using advanced filtration systems. Environmental Earth Sciences **404**(1) (2022). https://doi.org/10.1088/1755-1315/404/1/012017
6. Rehulina, et al.: Application of machine learning for predictive analytics in aquaculture. Aquaculture **2022** (2022). https://doi.org/10.1016/j.aquaculture.2016.10.019
7. Rojas, T., Mula, J., Sanchis, R.: Quantitative modelling approaches for lean manufacturing under uncertainty. Int. J. Prod. Res. **62**(16), 5989–6015 (2024). https://doi.org/10.1080/00207543.2023.2293138
8. Davis, R., Smith, K., Lee, P.: A study on the effects of temperature and salinity on shrimp larvae. J. World Aquaculture Society **49**(2) (2018). https://doi.org/10.1111/jwas.12818
9. Williams, T., Brown, J.: The impact of feed composition on the growth rates of shrimp. Aquaculture **2024** (2024). https://doi.org/10.1016/j.aquaculture.2024.740867
10. Johnson, A., Thompson, L.: Behavioral responses of shrimp to different salinity levels. J. Applied Ichthyology **18**(1) (2002). https://doi.org/10.1111/j.1749-7345.2002.tb00019.x

Influence of Injection Molding Parameters on the Orientation of Polypropylene Lamellar Structures Using Wide-Angle X-ray Scattering (WAXS) Technique

Daysi Baño-Morales[1,2](✉) 🄳, Andrea Fernández-Gorgojo[1] 🄳,
and Antonio Vizán-Idiope[1] 🄳

[1] Departamento de Ingeniería Mecánica, Universidad Politécnica de Madrid, Madrid, Spain
`daysi.bano@alumnos.upm.es`
[2] Grupo de Ingeniería Automotriz, Movilidad y Transporte (GiAUTO), Carrera de Ingeniería
Automotriz, Universidad Politécnica Salesiana, Campus Sur, Quito, Ecuador

Abstract. The optimization of injection molding process variables is important to minimize the number of final parts with defects, the cost and injection time. Among all the defects, shrinkage stands out for its important impact, since it causes the injected part to undergo dimensional changes during the cooling and post-processing stage. One of the factors influencing shrinkage is molecular orientation, as the injection phase imparts a predominant orientation to the polymer chains due to flow.

This study presents the influence of polymer flow along the part on molecular orientation by varying injection cycle parameters, such as mass temperature, pressure, and injection time. The crystalline lamella orientation in a semicrystalline polymer like polypropylene (PP) was analyzed using wide-angle X-ray scattering (WAXS) technique. The WAXS images indicate the periodicity of the lamellar structures in which polymer chains of crystalline and amorphous phases are stacked. The results showed that molecular orientation is higher at the entry of the specimen and decreases along its length. Regarding the manufacturing process parameters, mass temperature is the most significant factor in molecular orientation, promoting its alignment.

Keywords: Injection molding · polypropylene · molecular orientation · shrinkage · wide-angle X-ray scattering (WAXS)

1 Introduction

Injection molding is one of the most important processes for producing plastics with complex shapes and geometries [1]. The injection molding process consist of melting the initially solid polymer and injecting it at high pressure into a mold with the desired shape. Once the cavity is filled, the holding stage begins to compensate for the shrinkage that occurs during cooling. Cooling times have an influence on contractions and residual

© The Author(s) 2025
E. M. Inga Ortega et al. (Eds.): CITIS 2024, LNNS 1331, pp. 500–511, 2025.
https://doi.org/10.1007/978-3-031-87065-1_45

stresses. Finally, the mold opens and ejects the produced part [2]. This manufacturing process employs many parameters, such as the mass and mold temperatures, injection pressure and speed, holding time and pressure, among others [3]. Therefore, optimizing these process variables is crucial to minimize the number of defects in the final parts. A very significant defect in injection molding is differential shrinkage, which occurs when the injected part undergoes a dimensional change during the cooling stage [4, 5]. This type of defect leads to other issues, such as sink marks or voids due to incomplete filling, which affect the quality of the final product [6].

One factor influencing shrinkage is molecular orientation, since during the injection stage it gives a predominant orientation to the polymer chains due to flow-induced crystallization [7]. Polymer molecule orientation results in variations in the physical and mechanical properties of the material, leading to lower strength in the transverse direction compared to the longitudinal direction of the molecular chains [8].

Shrinkage is highly sensitive to part and mold design, the polymer used, and changes in process parameters [1, 9]. The manufacturing cycle parameters that have the greatest impact on shrinkage are mass temperature, injection pressure, injection time, and holding pressure [10, 11]. Increasing the mass temperature decreases the polymer viscosity, extending the solidification time and thus promoting molecular orientation. Conversely, excessive injection pressure promotes differences in molecular orientation and the formation of residual stresses, leading to part deformation [1].

Molecular orientation occurs during the mold filling phase when the molten polymer enters the mold. Polymer chains stretch due to the velocity gradient associated with laminar flow behavior. The flattened shape of most molten polymer velocity profiles (pseudoplastic behavior) causes regions with higher orientation -in the flow direction- to be located near the mold walls, while molecules in the core of the part remain randomly distributed [12].

Molecular orientation can be obtained through various techniques, such as X-ray scattering techniques: wide-angle X-ray scattering (WAXS) and small-angle X-ray scattering (SAXS). As shown in Fig. 1, the first technology reveals the structure at the atomic and molecular level ($\theta > 10°$) on a scale of 1 to 100 nm. The second technique is suitable for mesoscale structures ($\theta < 10°$) and analyzes the mesoscale structure that arises from the aggregation of atoms and molecules [13].

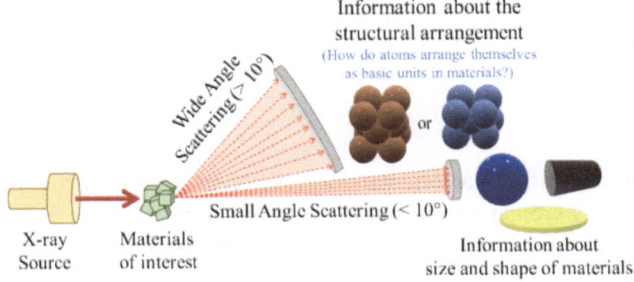

Fig. 1. Diagram illustrating small-angle and wide-angle X-ray scattering [14].

Some researchers have highlighted the usefulness of X-ray scattering techniques in identifying molecular orientation. For instance, Kantz et al. analyzed the crystalline orientation and morphology in the surface and intermediate layers of both a polypropylene homopolymer and a copolymer, employing optical microscopy and X-ray diffraction. The study revealed that the thickness of the surface layer varies inversely with the polymer mass temperature, while the thickness of the intermediate layer exhibits a complex relationship with injection pressure [15]. On the other hand, Zhu and Graham investigated the laminar and molecular orientation of isotactic polypropylene produced via injection molding, using WAXS and SAXS techniques. They concluded that, for thick materials, the reduction in lamellar orientation is associated with a constant degree of molecular orientation [13].

In this research, the effect of mass temperature, injection pressure, and injection time on molecular orientation is examined in isotactic polypropylene homopolymer, using the WAXS technique. Two specimens obtained near the gate and at the middle of the sample were analyzed.

2 Materials and Methods

2.1 Materials and Experimental Unit

The material used was an *ExxonMobil PP1013H1* polypropylene homopolymer. The main characteristics are listed in Table 1:

Table 1. Polypropylene PP-1013H1 properties.

Characteristic	Value	Unit
Melt Mass-flow Rate (MFR)	7.5	g/ 10 min
Density	0.9	g/cm^3
Melting Temperature	160	°C
Peak Crystallization Temperature	112	°C

The parts were obtained on the *BOY XS* plastic injection machine. Its main characteristics are summarized in Table 2.

2.2 Methods

In order to evaluate the molecular orientation at different manufacturing conditions, WAXS measurements were carried out on the *AntonPaar SA-XSpoint* 5.0 equipment at an X-ray wavelength of 0.15418 nm and a sample-detector distance of 64.199 mm. Studies were carried out with *SAXSanalysis* 4.30.0 software.

Based on the previous studies shown in the *Introduction section*, two different mass temperatures were analyzed, 200 °C and 220 °C; three different injection pressures, 240 bar, 250 bar and 260 bar; and two injection times, 2 s and 4 s. The configurations are

Table 2. Principal characteristics of the *BOY XS* injection molding machine.

Characteristic	Value	Unit
Clamping force (max – min)	100 – 10	kN
Screw diameter	14	mm
Injection force	35.4	kN
Max. Stroke volume (theoretical)	6.1	cm^3

Table 3. Injection molding parameters used in the research.

Sample	T_{mass} [°C]	P_{inj} [bar]	t_{inj} [s]
P_A	200	240	2
P_B	200	250	2
P_C	200	260	2
P_D	220	240	2
P_E	220	250	2
P_F	220	260	2
P_G	220	240	4
P_H	220	250	4
P_I	220	260	4

summarized in Table 3. The mold cooling system remained closed and a mold clamping force of 50 kN was maintained.

Figure 2 illustrates the design of the analyzed samples. For the WAXS tests, longitudinal sheets of 4x4x1 mm were cut perpendicular to the injection direction, one near the gate (E) and another at the midpoint of the specimen (M), as shown in Fig. 3.

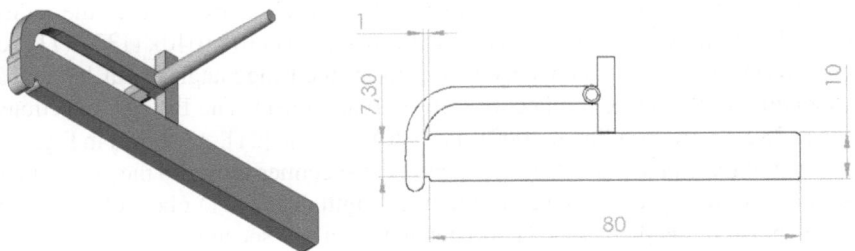

Fig. 2. Cavity design: sprue, runner, cold slug well, gate and sample (units in mm).

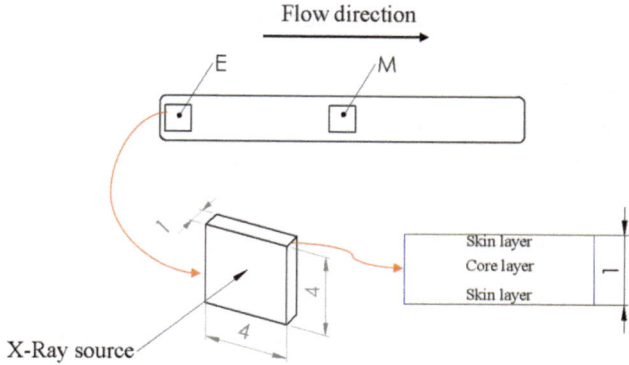

Fig. 3. Dimensions of the WAXS test specimen (units in mm), one near the gate (E) and another at the midpoint of the specimen (M).

3 Results and Discussion

Polypropylene often exhibits three different crystalline phases, α -monoclinic phase-, β -trigonal phase-, and γ -triclinic phase- [16]. The α phase is characterized by the presence of transverse crystals along with the main structure of radial lamellae, known as the "cross-linked" lamellar morphology. On the other hand, the β phase consists exclusively of radial lamellae and is only formed under specific conditions [17]. The β phase can be formed through the addition of nucleating agents, by the cooling rate during crystallization, or by orientation in the cutting zone [18, 19]. It is commonly found in samples subjected to mechanical deformation, such as extruded or injection-molded products, as these processes promote local crystallization and the nucleation of β crystals. The γ phase preferably forms under pressures exceeding 200 MPa when the polymer is fully crystallized [20].

Regarding the crystalline structure of the homopolymer PP obtained from the WAXS patterns at the entrance (Fig. 4a) and at center (Fig. 4b) of the sample, the specimen located in the center is mainly composed of the α form. Whilst the specimen near the gate is mainly composed of the α form, although some coexistence with the β form is also observed, as seen in Fig. 4c.

The Bragg reflections at 14°, 17°, 18.5°, 21° and 22° correspond to the indexed planes of the monoclinic crystals of isotactic PP, (α-form) (110), (040), (130), (111) and (131) + (041) and for the trigonal crystals (β-form) the Bragg angles with 16° and 21° corresponding to the indexed reflections of (300) and (301). The Braggs reflections at 25.5° and 28.5° corresponding to (060) and (220) α -phase [21]. As shown in Fig. 4, the (300) plane of the β phase is slightly evident in the specimen closer to the gate, whereas it disappears in the specimen taken at the mid-length. The (111) plane of the α phase and (301) plane of the β phase at 21° do not appear in any specimen.

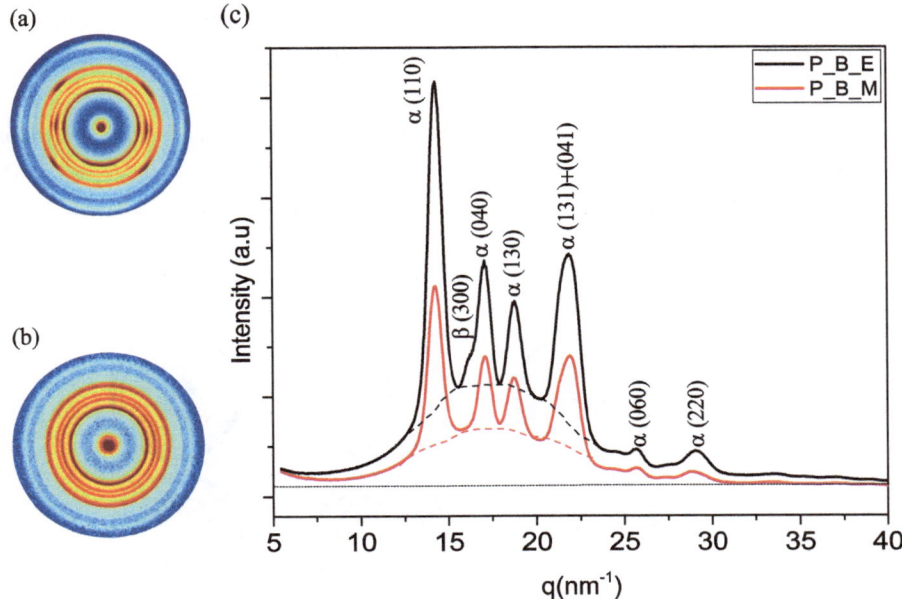

Fig. 4. Polypropylene WAXS pattern for P_B sample (a) close to the gate (P_B_E) and at mid-length (P_B_M), with $T_{mass} = 200\ °C$, $P_{inj} = 250$ bar, y $t_{inj} = 2$ s, (c) PP XRD pattern for P_B sample close to the gate (black curve) and at mid-length (red curve).

Tables 4, 6, and 7 show WAXS patterns under different injection molding conditions analyzed as the schematic diagram of WAXS pattern of Fig. 5.

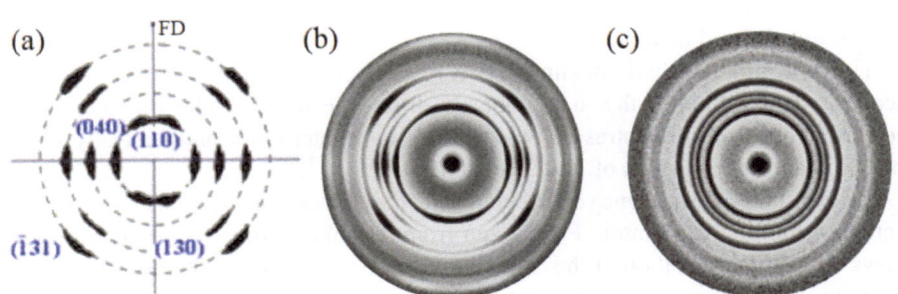

Fig. 5. (a) A schematic diagram of WAXS patterns taken from skin layer [21], WAXS pattern take from skin layer at $T_{mass} = 200\ °C$, $P_{inj} = 240$ bar y $t_{inj} = 2$ s (b) at the entry and (c) in the middle region of the specimen.

To compare the specimens obtained from the inlet zone with those from the mid-length zone, the WAXS images in Table 4 are presented. There is a high azimuthal dependence in the (110), (040), (130), and (131) planes of the α phase and a slight appearance of the (300) plane for the β phase in the specimens near the gate, indicating a high degree of orientation in the flow direction. In contrast, the WAXS images obtained from the specimens taken at mid-length show an isotropic scattering pattern, indicating random orientation of the lamellar structures in the plane.

Table 4. Wide-angle X-ray scattering patterns taken from various position around the length of the sample at $T_{mass} = 200$ °C, $t_{inj} = 2$ s and different injection pressure

Sample	P_A	P_B	P_C
	$P_{inj} = 240$ bar	$P_{inj} = 250$ bar	$P_{inj} = 260$ bar
Layer	Skin	Skin	Skin
Entry region			
Middle region			

A relative level of laminar orientation is obtained from the full width at half maximum (FWHM) of the azimuthal profiles. To analyze the dependence of the injection pressure on the orientation of the molecules, the melting temperature and injection time were kept constant. Figure 6 shows the XRD pattern of polypropylene for samples P_A, sample P_B and sample P_C evaluated at the entrance and at half the length of the specimen. Furthermore, the FWHM of azimuthal (040) obtained by a Gaussian fit is shown in Fig. 6a, summarized in Table 5.

The full width at half maximum (FWHM) decreases as the injection pressure increases, which implies that it promotes a difference in molecular orientation, and the formation of residual stresses, which leads to greater deformation. These results corroborate the conclusions of Portale et al. [20].

Table 6 shows the influence of orientation on the skin or core layer taken from a cross section of the sample. Sample P_D, taken from the middle zone (core) near the gate, shows a random orientation of the crystallites, so no orientation in the flow direction is observed. In contrast, samples P_E and P_F, taken from the surface near the gate, are not only highly crystalline but also have crystallites aligned in the flow direction. This is evident in Fig. 7a sample, where sample P_D has a lower peak intensity than samples P_E and P_F.

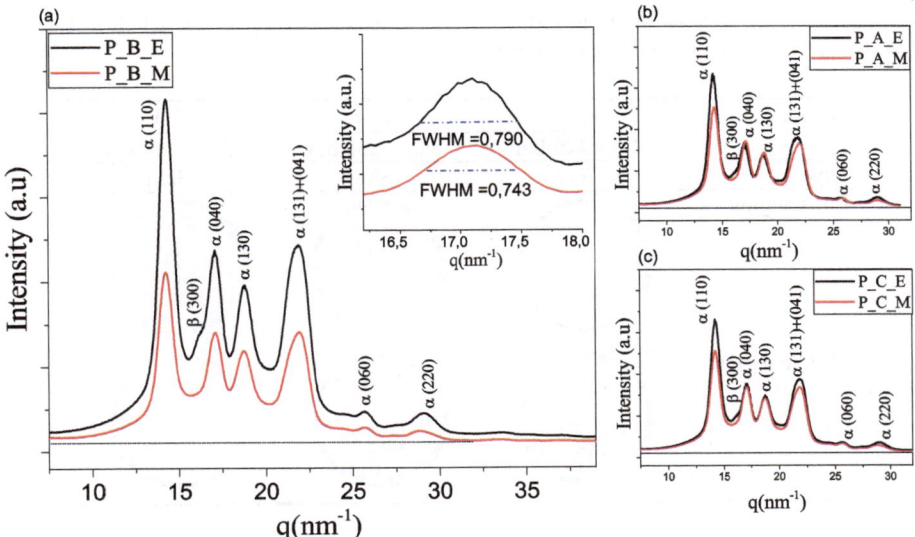

Fig. 6. PP XRD pattern for (a) P_B sample, (b) P_A sample, (c) P_C sample, close to the gate (black curve) and at mid-length (red curve).

Table 5. FWHM at $T_{mass} = 200\,°C$, $t_{inj} = 4$ s and different injection pressure

Samples	T_{inj} [°C]	P_{inj} [bar]	t_{inj} [s]	FWHM [nm^{-1}]	
				Entry	Middle
P_A	200	240	2	0.862	0.747
P_B	200	250	2	0.790	0.743
P_C	200	260	2	0.767	0.765

Table 7 shows the influence of the temperature of the molten polymer and the injection time on the orientation. Specimen P_B at $T_{mass} = 200\,°C$ has a greater degree of crystalline orientation than the specimen P_E at $T_{mass} = 220\,°C$. The planes (110), (040), (130), (131) are clearly observed in the specimen P_B as shown in Fig. 7b, however in the specimen P_E the intensity of the planes decreases and the rings become uniform, which corroborates the results of Kantz et al. [15].

The results presented in Fig. 8, which show the (110) reflection because it appears in the different ranges of distance, conclude that the bimodal characteristic is significant at $T_{mass} = 200\,°C$, and decreases with increasing temperature to $T_{mass} = 220\,°C$ [15, 22].

Table 6. WAXS patterns taken from various position around the length of the sample at $T_{mass} = 220$ °C, $t_{inj} = 2$ s and different injection pressure

Sample	P_D	P_E	P_F
	$P_{inj} = 240$ bar	$P_{inj} = 250$ bar	$P_{inj} = 260$ bar
Layer	Core	Skin	Skin
Entry region			
Middle region			

Table 7. Wide-angle X-ray scattering patterns taken in entry and middle region for $P_{inj} = 250$ bar and different melt temperature and injection time

Sample	P_B	P_E	P_H
	$T_{mass} = 200$ °C, $t_{inj} = 2$ s	$T_{mass} = 220$°C, $t_{inj} = 2$ s	$T_{mass} = 220$°C, $t_{inj} = 4$ s
Layer	Skin	Skin	Skin
Entry region			
Middle region			

With respect to the injection time, it is observed that as the injection time increases, keeping the temperature and pressure constant, the molecular orientation increases, for example, according to Fig. 8-above, the height of the peak of the sample P_H is greater than in the sample P_E. This is corroborated by the WAXS graphs of the P_E and P_H specimens in Fig. 7b since the azimuthal rings are observed with greater intensity in the P_H specimen.

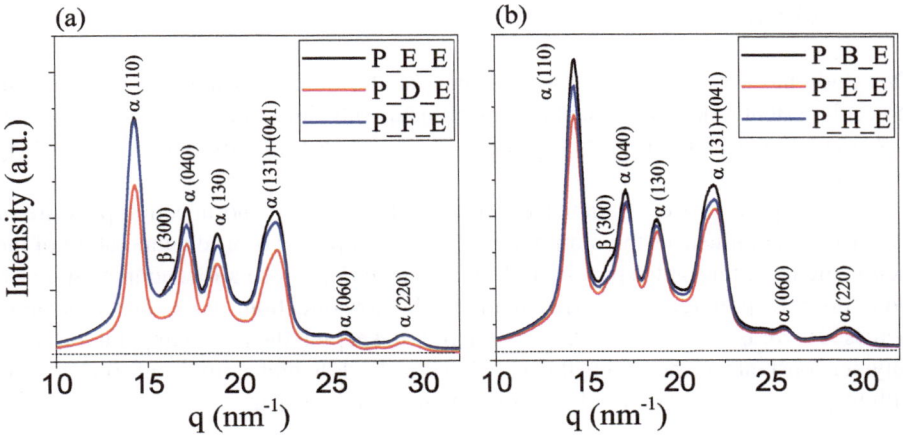

Fig. 7. PP XRD pattern for (a) samples at $T_{mass} = 220$ °C, $t_{inj} = 2$ s and different injection pressure, (b) samples at $P_{inj} = 250$ bar and different melt temperature and injection time

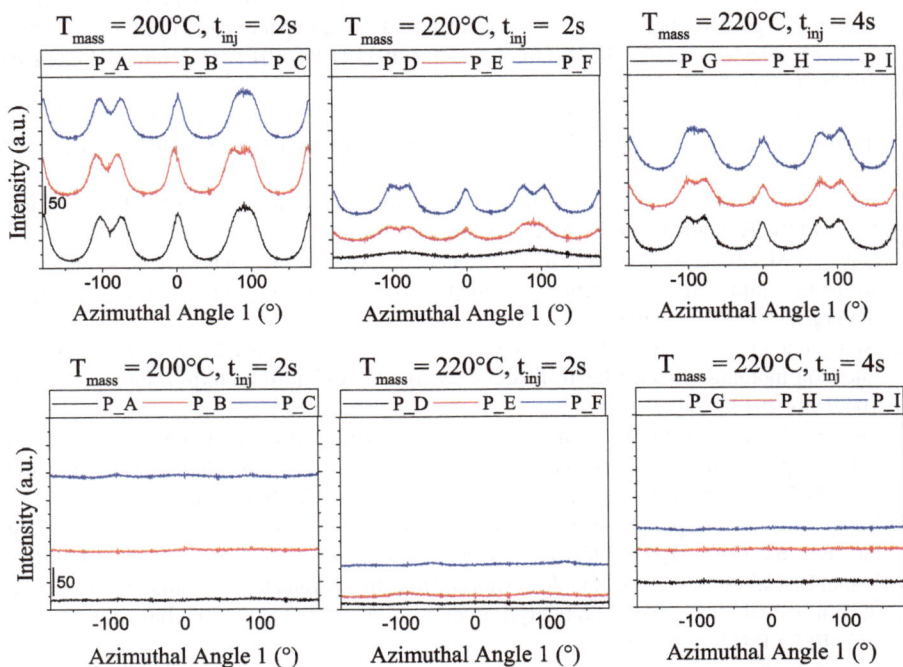

Fig. 8. Distributions of azimuthal profiles of (110) at $P_{inj} = 240$ bar (black line), $P_{inj} = 250$ bar (red line) and $P_{inj} = 260$ bar (blue line) in different melt temperature and injection time, (above) at the entry and (below) in the middle region of the specimen. The graphs illustrate how intensity patterns change with temperature and time at the entry (above) and in the middle region (below) of the specimen.

4 Conclusions

This study has shown that the WAXS technique allows to determine the predominant molecular orientation of the specimens manufactured by injection molding. The molecular orientation is greatest at the entrance of the piece and decreases longitudinally along the sample.

Regarding the process parameters, the mass temperature and injection pressure of polypropylene influences the morphology of the final products. The temperature of the melt is the most relevant aspect in molecular orientation, since a low temperature promotes ordering preferentially perpendicular to the flow direction, and as the temperature increases there is isotropic ordering. Furthermore, keeping the mass temperature of the polymer constant and increasing the injection time leads to greater molecular orientation with respect to the direction of flow throughout the molded part.

References

1. Zhao, N.Y., Lian, J.Y., Wang, P.F., Bin Xu, Z.: Recent progress in minimizing the warpage and shrinkage deformations by the optimization of process parameters in plastic injection molding: a review. Int. J. Adv. Manuf. Technol. **120**(1–2), 85–101 (2022). https://doi.org/10.1007/s00170-022-08859-0

2. Moayyedian, M., Abhary, K., Marian, R.: The analysis of defects prediction in injection molding. World Acad. Sci. Eng. Technol. Int. J. Mech. Mechatronics Eng. **10**(12), 1813–1816 (2016)

3. Kashyap, S., Datta, D.: Process parameter optimization of plastic injection molding: a review. Int. J. Plast. Technol. **19**(1), 1–18 (2015). https://doi.org/10.1007/s12588-015-9115-2

4. Hatta, N.M., Zain, A.M., Shayfull, Z., Sallehuddin, R.: AI approaches: Recent studies on shrinkage optimisation in injection moulding process. AIP Conf. Proc. **2030** (2018). https://doi.org/10.1063/1.5066793

5. Annicchiarico, D., Alcock, J.R.: Review of factors that affect shrinkage of molded part in injection molding. Mater. Manuf. Process. **29**(6), 662–682 (2014). https://doi.org/10.1080/10426914.2014.880467

6. Shen, C., Wang, L., Cao, W., Qian, L.: Investigation of the effect of molding variables on sink marks of plastic injection molded parts using Taguchi DOE technique. Polym. Plast. Technol. Eng. **46**(3), 219–225 (2007). https://doi.org/10.1080/03602550601152887

7. Isayev, A.: Anisotropic shrinkage in injection molding of various polyesters. Polym. Eng. Fac. Res. **52**, 1190–1194 (2006). https://ideaexchange.uakron.edu/polymerengin_ideas/134/

8. Wang, B., Cai, A.: Influence of mold design and injection parameters on warpage deformation of thin-walled plastic parts. Polimery/Polymers **66**(5), 283–292 (2021). https://doi.org/10.14314/POLIMERY.2021.5.1

9. Liu, W., et al.: Integration optimization of molding and service for injection-molded product. Int. J. Adv. Manuf. Technol. **84**(9), 2019–2028 (2016). https://doi.org/10.1007/s00170-015-7862-z

10. Usman Jan, Q.M., Habib, T., Noor, S., Abas, M., Azim, S., Yaseen, Q.M.: Multi response optimization of injection moulding process parameters of polystyrene and polypropylene to minimize surface roughness and shrinkage's using integrated approach of S/N ratio and composite desirability function. Cogent Eng. **7**(1), 1781424 (2020). https://doi.org/10.1080/23311916.2020.1781424

11. Xu, Y., Zhang, Q., Zhang, W., Zhang, P.: Optimization of injection molding process parameters to improve the mechanical performance of polymer product against impact. Int. J. Adv. Manuf. Technol. **76**(9), 2199–2208 (2015). https://doi.org/10.1007/s00170-014-6434-y
12. Malloy, R.A.: Plastic Part Design for Injection Molding (2010)
13. Zhu, P.W., Edward, G.: Orientational distribution of parent-daughter structure of isotactic polypropylene: a study using simultaneous synchrotron WAXS and SAXS. J. Mater. Sci. **43**(19), 6459–6467 (2008). https://doi.org/10.1007/s10853-008-2979-1
14. Asgar, H.: Multi-scale x-ray scattering to address fundamental research questions. Energy Frontier Research Center (2022). https://www.energyfrontier.us/content/multi-scale-x-ray-scattering-address-fundamental-research-questions
15. Kantz, M.R., Newman, H.D., Jr., Stigale, F.H.: The skin-core morphology and structure–property relationships in injection-molded polypropylene. J. Appl. Polym. Sci. **16**(5), 1249–1260 (1972)
16. Krajenta, J., Pawlak, A.: Crystallization of the β-form of polypropylene from the melt with reduced entanglement of macromolecules. Polymers **16**(12) (2024). https://doi.org/10.3390/polym16121710
17. Padden, F.J., Jr., Keith, H.D.: Spherulitic crystallization in polypropylene. J. Appl. Phys. **30**(10), 1479–1484 (1959). https://doi.org/10.1063/1.1734985
18. Keith, H.D., Padden, F.J., Jr., Walter, N.M., Wyckoff, H.W.: Evidence for a second crystal form of polypropylene. J. Appl. Phys. **30**(10), 1485–1488 (1959). https://doi.org/10.1063/1.1734986
19. Tashiro, K., Yamamoto, H., Funaki, K., Masunaga, H., Miyake, Y.: Three representative types of WAXD/SAXS patterns to establish the bimodal structure concept of stacked lamellae in isotactic polypropylene spherulites. Polym. J. **56**(5), 491–503 (2024). https://doi.org/10.1038/s41428-024-00893-x
20. Portale, G., et al.: Polymer crystallization studies under processing-relevant conditions at the SAXS/WAXS DUBBLE beamline at the ESRF. J. Appl. Crystallogr. **46**(6), 1681–1689 (2013). https://doi.org/10.1107/S0021889813027076
21. Machado, G., et al.: Crystalline properties and morphological changes in plastically deformed isotatic polypropylene evaluated by X-ray diffraction and transmission electron microscopy. Eur. Polym. J. **41**(1), 129–138 (2005). https://doi.org/10.1016/j.eurpolymj.2004.08.011
22. Björn, L., et al.: Scanning small-angle x-ray scattering of injection-molded polymers: anisotropic structure and mechanical properties of low-density polyethylene. ACS Appl. Polym. Mater. **5**(8), 6429–6440 (2023). https://doi.org/10.1021/acsapm.3c01007

Author Index